Haim Abramovich
Intelligent Materials and Structures

Also of Interest

Advanced Aerospace Materials.
Aluminum-Based and Composite Structures
Abramovich, 2019
ISBN 978-3-11-053756-7, e-ISBN 978-3-11-053757-4

Advanced Materials
van de Ven, Soldera (Eds.), 2019
ISBN 978-3-11-053765-9, e-ISBN 978-3-11-053773-4

Nanoscience and Nanotechnology.
Advances and Developments in Nano-sized Materials
Van de Voorde (Ed.), 2018
ISBN 978-3-11-054720-7, e-ISBN 978-3-11-054722-1

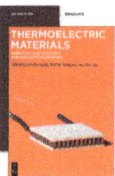

Thermoelectric Materials.
Principles and Concepts for Enhanced Properties
Kurosaki, Takagiwa, Shi 2020
ISBN 978-3-11-059648-9, e-ISBN 978-3-11-059652-6

Haim Abramovich

Intelligent Materials and Structures

2nd Edition

DE GRUYTER

Authors
Prof. Dr. Haim Abramovich
Aerospace Structural Laboratory
Technion-Israel Inst. of Technology
Technion City
32000 Haifa
Israel

ISBN 978-3-11-072669-5
e-ISBN (PDF) 978-3-11-072670-1
e-ISBN (EPUB) 978-3-11-072622-0

Library of Congress Control Number: 2021938034

Bibliographic information published by the Deutsche Nationalbibliothek
The Deutsche Nationalbibliothek lists this publication in the Deutsche Nationalbibliografie;
detailed bibliographic data are available on the Internet at http://dnb.dnb.de.

© 2021 Walter de Gruyter GmbH, Berlin/Boston
Cover image: ThomasVogel/iStock/Getty Images Plus
Typesetting: Integra Software Services Pvt. Ltd.
Printing and binding: CPI books GmbH, Leck

www.degruyter.com

Preface

The first edition of the book *Intelligent Materials and Structures* was published in 2016 and it was aimed at dissemination of advanced engineering ideas using relatively new multifunctional materials. The book was written to provide students and scholars a good understanding of the new emerging interdisciplinary topic of smart/intelligent structures. It served as an introductory book for graduate students wishing to study about intelligent materials and structures, enabling them to acquire the necessary physical and mathematical tools to understand the various perplexing aspects of this new engineering field.

The second edition of the book aims at updating the content of the first edition based on additional research performed by the author on smart structures and innovative research published in the literature during the past 5 years.

The second edition of the book contains eight chapters from the first edition, namely, an extended introductory chapter on the main intelligent materials and structures like piezoelectric materials, shape memory alloys (SMA), electrorheological and magnetorheological fluids and electrostrictive and magnetostrictive materials; laminated composite materials, including classical lamination theory and first-order shear deformation theory; piezoelectricity with its constitutive equations and various beams and plate models; SMA and pseudoelasticity; electrorheology and magnetorheology; electrostriction and magnetostriction; applications of intelligent materials in structures and devices of aerospace and medicine sectors and a new look on piezoelectric-based motors. The eight chapters of the book present basic equations for energy harvesting using piezoelectric- and electromagnetic-based devices together with their relevant literature.

Two new chapters have been added to the second edition of the book: Chapter 9 presents the introduction to fiber optic, while Chapter 10 presents miscellaneous topics like enhanced flexural behavior of plates equipped with SMA wires, piezoelectric fiber composites, acoustic energy harvesting, harvesting using SMA and road traffic harvesting using piezoelectric transducers.

Besides those new chapters, a new appendix dealing with the definition of the piezoelectric coupling coefficient was added to Chapter 3, while Chapter 8 was upgraded by two new sub-chapters dealing with bimorph electrical power under vibration excitation and a piezoelectric harvester with enhanced frequency bandwidth.

It is hoped that the second edition of the book will present a wider view of the various smart structures, enabling readers to become more familiar with the new engineering topic of intelligent materials and structures.

The author wishes to thank his wife Dorit and his children, Chen, Oz, Shir and Or, for their support, understanding, love and devotion. Their continuous support throughout the writing period enabled the publishing of this book.

Haifa, April 2021
Haim Abramovich

https://doi.org/10.1515/9783110726701-202

Contents

1 Introduction to Intelligent Materials and Structures

1.1 Types of Intelligent Materials

The aim of this chapter is to present the reader with the general topic of smart intelligent structures. The general area of intelligent materials incorporated into smart structures had been investigated for some decades; however since 2000, the number of published manuscripts in various journals had tremendously increased with applications in almost all areas of engineering.

Smart structures (also carrying the name intelligent structures, mainly in the aerospace field) are normally represented as a minor subset of a much larger field of research, as schematically shown in Figure 1.1 (see [1–2]).

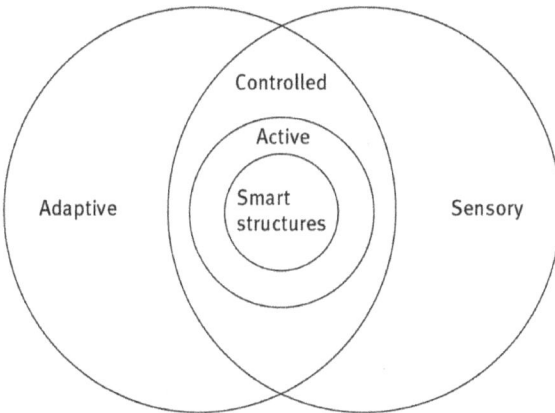

Figure 1.1: Smart structures as a subset of active and controlled structures [1].

According to Wada et al. [1], structures with actuators distributed throughout them are defined as *adaptive structures*. Typical examples of such adaptive structures are wings of conventional aircraft which possess articulated leading and trailing edge control surfaces. Accordingly, those structures that contain sensors distributed throughout them are called *sensory structures*. These sensory structures have transducers that might detect displacement strains, thus leading to the monitoring of mechanical properties, electromagnetic properties, temperature or the presence or accumulation of damage. The overlap structures that contain both actuators and sensors, and those that implicitly contain a closed-loop control system aimed at connecting the actuators and the sensors, are defined as *controlled structures*. A subset of the controlled structures are *active structures*. Active structures are distinguished from controlled structures by highly distributed actuators which have structural functionality and are part

https://doi.org/10.1515/9783110726701-001

of the load-bearing system [2]. Within this hierarchy, *smart structures* are only a subset of active structures that have highly distributed actuator and sensor systems with structural functionality, and in addition, distributed control functions and computing capabilities. Therefore, the term *smart structures* stands for structures that are capable of sensing and reacting to their environment in a predictable and desired manner, through the integration of various elements, such as sensors, actuators, power sources, signal processors, and communication network. In addition to carrying mechanical loads, smart structures may alleviate vibration, reduce acoustic noise, monitor their own condition and environment, automatically perform precision alignments or change their shape or mechanical properties on command (Figure 1.2). Another definition frequently appearing in the literature states that a smart structure is a system containing multifunctional parts that can perform sensing, control and actuation; it is a primitive analog of a biological body. As stated in [3], from this analogy of the bionic system of humans and animals, it can be seen that the following mechanisms may be essential for any material to be made intelligent:

1. A sensing tool to perceive the external stimuli (like a skin which senses thermal gradients or an eye that senses optical signals), to be termed as a *sensor*.
2. A communication network by which the sensed signal would be transmitted to a decision-making mechanism (like the nervous system in humans and animals) and a decision-making device (like the human brain), to be called *control*.
3. An actuating device that could be inherent in the material or externally coupled with it (like stiffening of muscles in humans or animals to resist deformation due to external loading), having the function of *actuator*.

All of these devices would need to be active in real-time applications for the material to respond intelligently and in an optimum time interval.

Smart materials are used to construct these smart structures, which can perform both sensing and actuation functions. Therefore, smart or intelligent materials are designed materials that have one or more properties that can be significantly changed in a controlled fashion by external stimuli, such as stress, temperature, moisture, pH and electric or magnetic fields.

Another interesting side of smart or intelligent structures is their interdisciplinary characteristics that require the involvement of a few engineering fields, such as aerospace or mechanical engineering (or any other applicative engineering field), electrical engineering, computer engineering, material science, applied physics, and system engineering. This is why this field is so attractive to many people coming from different fields of engineering and science, leading to many research studies being published in the literature. Although the scientists would publish their work in journals connected to their fields, three dedicated journals appeared to cover the newly emerged field of smart structures.

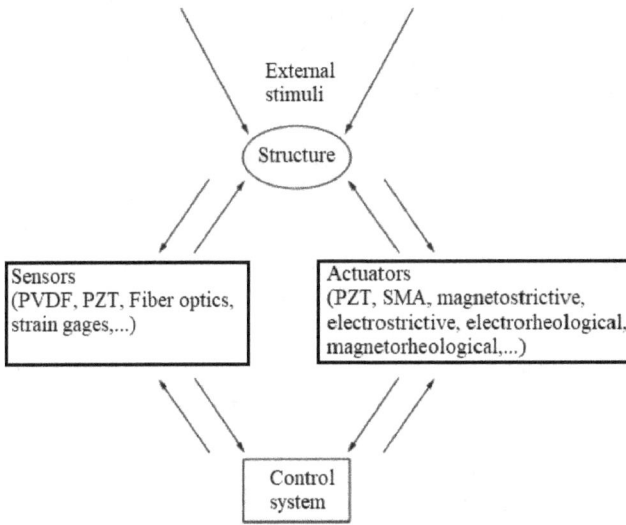

Figure 1.2: A schematic drawing of a smart structure.

- *Smart Materials and Structures,*[1] a multidisciplinary journal dedicated to technical advances in (and applications of) smart materials, systems, and structures, including intelligent systems, sensing and actuation, adaptive structures, and active control.
- *Journal of Intelligent Material Systems and Structures*[2] (JIMSS) is an international peer-reviewed journal that publishes the original research of the highest quality. JIMSS reports on the results of experimental or theoretical work on any aspect of intelligent material systems and/or structure research are also called smart structure, smart materials, active materials, adaptive structures and adaptive materials.
- *Smart Structures and Systems*[3] aims at providing a major publication channel for researchers in the general area of smart structures and systems. Typical subjects considered by the journal include sensors/actuators(materials/devices/informatics/networking), structural health monitoring and control and diagnosis/prognosis.

Beside those dedicated journals, one should note the open access journals, *Actuators, Sensors and Materials,*[4] with many manuscripts on smart structures topics.

---

1 Published by IOP Publishing (UK). http://iopscience.iop.org/0964-1726
2 Published by Sage Journals. http://jim.sagepub.com/
3 Published by Techno-Press. http://techno-press.org/?journal=sss&subpage=5#
4 Published by MDPI, www.mdpi.com

Books dedicated to smart structures and their applications were written [4–21] based on the experienced gained during the years. These include the early book of Ghandi and Thompson [4], through books dealing with various aspects of smart structures, like vibrations, dynamics and health monitoring [6–14] and ending with books carrying the general title of *Smart Structures*, covering all aspects of the topic [15–20]. Accompanying books deal with mechatronics[5] [13] and adaptronics[6] [21], the German paradigm for adaptive structures. It is worth noting that although the number and variety of books are relative high, only two can be used as undergraduate–graduate textbooks: the book authored by Leo [12] and the book co-authored by Chopra and Sirohi [20], which is more comprehensive, well posed and with a long list of references.

Piezoelectric materials can be considered the prime candidate to be included in a smart structure. The term *piezoelectricity* is also called the piezoelectric effect, and it is defined as the ability of certain materials, like quartz, certain ceramics and Rochelle salts to generate voltage when subjected to mechanical stresses or vibrations, or elongate or vibrate when subjected to a voltage. As stated in [22], the piezoelectric property that carries its name following a direct translation from Greek of the word *piezein* ("pressure electricity") was discovered by Pierre Curie and Jacques Curie in 1880 [23].

By analogy with temperature-induced charges in pyroelectric crystals, the Curie brothers observed electrification under mechanical pressure of certain crystals, including tourmaline, quartz, topaz, cane sugar and Rochelle salt. In a follow-up manuscript [24], the Curie brothers confirmed the second property of the piezoelectricity, what is called *the converse effect*. Since then, the term *piezoelectricity* is commonly used for more than a century to describe the ability of materials to develop electric displacement, denoted by the letter D, which is directly proportional to an applied mechanical stress, denoted by the Greek letter σ (Figure 1.3a). Following this definition, the electric charge appeared on the electrodes reverses its sign if the stress is changed from tensile to compressive. As follows from thermodynamics, all piezoelectric materials are also subject to a converse piezoelectric effect (Figure 1.3b), i.e., they deform under an applied electric field. Again, the sign of the strain S (elongation or contraction) changes to the opposite one if the direction of electric field E is reversed. The shear piezoelectric effect (Figure 1.3c) is also possible, as it linearly couples shear mechanical stress or strain with the electric charge.

Notice, that to understand the physics of piezoelectricity, one has to deepen into the crystallographic principles of the effect (see the pioneering work of Voigt [25] in 1910 and the more recently work of Jaffe et al. [26] from 1971). Table 1.1 presents

---

5 Mechatronics is a multidisciplinary field of engineering that includes a combination of mechanical engineering, electrical engineering, telecommunication engineering, control engineering and computer engineering.
6 Adaptive structure technology, briefly called Adaptronics, is an innovative, new cross sectional technology for the optimization of structure systems.

(a) The direct piezoelectric effect

(b) The converse piezoelectric effect

(c) The shear piezoelectric effect

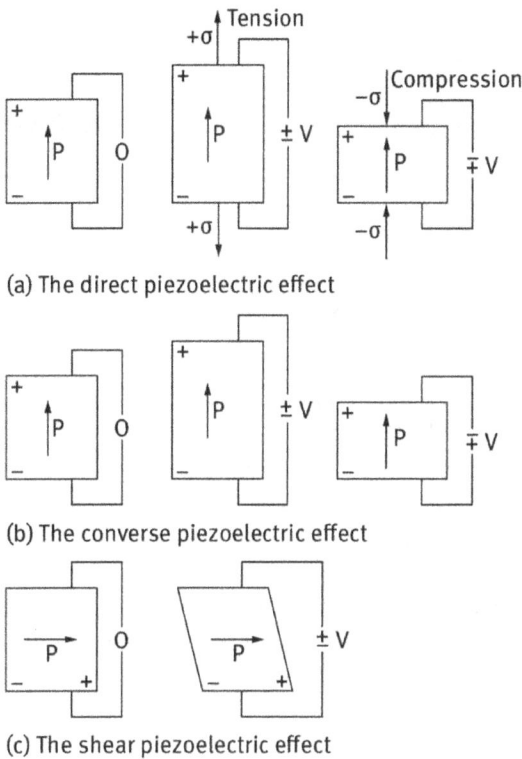

Figure 1.3: A schematic drawing of the direct (a), converse (b) and shear (c) piezoelectric effects.

typical values for main piezoelectric materials together with their symmetries as shown in [22]. The present book does not contain this part of the piezoelectricity and the reader interested is referred to Ref. [26] and to many other manuscripts dealing with the crystallography of piezoelectric. One should note that the invention of piezoelectric ceramics (see [26]) led to an astonishing performance of piezoelectric materials with industrial applications in many areas yielding a multibillion-dollar industry with a wide range of applications and uses.

The most widely used piezoelectric material is probably the PZT (lead zirconate titanate) [27]. The PZT ceramics shows both the direct piezoelectric effect (generating a voltage when it is strained) and the converse piezoelectric effect (it becomes strained when placed in an electric field). Therefore, the PZT ceramics can be used both as sensor and actuator. Piezoelectric ceramics are polycrystalline in nature and do not present any piezoelectric characteristics in their original state. To induce piezoelectric effects in PZT type ceramics, these materials undergo poling under their Curie temperature at high dc electrical fields of $2\,kV/mm$ (notice that this value is only an averaged one), resulting in an alignment of the polar axis of unit cells parallel to the applied field and thus yielding a permanent polarization of the

Table 1.1: Physical properties of major piezoelectric materials together with their symmetries.

Parameter				Materials			
	Quartz	BaTiO$_3$	PbTiO3:Sm	PZT 5 H	LF4T	PZN-8%PT [001]	PZN-8%PT [111]
Symmetry	32	4 mm	4 mm	3 m/4 mm	mm2/4 mm	3 m/4 mm	3 m/4 mm
d_{33} (pC/N)	2.3	190	65	593	410	2500	84
d_{31} (pC/N)	0.09	0.38	0	− 274	− 154	−1400	−20
$\varepsilon_{33}^T/\varepsilon_0$	5	1700	175	3400	2300	7000	1000
T_c (°C)	−	120	355	193	253	160	160

material. Another important fact associated with the polarization process of the ceramics is its permanently mechanical deformation due to the reorientation of its domains. One has to differentiate between the longitudinal, transversal and shear effects.
The longitudinal effect deals with the active strain in the polarization direction and
parallel to the electric field while the transversal effect is due to the resulting Poisson
in plane strain. The direction of a shear strain is parallel to the polarization direction
and vertical to the electric field. From the application point of view, it is customary that
only the longitudinal and transversal effects are being used while the shear effect is
being neglected. One should note the following facts: the maximum strain of PZT is
relatively small (0.12–0.18%) and it is limited by saturation effects and depolarization [28] and that the PZT ceramics exhibits hysteresis of 2% for very small signals and
reaches 10–15% at nominal voltage [28]. The density of PZT is typically 7.6g/cm^3 [29].

From a structural point of view, due to polarization process the PZT is not only
electrical but also mechanical anisotropic. The longitudinal Young's modulus for
PZT is typically is 50–70 GPa while in the transversal direction the values is only
35–49 GPa (for further information, see also [30]).

Another constraint is the fact that the piezoelectric effect depends also on the
ambient temperature. Below 260 K, it decreases with reducing temperature by a factor of approximately 0.4%/K [31]. The upper temperature is limited by its Curie temperature. PZT actuators can reliably be driven up to 70% of their Curie temperature
(between 150 and 350 °C) [30].

Temperatures above the Curie temperature would cause depolarization. High
mechanical stress can also depolarize a PZT ceramics. Normally, it is accustomed
to limit the applied stress to about 20–30% of its mechanical compression load
limit (200–300 MPa) [28]. Another important issue characterizing the PZT ceramics
is the fact that depolarization might occur when applying an electrical voltage in

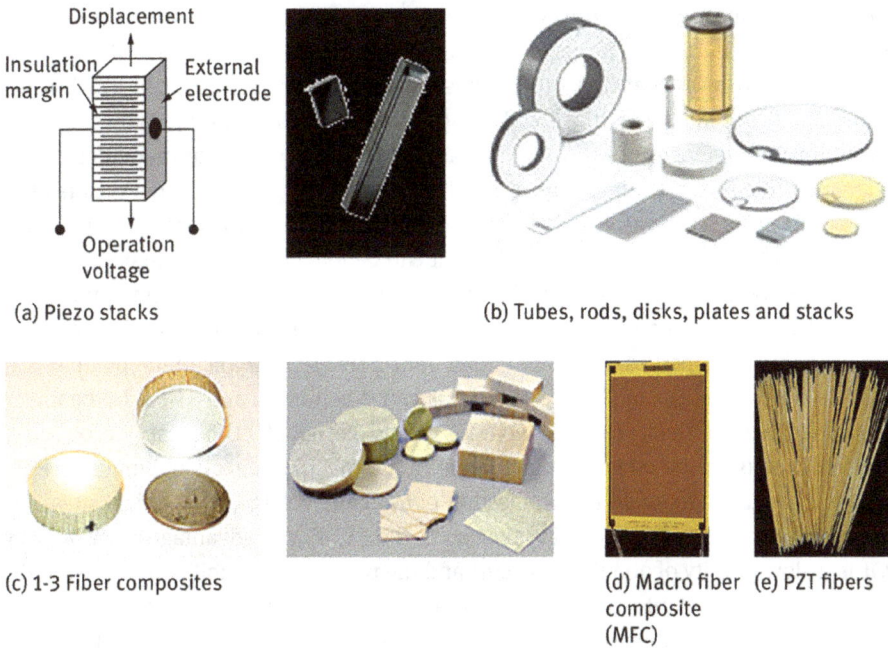

(a) Piezo stacks

(b) Tubes, rods, disks, plates and stacks

(c) 1-3 Fiber composites

(d) Macro fiber (e) PZT fibers
composite
(MFC)

Figure 1.4: Typical forms of PZT ceramics (a) stacks, (b) tubes, rods, disks, plates and stacks, (c) 1–3 Fiber composite, (d) MFC patches and (e) PZT fibers. Sources: NASA Smart Materials Corp. (a, c–e); PI ceramic (b).

the opposite direction to the polarization one which is larger than its coercivity voltage.[7]

Piezoelectric ceramics can be found in the form of thin sheets (patches), tubes, short rods, disks, fibers/stripes or stacked to form discrete piezostack actuators (see Figure 1.4).

Polyvinylidene fluoride (PVDF) and its copolymers with trifluoroethylene (TrFE) and tetrafluoroethylene (TFE) which are semicrystalline fluoropolymers [31] represent another way of producing piezoelectric materials based on polymers (see Figure 1.5). Like PZT, the PVDF can be used as sensor and actuator. Their electromechanical material behavior shows, like in the case of PZT, the three effects, namely, longitudinal, transversal and shear strains, whereas technically only the transversal effect is normally being used. The significant difference between PVDF and PZT is connected with their electromechanical material behavior namely in the direction of the electric field PVDF would contract instead of elongating like PZT. In plane, PVDF elongates whereas PZT

---

7 Electric coercivity is the ability of a ferroelectric material to withstand an external electric field without becoming depolarized.

contracts. Comparing the values of the piezoelectric constants (d_{31}, d_{32}, d_{33}) the PZT ceramics has 10–20 times larger piezoelectric strain constants than PVDF [31]. The active strain with 0.1% at up to 100 kHz is of the order of magnitude of PZT but requires significant larger electrical fields of 10–20 kV/mm. The amorphous region has a glass transition temperature ($\sim -40\,°C$) that dictates the mechanical properties of the polymer, while the crystallites have a melting temperature ($\sim 180\,°C$) that dictates the upper limit of its temperature. However, the piezoelectric effect is limited by the relatively low Curie temperature ($\sim 100\,°C$) and it can be reliably used at a maximum temperature of about 60–80 °C. Electrical poling is done using an electric field of the order of 50–80 kV/mm. Depending on whether stretching is uniaxial or biaxial, the electrical and mechanical properties are either highly anisotropic or isotropic in the plane of the polymer sheet. The in-plane Young's modulus varies between 2 and 3 GPa. In the direction of the electric field, Young's modulus would be about 1 GPa. The piezoelectric strain constants on the other hand are larger in its stretching direction. Similar to PZT, care must be taken to avoid the application of too large voltage, mechanical stress, or temperature to prevent the depoling process of the PVDF. The advantage of PVDF over PZT is its low density of about 1.47 g/cm^3 and the processing flexibility because tough, readily manufactured into large areas, and can be cut and formed into complex shapes. Major disadvantage is their low stiffness significantly reducing authority over the structure limiting PVDF to applications with low requirements concerning the forces.

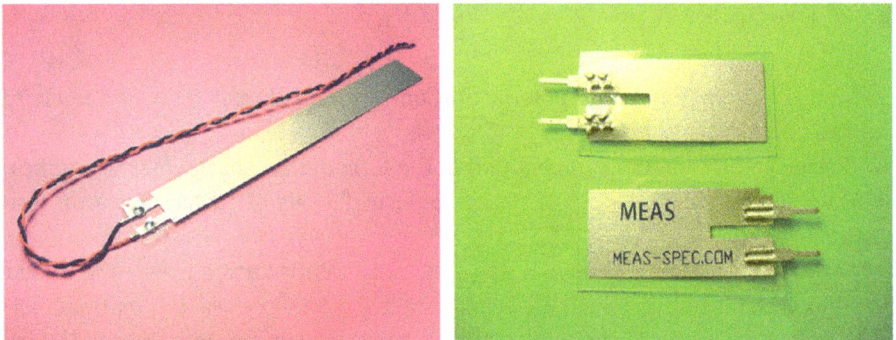

(a) Large PVDF film tab   (b) PVDF patch with connectors

Figure 1.5: Typical PVDF forms: (a) large film tab and (b) small film tabs.

Another important property of ferroelectric and ceramic (including piezoelectricity) materials, like PZT and PVDF is its pyroelectric effect,[8] which is the ability to generate voltage across its electrodes due to changes in the ambient temperature (both heating and cooling). The changes in the temperature lead to slight movement of

---

8 Pyroelectricity – from the Greek word *pyr* which means fire, and the word *electricity*.

the atoms within the crystal structure, leading to spontaneous polarization. One should note that the temperature should be time dependent, and the pyroelectric coefficient is defined according to the following relationship:

$$p_i \left[\frac{c}{m^2 K} \right] = \frac{\partial P_{S,i}}{\partial T}, \quad i = 1, 2, 3 \tag{1.1}$$

where p_i is the pyroelectric vector, P is the polarization vector and T is temperature.

Neglecting dielectric losses [36], the pyroelectric equations can be written as [36]

$$\dot{D} = \varepsilon \dot{E} + p \dot{T} \tag{1.2}$$

$$\dot{S} = p \dot{E} + \frac{c \dot{T}}{T} \tag{1.3}$$

where D, E, T and S are the electric induction, electric field, temperature and entropy, respectively. ε, p and c are the dielectric permittivity, the pyroelectric coefficient and the heat capacity, respectively. Note that ($\dot{}$) represents derivation with respect to time.

Based on the pyroelectric effect, infrared sensors and harvesting devices are designed to capture the heat time dependent energy and transform it into electrical energy as was suggested in [33–41]. Typical applications using the pyroelectric effect are shown in Figure 1.6.

An additional class of smart materials is the electrostrictive[9] one [42–51]. The electrostrictive material would have an induced strain proportionally to the square of the applied voltage (or of the applied electric field) and produce electricity when stretched. These materials are primarily used as precision control systems such as vibration control and acoustic regulation systems in engineering, vibration damping in floor systems and dynamic loading in building construction.

Electrostriction is a basic electromechanical phenomenon in all insulators or dielectrics [42]. It describes the electric field/polarization-induced strain (S_{ij}) to be proportional to the square of the electric field (E_i) or polarization (P_i), and is expressed by the following two equations:

$$S_{ij} = Q_{ijkl} P_k P_l \tag{1.4}$$

$$S_{ij} = M_{ijkl} E_k E_l \tag{1.5}$$

where Q_{ijkl} and M_{ijkl} are the electrostrictive coefficients. As the electrostriction is a fourth-rank tensor property it can be observed in all crystal symmetries [44]. The electrostrictive materials are not poled like the piezoelectric materials, and present a lower hysteresis than the piezoelectric ones; however, as they are more capacitive in nature,

---

9 The electrostrictive effect causes dimensional change of the material under the influence of applied an electric field.

(a) Thermal IR detector

(b) Pyroelectric waste heat generator concept

(c) Hybrid piezoelectric-pyroelectric nano-generator concept – from [40]

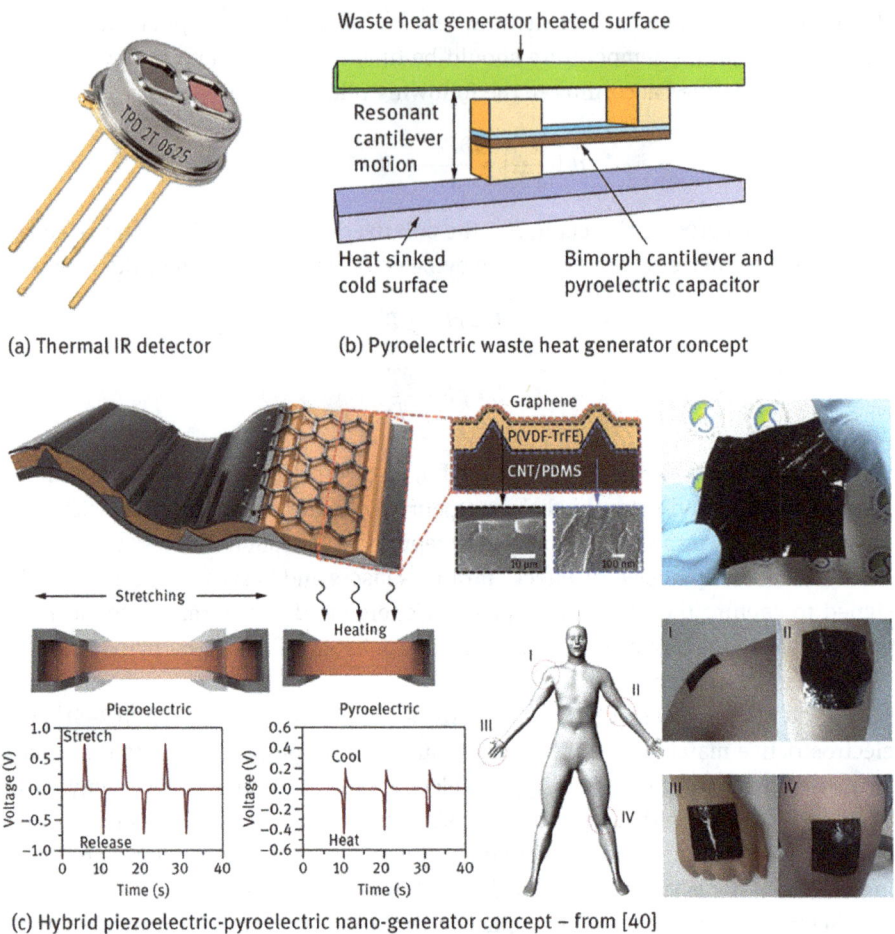

Figure 1.6: Pyroelectric applications: (a) IR detector, (b) waste heat generator concept [40] and (c) hybrid nanogenerator [40]. Sources: Warsash Scientific (a).

a larger driving current is required. They are also highly nonlinear since they respond to the square of the applied voltage/electric field. Electrostrictive ceramics, based on a class of materials are known as relaxor ferroelectrics, which show strains comparable to piezoelectric materials and have already found applications in many commercial systems [51]. The most prominent electrostrictive materials are lead magnesium niobate-lead titanate (PMN-PT)[10] and lead lanthanum zirconate titanate (PLZT). Those materials have a Young's modulus of about 700 GPa, are very brittle, possess a fast response

---

10 They are known as relaxor ferroelectrics possessing high relative permittivities (20,000–35,000) and high electrostrictive coefficients.

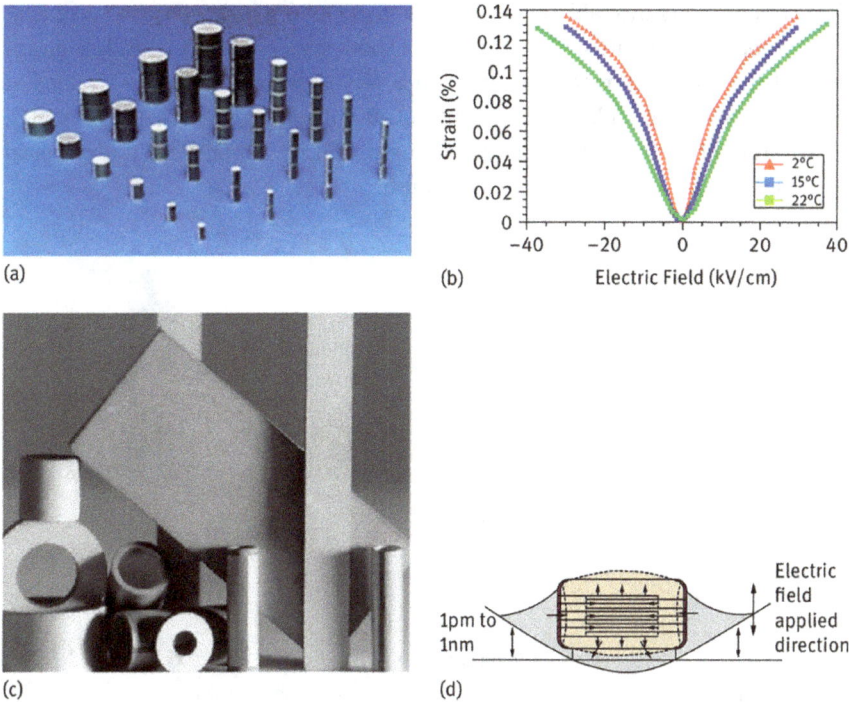

Figure 1.7: Electrostrictive materials: (a) multilayer PMN actuators, (b) PMN-15 behavior, (c) positioners and (d) deformation of a board. Sources: AOAXinetics (a), TRS Technologies (b), Piezomechanik GmbH (c) and MURATA (d).

time and low hysteresis loop, are suitable for frequencies up to 50 kHz and can be used only as actuators. Typical illustrations of electrostrictive materials can be found in Figure 1.7.

A property similar to electrostriction described above is the magnetostriction, which is the material property that causes a material to change its length when subjected to an electromagnetic field [52], also known as the Joule effect [53]. Another magnetostriction property, also known as the Villari effect, would cause materials to generate electromagnetic fields (will induce a change in the magnetic flux density) when they are deformed by an external force. The third magnetostrictive effect, also called the Barrett effect [54], states that the volume of a material will change in response to the application of a magnetic field. Magnetostrictive materials can thus be used for both sensing and actuation. Terfenol-D[11] (an alloy of terbium and iron

---

11 Terfenol-D, an alloy of the formula $Tb_xDy_{1-x}Fe_2$ ($x \sim 0.3$),s is a magnetostrictive material. It was initially developed in the 1970s by the Naval Ordnance Laboratory in USA. The technology for

(Fe)) is a commercially available magnetostrictive material that has "giant" magneto-striction at room temperature and can be used as both sensor and actuator. It is one of the rarest of the rare earth materials, and hence very expensive. It can be strained at large strains (2%) and has a Young's modulus of 200 GPa. It can generate large actuating forces (order of kNs) and possess low hysteresis. Figure 1.8 presents a few typical images of magnetostrictive materials and their use.

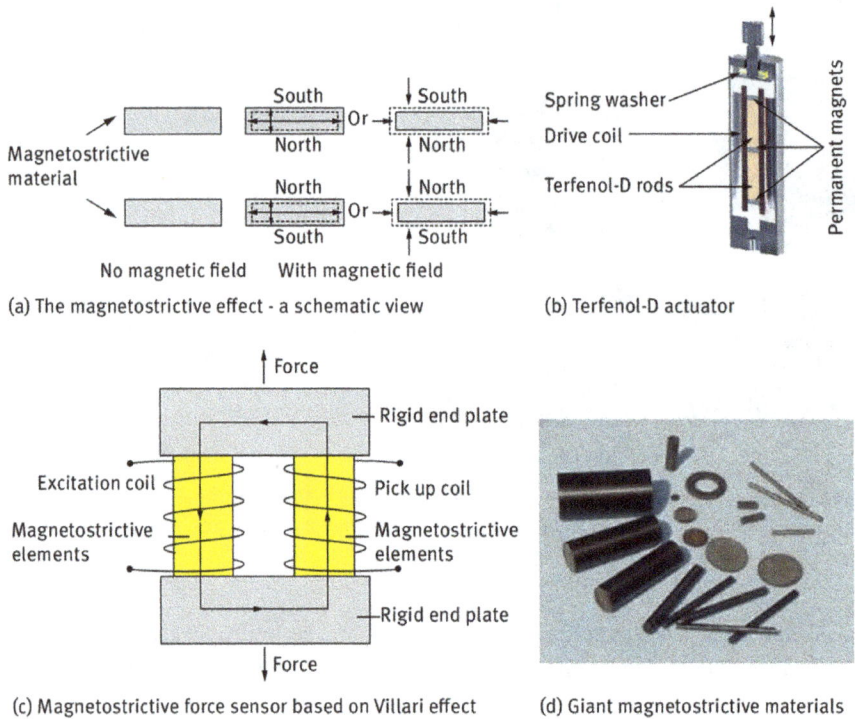

(a) The magnetostrictive effect - a schematic view

(b) Terfenol-D actuator

(c) Magnetostrictive force sensor based on Villari effect

(d) Giant magnetostrictive materials

Figure 1.8: Magnetostrictive materials: (a) the effect, (b) Terfenol-D actuator, (c) force sensor and (d) giant materials. Sources: John Fuchs: http://www.ctgclean.com/tech-blog/2011/12/ultrasonics-transducers-magnetostrictive-effect/ (posted on Dec. 30 2011) (a), [55] (b), [538] (c), GANSU TIANXING Rare Earth Functional Materials Co. Ltd. (d).

Another class of intelligent materials is the electroactive (see Figure 1.9a[12]) and magnetoactive polymers (MAPs) (Figure 1.9b). Electroactive polymers, or EAPs [56],

---

manufacturing the material efficiently was developed in the 1980s at Ames Laboratory under a U.S. Navy–funded program. It is named after terbium, iron (Fe), Naval Ordnance Laboratory (NOL) and the D comes from dysprosium.

12 Taken from www.hizook.com/blog/2009/12/28/electroactive-polymers-eap-artificial-muscles-epam-robot-applications#

are polymers that exhibit a change in size or shape when stimulated by an electric field. The most common applications of this type of material are in actuators and sensors. A typical characteristic property of an EAP is that it undergoes a large amount of deformation while sustaining large forces.

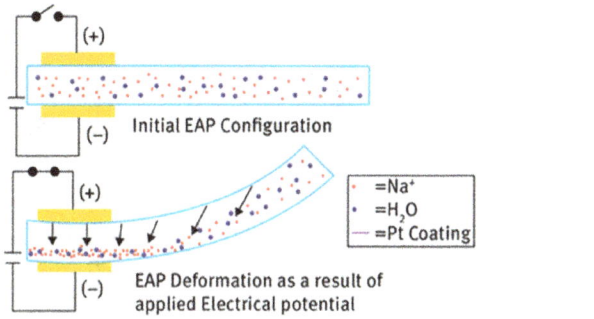

(a) Schematic electroactive activation

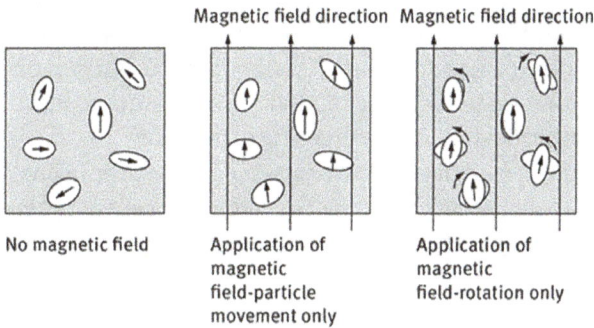

(b) Schematic magnetoactive activation

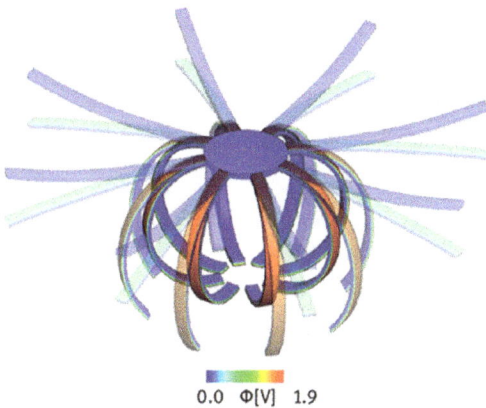

(c) Electroactive polymer based starfish gripper

Figure 1.9: Electroactive (EAP) and magnetoactive polymer (MAP): (a) EAP activation, (b) MAP activation (from [57]) and (c) Starfish gripper (from [58]).

The majority of smart-material-based actuators are made up of ceramic piezoelectric materials. While these materials are able to withstand large forces, they commonly only deform a fraction of a percent. In the late 1990s, it has been demonstrated that some EAPs can exhibit up to 32% strain, which is much more than any ceramic actuator [56]. One of the most common applications of EAPs is in the field of robotics in the development of artificial muscles; thus, an EAP is often referred to as an artificial muscle.

MAPs are polymer-based composites that respond to magnetic fields with large deformation or tunable mechanical properties. While a variety of these materials exist, most are composites of a soft polymer matrix with a filler of magnetic particles. The multiphysics interactions in MAPs give them two very attractive features. First, they respond to a magnetic field with variable mechanical properties (e.g., stiffness). Second, their shape and volume may be significantly changed in a magnetic field. Both features could be tuned by engineering the microstructure of the composites. Potential applications of MAPs include sensors, actuators, bio-medicine and augmented reality.

One should note that by embedding ferrous particles in a polymer matrix a solid elastomer is formed. Unlike in MRF (magnetorheological fluids – see discussion about it further in this chapter), the particles in an EAP or MAP have very restricted movement. The application of a magnetic field results in hardening of the MAP.

A different class of intelligent materials is the shape memory alloys (SMAs), which is a unique class of metal alloys that can recover apparent permanent strains when they are heated above a certain temperature and present a pseudoelasticity (PE) effect with large strains (see [58–61]). The SMAs have two stable phases: the high-temperature phase called austenite, and having a cubic crystal structure and the low-temperature phase called martensite with monoclinic crystal structure. In addition, the martensite state can be in one of two forms: twinned and detwinned, as presented schematically in Figure 1.10. A phase transformation that occurs between these two phases upon heating/cooling is the basis for the unique properties of the SMAs which are shape memory effect (SME) and PE.

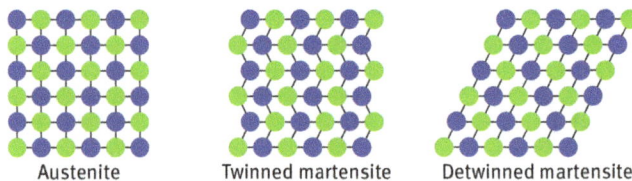

Austenite   Twinned martensite   Detwinned martensite

Figure 1.10: Schematic drawings of the austenite phase, the twinned martensite and the detwinned martensite.

The transformation from one phase to other can be done either by heating, or by mechanical loading or by a combination of both temperature and mechanical loading.

For the case of cooling in the absence of applied load the material transforms from austenite into twinned martensite with no observable macroscopic shape changes. Heating the material in the martensitic phase, a reverse phase transformation takes place yielding the austenite phase (see Figure 1.11).

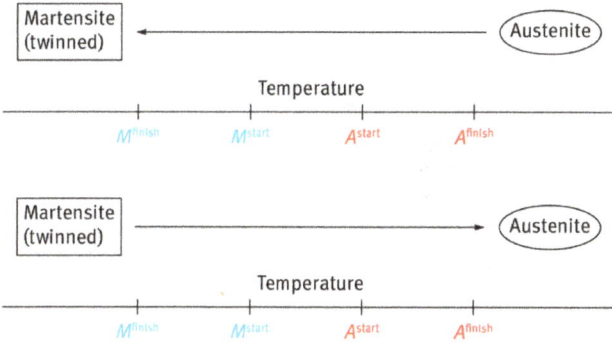

Figure 1.11: SMA – temperature-induced phase transformation without mechanical loading.

Figure 1.11 shows four characteristic temperatures: martensitic start temperature (M^{start}) which is the temperature at which the material starts transforming from austenite to martensite; martensitic finish temperature (M^{finish}), at which the transformation is complete and the material is fully in the martensitic phase; austenite start temperature (A^{start}) at which the reverse transformation (austenite to martensite) starts; and austenite finish temperature (A^{finish}) at which the reverse-phase transformation is completed and the material is the austenitic phase.

If a mechanical load is applied to the material while being at a state of twinned martensite (at low temperature) it is possible to obtain a detwinned martensite. When the load is released, the material remains deformed. Now if the material is heated to a temperature above A^{finish} will result in reverse-phase transformation (martensite to austenite) and will lead to a complete shape recovery (SME), as schematically shown in Figure 1.12.

When a mechanical load is applied in the austenitic phase and the material is cooled, the phase transformation will result in a detwinned martensite. Thus, very large strains (on the order of 5–8%) will be observed. Reheating the material will result in complete shape recovery. The above-described loading path is shown in Figure 1.13.[13] Note that the transformation temperatures in this case strongly depend on the magnitude of the applied load. Higher values of the applied load will lead to higher values of the transformation temperatures. Usually, a linear relationship between the applied mechanical load and the transformation temperatures is assumed, as shown in Figure 1.13.

---

13 From smart.tamu.edu/overview/overview.html

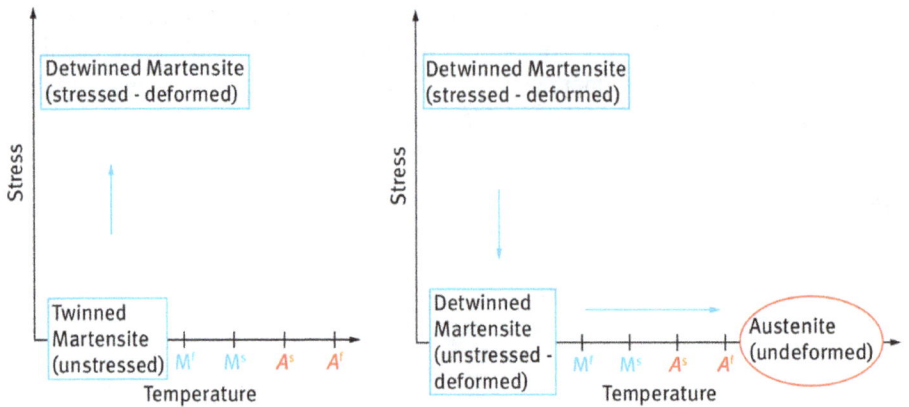

Figure 1.12: SMA – stress-induced phase transformation and the SME.

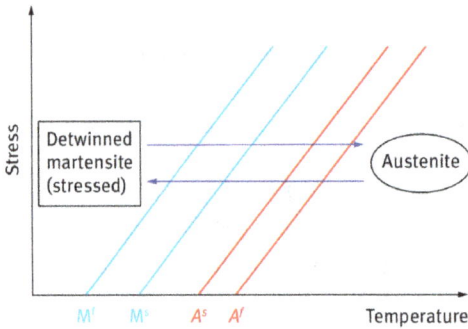

Figure 1.13: SMA – stress-induced phase transformation – austenite to detwinned martensite and opposite.

Figure 1.14 presents the stress–strain curves[14] for a typical SMA at various temperatures. Note the difference between the pseudoelastic effect (c) and the memory effect (b).

One of the most known SMA materials is NITINOL (sometimes called also nickel titanium) a metal alloy composed of nickel and titanium. The term NITINOL is derived from its composition and its place of discovery: (Nickel Titanium-Naval Ordnance Laboratory). Some of its properties are shown in Table 1.2 [61]. NITINOL has a large use in the medical sector (see Figure 1.15) and also in mechanical, civil and aeronautical sectors by providing reliable advanced actuators and springs.

The last topic to be presented, regarding intelligent materials, is electrorheological (ER) and magnetorheological (MR) fluids.

---

14 *From:* http://heim.ifi.uio.no/~mes/inf1400/COOL/RobotProsjekt/Flexinol/ShapeMemoryAlloys.htm

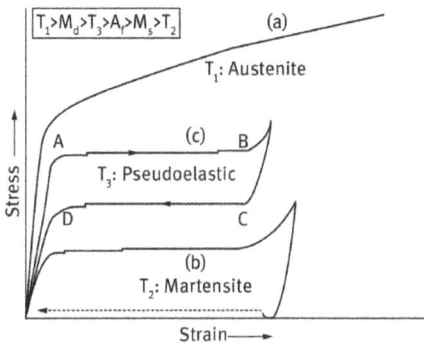

Figure 1.14: SMA – typical stress–strain curves at different temperatures relative to the phase transformation, showing (a) austenite, (b) martensite and (c) pseudoelastic behavior.

Table 1.2: Nitinol – mechanical and electrical properties.

Property	Martensite phase	Austenite phase
Density (g/cm^3)	6.45	6.45
Elastic modulus (GPa)	28–40	75–83
Yield stress (MPa)	70–140	195–690
Poisson's ratio	0.33	0.33
Coefficient of thermal expansion ($\times 10^{-6}/$)	6.6	11
Electrical resistance ($\times 10^{-6}\Omega$cm)	76	82
Thermal conductivity (W/cmK)	0.086	0.18

ER fluids are a class of materials whose rheological characteristics can be controlled by application of an electrical field. A typical ER fluid consists of colloidal dispersions of dielectric particles in a liquid with low dielectric constant. When the electric field is applied, the effective viscosity of the ER fluid increases dramatically. Above a critical value, the ER fluid might turn into a gel-type solid whose shear stress would increase as the electrical field is strengthened further. From the structural point of view, it was observed that the dielectric particles within the ER fluid will tend to form chains spanning between the two electrodes of the device (see Figure 1.16a and b). With an increase in the applied electric field, those chains would aggregate to form thick columns, yielding a dramatically increase in the apparent viscosity of the ER fluid that was brought to the world through a patent obtained by Winslow in 1947 [62] followed up by a manuscript in 1950 [63]. His patent is known as a functional fluid whose yield stress can be changed by the applied voltage. The property of ER fluid is such that it can be changed by base oil and characteristics, as well as size and density of particles dispersed in the fluid [64–65]. Extensive research had been performed to improve the performance and the stability of the ER fluid and to its applications as mechanical components such as suspensions, absorbers, engine mount, clutch, break and valve

When cold, it is
bent to any any
shape

Hot
water

The original
wire

Dropped in hot
water it returns to
its original shape

Nitinol Wire

(a) A simple experiment

(b) Protage® GPS™ Self-Expanding Nitinol
Biliary Stent System

504.4 mm

Nitinol helical spring

82.6 mm

Screw   Hardened
cap   disk

Shaft nut

Nitinol belleville washers

(c) SMA Devices: Helical spring and belleville washer

(d) Nitinol smart spring

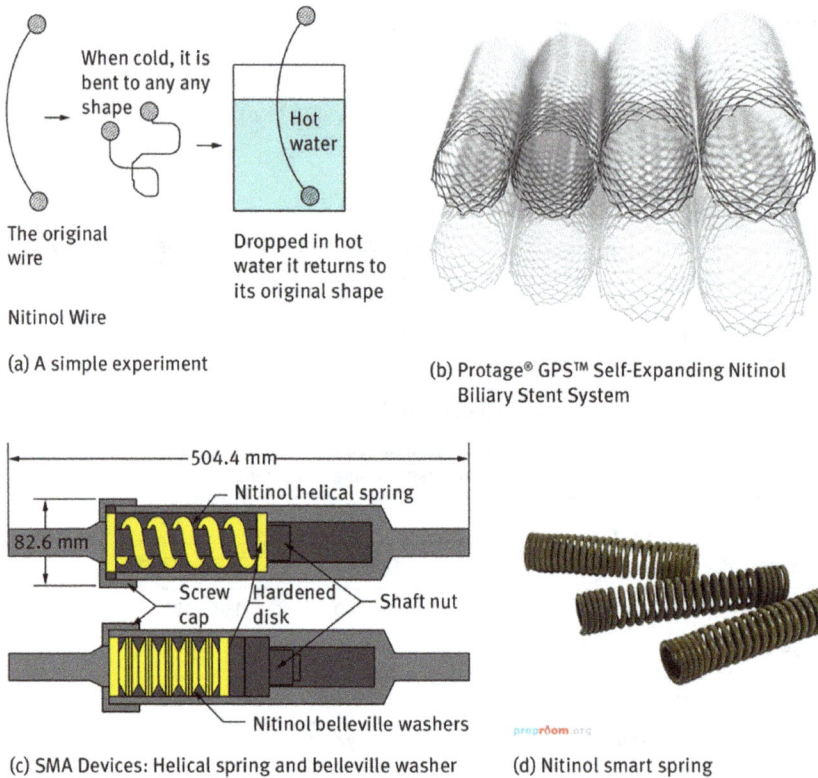

Figure 1.15: NITINOL applications and uses. Sources: www.talkingelectronics.com/projects/Niti
nol (a), M&J Co. Ltd, www.twmt.url.tw (b), www.neesrcr.gatech.edu (c), www.preproom.org (d).

(see Figure 1.16d). When activated, an ER fluid would behave like a Bingham plastic
model,[15] with a yield point which is determined by the electric field strength. After
the yield point is reached, the fluid shears as a fluid, namely, the incremental shear
stress is proportional to the rate of shear contrarily to a Newtonian-type fluid where
no yield point exists and the stress is directly proportional to the shear strain.

Although devices based on ER fluids are already on market, two major problems
still prevent their big breakthrough: the tendency of the particles suspensions to
settle out at the bottom of the container, thus preventing the formation of the chains
as a result of the application of an electrical field; the second problem is connected
to the breakdown voltage of air, being around 3 kV/mm, near the needed electric
field to activate an ER-fluid-based device. To answer the first problem, advanced research had been performed to match the densities of the particles and their respective

---

15 A Bingham (named after Eugene Cook Bingham 1878–1945) plastic is a viscous-plastic material
that behaves as a rigid body at low stresses but flows as a viscous fluid at high stress [66].

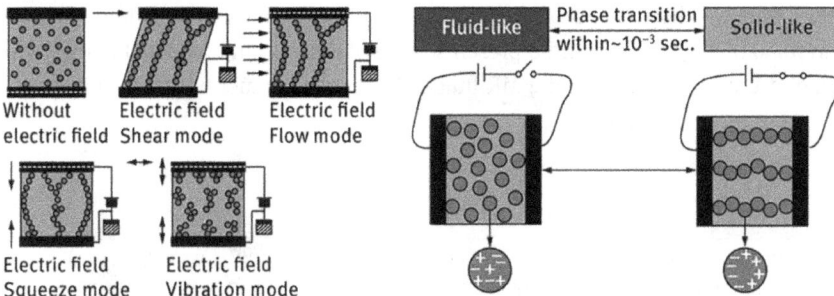

(a) Schematic mechanism of ER

Without electric field

Electric field Shear mode

Electric field Flow mode

Electric field Squeeze mode

Electric field Vibration mode

Fluid-like | Phase transition within~10^{-3} sec. | Solid-like

(b) Schematic diagram for ER activation

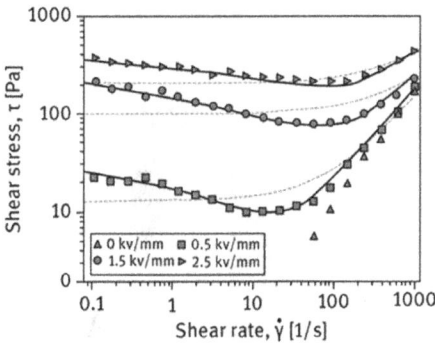

(c) Flow curves PANI/BaTiO$_3$ based ER fluid

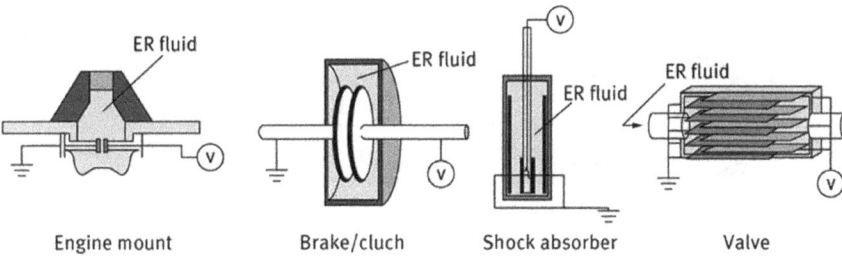

$\blacktriangle$ 0 kv/mm $\blacksquare$ 0.5 kv/mm $\bullet$ 1.5 kv/mm $\blacktriangleright$ 2.5 kv/mm

ER fluid

ER fluid

ER fluid

ER fluid

Engine mount   Brake/cluch   Shock absorber   Valve

(d) ER fluid applications

Figure 1.16: ER fluid: (a) schematic mechanism, (b) schematic activation, (c) shear rate versus shear stress – PANI/BaTiO$_3$-based fluid and (d) applications and uses.

liquid, and/or using nanoparticles. The second problem is tackled by insulating the ER device, and preventing it from the direct contact of air.

Similar to ER fluids, MR fluids is a type of smart rheological fluid that will change its apparent viscosity as a result of the application of a magnetic field. When subjected to a magnetic field, the fluid greatly increases its apparent viscosity, to the point of becoming a viscoelastic solid. The fluid yield stress can be controlled very accurately by varying the magnetic field intensity, using, for instance, an electromagnet, and thus enabling the control of the force being generated by the MR fluid to the designated receiver. Extensive discussions of the physics and applications of MR fluids can be found in [67–70]. Figure 1.17a and b presents some of the working modes (Figure 1.17a) and some of the applications using MR fluids. As previously mentioned for ER fluids,

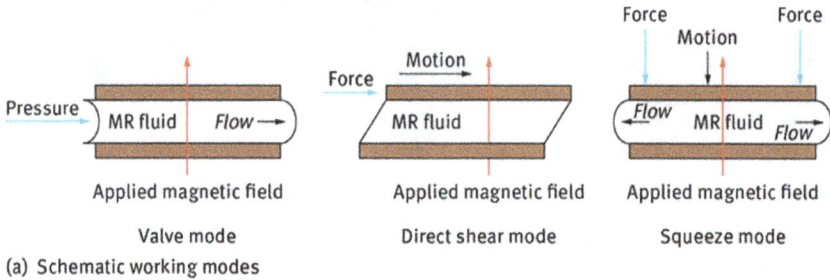

(a) Schematic working modes

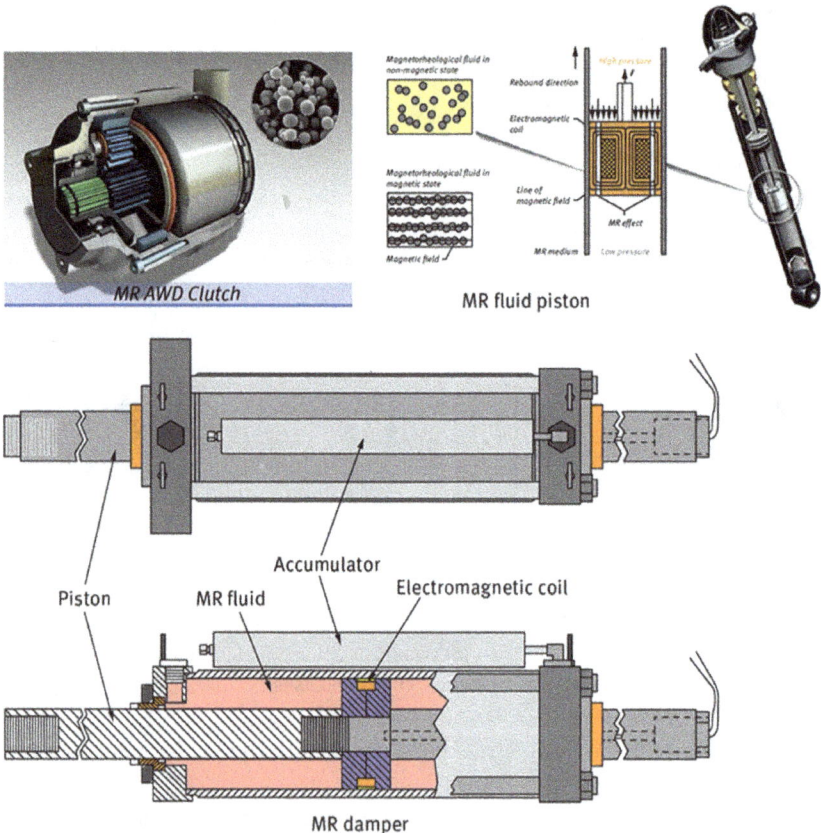

(b) Applications and uses

Figure 1.17: MR fluid: (a) schematic working modes and (b) applications and uses. Sources: scmero.ulb.ac.be/project (a); www.atzoniline.com, www.autoserviceprofessional.com, NEES Project Warehouse (b).

MR fluid particles are primarily on the micrometer scale and are too dense for Brownian-type motion to keep them suspended (in the lower density carrier fluid), and thus might settle over time because of the inherent density difference between the particle and its carrier fluid. As mentioned previously, this problem is dealt in the same way as for the ER fluid, or by maintaining a constant magnetic field, to prevent the settling of the suspended particles and their carrying liquid.

References

[1] Wada, B. K., Fanson, J. L. and Crawley, E. F., Adaptive structures, Journal of Intelligent Material Systems and Structures 1, April 1990, 157–174.
[2] Lazarus, K. B. and Napolitano, K. L., Smart structures, An overview, AD-A274 147, WL-TR-93-3101, Flight Dynamics Directorate Wright Laboratory Air Force, Material Command Wright-Patterson Air Force Base, Ohio 45433–7552, September 1993, 60 pp.
[3] Iyer, S. S. and Haddad, Y. M., Intelligent materials-An overview, International Journal of Pressure Vessels and Piping 58, 1994, 335–344.
[4] Gandhi, M. V. and Thompson, B. S., Smart Materials and Structures, 1st edn, London, New York, Chapman & Hall, 1992, 309.
[5] Culshaw, B., Smart Structures and Materials, Artech House, 1996, 207.
[6] Preumont, A., Vibration Control of Active Structures, Dordrecht, the Netherlands, Kluwer Academic Publishers, 1997, 259.
[7] Adeli, H. and Saleh, M., Control, Optimization, and Smart Structures: High-Performance Bridges and Buildings of the Future, John Wiley & Sons, May 3 1999, 265.
[8] Srinivasan, A. V. and McFarland, D. M., Smart Structures: Analysis and Design, 1st edn, Cambridge University Press, 2001, 223.
[9] Watanabe, K. and Ziegler, F., Dynamics of Advanced Materials and Smart Structures, Springer Science & Business Media, July 31 2003, 469.
[10] Wadhawan, V. K., Blurring the distinction between the living and nonliving, Monographs on Physics and Chemistry of Materials 65, Oxford University Press, 18 October 2007, 368.
[11] Bandyopadhyay, B., Manjunath, T. C. and Umapathy, M., Modeling, Control and Implementation of Smart Structures: A FEM-State Space Approach, Springer, April 22 2007, 258.
[12] Leo, D. J., Engineering Analysis of Smart Material Systems, John Wiley and Sons, Inc., 2007, 556.
[13] Cetikunt, S., Mechatronics, John Wiley and Sons, Inc, 2007, 615.
[14] Giurgiutiu, V., Structural Health Monitoring: With Piezoelectric Wafer Active Sensors, 1st edn, Elsevier, Academic Press, 2008, 747.
[15] Gaudenzi, P., Smart Structures: Physical behavior, Mathematical Modeling and Applications, John Wiley and Sons, Inc, 2009, 194.
[16] Pryia, S. and Inman, D. J., (eds.), Energy Harvesting Technologies, Springer Science & Business Media, 2009, 517.
[17] Carrera, E., Brischetto, S. and Nali, P., Plates and Shells for Smart Structures: Classical and Advanced Theories for Modeling and Analysis, John Wiley & Sons Ltd., September 2011, 322.
[18] Holnicki-Szulc, J. and Rodellar, J., Smart Structures: Requirements and Potential Applications in Mechanical and Civil Engineering, Springer Science & Business Media, December 6 2012, 391.
[19] Vepa, R., Dynamics of Smart Structures, 1st edn, Wiley, April 2010, 410.

[20] Chopra, I. and Sirohi, J., Smart Structures Theory, New York, Cambridge University Press, 2013, 905.

[21] Janocha, H., (ed.), Adaptronics and Smart Structures: Basics, Materials, Design, and Applications, Springer Science & Business Media, November 11 2013, 438.

[22] Kholkin, A. L., Pertsev, N. A. and Goltsev, A. V., Piezoelectricity and Crystal Symmetry, Ch. 2 from Piezoelectric and Acoustic Materials for Transducer Applications, Safari, A. and Akdogan, E. K., (eds.), Springer, 2008, 482.

[23] Curie, P. and Curie, J., Developpement par pression, de l'electricite polaire dans les cristaux hemiedres a faces inclines, Comptes Rendus (France) 91, 1880, 294–295.

[24] Curie, P. and Curie, J., Contractions et dilatations produites par des tensions electriques dans les cristaux hemiedres a faces inclines, Comptes Rendus (France) 93, 1881, 1137–1140.

[25] Voigt, W., Lerbuch der Kristallphysik, Leipzig–Berlin, Teubner, 1910.

[26] Jaffe, B., Cook, W. R. and Jaffe, H., Piezoelectric Ceramics, New York, Academic Press, 1971.

[27] Monner, H. P., Smart materials for active noise and vibration reduction, Novem – Noise and Vibration: Emerging Methods, Saint-Raphaël, France, 18–21 April 2005.

[28] www.piceramic.com/pdf/PIC_Tutorial.pdf.

[29] Giurgiutiu, V., Pomirleanu, R. and Rogers, C. A., Energy-Based Comparison of Solid-State Actuators, University of South Carolina, Report # USC-ME-LAMSS-2000-102, March 1, 2000.

[30] http://www.piceramic.com/pdf/material.pdf.

[31] Kawai, H., The Piezoelectricity of Poly(vinylidene fluoride), The Japan Society of Applied Physics 8, 1969, 975.

[32] Damjanovic, D., Ferroelectric, dielectric and piezoelectric properties of ferroelectric thin films and ceramics, Reports on Progress in Physics 61, 1998, 1267–1324.

[33] García, C., Arrazola, D., Aragó, C. and Gonzalo, J. A., Comparison of symmetric and asymmetric energy conversion cycles in PZT plates, Ferroelectrics Letters Section 28(1–2), 2000, 43–47.

[34] Navid, A. and Pilon, L., Pyroelectric energy harvesting using Olsen cycles in purified and porous poly(vinylidene fluoride-trifluoroethylene) [P(VDF-TrFE)] thin films, IOP Publishing, Smart Materials and Structures 20, 2011, 025012, 18 August 2010, 9. doi:10.1088/0964-1726/20/2/025012.

[35] Xie, J., Mane, P. P., Green, C. W., Mossi, K. M. and Leang, K. K., Energy harvesting by pyroelectric effect using PZT, Proceedings of SMASIS08 ASME Conference on Smart Materials, Adaptive Structures and Intelligent Systems, October 28–30, 2008, Ellicott City, Maryland, USA.

[36] Sebald, G., Guyomar, D. and Agbossou, A., On thermoelectric and pyroelectric energy harvesting, Smart Materials and Structures 18, 2009, 125006. doi:10.1088/0964-1726/18/12/125006.

[37] Xie, J., Experimental and numerical investigation on pyroelectric energy scavenging, 2009, VCU Theses and Dissertations, Paper 2041.

[38] Lee, F., Experimental and analytical studies on pyroelectric waste heat energy conversion, Master of Science Thesis in Mechanical Engineering, 2012, University of California, Los Angeles, USA, 138 pp.

[39] Erturn, U., Green, C., Richeson, M. L. and Mossi, K., Experimental analysis of radiation heat-based energy harvesting through pyroelectricity, Journal of Intelligent Material Systems and Structures 25(September), 2014, 1838–1849.

[40] Bowen, C. R., Taylor, J., LeBoulbar, E., Zabek, D., Chauhan, A. and Vaish, R., Pyroelectric materials and devices for energy harvesting applications, Energy and Environmental Sciences 7, 2014, 3836–3856.

[41] Navid, A., Lynch, C. S. and Pilon, L., Synthesis and characterization of commercial, purified, and Porous Vinylidene Fluoride-Trifluoroethylene P(VDF-TrFE) thin films, Smart Materials and Structures 19, 2010, 055006(1–13).

[42] Li, F., Jin, L., Xu, Z. and Zhang, S., Electrostrictive effect in ferroelectrics: An alternative approach to improve piezoelectricity, Applied Physics Reviews 1, 2014, 011103, AIP Publishing LLC. doi:http://dx.doi.org/10.1063/1.4861260.

[43] Surowiak, Z., Fesenko, E. G. and Skulski, R., Dielectric and electrostrictive properties of ferroelectric relaxors, Archives of Acoustics 24(3), 1999, 391–399.

[44] Newnham, R. E., Properties of Materials: Anisotropy, Symmetry, Structure, OUP, Oxford, November 11, 2004, 390 pp.

[45] Uchino, K., Nomura, S., Cross, L. E., Jang, S. J. and Newnham, R. E., Electrostrictive effect in lead magnesium niobate single crystals, Journal of Applied Physics 51, 1980, 1142–1145.

[46] Uchino, K., Nomura, S., Cross, L. E., Newnham, R. E. and Jang, S. J., Review Electrostrictive effect in perovskites and its transducer applications, Journal of Materials Science 16, 1981, 569–578.

[47] Cross, L. E., Jang, S. J., Newnham, R. E., Nomura, S. and Uchino, K., Large electrostrictive effects in relaxor ferroelectrics, Ferroelectrics 23, 1980, 187–191.

[48] Nomura, S., Kuwata, J., Jang, S. J., Cross, L. E. and Newnham, R. E., Electrostriction in Pb (Zn1/3Nb2/3)O3, Material Research Bulletin 14, 1979, 769–774.

[49] Jang, S. J., Uchino, K., Nomura, S. and Cross, L. E., Electrostrictive behavior of lead magnesium niobate based ceramic dielectrics, Ferroelectrics 27, 1980, 31–34.

[50] Uchino, K., Cross, L. E., Newnham, R. E. and Nomura, S., Electrostrictive effects in non-polar perovskites, Phase Transitions 1, 1980, 333–342.

[51] Zhang, Q. M., Bharti, V. and Zhao, X., Giant electrostriction and relaxor ferroelectrics behavior in electron-irradiated poly (vinylidene fluoride-tri-fluoroethylene) copolymer, Science 280, 1998, 2101–2104.

[52] Buschow, K. H. J. and De Boer, F. R., Physics of Magnetism and Magnetic Materials, Ch. 16, Vols 171–175, New York, Kluwer Academic/Plenum Publishers, 2003, 182.

[53] Joule, J. P., On the effects of magnetism upon the dimensions of iron and steel bars, The London, Edinburgh and Dublin philosophical magazine and journal of science (Taylor & Francis) 30, 1847, 225–241, Third Series: 76–87.

[54] Barrett, R. A. and Parsons, S. A., The influence of magnetic fields on calcium carbonate precipitation, Water Research 32, 1998, 609–612.

[55] Aljanaideh, O., Rakheja, S. and Su, C.-Y., Experimental characterization and modeling of rate-dependent asymmetric hysteresis of magnetostrictive actuators, Smart Materials and Structures 23, 2014, 035002, 12. doi:10.1088/0964-1726/23/3/035002.

[56] Bar-Cohen, Y., Electroactive polymers as artificial muscles – Capabilities, Potentials and Challenges, HANDBOOK ON BIOMIMETICS, Yoshihito Osada (Chief ed.), Section 11, in Chapter 8, "Motion" paper #134, publisher: NTS Inc., August 2000, 1–13. Han, Y., Mechanics of magneto-active polymers, Graduate Theses and Dissertations Paper 12929, Iowa State University, 2012, 112 pp.

[57] Zah, D. and Miehe, C., Variational-based computational homogenization of electro-magneto-active polymer composite at large strains, 11 th World Congress on Computational Mechanics (WCCM XI), 5th European Conference on Computational Mechanics (ECCM V), 6th European Conference on Computational Fluid dynamics (ECFD VI), July 20–25,204, Barcelona, Spain.

[58] Stoeckel, D., The shape memory effect – phenomenon, alloys and applications, Proceedings: Shape Memory Alloys for Power Systems, EPRI, 1995.

[59] Wei, Z. G., Sandström, R. and Miyazaki, S., Review shape-memory materials and hybrid composites for smart systems, Journal of Material Science 33, 1998, 3743–3762, 10 September 1997.

[60] Seelecke, S. and Müller, I., Shape memory alloy actuators in smart structures: Modeling and simulation, Applied Mechanics Reviews 57(1), January 2004, 23–46.

[61] Barbarino, S., Saavedra Flores, E. I., Ajaj, R. M., Dayyani, I. and Friswell, M. I., A review on shape memory alloys with applications to morphing aircraft, IOP Publishing, Smart Materials and Structures 23, 2014, 063001 (19 pp.), 1 August 2013. doi:10.1088/0964-1726/23/6/063001.

[62] Winslow, W. M., Method and means for translating electrical impulses into mechanical force, U.S. Patent 2,417,850, 25 March 1947.

[63] Winslow, W. M., Induced fibration of suspensions, Journal of Applied Physics 20(12), January 1950, 1137–1140.

[64] Ahn, Y. K., Yang, B.-S. and Morishita, S., Directionally controllable squeeze film damper using electro-rheological fluid, Journal of Vibration and Acoustics 124, January 2002, 105–109.

[65] Choi, H. J. and Jhon, M. S., Electrorheology of polymers and nanocomposites, Soft Matter 5, 2009, 1562–1567.

[66] Bingham, E. C., An investigation of the laws of plastic flow, Bulletin of the U.S. Bureau of Standards Bulletin 13, 1916, 309–353, scientific paper 278 (S278).

[67] Zhu, X., Jing, X. and Cheng, L., Magnetorheological fluid dampers: A review on structure design and analysis, Journal of Intelligent Material Systems and Structures 23(8), 2012, 839–873.

[68] Baranwal, D. and Deshmukh, T. S., MR-fluid technology and its application- A Review, International Journal of Emerging Technology and Advanced Engineering Website: www.ije tae.com (ISSN 2250–2459, ISO 9001: 2008Certified Journal 2 (12), December 2012).

[69] Sulakhe, V. N., Thakare, C. Y. and Aute, P. V., Review-MR fluid and its application, International Journal of Research in Aeronautical and Mechanical Engineering (IJRAME) 1(7), ISSN (Online), 2321–3051, November 2013, 125–133.

[70] Kciuk, S., Turczyn, R. and Kciuk, M., Experimental and numerical studies of MR damper with prototype magnetorheological fluid, Journal of Achievements in Materials and manufacturing Engineering 39(1), March 2010, 52–59.

1.2 Extended Review on Piezoelectric Materials, SMA, ER and MR Fluids and MS and ES Materials

The following extended review[16] does not claim to include all the references existing in the free literature, but it is aimed at presenting typical studies of the main topics presented in the book. References [1–268] belong to piezoelectric materials and their usage, [269–378] describe the research performed on SMA, [379–536] outline the research on ER and MR fluids, and [537–604] address magneto- and electrostrictive (MS and ES) materials and their applications.

1.2.1 Extended Review on Piezoelectricity

Piezoelectric materials had been and still are studied from all their scientific and engineering aspects. The long list of references provides the reader with valuable data regarding of what had been published in the literature. One should note that in

---

16 The references are brought in their chronological appearance in the literature and are roughly divided in parts relevant to what they contain.

Chapter 3 of the present book, additional references are included to further clarify the piezoelectric topic.

The present review of piezoelectric materials will start with polyvinylidene fluoride (PVDF) piezofilm which is characterized and its preliminary applications of which are discussed in [1–8]. The use of PVDF for vibration control of beams and plates is outlined in [9–10], while its use in space is discussed in [11–12]. Various applications of PVDF as sensors are presented in [13–16], while in [17–18] a hybrid array of PZT/PVDF is used for vibration control. The use of PVDF transducers for Lamb wave's measurements is studied in [14]. Various manufacturing concepts of PVDF and ways of improving its properties are discussed and presented in [19–35], while its application to harvest ambient energy is described in [38] where PVDF microbelts harvest energy from respiration yielding 1–2 mJ, and in [39] where the use of porous surface on the top of the conventional PVDF films increased the harvested energy.

Without doubt, the number of studies focusing on the static and dynamic behavior of beams with piezoelectric layers and patches using various beam, plate and shell theories is the largest among all the smart materials publications. A good introduction in this area is presented in related books and reviews presented in [40–54]. This starts with two of the basic books on piezoelectricity [40–41] and followed up with various reviews on the up to date state of the art on the ferroelectric ceramics [42–46]. The various theories applicable to piezoelectric laminates are reviewed in [47–49], while descriptions on the various usages of piezoelectric materials, like energy harvesters, or acting as actuators and suppressing vibrations and noise are outlined in [50–54]. The next group of references [55–105] addresses topics like: static and dynamic behavior of beam-like structures equipped with piezoelectric continuous layers and/or patches, the use of piezoelectric materials for damping vibrations either as a shunt or in closed loop control, how to control the piezolaminated beams and increasing the damping capabilities using what is called active constrained layer damping (ACLD). The ACLD configuration uses an active element, usually a piezoelectric layer, to augment the passive constraining layer (usually a viscoelastic layer), thereby enhancing the energy dissipation of the damping. The references include models to describe the piezoelectric actuations on beams (usually applying the basic Bernoulli–Euler theory), the use of cylindrical bending of laminated composite plates, which reduces the 2D problem to a 1D problem suitable to beam-like structures, advanced mechanical theories, like Timoshenko's theory and the coupled layerwise analysis of composite piezolaminated beams, sandwich-type beams with piezoelectric patches and the application of various control laws to provide active damping of beams. Most of the references mentioned above present theoretical/numerical calculations, while in some of them are accompanied with experimental results aimed at enhancing the various models developed in the studies. Another important subject is the behavior of helicopter rotors with piezoelectric patches aimed at inducing strains to morph the blade of the rotor thus providing control capabilities to modify the rotor shape according to the flight conditions. Typical studies are referenced in [106–110]. The sensing use of piezoelectric materials aimed at either detecting various impacts to close the control loop or to detect delaminations in laminated composite structures is covered in [111–124]. Theoretical and experimental results are brought and the fidelity of the theories used is discussed. The actuation of various structures by the application of piezoelectric layers, bimorph patches or stacks is highlighted in references [125–151]. The studies include self-sensing actuators, enhanced actuators capable of providing large strains, calculations of the power characteristics of the actuators, experimental and theoretical calculations of the piezoelectric-based actuators and their performances in various structures such as beams, plates and wings. One should note the all the above presented studies use exclusively the d_{31} strain coefficient for both actuation and sensing, with the larger coefficient d_{33} being applied for the stacks configurations. As can be seen in Chapter 3, which deals with the piezoelectric materials, there is a possibility to apply the shear-strain coefficient, d_{15}, to obtain actuation and sensing. This is highlighted in [152–162], where closed-form solutions for beams and plates using either shear-type transducers or both tension–compression (using the d_{31} coefficient) together with shear are used to enhance the damping of the structures or to enable energy harvesting. All the quoted studies present analytical/numerical calculations.

Another interesting application of the piezoelectric material is presented in the studies quoted in [163–181]. It was shown both analytically and numerically that for certain conditions, the bending stiffness of beam-like structures and/or plate structures can be increased by the use of piezoelectric layer or patches, in both static and dynamic cases. The piezoelectric materials, serving both as sensors and actuators, can control the buckling loads of the structures and even increase their stability margin by applying enough actuation to prevent the loss of the structure. In parallel, by enhancing the stiffness of the structure, the natural frequencies are also changed, leading to an increase while the structure is under tension (due to the piezoelectric-induced strains) or to a decrease in their values when the induced strains on the beam are compressive ones. The studies present both analytical/numerical predictions accompanied in some manuscripts by experimental results. References [182–186] address the elastic piezoelectric topic by applying Saint-Venant's principle[17] and obtaining elastic solutions to compound piezo beams.

The solutions for plates and shells with piezoelectric patches acting as sensors and/or actuators has been treated and presented in various manuscripts quoted in

---

17 Saint-Venant's principle (to honor a French elasticity theorist Adhémar Jean Claude Barré de Saint-Venant) states that "If the forces acting on a small portion of the surface of an elastic body are replaced by another statically equivalent system of forces acting on the same portion of the surface, this redistribution of loading produces substantial changes in the stresses locally but has a negligible effect on the stresses at distances which are large in comparison with the linear dimensions of the surface on which the forces are changed" (*Theory of Elasticity*, 2nd ed., by S. Timoshenko and J. N. Goodier, McGraw-Hill, 1934, p. 33.)

[187–223]. Various types of piezoelectric plates and shells are analyzed using Kirchh-off–Love and advanced theories like first-order shear deformation theory, to yield sensing and actuation of the piezoelectric parts of the structures that are mainly made from laminated composite layers (or sandwich), with active piezoelectric layers en-abling closed-loop control of their static and dynamic behavior. The manuscripts pres-ent numerical calculations in the form of predicted displacements, modal voltages, shape control of the structure and electrical characteristics of the adaptive structures.

Another wide published topic is the use of piezoelectric materials to harvest en-ergy from ambient vibrations and other parasitic energy sources and transform it to electrical energy aimed at providing various local users their needed electrical en-ergy. Typical references from a huge number of manuscripts published in the last 15 years are outlined in [224–249]. It starts with the way using vibrating cantilevered beam equipped with piezoelectric layer/patches matching the ambient frequency can extract mechanical energy from the vibration and using an adequate electrical circuit to transform to electrical energy and store it in either batteries or other sour-ces of storage like capacitors or supercapacitors. The numerous studies show nu-merical predictions of how much energy (or power) can be harvested, accompanied by well-designed tests to back up the theoretical predictions, and advanced electri-cal circuits aimed at increasing the electrical output. Various advanced designs in-clude ways of improving the devices to yield higher electrical energy to supply to the potential users.

In parallel to the conventional ceramic perovskite[18]-type piezoelectric material in the form of PZT (lead zirconate titanate) a lot of efforts have been directed to manufacture and predict the performance of what is called single crystals trans-ducers fabricated from single crystals of lithium niobate, synthetic quartz, or other materials. Although some of those materials are electrostrictive in their nature, like PNM-PT, these references were brought here, as most of them compare their proper-ties to PZT. These piezoelectric materials present superior piezoelectric properties, relative to PZT elements. Having a relative insensitivity to temperature and exhibit-ing very high electrical/mechanical energy converting factors made them very at-tractive for a variety of applications. This is shown in [250–262], where various models were developed to predict the output of devices built with single-crystal ma-terials, their usage as energy harvesters and their control.

Finally, the usage of piezoelectric materials as sensors in structural health mon-itoring (SHM) was addressed in many studies. A few typical examples are given in [263–268], where piezoelectric patches are used to sense the state of the structure and give advanced warning when it starts to lose its integrity, yielding a valuable tool to monitor the various critical parts within a structure (for example a bridge).

---

18 Perovskite (named after Russian mineralogist Lev Perovski) – a calcium titanium oxide mineral composed of calcium Titanate, with the chemical formula $CaTiO_3$.

The studies deal with the integration of the sensors and their connection to the general electric system which monitors the state of the structure.

1.2.2 Extended Review on Shape Memory Alloys

SMAs had been, and still are, researched from all their scientific and engineering aspects. The list of references [269–378] presents data related to SMAs which had been published in the literature. Note that in Chapter 4 of this book, additional references are included to clarify the SMA issue.

Although named SMA (Shape Memory Alloy), these alloys present two main characteristics: the first one is the SME which describes the capability of the alloy to undergo deformation at one temperature, then recover its original, undeformed shape by heating it above its transformation temperature. The second characteristic is called PE or sometimes SE (super elasticity). PE occurs just above the alloy's transformation temperature and no heating is necessary to recover the undeformed shape. This property exhibits enormous elastic strains, approx. 10–30 times that of normal metals, which are recovered without any plastic deformation.

To understand the pretty complicated behavior of a shape memory material, the readers are invited to start reading the various basic books and reviews presented in [269–293]. The books [269–288] cover the basics of the SMAs and include various examples of applications in the nonmedical and biomedical sectors. The reviews, present the up-to-date state of the art in scientific and technological aspects of SMA, including modeling, and the various applications based on SMA (see, e.g., the last reviews [289–292]). After being initiated in the SMA technology, some of the advanced applications are presented. References [294–301] present the way a structure can have enhanced stability, by inserting SMA wires, leading to increased stiffness, active vibration control and shape control of the structures. One of the major applications of SMA is the biomedical sector, where stents and other devices can be implanted in the human body (due to the compatibility of NITINOL[19] – an SMA, with the human body) and thus performing advanced rehabilitation-based tasks. The various applications of NITINOL in the biomedical sector are described in typical references [302–312].

The introduction of SMA into rods, columns, shafts and actuators is investigated in a large number of publications, with [313–329] being only typical ones. Constitutive models, together with experimental validation of the proposed models, enable engineers to design and manufacture structures with attached SMA wires.

The introduction of SMA into much advanced structures like plates and shells is outlined in [330–334], where basic models are developed and the characterization of the static, dynamic and buckling behavior of those structures is performed.

---

19 NITINOL = Nickel Titanium-Naval Ordnance Laboratory – Ni = nickel, Ti = titanium.

A long list of publications is devoted to the complicated issue of modeling the transformation from martensite to austenite phase and back to martensite to be able to predict correctly the behavior of the SMA in both PE and memory effect tracks. Some of those publications are quoted in [335–363], and include advanced numerical tools including the use of Preisach model[20] [345] to capture the hysteresis curve of the SMA.

As it is known that the SMA wires can apply very large forces during their transformation from one phase to the other, their use in morphing wings are explored and characterized in [364–369], which form only a short list of publications for that application. Other investigations connected with SMA, like new shape memory alloy system, control issues, fatigue and cycled loading and even the use as sensors are dealt in detail in references [370–378].

1.2.3 Extended Review on Electrorheological and Magnetorheological Fluids

ER and MR fluids are other types of intelligent materials that had been investigated in depth from all their scientific and engineering aspects. References [379–536] bring data bring data published in the literature, regarding ER and MR fluids, their physical aspects and their applications as dampers. Chapter 5 of this book contains additional references to cover the topic of ER and MR fluids.

Both ER and MR fluids contain suspended particles in a fluid. The application of an electric field or a magnetic field would change its rheological behavior, thus providing a new class of fluids having a behavior of fluid/solid-like materials. These characteristics of those fluids were investigated in depth, providing a long list of publications. Books and reviews were written to provide knowledge of these ER and MR fluids [379–386]. The list of references includes two books [382–386] covering the various scientific and technological aspects of the fluids, accompanied by many reviews and state-of-the-art publications [379–387] aimed at providing the necessary know-how experience to design and build devices based on those two types of fluids.

The electrorheological phenomenon is investigated in many published studies, like [388–445] and covers all its aspects, starting from modeling, through understanding ways of designing appropriate devices by changing the properties of the fluid and culminating in real applications of this smart fluid mainly in the form of controlled advanced adaptive dampers. The use of both ER and MR fluids is presented in [446–473], where various performance comparisons between the two fluids are performed, their rheological properties are investigated, the capabilities of various developed models are evaluated and their structural applications are investigated.

References [474–536] present the investigations performed only on MR fluids, the various approaches to model the hysteresis curve (including Preisach model [500])

---

20 Preisach, F., Über die Magnetische Nachwirkung, Zeitschrift für Physik, Vol. 94, 1935, 277–302.

implemented in the various numerical approaches, their performance assessment as advanced dampers and their application in real structures, like, for example, the lag damper for helicopter rotor blades or brakes for automobile industry.

1.2.4 Extended Review on Magnetostrictive and Electrostrictive Materials

Magneto- and electrostrictive materials are another type of intelligent materials to be covered in this book. The list of typical references [537–604] describes that various investigations been performed on both magnetostrictive and electrostrictive materials, and electromagnetic and magnetic applications covering various scientific and engineering aspects. As noted before for the other types of intelligent materials, a dedicated chapter of this book (Chapter 6) presents additional references to further add to the knowledge of magnetostrictive and electrostrictive materials published in the literature.

Magnetostriction is a property of ferromagnetic materials that would change their shape or dimensions during the magnetization process. Equivalently, the electrostrictive property, appearing in all electrical nonconductors (dielectrics), causes them to change their shape under the application of an electric field. The transformation of either magnetic or electrical energy into kinetic energy, enable the use of those properties to build both actuators and sensors. With the appearance of TERFE-NOL-D[21] as a "giant" magnetostrictive actuator and other similar materials from one side, and the electrostrictive materials like PMN (lead, magnesium, niobate), PLZT (lead, lanthanum, zirconate, titanate) or PMN-PT (lead, magnesium, niobate-lead, titanate) from the other side, triggered the investigations of those new materials, yielding a long list of publications.

A few reviews can be found in the literature; for example [537–539], one deals with both electrostriction and magnetostriction [537], while the other two, review only the magnetostriction phenomenon [538–539]. Indeed the number of publications dealing with magnetostrictive materials is very large with only typical examples being outlined in [540–575]. Various researchers present modeling of the magnetostriction phenomenon, investigation of the TERFENOL-D and GALFENOL[22]/ALFENOL[23] properties, sensing and actuation using magnetostrictive materials, the effects of stress, and careful experiments aimed at the evaluation of the various devices using magnetostrictive materials. The investigation on electrostrictive-based devices is much more restricted and

---

21 TERFENOL-D: TER = terbium, FE = iron, NOL = Naval Ordnance Laboratory (today Naval Surface Warfare Center-Carderock Division (NSWC-CD)), D = dysprosium.
22 GALFENOL-D: GAL = gallium, FE = iron, NOL = Naval Ordnance Laboratory, today Naval Surface Warfare Center-Carderock Division (NSWC-CD).
23 ALFENOL-D: AL = aluminum, FE = iron, NOL = Naval Ordnance Laboratory, today Naval Surface Warfare Center-Carderock Division (NSWC-CD).

includes less publication. Some of those manuscripts are brought in [576–587], in which modeling of PMN is performed, the investigation of plate-like structures and shells with electrostrictive actuators is evaluated, properties of piezoelectric versus electrostrictive materials are outlined for performance assessment and tests are described to enhance the theoretical part of the research. Finally, references [588–604] contain a list of publications which address the general use of the electromagnetic and magnetic phenomena for vibration suppression, vibration control, magnetic constrained layer to induce enhanced damping in the structure and energy harvesting.

References – Piezoelectricity

PVDF-related references

[1]   Furukawa, T., Uematsu, Y., Asakawa, K. and Wada, Y., Piezoelectricity, pyroelectricity, and thermoelectricity of polymer films, Journal of Applied Polymer Science 12(12), December 1968, 2675–2689.
[2]   Murayama, N., Oikawa, T., Katto, T. and Nakamura, K., Persistent polarization in poly (vinylidene fluoride). 2. Piezoelectricity of Poly(vinylidene fluoride) thermoelectrets, Journal of Polymer Science. Part B, Polymer physics 13 (5), May 1975, 1033–1047.
[3]   Zimmerman, R. L. and Suchicital, C., Electric field-induced piezoelectricity in polymer film, Journal of Applied Polymer Science 19(5), May 1975, 1373–1379.
[4]   Shuford, R. J., Wilde, A. F., Ricca, J. J. and Thomas, G. R., Characterization of piezoelectric activity of stretched and poled poly(vinylidene fluoride) part I: effect of draw ratio and poling conditions, Polymer Engineering and Science 16(1), January 1976, 25–35.
[5]   Das-Gupta, D. K., Piezoelectricity and Pyroelectricity, Key Engineering Materials, Ferroelectric Polymers and Ceramic-Polymer Composites 92–93, February 1994.
[6]   Simpson, J. O., Welch, S. S. and St. Clair, T. L., Novel piezoelectric polyimides, Materials Research Society MRS Proceedings 413, 1995.
[7]   Hodges, R. V. and McCoy, L. E., Comparison of Polyvinylidene Fluoride (PVDF) gauge shock pressure measurements with numerical shock code calculations, Propellants Explosives Pyrotechnics 24(6), December 1999, 353–359.
[8]   Hodges, R. V., McCoy, L. E. and Toolson, J. R., Polyvinylidene Fluoride (PVDF) gauges for measurement of output pressure of small ordnance devices, Propellants Explosives Pyrotechnics 25(1), January 2000, 13–18.
[9]   Stöbener, U. and Gaul, L., Modal vibration control for PVDF coated plates, Journal of Intelligent Material Systems and Structures 11(4), April 2000, 283–293.
[10]   Audrain, P., Masson, P., Berry, A., Pascal, J.-C. and Gazengel, B., The use of PVDF strain sensing in active control of structural intensity in beams, Journal of Intelligent Material Systems and Structures 15(5), May 2004, 319–327.
[11]   Williams, R. B., Austin, E. M. and Inman, D. J., Limitations of using membrane theory for modeling PVDF patches on inflatable structures, Journal of Intelligent Material Systems and Structures 12(1), January 2001, 11–20.
[12]   Williams, R. B., Austin, E. M. and Inman, D. J., Local effects of PVDF Patches on inflatable space-based structures, Proceedings of the 42nd AIAA/ASME/ASCE/AHS/ASC Structures Structural Dynamics and Materials Conference, Seattle, April 2001.

[13] Liao, W. H., Wang, D. H. and Huang, S. L., Wireless monitoring of cable tension of cable-stayed bridges using PVDF piezoelectric films, Journal of Intelligent Material Systems and Structures 12, 2001, 331–339.

[14] Lee, Y.-C., Tein, Y. F. and Chao, Y. Y., A Point-Focus PVDF transducer for lamb wave measurements, Journal of Mechanics 18(1), March 2002, 29–33.

[15] Wang, F., Tanaka, M. and Chonan, S., Development of a PVDF piezopolymer sensor for unconstrained in-sleep cardiorespiratory monitoring, Journal of Intelligent Material Systems and Structures 14(3), March 2003, 185–190.

[16] Lee, Y. C., Yu, J. M. and Huang, S. W., Fabrication and characterization of a PVDF hydrophone array transducer, Key Engineering Materials 270–273, August 2004, 1406–1413.

[17] Chang, W.-Y., Cheng, Y.-H., Liu, S.-Y., Hu, Y.-C. and Lin, Y.-C., Dynamic behavior investigation of the piezoelectric PVDF for flexible fingerprint, Materials Research Society MRS Proceedings, 949, 2006.

[18] Lin, B. and Giurgiutiu, V., PVDF and PZT piezoelectric wafer active sensors for structural health monitoring, Proceedings of the ASME 2005 International Mechanical Engineering Congress and Exposition, 2005, 69–76.

[19] Lin, B. and Giurgiutiu, V., Modeling and testing of PZT and PVDF piezoelectric wafer active sensors, Smart Materials and Structures 15(4), 2006, 1085–1093.

[20] Luo, D., Guo, Y., Hao, H., Liu, H. and Ouyang, S., Fabrication of energy storage media BST/PVDF-PAN, Materials Research Society MRS Proceedings, Vol. 949, 2006.

[21] Nunes, J. S., Sencadas, V., Wu, A., Kholkin, A. L., Vilarinho, P. M. and Lanceros-Méndez, S., Electrical and microstructural changes of β-PVDF under different processing conditions by scanning force microscopy, Materials Research Society MRS Proceedings, Vol. 949, 2006.

[22] Ren, X. and Dzenis, Y., Novel continuous Poly(vinylidene fluoride) nanofibers, Materials Research Society MRS Proceedings, 920, 2006.

[23] Nasir, M., Matsumoto, H., Minagawa, M., Tanioka, A., Danno, T. and Horibe, H., Preparation of porous PVDF nanofiber from PVDF/PVP Blend by electrospray deposition, Polymer Journal 39(10), May 2007, 1060–1064.

[24] Mago, G., Kalyon, D. M. and Fisher, F. T., Membranes of polyvinylidene fluoride and PVDF nanocomposites with carbon nanotubes via immersion precipitation, Journal of Nanomaterials 2008, September 2007, article ID 759825.

[25] Vickraman, P., Aravindan, V., Srinivasan, T. and Jayachandran, M., Polyvinylidenefluoride (PVdF) based novel polymer electrolytes complexed With Mg(ClO4)2, The European Physical Journal Applied Physics 45 (1), January 2009, article ID 11101.

[26] Costa, L. M. M., Bretas, R. E. S. and Gregorio, R. Jr., Effect of solution concentration on the electrospray/electrospinning transition and on the crystalline phase of PVDF, Materials Sciences and Applications 1(4), October 2010, 247–252.

[27] Kim, Y.-J., Ahn, C. H., Lee, M. B. and Choi, M.-S., Characteristics of electrospun PVDF/SiO2 composite nanofiber membranes as polymer electrolyte, Materials Chemistry and Physics 127(1–2), May 2011, 137–142.

[28] Liu, F., Hashim, N. A., Liu, Y., Abed, M. R. M. and Li, K., Progress in the production and modification of PVDF membranes, Journal of Membrane Science 375(1–2), June 2011.

[29] Hou, M., Tang, X. G., Zou, J. and Truss, R., Increase the mechanical performance of polyvinylidene fluoride (PVDF), Advanced Materials Research 393–395, November 2011, 144–148.

[30] Gupta, A. K., Tiwari, A., Bajpai, R. and Keller, J. M., Short circuit thermally stimulated discharge current measurement on PMMA:PEMA:PVDF ternary blends, Scientific Research: Materials Sciences and Application 2(8), August 2011, 1041–1048.

[31] Li, Q., Zhou, B., Bi, Q.-Y. and Wang, X.-L., Surface modification of PVDF membranes with sulfobetaine polymers for a stably anti-protein-fouling performance, Journal of Applied Polymer Science 125(5), March 2012, 4015–4027.

[32] Klimiec, E., Zaraska, W., Piekarski, J. and Jasiewicz, B., PVDF sensors – research on foot pressure distribution in dynamic conditions, Advances in Science and Technology 79, September 2012, 94–99.

[33] Chiu, -Y.-Y., Lin, W.-Y., Wang, H.-Y., Huang, S.-B. and Wu, M.-H., Development of a piezoelectric polyvinylidene fluoride (PVDF) polymer-based sensor patch for simultaneous heartbeat and respiration monitoring, Sensors and Actuators A, Physical 189, January 2013, 328–334.

[34] Hartono, A., Satira, S., Djamal, M., Ramli, R., Bahar, H. and Sanjaya, E., Effect of mechanical treatment temperature on electrical properties and crystallite size of PVDF film, Advances in Materials Physics and Chemistry AMPC 3(1), January 2013, 71–76.

[35] Jain, A., Kumar, J. S., Srikanth, S., Rathod, V. T. and Mahapatra, D. R., Sensitivity of polyvinylidene fluoride films to mechanical vibration modes and impact after optimizing stretching conditions, Polymer Engineering and Science 53(4), April 2013, 707–715.

[36] Kang, G.-D. and Cao, Y.-M., Application and modification of poly(vinylidene fluoride) (PVDF) membranes – a review, Journal of Membrane Science 463, January 2014, 145–165.

[37] Saïdi, S., Mannaî, A., Bouzitoun, M. and Mohamed, A. B., Alternating current conductivity and dielectric relaxation of PANI: PVDF composites, The European Physical Journal Applied Physics 66 (1), April 2014, 10201, http://dx.doi.org/10.1051/epjap/2014130245.

[38] Sun, C., Shi, J., Bayerl, D. J. and Wang, X., PVDF microbelts for harvesting energy from respiration, Energy & Environmental Science, 4, July 2011, 4508–4512.

[39] Chen, D., Sharma, T. and Zhang, J. X. J., Mesoporous surface control of PVDF thin films for enhanced piezoelectric energy generation, Sensors and Actuators A 216, September 2014, 196–201.

Ferroelectric ceramics and piezoelectric reviews–related references

[40] Cady, W. G., Piezoelectricity, McGraw-Hill Book Company, Inc., New York, 1946, 842.

[41] Yaffe, B., Cook, W. R. Jr. and Jaffe, H., Piezoelectric Ceramics, Academic Press, London, 1971, 317.

[42] Gruver, R. M., Buessem, W. R., Dickey, C. W. and Anderson, J. W., State-of-the-art review on ferroelectric ceramic materials, Technical Report AFML-TR-66-164, Air Force Materials Laboratory Research and Technology Division, Airforce Systems Command, Wright-Patterson Air Force Base, Ohio, USA, May 1966, 223.

[43] Haertling, G. H., Ferroelectric ceramics: History and technology, Journal of the American Ceramic Society 82(4), January 1999, 797–818.

[44] Muralt, P., Ferroelectric thin films for micro-sensors and actuators: A review, Journal Micromechanics and Microengineering 10, 2000, 136–146.

[45] Niezrecki, C., Brei, D., Balakrishnan, S. and Moskalik, A., Piezoelectric actuation: state of the art, The Shock and Vibration Digest 44(4), July 2001, 269–280.

[46] Setter, N., Damjanovic, D., Eng, L., Fox, G., Gevorgian, S., Hong, S., Kingon, A., Kohlstedt, H., Park, N. Y., Stephenson, G. B., Stolitchnov, I., Taganstev, A. K., Taylor, D. V., Yamada, T. and Streiffer, S., Ferroelectric thin films: review of materials, properties, and applications, Journal of Applied Physics 100, 2006, article ID 051606, doi:10.1063/1.2336999.

[47] Gopinathan, S. V., Varadan, V. V. and Varadan, V. K., A review and critique of theories for piezoelectric laminates, Smart Materials and Structures 9(1), 2000, 24–48.

[48] Chopra, I., Review of state-of-art of smart structures and integrated systems, AIAA Journal 40(11), November 2002, 2145–2187.

[49] Gupta, V., Sharma, M. and Thakur, N., Mathematical modeling of actively controlled piezo smart structures: a review, Smart Structures and Systems An International Journal 8(3), September 2011, 275–302.

[50] Li, T., Ma, J., Es-Souni, M. and Woias, P., Advanced piezoelectrics: materials, devices, and their applications, Smart Materials Research 2012, March 2012, article ID 259275.

[51] Liu, X., Zhang, K. and Li, M., A survey on experimental characterization of hysteresis in piezoceramic actuators, Advanced Materials Research 694–697, May 2013, 1558–1564.

[52] Harne, R. L. and Wang, K. W., A Review of the recent research on vibration energy harvesting via bistable systems, Smart Materials and Structures 22(2), 2013, article ID 023001.

[53] Ramadan, K. S., Sameoto, D. and Evoy, S., A review of piezoelectric polymers as functional materials for electromechanical transducers, Smart Materials and Structures 23 (3), January 2014, article ID 033001.

[54] Aridogan, U. and Basdogan, I., A review of active vibration and noise suppression of plate-like structures with piezoelectric transducers, Journal of Intelligent Material Systems and Structures 26(12), August 2015, 1455–1476.

Static and Dynamic Behavior of Piezoceramic Beams, Damping, Control Issues and Active Constrained Layered Damping–Related References

[55] Baz, A., Static deflection control of flexible beams by piezo-electric actuators, NASA Technical Report No: N87-13788, September 1986.

[56] Baz, A., Poh, S. and Studer, P., Optimum vibration control of flexible beams by piezo-electric actuators, Proceedings of 6th Conference on the Dynamics & Control of Large Structures, Blacksburg, June 1987, 217–234.

[57] Crawley, E. F. and Anderson, E. H., Detailed models of piezoceramic actuation of beams, Journal of Intelligent Material Systems and Structures 1(1), January 1990, 4–25.

[58] Chandra, R. and Chopra, I., Structural modeling of composite beams with induced-strain actuation, AIAA Journal 31(9), September 1993, 1692–1701.

[59] Lagoudas, D. C. and Bo, Z., The cylindrical bending of composite plates with piezoelectric and SMA layers, Journal of Smart Materials and Structures 3, 1994, 309–317.

[60] Kim, S. J. and Jones, J. D., Influence of piezo-actuator thickness on the active vibration control of a cantilever beam, Journal of Intelligent Material Systems and Structures 6(5), September 1995, 610–623.

[61] Inman, D. J., Huang, S.-C. and Austin, E. M., Piezoceramic versus viscoelastic damping treatments, Proceedings of the 6th International Conference on Adaptive Structure Technology, November 1995, 241–252.

[62] Saravanos, D. A. and Heyliger, P. R., Coupled layerwise analysis of composite beams with embedded piezoelectric sensors and actuators, Journal of Intelligent Material Systems and Structures 6(3), 1995, 350–363.

[63] Chen, P. C. and Chopra, I., Induced strain actuation of composite beams and rotor blades with embedded piezoceramic elements, Smart Materials and Structures 5(1), February 1996, 35–48.

[64] Lee, H. J. and Saravanos, D. A., Coupled layerwise analysis of thermopiezoelectric smart composite beams, AIAA Journal 34(6), June 1996, 1231–1237.

[65] Zapfe, J. A. and Lesieutre, G. A., Iterative calculation of the transverse shear distribution in laminated composite beams, AIAA Journal 34(6), June 1996, 1299–1300.

[66] Park, C. and Chopra, I., Modeling piezoceramic actuation of beams in torsion, AIAA Journal 34(12), December 1996, 2582–2589.

[67] Lesieutre, G. A. and Lee, U., A finite element model for beams having segmented active constrained layers with frequency-dependent viscoelastic material properties, Smart Materials and Structures 5, 1996, 615–627.

[68] Smith, C. B. and Wereley, N. M., Transient analysis for damping identification in rotating composite beams with integral damping layers, Smart Materials and Structures 5(5), 1996, 540–550.

[69] Baz, A., Dynamic boundary control of beams using active constrained layer damping, Mechanical Systems and Signal Processing 11(6), November 1997, 811–825.

[70] Abramovich, H. and Pletner, B., Actuation and sensing of piezolaminated sandwich type structures, Composite Structures 38(1–4), 1997, 17–27.

[71] Baz, A., Dynamic boundary control of beams with active constrained layer damping, Journal of Mechanical Systems & Signal Processing 11(6), 1997, 811–825.

[72] Lesieutre, G. A., Vibration damping and control using shunted piezoelectric materials, Shock and Vibration Digest 30, 1998, 187–195.

[73] Liao, W. H., Actuator location for active/passive piezoelectric control systems, Proceedings of the Symposium on Image Speech Signal Processing and Robotics ISSPR '98, 1, 1998, 341–346.

[74] Abramovich, H. and Meyer-Piening, H.-R., Induced vibrations of piezolaminated elastic beams, Composite Structures 43(1), September 1998, 47–55.

[75] Abramovich, H., Deflection control of laminated composite beams with piezoceramic layers-closed form solutions, Composite Structures 43(3), November 1998, 217–231.

[76] Saravanos, D. A. and Heyliger, P. R., Mechanics and computational models for laminated piezoelectric beams plates and shells, Applied Mechanics Reviews 52(10), 1999, 305–320.

[77] Barboni, R., Mannini, A. and Gaudenzi, P., Optimal placement of PZT actuators for the control of beam dynamics, Smart Materials and Structures 9(1), February 2000, 110–120.

[78] Ang, K. K., Reddy, J. N. and Wang, C. M., Displacement control of timoshenko beams via induced strain actuators, Smart Materials and Structures 9, 2000, PII: S0964-1726(00)16310-4, 981–984.

[79] Shih, H.-R., Distributed vibration sensing and control of a piezoelectric laminated curved beam, Smart Materials and Structures 9(6), 2000, 761–766.

[80] Wang, Q. and Quek, S. T., Flexural vibration analysis of sandwich beam coupled with piezoelectric actuator, Smart Materials and Structures 9(1), 2000, 103–109.

[81] Smith, C., Analytical modeling and equivalent electromechanical loading techniques for adaptive laminated piezoelectric beams and plates, Master's Thesis at Virginia Polytechnic Institute and State University, USA, etd-02062001-101444, January 2001.

[82] Krommer, M., On the correction of the Bernoulli–Euler beam theory for smart piezoelectric beams, Smart Materials and Structures 10(4), August 2001, 668–680.

[83] Achuttan, A., Keng, A. K. K. and Ming, W. C., Shape control of coupled nonlinear piezoelectric beams, Smart Materials and Structures 10(5), October 2001, 914–924.

[84] Cai, C., Liu, G. R. and Lam, K. Y., A technique for modelling multiple piezoelectric layers, Smart Materials and Structures 10(4), August 2001, 689–694.

[85] Tong, L., Sun, D. and Atluri, S. N., Sensing and actuating behaviors of piezoelectric layers with debonding in smart beams, Smart Materials and Structures 10(4), August 2001, 724–729.

[86] Sun, D., Tong, L. and Atluri, S. N., Effects of piezoelectric sensor/actuator debonding on vibration control of smart beams, International Journal of Solids and Structures 38(50–51), December 2001, 9033–9051.

[87] Baz, A. and Ro, J., Vibration control of rotating beams with active constrained layer damping, Journal of Smart Materials and Structures 10(1), 2001, 112–120.

[88] Ray, M. and Baz, A., Control of nonlinear vibration of beams using active constrained layer damping, Journal of Vibration and Control 7, 2001, 539–549.

[89] Trindade, M. A., Benjeddou, A. and Ohayon, R., Finite element modelling of hybrid active-passive vibration damping of multilayer piezoelectric sandwich beams – part i: formulation, International Journal for Numerical Methods in Engineering 51(7), 2001, 835–854.

[90] Trindade, M. A., Benjeddou, A. and Ohayon, R., Finite element modelling of hybrid active-passive vibration damping of multilayer piezoelectric sandwich beams – part ii: system analysis, International Journal for Numerical Methods in Engineering 51(7), 2001, 855–864.

[91] Trindade, M. A., Benjeddou, A. and Ohayon, R., Piezoelectric active vibration control of damped sandwich beams, Journal of Sound and Vibration 246(4), 2001, 653–677.

[92] Wang, G. and Wereley, N. M., Spectral finite element analysis of sandwich beams with passive constrained layer damping, ASME Journal of Vibration and Acoustics 124(3), July 2002, 376–386.

[93] Abramovich, H. and Livshits, A., Flexural vibrations of piezolaminated slender beams: a balanced model, Journal of Vibration and Control 8(8), August 2002, 1105–1121.

[94] Waisman, H. and Abramovich, H., Variation of natural frequencies of beams using the active stiffening effect, Composites Part B: Engineering 33(6), September 2002, 415–424.

[95] Yocum, M. and Abramovich, H., Static behavior of piezoelectric actuated beams, Computers & Structures 80(23), September 2002, 1797–1808.

[96] Waisman, H. and Abramovich, H., Active stiffening of laminated composite beams using piezoelectric actuators, Composite Structures 58(1), October 2002, 109–120.

[97] Gao, J. X. and Liao, W. H., Damping characteristics of beams with enhanced self-sensing active constrained layer treatments under various boundary conditions, ASME Journal of Vibration and Acoustics 127(2), 2005, 173–187.

[98] Gao, J. X. and Liao, W. H., Vibration analysis of simply supported beams with enhanced self-sensing active constrained layer damping treatments, Journal of Sound and Vibration 280(1–2), 2005, 329–357.

[99] Park, C. H. and Baz, A., Vibration control of beams with negative capacitive shunting of interdigital electrode piezoceramics, Journal of Vibration and Control 11(3), 2005, 331–346.

[100] Edery-Azulay, L. and Abramovich, H., Active damping of piezo-composite beams, Composite Structures 74(4), August 2006, 458–466.

[101] Edery-Azulay, L. and Abramovich, H., The integrity of piezo-composite beams under high cyclic electro-mechanical loads- experimental results, Smart Materials and Structures 16(4), July 2007, 1226–1238.

[102] Adhikari, S. and Friswell, M. I., Shaped modal sensors for linear stochastic beams, Journal of Intelligent Material Systems and Structures 20(18), December 2009, 2269–2284.

[103] Wang, T. R., Structural responses of surface-mounted piezoelectric curved beams, Journal of Mechanics 26 (4), December 2010, 439–451. Kim, J. S. and Wang, K. W., An Asymptotic Approach for the Analysis of Piezoelectric Fiber Composite Beams, Smart Materials and Structures 20, 2011, doi:10.1088/0964-1726/20/2/025023.

[104] Bachmann, F., Bergamini, A. and Ermanni, P., Optimum piezoelectric patch positioning – a strain energy–based finite element approach, Journal of Intelligent Material Systems and Structures 23(14), 2012, 1575–1591.

[105] Koroishi, E. H., Molina, F. A. L., Faria, A. W. and Steffen, V. Jr., Robust optimal control applied to a composite laminated beam, Journal of Aerospace Technology and Management 7(1), January–March 2015, 70–80.

Piezoelectric Applications in Rotor-Related References

[106] Giurgiutiu, V., Chaudhry, Z. A. and Rogers, C. A., Active control of helicopter rotor blades with induced strain actuators, Proceedings of the 35th Structures Structural Dynamics and Materials Conference, Hilton Head, 1994, 288–297.

[107] Giurgiutiu, V., Chaudhry, Z. and Rogers, C. A., Engineering feasibility of induced strain actuators for rotor blade active vibration control, Journal of Intelligent Material Systems and Structures 6(5), 1995, 583–597.

[108] Chen, P. and Chopra, I., Hover testing of smart rotor with induced-strain actuation of blade twist, AIAA Journal 35(1), January 1997, 6–16.

[109] Chen, P. and Chopra, I., Wind tunnel test of a smart rotor model with individual blade twist control, Journal of Intelligent Material Systems and Structures 8(5), May 1997, 414–423.

[110] Koratkar, N. A. and Chopra, I., Testing and validation of a Froude scaled helicopter rotor model with piezo-bimorph actuated trailing-edge flaps, Journal of Intelligent Material Systems and Structures 8(7), July 1997, 555–570.

Sensing (Including Delamination) Using Piezoelectric Material–Related References

[111] Pletner, B. and Abramovich, H., Adaptive suspensions of vehicles using piezoelectric sensors, Journal of Intelligent Material Systems and Structures 6(6), November 1995, 744–756.

[112] Sirohi, J. and Chopra, I., Fundamental understanding of piezoelectric strain sensors, Journal of Intelligent Material Systems and Structures 11(April), 2000, 246–257.

[113] Giurgiutiu, V. and Zagrai, A. N., Characterization of piezoelectric wafer active sensors, Journal of Intelligent Material Systems and Structures 11(12), 2000, 959–976.

[114] Tabellout, M., Raquois, A., Emery, J. R. and Jayet, Y., The inserted piezoelectric sensor method for monitoring thermosets cure, The European Physical Journal Applied Physics 13(2), February 2001, 107–113.

[115] Giurgiutiu, V., Bao, J. and Zhao, W., Piezoelectric wafer active sensor embedded ultrasonics in beams and plates, Experimental Mechanics 43(4), 2003, 428–449.

[116] Baz, A., Poh, S., Lin, S. and Chang, P., Distributed sensing of rotating beams, Proceedings of the 1st International Workshop on Smart Materials and Structures Technology, Honolulu, January 2004.

[117] Chrysohoidis, N. A. and Saravanos, D. A., Assessing the effects of delamination on the damped dynamic response of composite beams with piezoelectric actuators and sensors, Smart Materials and Structures 13(4), May 2004, 733–742. doi:10.1088/0964-1726/13/4/01.

[118] Plagianakos, T. S. and Saravanos, D. A., Coupled high-order shear layerwise analysis of adaptive sandwich composite beams with piezoelectric actuators and sensors, AIAA Journal 43(4), April 2005, 885–894.

[119] Chrysochoidis, N. A. and Saravanos, D. A., Generalized layerwise mechanics for the static and modal response of delaminated composite beams with active piezoelectric sensors, International Journal of Solids and Structures 44(25–26), 2007, 8751–8768. doi:10.1016/j.ijsolstr.2007.07.004.

[120] Chrysochoidis, N. A. and Saravanos, D. A., High frequency dispersion characteristics of smart delaminated composite beams, Journal of Intelligent Material Systems and Structures 20(9), June 2009. doi:10.1177/1045389X09102983.

[121] Abramovich, H., Burgard, M., Edery-Azulay, L., Evans, K. E., Hoffmeister, M., Miller, W., Scarpa, F., Smith, C. W. and Tee, K. F., Smart tetrachiral and hexachiral honeycomb: sensing and impact detection, Composites Science and Technology 70(7), July 2010, 1072–1079.

[122] Lin, B., Giurgiutiu, V., Pollock, P., Xu, B. and Doane, J., Durability and survivability of piezoelectric wafer active sensors on metallic structure, AIAA journal 48(3), 2010, 635–643.

[123] Martinez, M. and Artemev, A., A novel approach to a piezoelectric sensing element, Journal of Sensors 2010, 2010, article ID 816068. doi:10.1155/2010/816068.

[124] Gresil, M. and Giurgiutiu, V., Guided wave propagation in composite laminates using piezoelectric wafer active sensors, Aeronautical Journal 117(1196), 2013.

Piezoelectric Actuator Application–Related References

[125] Dosch, J. J., Inman, D. J. and Garcia, E., A self sensing piezoelectric actuator for collocated control, Journal of Intelligent Material Systems and Structures 3(1), 1992, 166–185.

[126] Giurgiutiu, V. and Rogers, C. A., Large-amplitude rotary induced-strain (laris) actuator, Journal of Intelligent Material Systems and Structures 8(1), 1997, 41–50.

[127] Giurgiutiu, V. and Rogers, C. A., Power and energy characteristics of solid-state induced-strain actuators for static and dynamic applications, Journal of Intelligent Material Systems and Structures 8(9), 1997, 738–750.

[128] Kowbel, W., Xia, X., Withers, J. C., Crocker, M. J. and Wada, B. K., PZT/PVDF composite for actuator/sensor application, Materials Research Society MRS Proceedings 493, 1997.

[129] Lesieutre, G. A. and Davis, C. L., Can a coupling coefficient of a piezoelectric actuator be higher than those of its active material, Journal of Intelligent Materials Systems and Structures 8, 1997, 859–867.

[130] Pietrzakowski, M., Dynamic model of beam-piezoceramic actuator coupling for active vibration control, mechanika teoretyczna i stosowana, Journal of Theoretical and Applied Mechanics 35(1), 1997, 3–20.

[131] Chandra, R. and Chopra, I., Actuation of trailing edge flap in a wing model Using Piezostack Device, Journal of Intelligent Material Systems and Structures 9(10), October 1998, 847–853.

[132] Moskalik, A. J. and Brei, D., Force-deflection behavior of piezoelectric C-block actuator arrays, Smart Materials and Structures 8(5), 1999, 531–543.

[133] Sirohi, J. and Chopra, I., Fundamental behavior of piezoceramic sheet actuators, Journal of Intelligent Material Systems and Structures 11(1), January 2000, 47–61.

[134] Tylikowski, A., Influence of bonding layer on piezoelectric actuators of an axisymmetrical annular plate, Journal of Theoretical and Applied Mechanics 3(38), May 2000, 607–621.

[135] Benjeddou, A., Trindade, M. A. and Ohayon, R., Piezoelectric Actuation mechanisms for intelligent sandwich structures, Smart Materials and Structures 9(3), 2000, 328–335.

[136] Galante, T., Frank, J., Bernard, J., Chen, W., Lesieutre, G. A. and Koopmann, G. H., A high-force high-displacement piezoelectric inchworm actuator, Journal of Intelligent Materials Systems and Structures 10(12), 2000, 962–972.

[137] Lee, T. and Chopra, I., Design of piezostack-driven trailing-edge flaps for helicopter rotors, Journal Smart Material and Structures 10(1), February 2001, 15–24.

[138] Monturet, V. and Nogarede, B., Optimal dimensioning of a piezoelectric bimorph actuator, The European Physical Journal Applied Physics 17(2), February 2002, 107–118.

[139] Yocum, M., Abramovich, H., Grunwald, A. and Mall, S., Fully reversed electromechanical fatigue behavior of composite laminate with embedded piezoelectric actuator/sensor, Smart Materials and Structures 12(4), June 2003, 556–564.

[140] Abramovich, H., Piezoelectric actuation for smart sandwich structures – closed form solutions, Journal of Sandwich Structures & Materials 5(4), October 2003, 377–396.

[141] Law, W. W., Liao, W. H. and Huang, J., Vibration control of structures with self-sensing piezoelectric actuators incorporating adaptive mechanisms, Smart Materials and Structures 12, 2003, 720–730.

[142] Lesieutre, G. A., Rusovici, R., Koopmann, G. H. and Dosch, J. J., Modeling and characterization of a piezoceramic inertial actuator, Journal Sound and Vibration 261(1), 2003, 93–107.

[143] Barrett, R., McMurtry, R., Vos, R., Tiso, P. and De Breuker, R., Post-Buckled Precompressed (PBP) elements: a new class of flight control actuators enhancing high-speed autonomous VTOL MAVs, smart structures and materials 2005: industrial and commercial applications of smart structures technologies, edited by Edward V. White, Proceedings of SPIE Vol. 5762, SPIE, Bellingham, WA, 2005, doi:10.1117/12.599083, 2005.

[144] Abramovich, H., Weller, T. and Yeen-Ping, S., Dynamics response of a high aspect ratio wing equipped With PZT patches-a theoretical and experimental study, Journal of Intelligent Materials Systems and Structures 16(11–12), December 2005, 919–923.

[145] Parsons, Z. and Staszewski, W. J., Nonlinear acoustics with low-profile piezoceramic excitation for crack detection in metallic structures, Smart Materials and Structures 15, 2006, 1110–1118.

[146] Nir, A. and Abramovich, H., Design, analysis and testing of a smart fin, Composite Structures 92(4), March 2010, 863–872.

[147] Haller, D., Paetzold, A., Losse, N., Neiss, S., Peltzer, I., Nitsche, W., King, R. and Woias, P., Piezo-Polymer-composite unimorph actuators for active cancellation of flow instabilities across airfoil, Journal of Intelligent Material Systems and Structures 22(5), March 2011, 461–474.

[148] Wereley, N. M., Wang, G. and Chaudhuri, A., Demonstration of uniform cantilevered beam bending vibration using a pair of piezoelectric actuators, Journal of Intelligent Material Systems and Structures 22(4), 2011, 307–316. doi:10.1177/1045389X10379661.

[149] Pan, C. L. and Liao, W. H., A new two-axis optical scanner actuated by piezoelectric bimorphs, International Journal of Optomechatronics 6(4), 2012, 336–349. doi:10.1080/15599612.2012.721867.

[150] Arrieta, A. F., Bilgen, O., Friswell, M. I. and Ermanni, P., Modelling and configuration control of wing-shaped Bi-stable piezoelectric composites under aerodynamic loads, Aerospace Science and Technology 29(1), August 2013, 453–461.

[151] Davis, J., Kim, N. H. and Lind, R., Control of the flexural axis of a wing with piezoelectric actuation, Journal of Aircraft 52(2), March-April 2015, 584–594.

Piezoelectric Shear-Type Actuator–Related References

[152] Aldraihem, O. J. and Khdeir, A. A., Smart beams with extension and thickness-shear piezoelectric actuators, Smart Materials and Structures 9, 2000, PII: S0964-1726(00)07958-1.

[153] Przbylowicz, P. M., An application of piezoelectric shear effect to active damping of transverse vibration in beams, Journal of Theoretical and Applied Mechanics 3(38), 2000, 573–589.

[154] Vel, S. S. and Batra, R. C., Exact solution for rectangular sandwich plates with embedded piezoelectric shear actuators, AIAA Journal 39(7), July 2001, 1363–1373.

[155] Benjeddou, A., Gorge, V. and Ohayon, R., Use of piezoelectric shear response in adaptive sandwich shells of revolution – part 1: theoretical formulation, Journal of Intelligent Material Systems and Structures 12(4), 2001, 235–245.

[156] Benjeddou, A., Gorge, V. and Ohayon, R., Use of piezoelectric shear response in adaptive sandwich shells of revolution – part 2: finite element implementation, Journal of Intelligent Material Systems and Structures 12(4), 2001, 247–257.

[157] Vel, S. S. and Batra, R. C., Analysis of piezoelectric bimorphs and plates with segmented actuators, Thin Wall Structure 39, 2001, 23–44.

[158] Edery-Azulay, L. and Abramovich, H., Piezoelectric actuation and sensing mechanisms-closed form solutions, Composite Structures 64(3–4), June 2004, 443–453.

[159] Senthil, S. V. and Baillargeon, B. P., Active vibration suppression of smart structures using piezoelectric shear actuators, Proceedings of the 15th International Conference on Adaptive Structures and Technologies, ICAST, Bar Harbor, October 2004.

[160] Edery-Azulay, L. and Abramovich, H., Augmented damping of a piezo-composite beam using extension and shear piezoceramic transducers, Composite B: Engineering 37(4–5), 2006, 320–327.

[161] Sawano, M., Tahara, K., Orita, Y., Nakayama, M. and Tajitsu, Y., New design of actuator using shear piezoelectricity of a chiral polymer, and prototype device, Polymer International 59(3), August 2009, 365–370.

[162] Aladwani, A., Aldraihem, O. and Baz, A., Single degree of freedom shear-mode piezoelectric energy harvester, ASME Journal of Vibration and Acoustics 135(5), 2013, 051011, Paper No. VIB-11-1250. doi:10.1115/1.4023950.

Enhanced Stability of Beams, Plates and Shells with Piezoelectric Layer/Patch–Related References

[163] Meressi, T. and Paden, B., Buckling control of a flexible beam using piezoelectric actuators, Journal of Guidance Control, and Dynamics 16(5), 1993, 977–980.

[164] Alghamdi, A. A. A., Adaptive imperfect column with piezoelectric actuators, Journal of Intelligent Materials Systems and structures 12, March 2001, 183–189. doi:10.1106/ua0k-qwxq-p8kl-g3k2.

[165] Rao, G. V. and Singh, G., A smart structures concept for the buckling load enhancement of columns, Smart Materials and Structures 10(4), August 2001, 843–845.

[166] Batra, R. C. and Geng, T. S., Enhancement of the dynamic buckling load for a plate by using piezoceramic actuators, Smart Materials and Structures 10(5), October 2001, 925–933.

[167] Varelis, D. and Saravanos, D. A., Nonlinear coupled mechanics and buckling analysis of composite plates with piezoelectric actuators and sensors, Smart Materials and Structures 11(3), June 2002, 330–336.

[168] Mukherjee, A. and Chaudhuri, A. S., Active control of dynamic instability of piezolaminated imperfect columns, Smart Materials and Structures 11(6), 2002, 874–879.

[169] Wang, Q. and Varadan, V. K., Transition of the buckling load of beams by the use of piezoelectric layers, Smart Materials and Structures 12(5), 2003, 696–702.

[170] Fridman, Y. and Abramovich, H., Enhanced structural behavior of flexible laminated composite beams, Composite Structures 82(1), January 2008, 140–154.

[171] Zehetner, C. and Irschik, I., On the static and dynamic stability of beams with an axial piezoelectric actuation, Smart Structures and Systems 4(1), January 2008, 67–84.

[172] Sridharan, S. and Kim, S., Piezoelectric control of columns prone to instabilities and nonlinear modal interaction, Smart Materials and Structures 17(3), 2008, article ID 035001.

[173] Sridharan, S. and Kim, S., Piezoelectric control of stiffened panels subject to interactive buckling, International Journal of Solids and Structures 46(6), 2009, 1527–1538.

[174] De Faria, A. R. and Donadon, M. V., The use of piezoelectric stress stiffening to enhance buckling of laminated plates, Latin American Journal of Solids and Structures 7(March), 2010, 167–183.

[175] Wang, Q. S., Active buckling control of beams using piezoelectric actuators and strain gauge sensors, Smart Materials and Structures 19(62010), May 2010, article ID 065022.

[176] Enss, G. C., Platz, R. and Hanselka, H., An approach to control the stability in an active load-carrying beam-column by one single piezoelectric stack actuator, Proceedings of International Conference on Noise and Vibration (ISMA) 2010 including International Conference on Uncertainty in Structural Dynamics (USD) 2010, Leuven Belgium, 535–546.

[177] Abramovich, H., A new insight on vibrations and buckling of a cantilevered beam under a constant piezoelectric actuation, Composite Structures 93(2), January 2011, 1054–1057.

[178] Qishan, W., Active vibration and buckling control of piezoelectric smart structures, Ph.D. Thesis submitted to Civil Engineering and Applied Mechanics Dep., McGill University, Montreal, Quebec, Canada, 2012.

[179] Wluka, P. and Kubiak, T., Stability of cross-ply composite plate with piezoelectric actuators, Stability of Structures XIII-th Symposium, Zakopane, Poland, 2012, 667–686.

[180] Zenz, G. and Humer, A., Experimental investigations to enhance the buckling load of slender beams, 7th ECCOMAS Thematic Conference on Smart Structures and Materials, SMART 2015, Araúo, A. L., Mota Soares, C. A., et al. (eds.), 2015.

[181] Abramovich, H., Axial stiffness variation of thin walled laminated composite beams using piezoelectric patches – a new experimental insight, 26th International Conference on Adaptive Structures and Technologies at Kobe, Japan, on October 14–16, 2015 (ICAST2015).

Saint-Venant's Principle and Piezoelasticity-Related References

[182] Ruan, X., Danforth, S. C., Safari, A. and Chou, T.-W., Saint-venant end effects in piezoceramic materials, International Journal of Solids and Structures 37(19), May 2000, 2625–2637.

[183] Rovenski, V., Harash, E. and Abramovich, H., Saint-Venant's problem for homogeneous piezoelastic beams, Journal of Applied Mechanics 74(6), December 2006, 1095–1103.

[184] Rovenski, V. and Abramovich, H., Behavior of piezoelectric beams under axially non-uniform distributed loading, Journal of Elasticity 88(3), September 2007, 223–253.

[185] Rovenski, V. and Abramovich, H., Saint Venant's problem for compound piezoelastic beams, Journal of Elasticity 96, April 2009, 105–127.

[186] Krommer, M., Berik, P., Vetyukov, Y. and Benjeddou, A., Piezoelectric d_{15} Shear-response-based torsion actuation mechanism: an exact 3D Saint-Venant type solution, International Journal of Smart and Nano Materials 3 (2), June 2012, 82–102.

Plates and Shells with Piezoelectric Layers and/or Patch-Related References

[187] Barrett, R., Active Plate and Wing research Using EDAP Elements, Smart Materials and Structures 1(3), September 1992, 214–226.

[188] Shah, D. K., Joshi, S. P. and Chan, W. S., Static structural response of plates with piezoceramic layers, Smart Materials and Structures 2, 1993, 172–180.

[189] Heyliger, P., Ramirez, G. and Saravanos, D. A., Coupled discrete-layer finite elements for laminated piezoelectric plates, Communications in Numerical Methods in Engineering 10, 1994, 971–981.

[190] Heyliger, P. R. and Saravanos, D. A., Exact free-vibration analysis of laminated plates with embedded piezoelectric layers, Journal of Acoustical Society of America 98(3), 1995, 1547–1557.

[191] Miller, S. E. and Abramovich, H., A Self-Sensing Piezolaminated Actuator model for shells using a first order Shear deformation theory, Journal of Intelligent Material Systems and Structures 6(5), 1995, 624–638.

[192] Mitchell, J. A. and Reddy, J. N., A refined hybrid plate theory for composite laminates with piezoelectric laminae, International Journal of Solids and Structure 32(16), 1995, 2345–2367.

[193] Heyliger, P. R., Pei, K. C. and Saravanos, D. A., Layerwise mechanics and finite element model for laminated piezoelectric shells, AIAA Journal 34(11), 1996, 2353–2360.

[194] Miller, S. E., Oshman, Y. and Abramovich, H., Modal control of piezolaminated anisotropic rectangular plates: part 1 – modal transducer theory, AIAA Journal 34(9), September 1996, 1868–1875.

[195] Miller, S. E., Oshman, Y. and Abramovich, H., Modal control of piezolaminated anisotropic rectangular plates: part 2 – control theory, AIAA Journal 34(9), September 1996, 1876–1884.

[196] Baz, A., Ro, J., Vibration control of plates with active constrained layer damping, Journal of Smart Materials & Structures 5, 1996, 272–280.

[197] Lin, C.-C., Hsu, C.-Y. and Huang, H.-N., Finite element analysis on deflection control of plate with piezoelectric actuators, Composite Structures 35(4), 1996, 423–433.

[198] Kaljevic, I. and Saravanos, D. A., Steady-state response of acoustic cavities bounded by piezoelectric composite shell structures, Journal of Sound and Vibration 205(3), July 1997, 459–476.

[199] Pletner, B. and Abramovich, H., A Consistent Methodology for the modeling of piezolaminated shells, AIAA Journal 35(8), August 1997, 1316–1326.

[200] Saravanos, D. A., Mixed laminate theory and finite element for smart piezoelectric composite shell structures, AIAA Journal 35(8), August 1997, 1327–1333.

[201] Miller, S. E., Abramovich, H. and Oshman, Y., Selective modal transducers for anisotropic rectangular plates: experimental validation, AIAA Journal 35(10), October 1997, 1621–1629.

[202] Lee, H. J. and Saravanos, D. A., Generalized finite element formulation for smart multilayered thermal piezoelectric plates, International Journal of Solids and Structures 34(26), 1997, 3355–3371.

[203] Shields, W., Ro, J. and Baz, A., Control of sound radiation from a plate into an acoustic cavity using active piezoelectric damping composites, Journal of Smart Materials & Structures 7, 1998, 1–11.

[204] Hong, C. H. and Chopra, I., Modeling and validation of induced strain actuation of composite coupled plates, AIAA Journal 37(3), March 1999, 372–377.

[205] Saravanos, D. A., Damped vibration of composite plates with passive piezoelectric-resistor elements, Journal of Sound and Vibration 221(5), April 1999, 867–885.

[206] Miller, S. E., Oshman, Y. and Abramovich, H., Selective Modal transducers for piezolaminated anisotropic shells, Journal of Guidance Control and Dynamics 22 (3), May–June 1999, 455–466.

[207] Zhang, X. D. and Sun, C. T., Analysis of a Sandwich plate containing a piezoelectric core, Smart Materials and Structures 8, 1999, 31–40.

[208] Chee, C. Y. K., Static shape control of laminated composite plate smart structure using piezoelectric actuators, PhD Thesis at The University of Sydney Aeronautical Engineering, Australia, thesis ID 2123/709, 2000.

[209] Saravanos, D. A., Passively damped laminated piezoelectric shell structures with integrated electric networks, AIAA Journal 38(7), July 2000, 1260–1268.

[210] Chee, C. Y. K., Tong, L. and Steven, G. P., A mixed model for adaptive composite plates with piezoelectric for anisotropic actuation, Computers & Structures 77(3), 2000, doi:10.1016/S0045-7949(99)00225-4, 253–268.

[211] Abramovich, H. and Meyer-Piening, H.-R., Actuation and sensing of soft core sandwich plates with a built-in adaptive layer, Journal of Sandwich Structures & Materials 3 (1), January 2001, doi:10.1106/kx19-fa1t-x1b8-56gq, 75–86.
[212] Miller, S. E., Oshman, Y. and Abramovich, H., A Selective Modal Control Theory for piezolaminated anisotropic shells, Journal of Guidance Control and Dynamics 24 (4), July–August 2001, 844–852.
[213] Chee, C. K., Tong, L. and Steven, G. P., Static shape control of composite plates using a curvature-displacement based algorithm, International Journal of Solids and Structures 38, 2001, 6381–6403.
[214] Saravanos, D. A. and Christoforou, A. P., Low-energy impact of adaptive cylindrical laminated piezoelectric-composite shells, International Journal of Solids and Structures 39(8), May 2002, 2257–2279.
[215] Tzou, H. S. and Wang, D. W., Micro-sensing Characteristics and modal voltages of piezoelectric laminated linear and nonlinear toroidal shells, Journal of Sound & Vibration 254(2), 2002, 203–218.
[216] Wu, C. Y., Chang, J. S. and Wu, K. C., Analysis of wave propagation in infinite piezoelectric plates, Journal of Mechanics 21(2), June 2005, 103–108.
[217] Edery-Azulay, L. and Abramovich, H., A Reliable Plain Solution for rectangular plates with piezoceramic patches, Journal of Intelligent Material Systems and Structures 18(5), May 2007, 419–433.
[218] Larbi, W., Deü, J.-F. and Ohayon, R., Vibration of axisymmetric composite piezoelectric shells coupled with internal fluid, International Journal for Numerical Methods in Engineering 71(12), 2007, 1412–1435.
[219] Ghergu, M., Griso, G., Mechkour, H. and Miara, B., Homogenization of thin piezoelectric perforated shells, ESAIM: Mathematical Modeling and Numerical Analysis 41 (5), September–October 2007, 875–895.
[220] Edery-Azulay, L. and Abramovich, H., Piezolaminated plates-highly accurate solutions based on the extended kantorovich method, Composite Structures 84(3), July 2008, 241–247.
[221] Chen, Z. G., Hu, Y. T. and Yang, J. S., Shear horizontal piezoelectric waves in a piezoceramic plate imperfectly bonded to two piezoceramic half-spaces, Journal of Mechanics 24(3), September 2008, 229–239.
[222] Li, H., Chen, Z. B. and Tzou, H. S., Distributed actuation characteristics of clamped-free conical shells using diagonal piezoelectric actuators, Smart Materials and Structures 19(11), 2010, article ID 115015.
[223] Messina, A. and Carrera, E., Three-dimensional free vibration of multi-layered piezoelectric plates through approximate and exact analyses, Journal of Intelligent Material Systems and Structures 26(5), March 2015, 489–504.

Piezoelectric-Based Energy Harvesting–Related References

[224] Sodano, H. A., Magliula, E., Park, G. and Inman, D. J., Electric power generation using piezoelectric devices, Proceedings of the 13th International Conference on Adaptive Structures and Technologies, October 2002, 153–161.
[225] Ottman, G., Bhatt, A., Hofmann, H. and Lesieutre, G. A., Adaptive piezoelectric energy harvesting circuit for wireless remote power supply, IEEE Transactions on Power Electronics 17(5), 2002, 669–676.

[226] Sunghwan, K., Low Power Energy Harvesting with Piezoelectric Generators, Ph.D. Thesis submitted to the School of Engineering at the University of Pittsburgh, PA, USA, December 2002, 136.

[227] Eggborn, T., Analytical Model to Predict Power Harvesting in Piezoelectric Material, Master's Thesis at the Virginia Polytechnic Institute and State University, Blacksburg, Virginia, May 2003, 94.

[228] Lesieutre, G. A., Hofmann, H. and Ottman, G., Damping as a result of piezoelectric energy harvesting, Journal of Sound and Vibration 269, 2004, 991–1001.

[229] Sodano, H. A., Inman, D. J. and Park, G., Comparison of piezoelectric energy harvesting devices for recharging batteries, Journal of Intelligent Material Systems and Structures 16(10), October 2005, 799–807.

[230] Ng, T. H. and Liao, W. H., Sensitivity Analysis and Energy Harvesting for a Self-Powered Piezoelectric Sensor, Journal of Intelligent Material Systems and Structures 16(10), 2005, 785–797.

[231] Sodano, H. A., Lloyd, J. and Inman, D. J., An experimental comparison between several active composite actuators for power generation, Smart Materials and Structures 15, 2006, 1211–1216.

[232] Guan, M. J. and Liao, W. H., On the efficiencies of piezoelectric energy harvesting circuits towards storage device voltages, Smart Materials and Structures 16, 2007, 498–505.

[233] Erturk, A. and Inman, D. J., On mechanical modeling of cantilevered piezoelectric vibration energy harvesters, Journal of Intelligent Material Systems and Structures 19(11), 2008, 1311–1325.

[234] Kauffman, J. L. and Lesieutre, G. A., A low-order model for the design of piezoelectric energy harvesting devices, Journal of Intelligent Material Systems and Structures 20, March 2009, 495–504.

[235] Liang, J. R. and Liao, W. H., Piezoelectric energy harvesting and dissipation on structural damping, Journal of Intelligent Material Systems and Structures 20(5), 2009, doi:10.1177/1045389X08098194, 515–527.

[236] Erturk, A., Tarazaga, P. A., Farmer, J. R. and Inman, D. J., Effect of strain nodes and electrode configuration on piezoelectric energy harvesting from cantilevered beams, ASME Journal of Vibration and Acoustics 131 (1), January 2009, 011010, doi:10.1115/1.2981094.

[237] Friswell, M. I. and Adhikari, S., Sensor shape design for piezoelectric cantilever beams to harvest vibration energy, Journal of Applied Physics 108 (1), July 2010, 014901, http://dx.doi.org/10.1063/1.3457330.

[238] Rödig, T., Schönecker, A. and Gerlach, G., A survey on piezoelectric ceramics for generator applications, Journal of the American Ceramic Society 93(4), April 2010, 901–912.

[239] Dietl, J. M., Wickenheiser, A. M. and Garcia, E., A Timoshenko Beam Model for cantilevered piezoelectric energy harvesters, Smart Materials and Structures 19(5), 2010, article ID 055018.

[240] Erturk, A. and Inman, D. J., Broadband vibration energy harvesting using bistable beams and plates, Proceedings of the American Ceramic Society Symposium: ACerS Electronic Materials and Applications, Orlando, January 2011.

[241] Abramovich, H., Tsikhotsky, E. and Klein, G., An experimental determination of the maximal allowable stresses for high power piezoelectric generators, Journal of Ceramic Science and Technology 4(3), 2013, 131–136.

[242] Abramovich, H., Tsikhotsky, E. and Klein, G., An experimental investigation on PZT behavior under mechanical and cycling loading, Journal of the Mechanical Behavior of Materials 22(3–4), 2013, 129–136.

[243] Leinonen, M., Palosaari, J., Juuti, J. and Jantunen, H., Combined Electrical and Electromechanical simulations of a piezoelectric cymbal harvester for energy harvesting from walking, Journal of Intelligent Material Systems and Structures 25(4), March 2014, 391–400.

[244] Xiaomin, X., Luqi, C., Xiaohong, W. and Qing, S., Study on electric-mechanical hysteretic model of macro-fiber composite actuator, Journal of Intelligent Material Systems and Structures 25(12), August 2014, 1469–1483.

[245] Zhang, Y., Cai, S. C. S. and Deng, L., Piezoelectric-based energy harvesting in bridge systems, Journal of Intelligent Material Systems and Structures 25(12), August 2014, 1414–1428.

[246] Bilgen, O., Friswell, M. I., Ali, S. F. and Litak, G., Broadband vibration energy harvesting from a vertical cantilever piezocomposite beam with tip mass, International Journal of Structural Stability and Dynamics 15 (2), March 2015, Paper 1450038.

[247] Vijayan, K., Friswell, M. I., Khodaparast, H. H. and Adhikari, S., Non-linear energy harvesting from coupled impacting beams, International Journal of Mechanical Sciences 96–97, June 2015, 101–109.

[248] Wang, J., Shi, Z., Xiang, H. and Song, G., Modeling on energy harvesting from a railway system using piezoelectric transducers, Smart Materials and Structures 24(10), October 2015, article ID 105017.

[249] Ansari, M. H. and Karami, M. A., Energy harvesting from controlled buckling of piezoelectric beams, Smart Materials and Structures 24(11), November 2015, article ID 115005.

Single Crystal–Related References

[250] Rusovici, R., Dosch, J. J. and Lesieutre, G. A., Design of a single-crystal piezoceramic vibration absorber, Journal of Intelligent Materials Systems and Structures 13(11),November 2002, 705–712.

[251] Tressler, J. F., A comparison of single crystal versus ceramic piezoelectric materials for acoustic applications, The Journal of the Acoustical Society of America 113(4), April 2003, doi: org/10.1121/1.4780737, 2311.

[252] Lloyd, J. M., Williams, R. B., Inman, D. J. and Wilkie, W. K., An analytical model of the mechanical properties of single-crystal macro fiber composite actuators, Proceedings of the SPIE's 11th Annual International Symposium on Smart Structures and Materials, San Diego, Paper no. 5387–08, March 2004.

[253] Rusovici, R. and Lesieutre, G. A., Design of a single-crystal piezoceramic-driven, synthetic-jet actuator, Proceeding of SPIE's Symposium on Smart Structures and Materials, SPIE 5390, March 2004, doi:10.1117/12.539576.

[254] Wilkie, W. K., Inman, D. J., Lloyd, J. M. and High, J. W., Anisotropic piezocomposite actuator incorporating machined PMN-PT single crystal, Journal of Intelligent Material Systems and Structures 17(1), January 2006, 15–28.

[255] Erturk, A., Lee, H. Y. and Inman, D. J., Investigation of soft and hard ceramics and single crystals for resonant and off-resonant piezoelectric energy harvesting, Proceedings of the 3rd ASME Conference on Smart Materials Adaptive Structures and Intelligent Systems SMASIS, Philadelphia, 2010.

[256] Bilgen, O., Karami, M. A., Inman, D. J. and Friswell, M. I., Actuation characterization of cantilevered unimorph beams with single crystal piezoelectric materials, Smart Materials and Structures 20(5), May 2011, article ID 055024.

[257] Karami, M. A., Bilgen, O., Inman, D. J. and Friswell, M. I., Experimental and analytical parametric study of single crystal unimorph beams for vibration energy harvesting, Transactions on Ultrasonics Ferroelectrics and Frequency Control 58(7), July 2011, 1508–1520.

[258] Bilgen, O., Wang, Y. and Inman, D. J., Electromechanical comparison of cantilevered beams with multifunctional piezoceramic devices, Mechanical Systems and Signal Processing 27, February 2012, 763–777.

[259] Anton, S. R., Erturk, A. and Inman, D. J., Bending strength of piezoelectric ceramics and single crystals for multifunctional load-bearing applications, IEEE Transactions on Ultrasonics Ferroelectrics and Frequency Control 59(6), June 2012, 1085–1092.

[260] Zhou, Q., Lamb, K. H., Zheng, H., Qiu, W. and Shung, K. K., Piezoelectric single crystal ultrasonic transducers for biomedical applications, Progress in Materials Science 66, October 2014, 87–111.

[261] Jiang, X., Kim, J. and Kim, K., Relaxor-PT-single crystal piezoelectric sensors, Crystals 4, 2014, doi:10.3390/cryst4030351, 351–376.

[262] Patel, S. and Vaish, R., Design of PZT-PT functionally graded piezoelectric material for low-frequency actuation applications, Journal of Intelligent Material Systems and Structures 26(3), February 2015, 321–327.

Piezoelectric Patch–Based Structural Health Monitoring (SHM)-Related References

[263] Park, G., Kabeya, K., Cudney, H. H. and Inman, D. J., Removing effects of temperature changes from piezoelectric impedance-based qualitative health monitoring, Proceedings of SPIE Conference on Sensory Phenomena and Measurement Instrumentation for Smart Structures and Materials, SPIE Vol. 3330, March 1998, 103–114.

[264] Schulz, M., Pai, P. F. and Inman, D. J., Health Monitoring and Active control of composite structures using piezoceramic patches, Composites Part B: Engineering 30(7), 1999, 713–725.

[265] Nothwang, W. D., Hirsch, S. G., Demaree, J. D., Hubbard, C. W., Cole, M. W. and Lin, B., Direct integration of thin film piezoelectric sensors with structural materials for structural health monitoring, Integrated Ferroelectrics An International Journal 83(1), 2006, 139–148.

[266] Grisso, B., Advancing autonomous structural health monitoring, Ph.D. Thesis at the Virginia Polytechnic Institute and State University, etd-12062007-105329, November 2007.

[267] Giurgiutiu, V., Piezoelectric wafer active sensors for structural health monitoring of composite structures using tuned guided waves, Journal of Engineering Materials and Technology 133(4), 2011, article ID 041012.

[268] Wang, R. L., Gu, H. and Song, G., Active sensing based bolted structure health monitoring using piezoceramic transducers, International Journal of Distributed Sensor Networks 2013, July 2013, article ID 583205.

References – Shape Memory Alloys

SMA Books and Review–Related References

[269] Perkins, J. (ed.), Shape Memory Effects in Alloys, Springer US, 1975, ISBN: 978-1-4684-2211-5.

[270] Achenbach, M. and Muller, I., Simulation of material behavior of alloys with shape memory, Archives of Mechanics 37(6), 1985, 573–585.

[271] Funakubo, H. (ed.), Shape Memory Alloys, Gordon and Breach Science Publication, New York, NY, USA, 1987.

[272] Tadaki, T., Otsuka, K. and Shimizu, K., Shape memory alloys, Annual Review of Material Science 18, 1988, 25–45.

[273] Achenbach, M., A model for an alloy with shape memory, International Journal of Plasticity 5, 1989, 371–395.

[274] Duerig, T., Melton, K., Stöckel, D. and Wayman, C. M., Engineering Aspects of Shape Memory Alloys, Elsevier, 1990, ISBN: 978-0-7506-1009-4, 499.

[275] Wayman, C. M., Shape memory and related phenomena, Progress in Materials Science 36, 1992, 203–224.

[276] Wayman, C. M., Shape memory alloys, MRS Bulletin 18(4), April 1993, 49–56.

[277] Stöckel, D., The shape memory effect – phenomenon, alloys and applications, Proceedings of the Electric Power Research Institute EPRI on Shape Memory Alloys for Power Systems, 1995.

[278] Birman, V., Review of mechanics of shape memory alloy structures, Applied Mechanics Reviews 50(11), November 1997, 629–645.

[279] Wei, Z. G., Sandström, R. and Miyazaki, S., Review shape-memory materials and hybrid composites for smart systems, Journal of Material Science 33, 1998, 3743–3762.

[280] Humbeeck, J. V., Non-medical applications of shape memory alloys, Materials Science and Engineering 273–275, December 1999, 134–148.

[281] Otsuka, K. and Ren, X., Recent developments in the research of shape memory alloys, Intermetallics 7(5), 1999, 511–528.

[282] Wu, M. H. and Schetky, L. M., Industrial applications for shape memory alloys, Proceedings of the International Conference on Shape Memory and Superelastic Technologies, Pacific Grove, California, USA, 2000, 171–182.

[283] Humbeeck, J. V., Shape memory alloys: a material and a technology, Advanced Engineering Materials 3(11), November 2001, 837–850.

[284] Huang, W., On the selection of shape memory alloys for actuators, Materials & Design 23(1), February 2002, 11–19.

[285] Otsuka, K. and Kakeshita, T., Science and technology of shape-memory alloys: new developments, MRS Bulletin, February 2002, 91–100.

[286] Frecker, M. I., Recent advances in optimization of smart structures and actuators, Journal of Intelligent material systems and structures 14, April–May 2003, 207–216.

[287] Seelecke, S. and Müller, I., Shape memory alloy actuators in smart structures: modeling and simulation, Applied Mechanics Reviews 57(1), January 2004, doi:10.1115/1.1584064, 23–46.

[288] Yoneyama, T. and Miyazaki, S. (eds.), Shape Memory Alloys for Biomedical Applications, 1st edn., Woodhead Publishing, 28 November 2008, 352.

[289] Shabalovskaya, S., Anderegg, J. and Humbeeck, J. V., Critical overview of nitinol surfaces and their modifications for medical applications, Acta Biomaterialia 4, 2008, 447–467.

[290] Ozbulut, O. E., Hurlebaus, S. and Desroches, R., Seismic response control using shape memory alloys: a review, Journal of Intelligent Material Systems and Structures 22(14), August 2011, doi:10.1177/1045389X11411220, 1531–1549.

[291] Barbarino, S., Saavedra Flores, E. I., Ajaj, R. M., Dayyani, I. and Friswell, M. I., A review on shape memory alloys with applications to morphing aircraft, Smart Materials Structures 23(6), August 2014, doi:10.1088/0964-1726/23/6/063001.

[292] Jani, J. M., Leary, M., Subic, A. and Gibson, M. A., A review of shape memory alloy research, Applications and Opportunities, Materials & Design 56, March 2014, 1078–1113.

[293] Lecce, L. and Concilio, A. (eds.), Shape Memory Alloy Engineering for Aerospace Structural and Biomedical Applications, Elsevier, 2015, 421.

Enhanced Buckling and Vibrations Using SMA-Related References

[294] Baz, A. and Tampe, L., Active Control of Buckling of Flexible Beams, Proceeding of the ASME Conference on Failure Prevention and Reliability, Montreal, September 1989, 211–218.

[295] Baz, A., Poh, S., Ro, J., Mutua, M. and Gilheany, J., Active Control of Nitinol-Reinforced Composite Beam, in Intelligent Structural Systems, Tzou, H. S. and Anderson, G. L. (eds.), Dordrecht, Springer Science + Business Media, 1992, 169–212.

[296] Baz, A., Imam, K. and McCoy, J., Active Vibration Control of Flexible Beams Using Shape Memory Actuators, Journal of Sound and Vibration 140(3), August 1990, 437–456.

[297] Lagoudas, D. C. and Tadjbakhsh, J. G., Active Flexible Rods with Embedded SMA Fibers, Smart Materials and Structures 1, 1992, 162–167.

[298] Brinson, L. C., Huang, M. S., Boller, C. and Brand, W., Analysis of Controlled Beam Deflections Using SMA Wires, Journal of Intelligent Material Systems and Structures 8, January 1997, 12–25.

[299] Birman, V., Theory and Comparison of the Effect of Composite and Shape Memory Alloy Stiffeners on Stability of Composite Shells and Plates, International Journal of Mechanical Science 39(10), 1997, 1139–1149.

[300] Baz, A., Chen, T. and Ro, J., Shape Control of NITINOL-Reinforced Composite Beams, Composites Part B: Engineering 31(8), 2000, 631–642.

[301] Tsai, X.-Y. and Chen, L.-W., Dynamic Stability of a Shape Memory Alloy Wire Reinforced Composite Beam, Composite Structures 56(3), May–June, 2002, 235–241.

Medical Applications of SMA-Related References

[302] Ikuta, K., Tsukumoto, M. and Hirose, S., Shape Memory Alloy Servo Actuator System with Electric Resistance Feedback and Application for Active Endoscope, Proceedings of the IEEE International Conference on Robotics and Automation, April 1988, doi:10.1109/ROBOT.1988.12085, 427–430.

[303] Pelton, A. R., Stöckel, D. and Duerig, T. W., Medical Uses of Nitinol, Proceedings of the International Symposium on Shape Memory Materials, Kanazawa, Japan, May 1999, Material Science Forum, Vols. 327–328, 2000, 63–70.

[304] Filip, P, Titanium-Nickel Shape Memory Alloys in Medical Applications. In: Titanium in Medicine, Material Science, Surface Science, Engineering, Biological Responses and Medical Applications, Brunette, D. M., Tengvall, P. Textor, M. and Thomsen, P, (eds), Part II, 2001, 53–86.

[305] Duerig, T., Stöckel, D. and Johnson, D., SMA-Smart Materials for Medical Applications, Proceedings of SPIE 4763, 2002, 7–15.

[306] Machado, L. G. and Savi, M. A., Medical Applications of Shape Memory Alloys, Brazilian, Journal of Medical and Biological Research 36(2003), 683–691.

[307] Morgan, N. B., Medical Shape Memory Alloy Applications – the Market and its Product, Material Science and Engineering:A 378(1–2), July 2004, 16–23.

[308] Kleinstreuer, C., Li, Z., Basciano, C. A., Seelecke, S. and Farber, M. A., Computational Mechanics of Nitinol Stents Grafts, Journal of Biomechanics 41(11), 2008, 2370–2378.

[309] De Miranda, R. L., Zamponi, C. and Quandt, E., Fabrication of TiNi Thin Film Stents, Smart Materials and Structures 18(10), October 2009, article ID 104010.

[310] Schaffer, J., Mechanical Conditioning of Superelastic Nitinol Wire for Improved Fatigue Resistance, Journal of ASTM International 7(5), 2010, 1–7.

[311] Petrini, L. and Migliavacca, F., Biomedical Applications of Shape Memory Alloys, Journal of Metallurgy, 2011 article ID 501483, 15.

[312] Zainal, M. A., Sahlan, S. and Ali, M. S. M., Micromachined Shape Memory Alloy Microactuators and Their Application in Biomedical Devices, Micromachines 6, 2015, 879–901.

Beams, Columns, Rods and Actuators with SMA-Related References

[313] Baz, A. and Ro, J., Thermo-Dynamic Characteristics of Nitinol-Reinforced Composite Beams, Composites Engineering 2(5–7), 1992, 527–542.

[314] Baz, A. and Chen, T., Torsional Stiffness of NITINOL-Reinforced Composite Drive Shafts, Composite Engineering 3(12), 1993, 1119–1130.

[315] Boyd, J. G. and Lagoudas, D. C, Thermomechanical Response of Shape Memory Composites, Journal of Intelligent Material Systems and Structures 5(3), May 1994, 333–346.

[316] Zhang, C. and Zee, R. H., Development of Ni-Ti Based Shape Memory Alloys for Actuation and Control, Proceedings of the 31st Intersociety Energy Conversion Engineering Conference IECEC 1, 1996, doi:10.1109/IECEC.1996.552877, 239–244.

[317] Chen, Q. and Levy, C., Active Vibration Control of Elastic Beam by Means of Shape Memory Alloy Layers, Smart Materials and Structures 5(4), August 1996, 400–406.

[318] Dolce, M., Cardone, D. and Marnetto, R., Implementation and Testing of Passive Control Devices Based on Shape Memory Alloys, Earthquake Engineering & Structural Dynamics 29(7), July 2000, 945–968.

[319] Prahlad, H. and Chopra, I., Experimental Characterization of Ni-Ti Shape Memory Alloy Wires Under Uniaxial Loading Conditions, Journal of Intelligent Material Systems and Structures 11(4), April 2000, 263–271.

[320] Jonnalagadda, K. D., Sottos, N. R., Qidwai, M. A. and Lagoudas, D. C, Insitu-Displacement Measurements and Theoretical Prediction of Embedded SMA Actuation, Journal of Smart Materials and Structures 9, 2000, 701–710.

[321] Pae, S., Lee, H., Park, H. and Hwang, W., Realization of Higher-Mode Deformation of Beams Using Shape Memory Alloy Wires and Piezoceramics, Smart Materials and Structures 9(6), December 2000, 848–854.

[322] Epps, J. J. and Chopra, I., In-Flight Tracking of Helicopter Rotor Blades Using Shape Memory Alloy Actuators, Structures Smart Material and Structures 10(1), February 2001, 104–111.

[323] Prahlad, H. and Chopra, I., Comparative Evaluation of Shape Memory Alloy Constitutive Models with Experimental Data, Journal of Intelligent Material Systems and Structures 12(6), June 2001, 383–397.

[324] Lammering, R. and Schmidt, I., Experimental Investigation on the Damping Capacity of NiTi Components, Smart Materials and Structures 10(5), October 2001, 853–859.

[325] Matsuzaki, Y., Naito, H., Ikeda, T. and Funami, K., Thermo-Mechanical Behavior Associated with Pseudoelastic Transformation of Shape Memory Alloys, Smart Materials and Structures 10(5), October 2001, 884–892.

[326] Chandra, R., Active Shape Control of Composite Blades Using Shape Memory Alloys, Smart Materials and Structures 10(5), October 2001, 1018–1024.

[327] Mehrabi, R., Kadkhodaei, M., Andani, M. T. and Elahinia, M., Microplane Modeling of Shape Memory Alloy Tubes Under Tension, Torsion and Proportional Tension-Torsion Loading, Journal of Intelligent Material Systems and Structures 26(2), January 2015, 144–155.

[328] Malukhin, K. and Ehmann, K., Model of a NiTi Shape Memory Alloy Actuator, Journal of Intelligent Material Systems and Structures 26(4), March 2015, 386–399.

[329] Lacasse, S., Terriault, P., Simoneau, C. and Brailovski, V., Design, Manufacturing and Testing of an Adaptive Panel With Embedded Shape Memory Actuators, Journal of Intelligent Material Systems and Structures 26(15), October 2015, 2055–2072.

SMA Reinforced Plates and Shells–Related References

[330] Ro, J. and Baz, A., NITINOL-Reinforced plates: Part I. Thermal Characteristics, Composite Engineering 5(1), 1995, 61–75.
[331] Ro, J. and Baz, A., NITINOL-Reinforced plates: Part II. Static and buckling Characteristics, Composite Engineering 5(1), 1995, 77–90.
[332] Ro, J. and Baz, A., NITINOL-Reinforced plates: Part III. Dynamic characteristics, Composite Engineering 5(1), 1995, 91–106.
[333] Park, J.-S., Kim, J.-H. and Moon, S.-H., Vibration of Thermally Post-Buckled Composite Plates Embedded with Shape Memory Alloy Fibers, Composite Structures 63, 2004, 179–188.
[334] DeHaven, J. G. and Tzou, H. S., Forced Response of Cylindrical Shells Coupled with Nonlinear SMA Actuators Regulated by Sinusoidal and Saw-Tooth Temperature Profiles, Journal of Engineering Mathematics 61, 2008, doi:10.1007/s10665-008-9231-5.

Characterization and Modeling of SMA-Related References

[335] Boyd, J. G. and Lagoudas, D. C., A Thermodynamical Constitutive Model for Shape Memory Materials, Part I: The Monolithic Shape Memory Alloy, International Journal of Plasticity 12, 1996, 805–842.
[336] Boyd, J. G. and Lagoudas, D. C., A Thermodynamical Constitutive Model for Shape Memory Materials, Part II, The SMA Composite Material, International Journal of Plasticity 12, 1996, 843–873.
[337] Bo, Z. and Lagoudas, D. C., Thermomechanical Modeling of Polycrystalline SMAs Under Cyclic Loading, Part I: Theoretical Derivations, International Journal of Engineering Science 37(9), July 1999, 1089–1140.
[338] Bo, Z. and Lagoudas, D. C., Thermomechanical Modeling of Polycrystalline SMAs Under Cyclic Loading, Part II: Material Characterization and Experimental Results for a Stable transformation Cycle, International Journal of Engineering Science 37(9), July 1999, 1141–1173.
[339] Bo, Z. and Lagoudas, D. C., Thermomechanical Modeling of Polycrystalline SMAs Under Cyclic Loading, Part III: Evolution of Plastic Strains and Twoway Shape Memory Effect, International Journal of Engineering Science 37(9), July 1999, 1175–1203.
[340] Bo, Z. and Lagoudas, D. C., Thermomechanical Modeling of Polycrystalline SMAs Under Cyclic Loading, Part IV: Modeling of Minor Hysteresis Loops, International Journal of Engineering Science 37(9), July 1999, 1205–1249.
[341] Auricchio, F. and Sacco, E., A one-dimensional model for superelastic shape-memory alloys with different elastic properties between austenite and martensite, International Journal Non-Linear Mechanics 32(6), November 1997, 1101–1114.

[342] Lagoudas, D. C., Moorthy, D., Qidwai, M. A. and Reddy, J. N., Modeling of the Thermomechanical Response of Active Composite Laminates with SMA Layers, Journal of Intelligent Material Systems and Structures 8, 1997, 476–488.

[343] Xu, G.-M., Lagoudas, D. C., Hughes, D. and Wen, J. T., Modeling of a Flexible Beam Actuated by Shape Memory Alloys Wires, Journal of Smart Materials and Structures 6, 1997, 265–277.

[344] Webb, G. V., Lagoudas, D. C. and Kurdila, A. J., Hysteresis Modeling of SMA Actuators for Control Applications, Journal of Intelligent Materials Systems and Structures 9, 1998, 432–448.

[345] Matsuzaki, Y., Funami, K. and Naito, H., Inner Loops of Pseudoelastic Hysteresis of Shape Memory Alloys: Preisach Approach, Smart Structures and Materials 2002, Proceedings of the SPIE, 4699, 2002, 355–364.

[346] Seelecke, S. Modeling the Dynamic Behavior of Shape Memory Alloys, International Journal of Non-Linear Mechanics, 37, 2002, 1363–1374.

[347] Prahlad, H. and Chopra, I., Development of Strain-Rate Dependent Model for Uniaxial Loading of SMA Wires, Journal of Intelligent Material Systems and Structures 14(5), July 2003, 429–442.

[348] Auricchio, F., Marfia, S. and Sacco, E., Modelling of SMA materials: Training and Two Way Memory Effects, Computers & Structures 81(24–25), September 2003, 2301–2317.

[349] Singh, K., Sirohi, J. and Chopra, I., An Improved Shape-Memory Alloy Actuator for Rotor Blade Tracking, Journal of Intelligent Material Systems and Structures 14(12), December 2003, 767–786.

[350] Lagoudas, D. C., Khan, M. M., Mayes, J. J. and Henderson, B. K., Pseudoelastic SMA Spring Elements for Passive Vibration Isolation, Part I: Modeling, Journal of Intelligent Material Systems and Structures 15, 2004, 415–441.

[351] Lagoudas, D. C., Khan, M. M., Mayes, J. J. and Henderson, B. K., Pseudoelastic SMA Spring Elements for Passive Vibration Isolation, Part II: Simulations and Experimental Correlations, Journal of Intelligent Material Systems and Structures 15, 2004, 443–470.

[352] Manzo, J., Garcia, E., Wickenheiser, A. M. and Horner, G., Design of a Shape-Memory Alloy Actuated Macro-Scale Morphing Aircraft Mechanism, Proceedings of SPIE Conference, San Diego, SPIE 5764, March 2005, 232–240.

[353] Prahlad, H. and Chopra, I., Modeling and Experimental Characterization of SMA Torsional Actuators, Journal of Intelligent Material Systems and Structures 18(1), January 2007, 29–38.

[354] Hartl, D. and Lagoudas, D. C., Aerospace Applications of Shape Memory Alloys, Journal of Aerospace Engineering 221, 2007, 535–552.

[355] Popov, P. and Lagoudas, D. C., A 3-D Constitutive Model for Shape Memory Alloys Incorporating Pseudoelasticity and Detwinning of Self-Accommodated Martensite, International Journal of Plasticity 23, 2007, 1679–1720.

[356] Liang, C. and Rogers, C. A., Design of Shape Memory Alloy Actuators, Journal of Mechanical Design 114 (2), June 2008, doi:10.1115/1.2916935.

[357] Churchill, C. B., Shaw, J. A. and Iadicola, M. A., Tips and Tricks for Characterizing Shape Memory Alloy Wire: Part 2-Fundamental Isothermal Responses, Experimental Techniques 33(1),January–February 2009, doi:10.1111/j.1747-1567.2008.00460.x, 51–62.

[358] Churchill, C. B., Shaw, J. A. and Iadicola, M. A., Tips and Tricks for Characterizing Shape Memory Alloy Wir: Part 3-Localization and Propagation Phenomena, Experimental Techniques 33 (5),September–October 2009, doi:10.1111/j.1747-1567.2009.00558.x, 70–78.

[359] Bertacchini, O. W., Characterization and Modeling of Transformation Induced Fatigue of Shape Memory Alloy Actuators, Ph.D. Thesis at Texas A&M University, December 2009.

[360] Hartl, D., Modeling of Shape Memory Alloys Considering Rate-Independent and Rate-Dependent Irrecoverable Strains, Ph.D. Thesis at Texas A&M University, December 2009.

[361] Hartl, D. J. and Lagoudas, D. C., Constitutive Modeling and Structural Analysis Considering Simultaneous Phase Transformation and Plastic Yield in Shape Memory Alloys, Smart Materials and Structures 18(10), 2009, article ID 104017.

[362] Churchill, C. B., Shaw, J. A. and Iadicola, M. A., Tips and Tricks for Characterizing Shape Memory Alloy Wire: Part 4-Thermo-Mechanical Coupling, Society for Experimental Mechanics 34 (2),March-April 2010, 63–80.

[363] Furst, S. J. and Seelecke, S., Modeling and Experimental Characterization of the Stress, Strain, and Resistance of Shape Memory Alloy Actuator Wires with Controlled Power Input, Journal of Intelligent Material Systems and Structures 23(11), July 2012, 1233–1247.

Morphing with SMA-Related References

[364] Strelec, J. K. and Lagoudas, D. C., Fabrication and testing of a shape memory alloy actuated reconfigurable wing, Proceeding of the SPIE Conference on Smart Structures and Materials, SPIE 4701, July 2002, doi:10.1117/12.474664, 267–280.

[365] Strelec, J., Design and Implementation of a Shape Memory Alloy Actuated Reconfigurable Wing, Master's Thesis at the Texas A&M University, 2002.

[366] Strelec, J. K., Lagoudas, D. C., Khan, M. A. and Yen, J., Design and implementation of a shape memory alloy actuated reconfigurable wing, Journal of Intelligent Material Systems and Structures 14, 2003, 257–273.

[367] Peng, F., Jiang, X.-X., Hu, Y.-R. and Ng, A., Application of shape memory alloy actuators in active shape control of inflatable space structures, Proceedings of the IEEE Aerospace Conference, March 2005, doi:10.1109/AERO.2005.1559577.

[368] Oehler, S., Developing methods for designing shape memory alloy actuated morphing aerostructures, Master's Thesis at the Texas A&M University, 2012.

[369] Naghashian, S., Fox, B. L. and Barnett, M. R., Actuation curvature limits for a composite beam with embedded shape memory alloy wires, Smart Materials and Structures 23(6), June 2014, article ID 065002.

General SMA Applications and Uses–Related References

[370] Wuttig, M., Li, J. and Craciunescu, C., A new ferromagnetic shape memory alloy system, Scripta Materialia 44(10), May 2001, 2393–2397.

[371] Ishida, A. and Martynov, V., Sputter-deposited shape-memory alloy thin films: properties and applications, MRS Bulletin, February 2002, 111–114.

[372] Ma, N. and Song, G., Control of shape memory alloy actuator using pulse width modulation, Smart Materials and Structures 12(5), 2003, 712–719.

[373] Des Roches, R., McCormick, J. and Delemont, M., Cyclic properties of superelastic shape memory alloy wires and bars, Journal of Structural Engineering 130(1), January 2004, 38–46.

[374] Nagai, H. and Oishi, R., Shape memory alloys as strain sensors in composites, Smart Materials and Structures 15(2), 2006, 493–498.

[375] Schick, J., Transformation Induced fatigue of Ni-Rich NiTi shape memory alloy actuators, Master's Thesis at the Texas A&M University, December 2009.

[376] Lagoudas, D. C., Miller, D. A., Rong, L. and Kumar, P. K., Thermomechanical fatigue of shape memory alloys, Smart Materials and Structures 18, 2001, article ID 085021.

[377] Lan, C.-C. and Fan, C.-H., An accurate self-sensing method for the control of shape memory alloy actuated flexures, Sensors and Actuators 163(1), doi:10.1016/j.sna.2010.07.018, April 2010, 323–332.

[378] Moussa, M. O., Moumni, Z., Doare, O., Touze, C. and Zaki, W., Non-linear dynamic thermomechanical behavior of shape memory alloys, Journal of Intelligent Material Systems and Structures 23(14), 2012, 1593–1611.

References – Electrorheological and Magnetorheological Fluids

Reviews and Books on Electrorheological and Magnetorheological Fluid–Related References

[379] Jordan, T. C. and Shaw, M. T., Electrorheology, Material Research Society MRS Bulletin 16(8), August 1991, 38–43.

[380] Stanway, R., Sprostonz, J. L. and El-Wahed, A. K., Applications of electrorheological fluids in vibration control: a survey, Smart Materials and Structures 5(4), August 1996, 464–482.

[381] Kamath, G. M. and Wereley, N. M., Modeling the damping mechanism in electro-rheological fluid based dampers, ASTM Metals Test Methods and Analytical Procedures, STP1304-EB, January 1997, 331–348.

[382] Tao, R. (ed.), Electrorheological fluids and magnetorheological suspensions, Proceedings of the 7th International Conference on Electrorheological Fluids and Magnetorheological Suspensions, Honolulu, Hawaii, USA, July 12–23, 1999, World Scientific Publishing, 850 pp.

[383] Jolly, M. R., Properties and applications of magnetorheological fluids, symposium LL-materials for smart systems III. In: Fogle, M. W., Uchino, K., Ito, Y. and Gotthardt, R., (eds), Materials Research Society (MRS) Proceedings, Vol. 604, 1999.

[384] Phulé, P. P., Magnetorheological (MR) fluids: principles and applications, Smart Material Bulletin 2001(2), February 2001, 7–10.

[385] Wen, W., Huang, X. and Sheng, P., Electrorheological fluids: structures and mechanism, Soft Matter 4, 2008, 200–210.

[386] Wereley, N. M. (ed.), *Magnetorheology: Advances and Applications*, RSC Smart Materials series, RSC Publishing, 26 November 2013, 396 pp.

[387] Kulkarni, A. N. and Patil, S. R., Magneto-Rheological (MR) and Electro-Rheological (ER) fluid damper: a review parametric study of fluid behavior, International Journal of Engineering Research and Applications 3(6),November-December 2013, 1879–1882.

Electrorheological-Related References

[388] Uejima, H., Dielectric mechanism and rheological properties of electro-fluids, Japanese Journal of Applied Physics 11(3), March 1972, 319–326.

[389] Davis, L. C., Finite-element analysis of particle-particle forces in electrorheological fluids, Applied Physics Letters 60(3), September 1991, 319–321.

[390] Halsey, T. C., Electrorheological fluids: structure formation relaxation and destruction, Materials Research Society MRS Proceedings, Vol. 248, 1991.

[391] Katsikopoulos, P. and Zukoski, C., Relaxation processes in the electrorheological response, Materials Research Society MRS Proceedings, Vol. 248, 1991.

[392] Davis, L. C., Polarization forces and conductivity effects in electrorheological fluids, Journal of Applied Physics 72, February 1992, 1334–1340.
[393] Choi, S.-B., Park, Y.-K. and Kim, J.-D., Vibration characteristics of hollow cantilevered beams containing an electrorheological fluid, International Journal of Mechanical Sciences 35(9), August 1992, 757–768.
[394] Coulter, J. P., Weiss, K. D. and Carlson, J. D., Engineering applications of electrorheological materials, Journal of Intelligent Material Systems and Structures 4(2), April 1993, 248–259.
[395] Powell, J. A., Modelling the Oscillatory Response of an Electrorheological Fluid, Smart Materials and Structures 3(4), March 1994, 416–438.
[396] Rajagopal, K. R., Yalamanchili, R. C. and Wineman, A. S., Modeling electrorheological materials through mixture theory, International Journal of Engineering Science 32(3), 1994, 481–500.
[397] Rajagopal, K. R. and Ruziicka, M., On the modeling of electrorheological materials Mechanics Research Communications 23(4), April 1996, 401–407.
[398] Wu, S., Lu, S. and Shen, J., Electrorheological suspensions, Polymer International 41(4), December 1996, 363–367.
[399] Gamota, D. R., Schubring, A. W., Mueller, B. L. and Filisko, F. E., Amorphous ceramics as the particulate phase in electrorheological materials systems, Journal of Materials Research 11(1), 1996, 144–155.
[400] Kamath, G. M., Hurt, M. K. and Wereley, N. M., Analysis and testing of bingham plastic behavior in semi-active electrorheological fluid dampers, Smart Materials and Structures 5(5), 1996, 576–590.
[401] Gordaninejad, F. and Bindu, R., A Scale Study of electrorheological fluid dampers, Journal of Structural Control 4(2), December 1997, 5–17.
[402] Davis, L. C., Time-dependent and nonlinear effects in electrorheological fluids, Journal of Applied Physics 81(4), 1997, 1985–1991.
[403] Kohl, J. G. and Tichy, J. A., Expressions for coefficients of electrorheological fluid dampers, Lubrication Science 10(2), February 1998, 135–1143.
[404] Conrad, H., Properties and design of electrorheological suspensions, Materials Research Society MRS Bulletin 23(8), August 1998, 35–42.
[405] Inoue, A., Ide, Y., Maniwa, S., Yamada, H. and Oda, H., ER Fluids Based on liquid-crystalline polymers, Materials Research Society MRS Bulletin 23(8), August 1998, 43–49.
[406] Wen, W., Tam, W. Y. and Sheng, P., Electrorheological fluids using bi-dispersed particles, Journal of Material Research 13(10), 1998, 2783–2786.
[407] Wu, C. W. and Conrad, H., Influence of mixed particle size on electrorheological response, Journal of Applied Physics 83(7), 1998, 3880–3884.
[408] Kohl, J. G., Tichy, J. A., Craig, K. C. and Malcolm, S. M., Determination of the bingham parameters of an electrorheological fluid in an axial flow concentric-cylinder rheometer, Tribotest Journal 5(3), March 1999, 221–224.
[409] Johnson, A. R., Bullough, W. A. and Makin, J., Dynamic Simulation and Performance of an electrorheological clutch based reciprocating mechanism, Smart Materials and Structures 8(5), October 1999, 591–600.
[410] Lee, C.-Y. and Cheng, C.-C., Complex moduli of electrorheological material under oscillatory shear, International Journal of Mechanical Sciences 42(3), March 2000, 561–573.
[411] Lee, C. Y. and Liao, W. C., Characteristics of an electrorheological fluid valve used in an inkjet print head, Smart Materials and Structures 9, March 2000, 839–847.
[412] Fukuda, T., Takawa, T. and Nakashima, K., Optimum vibration control of CFRP sandwich beam using electrorheological fluids and piezoceramic actuators, Smart Materials and Structures 9(1), February 2000, 121–125.

[413] Wen, W., Ma, H., Tam, W. Y. and Sheng, P., Frequency-induced structure variation in electrorheological fluids, Applied Physics Letters 77(23), December 2000, 3821–3823.

[414] Sakamoto, D., Oshima, N. and Fukuda, T., Tuned sloshing damper using electrorheological fluid, Smart Materials and Structures 10, January 2001, 963–969.

[415] Wang, B., Liu, Y. and Xiao, Z., Dynamical modelling of the chain structure formation in electrorheological fluids, International Journal of Engineering Science 39(4), March 2001, 453–475.

[416] Hao, T., Electrorheological fluids, Advanced Materials 13(24), December 2001, 1847–1857.

[417] Chen, S. H., Yang, G. and Liu, X. H., Response analysis of vibration systems with ER dampers, Smart Materials and Structures 10(5), October 2001, 1025–1030.

[418] Kang, Y. K., Kim, J. and Choi, S.-B., Passive and active damping characteristics of smart electrorheological composite beams, Smart Materials and Structures 10(4), 2001, 724–729.

[419] Yoshida, K., Kikuchi, M., Park, J.-H. and Yokota, S., Fabrication of micro electrorheological valves (ER Valves) by Micromachining and experiments, Sensors and Actuators 95(2–3), January 2002, 227–233.

[420] Hong, S.-R., Choi, S.-B. and Han, M.-S., Vibration control of a frame structure using electrorheological fluid mounts, International Journal of Mechanical Sciences 44, April 2002, 2027–2045.

[421] Noresson, V., Ohlson, N. G. and Nilsson, M., Design of electrorheological dampers by means of finite element analysis: theory and applications, Materials and Design 23(4), June 2002, 361–369.

[422] Sproston, J. L., El Wahed, A. K. and Stanway, R., The rheological characteristics of electrorheological fluids in dynamic squeeze, Journal of Intelligent Material Systems and Structures 13(10), October 2002, 655–660.

[423] Chen, S. H. and Yang, G., A method for determining locations of electrorheological dampers in structures, Smart Materials and Structures 12(2), April 2003, 164–170.

[424] Lindler, J. and Wereley, N. M., Quasi-steady bingham-plastic analysis of an electrorheological flow mode bypass damper with piston bleed, Smart Materials and Structures 12(3), 2003, 305–317.

[425] Barber, G. C., Jiang, Q. Y., Zou, Q. and Carlson, W., Development of a laboratory test device for electrorheological fluids in hydrostatic lubrication, Tribotest Journal 11(3), March 2005, 185–191.

[426] Phani, A. S. and Venkatraman, K., Damping characteristics of electrorheological fluid sandwich beams, Acta Mechanica 180(1), June 2005, 195–201.

[427] Choi, S. B., Choi, H. J., Choi, Y. T. and Wereley, N. M., Preparation and mechanical characteristics of poly-methylaniline based electrorheological fluid, Journal of Applied Polymer Science 96(5), 2005, 1924–1929.

[428] Lim, S. C., Park, J. S., Choi, S. B., Choi, Y. T. and Wereley, N. M., Design and analysis program of electrorheological devices for vehicle systems, International Journal of Vehicle Autonomous Systems 3(1), 2005, 15–33.

[429] Shulman, Z. P., Korobko, E. V., Levin, M. L., Binshtok, A. E., Bilyk, V. A. and Yanovsky, Yu. G., Energy dissipation in electrorheological damping devices, Journal of Intelligent Material Systems and Structures 17(4), April 2006, 315–320.

[430] Liu, L., Chen, X., Niu, X., Wen, W. and Sheng, P., Electrorheological fluid-actuated microfluidic pump, Applied Physics Letters 89(8), May 2006, 083.505–083.506.

[431] Yan, Q. S., Bi, F. F. and Wu, N. Q., Performance evaluation of electro-rheological fluid and its applications to micro-parts fine-machining, Journal Key Engineering Materials 315–316, July 2006, 352–356.

[432] Nikitczuk, J., Weinberg, B. and Mavroidis, C., Control of electrorheological fluid based resistive torque elements for use in active rehabilitation devices, Smart Materials and Structures 16(2), February 2007, 418–428.

[433] Ramkumar, K. and Ganesan, N., Vibration and damping of composite sandwich box column with viscoelastic/electrorheological fluid core and performance comparison, Materials and Design 30(8), September 2009, 2981–2994.

[434] Chen, Y. G. and Yan, H., The performance analysis of electro-rheological damper, Advanced Materials Research 179–180, January 2011, 443–448.

[435] Kaushal, M. and Joshi, Y. M., Self-similarity in electrorheological behavior, Soft Matter 7, May 2011, 9051–9060.

[436] El Wahed, A. K., The Influence of solid-phase concentration on the performance of electrorheological fluids in dynamic squeeze flow, Materials and Design 32(3), 2011, 1420–1426.

[437] Hoppe, R. H. W. and Litvinov, W. G., Modeling simulation and optimization of electrorheological fluids, Handbook of Numerical Analysis, Vol. 16, 2011, 719–793.

[438] Liu, Y. D. and Choi, H. J., Electrorheological fluids: smart soft matter and characteristics, Soft Matter 8, May 2012, 11.961–11.978.

[439] Krivenkov, K., Ulrich, S. and Bruns, R., Extending the operation range of electrorheological actuators for vibration control through novel designs, Journal of Intelligent Material Systems and Structures 23(12), 2012, 1323–1330.

[440] Mohammadi, F. and Sedaghati, R., Dynamic mechanical properties of an electrorheological fluid under large amplitude oscillatory shear strain, Journal of Intelligent Material Systems and Structures 23(10), 2012, 1093–1105.

[441] Jiang, B. and Shi, W. K., The model simulation analysis of electro-rheological fluid engine mounting system, Advanced Materials Research 694–697, May 2013, 338–343.

[442] Allahverdizadeha, A., Mahjooba, M. J., Malekib, M., Nasrollahzadeha, N. and Naeia, M. H., Structural Modeling, Vibration Analysis and Optimal viscoelastic layer characterization of adaptive sandwich beams with electrorheological fluid core, Mechanics Research Communications 51, July 2013, 15–22.

[443] Wang, Z., Gong, X., Yang, F., Jiang, W. and Xuan, S., Dielectric relaxation effect on flow behavior of electrorheological fluids, Journal of Intelligent Material Systems and Structures 26(10), May 2014, 1141–1149.

[444] Hoseinzadeh, M. and Rezaeepazhand, J., Vibration suppression of composite plates using smart electrorheological dampers, International Journal of Mechanical Sciences 84, July 2014, 31–40.

[445] Wang, Z., Gong, X., Yang, F., Jiang, W. and Xuan, S., Dielectric relaxation effect on flow behavior of electrorheological fluids, Journal of Intelligent Material Systems and Structures 26(10), July 2015, 1141–1149.

Electrorheological- and Magnetorheological-Related References

[446] Weiss, K. D., Carlson, J. D. and Nixon, D. A., Viscoelastic properties of magneto- and electrorheological fluids, Journal of Intelligent Material Systems and Structures 5(6), November 1994, 772–775.

[447] Phulé, P. P. and Ginder, J. M., The materials science of field-responsive fluids, Materials Research Society MRS Bulletin 23(8), August 1998, 19–22.

[448] Wereley, N. M., Pang, L. and Kamath, G. M., Idealized hysteresis modeling of electrorheological and magnetorheological dampers, Journal of Intelligent Material Systems and Structures 9(8), August 1998, 642–649.

[449] Wereley, N. M. and Pang, L., Nondimensional analysis of semi-active electrorheological and magnetorheological dampers using approximate parallel plate models, Smart Materials and Structures 7(5), 1998, 732–743.

[450] Rankin, P. J., Ginder, J. M. and Klingenberg, D. J., Electro- and magnetorheology, Current Opinion in Colloid & Interface Science 3(4), August 1998, 373–381.

[451] El Wahed, A. K., Sproston, J. L. and Schleyer, G. K., A comparison between electrorheological and magnetorheological fluids subjected to impulsive loads, Journal of Intelligent Material Systems and Structures 10(9), September 1999, 695–700.

[452] Lee, D. Y. and Wereley, N. M., Quasi-steady Herschel-Bulkley analysis of electro- and magneto-rheological flow mode dampers, Journal of Intelligent Material Systems and Structures 10(10), September 1999, 761–769.

[453] Yalcintas, M. and Dai, H., Magnetorheological and electrorheological materials in adaptive structures and their performance comparison, Smart Materials and Structures 8(5), October 1999, 560–573.

[454] Xu, Y. L., Qu, W. L. and Ko, J. M., Seismic response control of frame structures using magnetorheological/electrorheological dampers, Earthquake Engineering Structural Dynamics 29(5), April 2000, 557–575.

[455] Wang, Z., Fang, H., Lin, Z. and Zhou, L., Dynamic simulation studies of structural formation and transition in electro-magnetorheological fluids, International Journal of Modern Physics B 15(6–7), 2001, 842–850.

[456] Butz, T. and Von Stryk, O., Modelling and simulation of electro- and magnetorheological fluid dampers, ZAMM Journal of Applied Mathematics and Mechanics 82(1), 2002, 3–20.

[457] El Wahed, A. K., Sproston, J. L. and Schleyer, G. K., Electrorheological and magnetorheological fluids in blast resistant design applications, Materials and Design 23(4), June 2002, 391–404.

[458] Choi, Y.-T. and Wereley, N. M., Comparative analysis of the time response of electrorheological and magnetorheological dampers using nondimensional parameters, Journal of Intelligent Material Systems and Structures 13(7–8), July 2002, 443–451.

[459] Lee, D.-Y., Choi, Y.-T. and Wereley, N. M., Performance analysis of ER/MR impact damper systems using Herschel-Bulkley model, Journal of Intelligent Material Systems and Structures 13(7–8), July–August 2002, 525–531.

[460] Dimock, G, Yoo, J.-H. and Wereley, N. M., Quasi-steady Bingham biplastic analysis of electrorheological and magnetorheological dampers, Journal of Intelligent Material Systems and Structures 13(9), September 2002, 549–559.

[461] Widjaja, J. and Samali, B., Li, J., Electrorheological and magnetorheological duct flow in shear-flow mode using Herschel-Bulkley constitutive model, Journal of Engineering Mechanics 129(12), November 2003, 1475–1477.

[462] Sims, N. D., Holmes, N. J. and Stanway, R., A unified modelling and model updating procedure for electrorheological and magnetorheological vibration dampers, Smart Materials and Structures 13(1), December 2003, 100–121.

[463] Lindler, J., Choi, Y.-T. and Wereley, N. M., Double adjustable electrorheological and magnetorheological shock absorbers, International Journal of Vehicle Design 33(1–3), 2003, 189–206.

[464] Rosenfeld, N. C. and Wereley, N. M., Volume-constrained optimization of magnetorheological and electrorheological valves and dampers, Smart Materials and Structures 13(6), July 2004, 1303–1313.

[465] Gandhi, F. and Bullough, W. A., On the phenomenological modeling of electrorheological and magnetorheological fluid pre-yield behavior, Journal of Intelligent Material Systems and Structures 16(3), March 2005, 237–248.

[466] Choi, Y. T., Cho, J. U., Choi, S. B. and Wereley, N. M., Constitutive models of electrorheological and magnetorheological fluids using viscometers, Smart Materials and Structures 14(5), October 2005, 1025–1034.

[467] Choi, Y.-T., Bitman, L. and Wereley, N. M., Nondimensional eyring analysis of electrorheological and magnetorheological dampers, Journal of Intelligent Material Systems and Structures 16(5), 2005, 383–394.

[468] Yoo, J.-H. and Wereley, N. M., Nondimensional analysis of annular duct flow in ER/MR dampers, International Journal of Modern Physics Part B 19(7–9), 2005, 1577–1583.

[469] Han, Y. M., Nguyen, Q. H., Choi, S. B. and Kim, K. S., Hysteretic behaviors of yield stress in smart ER/MR materials: experimental results, Key Engineering Materials 326–328, December 2006, 1459–1462.

[470] Wereley, N. M., Nondimensional Herschel–Bulkley analysis of magnetorheological and electrorheological dampers, Journal of Intelligent Material Systems and Structures 19(3), March 2008, 257–268.

[471] Wereley, N., Quasi-steady Herschel-Bulkley analysis of magnetorheological and electrorheological dampers, Journal of Intelligent Material Systems and Structures 19(3), 2008, 257–268.

[472] Goldasz, J. and Sapinski, B., Nondimensional characterization of flow-mode magnetorheological/electrorheological fluid dampers, Journal of Intelligent Material Systems and Structures 23(14), September 2012, 1545–1562.

[473] Esteki, K., Bagchi, A. and Sedaghati, R., Dynamic analysis of electro- and magnetorheological fluid dampers using duct flow models, Smart Materials and Structures 23(3), 2014, article ID 035016.

Magnetorheological-Related References

[474] Phulé, P. P., Ginder, J. M. and Jatkar, A. D., Synthesis and properties of magnetorheological mr fluids for active vibration control, Materials Research Society MRS Proceedings 459, 1996.

[475] Carlson, J. D., Catanzarite, D. M. and St. Clair, K. A., Commercial magnetorheological fluid devices, International Journal of Modern Physics B 10(23&24), October 1996, 2857–2965.

[476] Ginder, J. M., Davis, L. C. and Elie, L. D., Rheology of magnetorheological fluids: models and measurements, International Journal of Modern Physics B 10(23&24), October 1996, 3293–3303.

[477] Ashour, O. N., Kinder, D., Giurgiutiu, V. and Rogers, C. A., Manufacturing and characterization of magnetorheological fluids, Proceedings of the SPIE Conference on Smart Structures and Materials: Smart Materials Technologies, SPIE 3040, 1997, doi:10.1117/12.267112, 174–184.

[478] Ginder, J. M., Behavior of magnetorheological fluids, Materials Research Society MRS Bulletin 23(8), August 1998, 26–29.

[479] Phulé, P. P., Synthesis of novel magnetorheological fluids, Materials Research Society MRS Bulletin 23(8), August 1998, 23–25.

[480] Jolly, M. R., Bender, J. W. and Carlson, J. D., Properties and applications of commercial magnetorheological fluids, Journal of Intelligent Material Systems and Structures 10(1), January 1999, 5–13.

[481] Kamath, G. M., Wereley, N. M., Jolly, M. R., Characterization of magnetorheological helicopter lag dampers, Journal of the American Helicopter Society 44(3), July 1999, 234–248.

[482] Wereley, N. M., Kamath, G. M. and Madhavan, V., Hysteresis modeling of semi-active magnetorheological helicopter lag dampers, Journal of Intelligent Material Systems and Structures 10(8), August 1999, 624–633.

[483] Wang, D. H. and Liao, W. H., Application of MR dampers for semi-active suspension of railway vehicles, Proceedings of the International Conference on Advances in Structural Dynamics, 2000, 1389–1396

[484] Snyder, R., Kamath, G. M. and Wereley, N. M., Characterization and analysis of magnetorheological damper behavior under sinusoidal loading, AIAA Journal 39(7), July 2001, 1240–1253.

[485] Bica, I., Damper with magnetorheological suspension, Journal of Magnetism and Magnetic Materials 241(2–3), March 2002, 196–200.

[486] Li, W. H., Chen, G., Yeo, S. H. and Du, H., Dynamic properties of magnetorheological materials, Key Engineering Materials 227, August 2002, 119–124.

[487] Bossis, G., Lacis, S., Meunier, A. and Volkova, O., Magnetorheological fluids, Journal of Magnetism and Magnetic Materials 252, November 2002, 224–228.

[488] Bossis, G., Volkova, O., Lacis, S. and Meunier, A., Magnetorheology: Fluids, structures and rheology. In: *Ferrofluids*, Vol. 594 of the series Lecture Notes in Physics, Springer-Verlag, 2002, 202–230. doi:10.1007/3-540-45646-5_11.

[489] Lai, C. Y. and Liao, W. H., Vibration control of a suspension system via a magnetorheological fluid damper, Journal of Vibration and Control 8, 2002, 527–547.

[490] Liao, W. H. and Lai, C. Y., Harmonic analysis of a magnetorheological damper for vibration control, Smart Materials and Structures 11(2), April 2002, 288–296.

[491] Yoo, J.-H. and Wereley, N. M., Design of a high-efficiency magnetorheological valve, Journal of Intelligent Material Systems and Structures 13(10), 2002, 679–687.

[492] Bossisa, G., Khuzira, P., Lacisb, S. and Volkova, O., Yield behavior of magnetorheological suspensions, Journal of Magnetism and Magnetic Materials 258–259, 2003, 456–458.

[493] Lam, A. H. F. and Liao, W. H., Semi-active control of automotive suspension systems with magnetorheological dampers, International Journal of Vehicle Design 33(1–3), 2003, 50–75.

[494] Liao, W. H. and Wang, D. H., Semiactive vibration control of train suspension systems via magnetorheological dampers, Journal of Intelligent Material Systems and Structures 14(3), 2003, 161–172.

[495] Yoo, J. H., Sirohi, J. and Wereley, N. M., A Magnetorheological piezo hydraulic actuator, Journal of Intelligent Material Systems and Structures 16(11–12), 2005, 945–954.

[496] Jean, P., Ohayon, R. and Le Bihan, D., Payload/Launcher vibration isolation: MR dampers modeling with fluid compressibility and inertia effects through continuity and momentum equations, International Journal of Modern Physics B 19(7–9), 2005, 1534–1541.

[497] Lau, Y. K. and Liao, W. H., Design and analysis of magnetorheological dampers for train suspension, Journal of Rail and Rapid Transit 219(4), 2005, 261–276.

[498] Brigley, M., Choi, Y.-T., Wereley, N. M. and Choi, S.-B., Magnetorheological isolators using multiple flow modes, Journal of Intelligent Material Systems and Structures 18(12), 2007, 1143–1148.

[499] Guerrero-Sanchez, C., Lara-Ceniceros, T., Jimenez-Regalado, E., Ras, M. and Schubert, U.S., Magnetorheological fluids based on ionic liquids, Advanced Materials 19(13), 1740–1747.

[500] Han, Y.-M., Choi, S.-B. and Wereley, N. M., Hysteretic behavior of magnetorheological fluid and identification using Preisach model, Journal of Intelligent Material Systems and Structures 18(8), 2007, 973–981.

[501] John, S., Chaudhuri, A. and Wereley, N. M., A magnetorheological actuation system: test and model, Smart Materials and Structures 17, March–April 2008, article ID 025023.

[502] Hong, S.-R., Wereley, N. M., Choi, Y.-T. and Choi, S.-B., Analytical and experimental validation of a nondimensional bingham model for mixed mode magnetorheological dampers, Journal of Sound and Vibration 312(3), May 2008, 399–417.

[503] Han, Y. M., Kim, C. J. and Choi, S. B., Design and control of magnetorheological fluid-based multifunctional haptic device for vehicle applications, Advanced Materials Research 47–50, June 2008, 141–144.

[504] Muc, A. and Barski, M., Homogenization numerical analysis & optimal design of MR fluids, Advanced Materials Research 47–50, June 2008, 1254–1257.

[505] Hu, W. and Wereley, N. M., Hybrid magnetorheological fluid elastomeric dampers for helicopter stability augmentation, Smart Materials and Structures 17(4), August 2008, article ID 045021.

[506] Sapiński, B. and Snamina, J., Modeling of an adaptive beam with MR fluid, Applied Mechanics and Materials 147–149, January 2009, 831–838.

[507] Mazlan, S. A., Issa, A., Chowdhury, H. A. and Olabi, A. G., Magnetic circuit design for the squeeze mode experiments on magnetorheological fluids, Materials and Design 30(6), June 2009, 1985–1993.

[508] Choi, Y.-T. and Wereley, N. M., Self-powered magnetorheological dampers, ASME Journal Vibration and Acoustics 31(4), 2009, article ID 044501.

[509] Zheng, L., Li, Y. N. and Baz, A., Fuzzy-sliding mode control of a full car semi-active suspension systems with MR dampers, Journal of Smart Structures & Systems 5(3), 2009, 261–278.

[510] Zheng, G. X., Huang, Y. J. and Gan, B., The application of MR fluid damper to the vibrating screen as the enhance of screening efficiency, Advanced Materials Research 97–101(March), 2010, 2628–2633.

[511] Park, B. J., Fang, F. F. and Choi, H. J., Magnetorheology: *materials and application*, Soft Matter 6, June 2010, 5246–5253.

[512] Han, K., Feng, Y. T. and Owen, D. R. J., Three-dimensional modelling and simulation of magnetorheological fluids, International Journal Numerical Methods in Engineering 84(11), November 2010, 1273–1302.

[513] Wang, D. H., Bai, X. X. and Liao, W. H., An integrated relative displacement self-sensing magnetorheological damper: prototyping and testing, Smart Materials and Structures 19(10), October 2010, article ID 105008.

[514] Snamina, J., Energy dissipation in three-layered plate with magnetorheological fluid, Solid State Phenomena 177, July 2011, 143–150.

[515] Mazlan, S. A., Ismail, I., Fathi, M. S., Rambat, S. and Anis, S. F., An experimental investigation of magnetorheological MR fluids under quasi-static loadings, Key Engineering Materials 495, November 2011, 285–288.

[516] Zhao, D. M. and Liu, X. P., Magnetorheological fluid test and application, Advanced Materials Research 395–398, November 2011, 2158–2161.

[517] DeVicente, J., Klingenberg, D. J. and Hidalgo-Alvareza, R., Magnetorheological fluids: a review, Soft Matter, 2011, 3701–3710.

[518] Kothera, C. S., Ngatu, G. T. and Wereley, N. M., Control evaluations of a magnetorheological fluid elastomeric (MRFE) lag damper for helicopter rotor blades, AIAA Journal of Guidance Control and Dynamics 34(4), 2011, 1143–1156.

[519] Wang, D. H. and Liao, W. H., Magnetorheological fluid dampers: a review of parametric modeling, Smart Materials and Structures 20(2), February 2011, article ID 023001.

[520] Wereley, N. M., Choi, Y.-T. and Singh, H., Adaptive energy absorbers for drop-induced shock mitigation, Journal of Intelligent Material Systems and Structures 22(6), 2011, 515–519.

[521] Choi, S. B. and Nguyen, Q. H., Selection of magnetorheological brakes via optimal design considering maximum torque and constrained volume, Smart Materials and Structures 21(1), December 2011, article ID 015012.

[522] Chen, C. and Liao, W.-H., A self-sensing magnetorheological damper with power generation, Smart Materials and Structures 21(2), January 2012, article ID 025014.

[523] Rodríguez-López, J., Segura, L. E. and De Espinosa Freijo, F. M., Ultrasonic velocity and amplitude characterization of magnetorheological fluids under magnetic fields, Journal of Magnetism and Magnetic Materials 324(2), January 2012, 222–230.

[524] Zhu, S. X., Liu, X. and Ding, L., Modeling and analysis of magnetorheological fluid damper under impact load, Advanced Materials Research 452–453, January 2012, 1481–1485.

[525] Asthana, C. B. and Bhat, R. B., A novel design of landing gear oleo strut damper using MR fluid for aircraft and UAV's, Applied Mechanics and Materials 225, November 2012, 275–280.

[526] Chen, C. and Liao, W. H., A self-sensing magnetorheological damper with power generation, Smart Materials and Structures 21(2), February 2012, article ID 025014.

[527] Wang, X. L., Hao, W. J. and Li, G. F., Study on magnetorheological fluid damper, Applied Mechanics and Materials 405–408, September 2013, 1153–1156.

[528] Rajamohan, V., Sundararaman, V. and Govindarajan, B., Finite element vibration analysis of a magnetorheological fluid sandwich beam, Procedia Engineering 64, 2013, 603–612.

[529] Yazid, I. I. M., Mazlan, S. A., Kikuchi, T., Zamzuri, H. and Imaduddin, F., Design of magnetorheological damper with a combination of shear and squeeze modes, Materials and Design 54, February 2014, 87–95.

[530] Singh, H. J., Hu, W., Wereley, N. M. and Glass, W., Experimental validation of a magnetorheological energy absorber optimized for shock and impact loads, Smart Materials and Structures 23(12), April 2014, article ID 125033.

[531] Ismail, I. and Aqida, S. N., Fluid-particle separation of magnetorheological (MR) fluid in MR machining application, Key Engineering Materials 611–612, May 2014, 746–755.

[532] Ashtiani, M., Hashemabadi, S. H. and Ghaffari, A., A review on the magnetorheological fluid preparation and stabilization, Journal of Magnetism and Magnetic Materials 374, January 2015, 716–730.

[533] Becnel, A. C., Sherman, S. G., Hu, W. and Wereley, N. M., Nondimensional scaling of magnetorheological rotary shear mode devices using the mason number, Journal of Magnetism and Magnetic Materials 380, April 2015, 90–97.

[534] Sherman, S. G., Becnel, A. C. and Wereley, N. M., Relating mason number to bingham number in magnetorheological fluids, Journal of Magnetism and Magnetic Materials 380(April), 2015, 98–104.

[535] Sapinski, B. and Goldasz, J., Development and performance evaluation of an MR squeeze-mode damper, Smart Materials and Structures 24(11), November 2015, article ID 115007.

[536] Morrilas, J. R., González, E. C. and De Vicente, J., Effect of particle aspect ratio in magnetorheology, Smart Materials and Structures 24(12), December 2015, article ID 125005.

References – Magnetostrictive and Electrostrictive Materials

Reviews on Magnetostrictive- and Electrostrictive-Related References

[537] Von Rensburg, R. J., Hunberstone, V., Close, J. A., Tavernert, A. W., Stevens, R., Greenough, R. D., Connor, K. P. and Gee, M. G., Review of Constitutive Description and measurement Methods for Piezoelectric, Electrostrictive and Magnetostrictive Materials, ©Crown Copyright

1994, ISSN 0959 2423, National Physical Laboratory, Teddington, Middlesex, United Kingdom, TW11 0LW, July 1994, 44 pp.

[538] Calkins, F. T., Flatau, A. B. and Dapino, M. J., An Overview of Magnetostrictive Sensor Applications, Proceedings of the Structural Dynamics and Mechanics Conference, AIAA Paper 99–1551, April 1999.

[539] Dapino, M. J., Magnetostrictive Materials, Encyclopedia of Smart Materials, Schwartz, M., (ed.), New York, John Wiley and Sons, 2002, 600–620.

Magnetostrictive-Related References

[540] Clark, A. E., Savage, H. T. and Spano, M. L., Effect of stress on magnetostriction and magnetization of single crystal, IEEE Transactions on Magnetics 20, 1984, 1443–1445.

[541] Butler, J. L., Butler, S. C. and Clark, A. E, Unidirectional magnetostrictive/piezoelectric hybrid transducer, Journal of the Acoustical Society of America 88(1), 1990, 7–11.

[542] Flatau, A., Hall, D. L. and Schlesselman, J. M., magnetostrictive active vibration control systems, Proceedings of the AIAA 30th aerospace science meeting: recent advances in adaptive and sensory materials and their applications, Paper No. 92–0490, 1992, 419–429.

[543] Hall, D. L. and Flatau, A. B., Nonlinearities Harmonics and Trends in Dynamic Applications of Terfenol-D, Proceedings of the SPIE Conference on Smart Structures and Intelligent Systems, SPIE 1917, February 1993, 929–939.

[544] Pratt, J. and Flatau, A. B., Collocated Sensing and Actuation using magnetostrictive materials, Proceedings of the SPIE Conference on Smart Structures and Intelligent Systems, 1917, February 1993, 952–961.

[545] Hall, D. L. and Flatau, A. B., One-dimensional analytical constant parameter linear electromagnetic-magnetomechanical models of a cylindrical magnetostrictive terfenol-D transducer, Proceedings of the Second International Conference on Intelligent Materials, July 1994, 605–616.

[546] Hall, D. L. and Flatau, A. B, Broadband performance of a magnetostrictive shaker, active control of noise and vibration, Journal of Intelligent Material Systems and Structures 6(1), January 1995, 109–116.

[547] Bothwell, C. M., Chandra, R. and Chopra, I., Torsional actuation with extension-torsional composite coupling and magnetostrictive actuators, AIAA Journal 33(4), April 1995, 723–729.

[548] Calkins, F. T. and Flatau, A. B., Transducer Based Measurement of Terfenol-D Material Properties, Proceedings of the SPIE Conference on Smart Structures and Integrated Systems, SPIE 2717, February 1996.

[549] Dapino, M. J., Calkins, F. T., Hall, D. L., Flatau, A. B., Measured Terfenol-D material properties under varied operating conditions, Proceedings of the SPIE Conference on Smart Structures and Integrated Systems, SPIE 2717, February 1996, 697–708.

[550] Calkins, F. T., Flatau, A. B. and Hall, D. L., Characterization of the Dynamic Material Properties of Magnetostrictive Terfenol-D, Proceedings of the 4th Annual Workshop: Advances in Smart Materials for Aerospace Applications, NASA Technical Report Id 19960047691, March 1996.

[551] Calkins, F. T. and Flatau, A. B, Experimental evidence for maximum efficiency operation of a magnetostrictive transducer, Journal of the Acoustical Society of America 99(4), April 1996, 2536–2536.

[552] Giurgiutiu, V. and Rogers, C. A., Power and energy characteristics of solid state induced strain actuators for static and dynamic applications, Journal of Intelligent Material Systems and Structures 8(9), September 1997, 738–750.

[553] Calkins, F. T., Dapino, M. J. and Flatau, A. B., Effect of Prestress on the Dynamic Performance of a Terfenol-D Transducer, Proceedings of the SPIE Conference on Smart Structures and Materials, SPIE 3041, March 1997, 293–304.

[554] Dapino, M. J., Calkins, F. T. and Flatau, A. B., Statistical Analysis of Terfenol-D Material Properties, Proceedings of the SPIE Conference on Smart Structures and Materials, SPIE 3041, March 1997, 256–267.

[555] Body, C., Reyne, G. and Meunier, G., Nonlinear finite element modelling of magneto-mechanical phenomena in giant magnetostrictive thin films, IEEE Transactions on Magnetics 33(2), 1997, 1620–1623.

[556] Snodgrass, J., Calkins, F. T., Dapino, M. J. and Flatau, A. B., Terfenol-D material property study, Journal of the Acoustical Society of America 101(5), May 1997, 3094–3094.

[557] Giurgiutiu, V. and Rogers, C. A., Power and energy characteristics of solid state induced strain actuators for static and dynamic applications, Journal of Intelligent Material Systems and Structures 8(9), September 1997, 738–750.

[558] Giurgiutiu, V., Jichi, F., Quattrone, R. F. and Berman, J. B., Experimental study of magnetostrictive tagged composite strain sensing response for structure health monitoring, Proceedings of the International Workshop of Structural Health Monitoring, 1999, 8–10.

[559] Dapino, M. J., Smith, R. C., Faidley, L. E. and Flatau, A. B., Coupled structural-magnetic strain and stress model for magnetostrictive transducers, Journal of Intelligent Material Systems and Structures 11(2000), 135–152.

[560] Butler, S. C. and Tito, F. A., A broadband hybrid magnetostrictive/ piezoelectric transducer array, Oceans, 2000 MTS/IEEE, 3, 2000, 1469–1475.

[561] Reddy, J. N. and Barbosa, J. I, On vibration suppression of magnetostrictive beams, Smart Materials and Structures 9(1), 2000, 49–58.

[562] Duenas, T. A. and Carman, G. P, Large magnetostrictive response of terfenol-d resin composites, Journal of Applied Physics 87(9), 2000, 4696–4701.

[563] Clark, A. E., Wun-Fogle, M., Restorff, J. B., Lograsso, T. A., Ross, A. R. and Schlagel, D. L., Magnetostrictive galfenol/alfenol single crystal alloys under large compressive stresses, Proceedings of the 7th International Conference on New Actuators, Borgmann, H. (ed.), Bremen, Germany, 2000, 111–115.

[564] Ludwig, A. and Quandt, E., Giant magnetostrictive thin films for applications in microelectromechanical systems, Journal of Applied Physics 87(9), 2000, 4691–4695.

[565] Flatau, A. and Kellogg, R. A., Magnetostrictive transducer performance characterization, The Journal of the Acoustical Society of America 109(5), 2001, 2434–2434.

[566] Giurgiutiu, V., Jichi, F., Berman, J. B. and Kamphaus, J. M., Theoretical and experimental investigation of magnetostrictive composite beams, Smart Materials and Structures 10(5), 2001, 934–945.

[567] Smith, R. C., Inverse compensation for hysteresis in magnetostrictive transducers, Mathematical and Computer Modelling 33, 2001, 285–298.

[568] Kakeshita, T. and Ullakko, K., Giant magnetostriction in ferromagnetic shape-memory alloys, MRS Bulletin, February 2002, 105–109.

[569] Smith, R. C., Dapino, M. J. and Seelecke, S., Free energy model for hysteresis in magnetostrictive transducers, Journal of Applied Physics 93, 2003, 458–466.

[570] Pomirleanu, R. and Giurgiutiu, V., High-field characterization of piezoelectric and magnetostrictive actuators, Journal of Intelligent Material Systems and Structures 15(3), 2004, 161–180.

[571] Dapino, M. J., On magnetostrictive materials and their use in adaptive structures, International Journal of Structural Engineering and Mechanics 17(3–4), 2004, 303–329.

[572] Tan, X. and Baras, J. S., Modeling and control of hysteresis in magnetostrictive actuators, Automatica 40(9), 2004, 1469–1480.

[573] Tzou, H. S., Chai, W. K. and Hanson, M., Dynamic actuation and quadratic magnetoelastic coupling of thin magnetostrictive shells, ASME Journal of Vibration and Acoustics 128(3), June 2006, 385–391.

[574] Chai, W. K., Tzou, H. S., Arnold, S. M. and Lee, H.-J., Magnetostrictive micro-actuations and modal sensitivities of thin cylindrical magnetoelastic shells, Journal of Pressure Vessel Technology 130(1), February 2008, article ID 011206.

[575] Chaudhuri, A., Yoo, J. H. and Wereley, N. M., Design, rtest and model of a hybrid magnetostrictive hydraulic actuator, Smart Materials and Structures 18(8), 2009, article ID 085019.

Electrostrictive-Related References

[576] Damjanovic, D. and Newnham, R. E., Electrostrictive and piezoelectric materials for actuator applications, Journal of Intelligent Material Systems and Structures 3, 1992, 190–208.

[577] Pilgrim, S. M., Massuda, M., Prodey, J. D. and Ritter, A. P., Electrostrictive Sonar Drivers for Flextensional Transducers, Transducers for Sonics and Ultrasonics, McCollum, M., Hamonic, B. F. and Wilson, O. B., (eds.), Lancaster, PA, Technomic, 1993.

[578] Hom, C. L. and Shankar, N., A Fully Coupled, Constitutive model for electrostrictive ceramic materials, Journal of Intelligent Material Systems and Structures 5, 1994, 795–801.

[579] Hom, C. L., Dean, P. D. and Winzer, S. R., Simulating electrostrictive deformable mirrors: i. nonlinear static analysis, Smart Materials and Structures 8, 1999, 691–699.

[580] Piquette, J. C. and Forsythe, S. E., Generalized material model for Lead Magnesium Niobate (PMN) and an associated electromechanical equivalent circuit, Journal of the Acoustical Society of America 104(1998), 2763–2772.

[581] Hom, C. L., Simulating electrostrictive deformable mirrors: II, Nonlinear Dynamic Analysis, Smart Materials and Structures 8, 1999, 700–708.

[582] Chaudhuri, A. and Wereley, N. M., Experimental validation of a hybrid electrostrictive hydraulic actuator analysis, ASME Journal of Vibration and Acoustics 132(2), 2010, article ID 021006.

[583] Pablo, F., Osmont, D. and Ohayon, R., A Plate Electrostrictive, Finite element – Part I: Modeling and variational formulations, Journal of Intelligent Material Systems and Structures 12(11), 2001, 745–759.

[584] Pablo, F., Osmont, D. and Ohayon, R., Modeling of plate structures equipped with current driven electrostrictive actuators for active vibration control, Journal of Intelligent Material Systems and Structures 14(3), 2003, 173–183.

[585] Tzou, H. S., Chai, W. K. and Arnold, S. M, Structronics and actuation of hybrid electrostrictive/ piezoelectric thin shells, ASME Journal of Vibration and Acoustics 128, February 2006, 79–87.

[586] Chai, W. K. and Tzou, H. S., Design and testing of a hybrid electrostrictive/piezoelectric polymeric beam with bang-bang control, Journal of Mechanical Systems and Signal Processing 21(2007), 417–429.

[587] John, S., Sirohi, J., Wang, G. and Wereley, N. M., Comparison of piezoelectric, magnetostrictive and electrostrictive hybrid hydraulic actuators, Journal Intelligent Material Systems and Structures 18(10), 2007, 1035–1048.

Electromagnetic- and Magnetic-Related References

[588] Baz, A, Magnetic constrained layer damping, Proceedings of the 11th Conference on Dynamics & Control of Large Structures, Blacksburg, May 1997, 333–344.

[589] Oh, J., Ruzzene, M. and Baz, A., Control of the dynamic characteristics of passive magnetic composites, Composites Part B: Engineering 30(7), October 1999, 739–751.

[590] Oh, J., Poh, S., Ruzzene, M. and Baz, A., Vibration control of beams using electromagnetic compressional damping treatment, ASME Journal of Vibration & Acoustics 122(3), 2000, 235–243.

[591] Ebrahim, A. and Baz, A., Vibration control of plates using magnetic constrained layer damping, Journal of Intelligent Material Systems and Structures 11(10), October 2000, 791–797.

[592] Baz, A. and Poh, S., Performance characteristics of magnetic constrained layer damping, Journal of Shock & Vibration 7(2), 2000, 18–90.

[593] Omer, A. and Baz, A., Vibration control of plates using electromagnetic damping treatment, Journal of Intelligent Material Systems & Structures 11(10), 2000, 791–797.

[594] Ruzzene, M., Oh, J. and Baz, A., Finite element modeling of magnetic constrained layer damping, Journal of Sound & Vibration 236(4), 2000, 657–682.

[595] Boyd, J., Lagoudas, D. C. and Seo, C., Arrays of micro-electrodes and electromagnets for processing of eectro-magneto-elastic multifunctional composite materials, Proceedings of SPIE's 10th Annual International Symposium on Smart Structures and Materials, San Diego, SPIE 5053, March 2003, 70–80.

[596] Sodano, H. A., Inman, D. J. and Belvin, W. K, Development of a new passive-active magnetic damper for vibration suppression, ASME Journal of Vibration and Acoustics 128(3), 2006, 318–327.

[597] Joyce, B., Development of an Electromagnetic Energy Harvester for Monitoring Wind Turbine Blades, Master's Thesis at the Virginia Polytechnic Institute and State University, 10919/36354, December 2011.

[598] Han, Y., Mechanics of magneto-active polymers, Ph.D. Thesis at the Iowa State University, Thesis No. 12929, 2012.

[599] Cottone, F., Basset, P., Vocca, H., Gammaitoni, L. and Bourouina, T., Bistable electromagnetic generator based on buckled beams for vibration energy harvesting, Journal of Intelligent Material Systems and Structures, October 2013, 1484–1495. doi:10.1177/1045389X13508330,.

[600] Hobeck, J. D. and Inman, D. J., Magnetoelastic Metastructures for Passive Broadband Vibration Suppression, Proceedings of SPIE's Smart Structures & NDE Conference, San Diego, Paper No. 9431–43, March 2015.

[601] Gonzalez-Buelga, A., Clare, L. R., Cammarano, A., Neild, S. A., Burrow, S. G. and Inman, D. J., An electromagnetic vibration absorber with harvesting and tuning capabilities, Structural Control and Health Monitoring 22(11), 2015. doi:10.1002/stc.1748.

[602] Shen, W. and Zhu, S., Harvesting energy via electromagnet damper: application to bridge stay cables, Journal of Intelligent Material Systems and Structures 26(1), January 2015, 156–171.

[603] Salas, E. and Bustamante, R., Numerical solution of some boundary value problems in nonlinear magnetoelasticity, Journal of Intelligent Material Systems and Structures 26(2), January 2015, 3–19.

[604] Farjoud, A. and Bagherpour, E. A., Electromagnet design for magnetorheological devices, Journal of Intelligent Material Systems and Structures 27(1), January 2016, 51–70.

1.3 Typical Properties of Intelligent Materials

The aim of the present chapter is to present the reader with data regarding mechanical and electrical properties of various materials defined as smart or intelligent materials. First, we shall present typical properties for piezoelectric materials as presented by PI Company from Germany[24] one of the known companies for piezoelectric materials and devices. Similar products can also be found in companies such as Morgan,[25] CeramTec AG,[26] APC International, Ltd.,[27] Noliac,[28] TRS Ceramics, Inc.,[29] and many other worldwide companies. Table 1.3 presents the properties for soft PZT materials; Table 1.4 exhibits the properties for hard PZT, while Table 1.5 shows the properties for lead-free materials as published by PI.

Table 1.3: Properties of soft PZT (source PI company).

Property	Symbol/unit	PIC151	PIC255/PIC252*	PIC155	PIC153	PIC152
Physical and dielectric properties						
Density	ρ (g/cm^3)	7.80		7.80	7.60	7.70
Curie temp.	T_c (°C)	250	350	345	185	340
Relative permittivity	$\varepsilon_{33}^T/\varepsilon_0$ [1]	2400	1750	1450	4200	1350
	$\varepsilon_{11}^T/\varepsilon_0$ [2]	1980	1650	1400		
Dielectric loss factor	$\tan\delta$ (10^{-3})	29	20	20	30	15
Electromechanical properties						
Coupling factor	k_p	0.62	0.62	0.62	0.62	0.48
	k_t	0.53	0.47	0.48		
	k_{31}	0.38	0.35	0.35		
	k_{33}	0.69	0.69	0.69		0.58
	k_{15}			0.66		
Piezoelectric charge coefficient	d_{31} (10^{-12} C/N)	−210	−180	−165		

24 www.physikinstrumente.com
25 www.morgantechnicalceramics.com
26 www.ceramtec.com/piezo-applications
27 www.americanpiezo.com
28 www.noliac.com
29 www.trsceramics.com

Table 1.3 (continued)

Property	Symbol/unit	PIC151	PIC255/PIC252*	PIC155	PIC153	PIC152
	d_{33} (10^{-12}C/N)	500	400	360	600	300
	d_{15} (10^{-12}C/N)		550			
Piezoelectric voltage coefficient	g_{31} (10^{-3}Vm/N)	−11.5	−11.3	−12.9		
	g_{33} (10^{-3}Vm/N)	22	25	27	16	25
Acoustomechanical properties						
Frequency coefficients	(N_p Hz m)	1950	2000	1960	1960	2250
	(N_1 Hz m)	1500	1420	1500		
	(N_3 Hz m)	1750		1780		
	(N_t Hz m)	1950	2000	1990	1960	1920
Elastic compliance coefficient	S_{11}^E (10^{-12}m^2/N)	15.0	20.7	19.7		
	S_{33}^E (10^{-12}m^2/N)	19.0		11.1		
Elastic stiffness coefficient	C_{33}^D (10^{10}N/m^2)	10.0		11.1		
Mechanical quality factor	Q_m	100	80	80	50	100
Temperature stability						
Temp. coeff. of ε_{33}^T [3]	$TK\varepsilon_{33}^T$ (10^{-3}/K)	6	4	6	5	2
Time stability (relative change of parameter per decade of time in %)						
Relative permittivity	C_ε		−1.0	−2.0		
Coupling factor	C_k		−1.0	−2.0		

*Material for the multilayer-tape technology. Matrix of coefficients on request; [1]Parallel to the polarization direction, [2] perpendicular to the polarization direction,[3] in the range of −20 up to +125.

Table 1.4: Properties of hard PZT (source PI Company).

Property	Symbol/unit	PIC181	PIC184[1]	PIC144[1]	PIC241	PIC300	PIC300
Physical and dielectric properties							
Density	ρ g/cm^3	7.80	7.75	7.95	7.80	7.80	5.50
Curie temp.	T_c (°C)	330	295	320	270	370	150
Relative permittivity	$\varepsilon_{33}^T/\varepsilon_0^{(1)}$	1200	1015	1250	1650	1050	950
	$\varepsilon_{11}^T/\varepsilon_0^{(2)}$	1500	1250	1500	1550	950	
Dielectric loss factor	$\tan\delta$ (10^{-3})	3	5	4	5	3	15
Electromechanical properties							
Coupling factor	k_p	0.56	0.55	0.60	0.50	0.48	0.30
	k_t	0.46	0.44	0.48	0.46	0.43	0.42
	k_{31}	0.32	0.30	0.30	0.32	0.25	0.18
	k_{33}	0.66	0.62	0.66	0.64	0.46	
	k_{15}	0.63	0.65		0.63	0.32	
Piezoelectric charge coefficient	d_{31} (10^{-12}C/N)	-120	-100	-110	-130	-80	-50
	d_{33} (10^{-12}C/N)	265	219	265	290	155	120
	d_{15} (10^{-12}C/N)	475	418		265	155	
Piezoelectric voltage coefficient	g_{31} (10^{-3}Vm/N)	-11.2	-11.1	-10.1	-9.8	-9.5	
	g_{33} (10^{-3}Vm/N)	25	24.4	25	21	16	-11.9
Acoustomechanical properties							
Frequency coefficients	(N_p Hz m)	2270	2195	2180	2190	2350	3150
	(N_1 Hz m)	1640	1590	1590	1590	1700	2300
	(N_3 Hz m)	2010	1930		1550	1700	2500
	(N_t Hz m)	2110	2035	2020	2140	2100	
Elastic compliance coefficient	S_{11}^E (10^{-12}m^2/N)	11.8	12.7	12.4	12.6	11.1	
	S_{33}^E (10^{-12}m^2/N)	14.2	14.0	15.5	14.3	11.8	

Table 1.4 (continued)

Property	Symbol/unit	PIC181	PIC184[1]	PIC144[1]	PIC241	PIC300	PIC300
Elastic stiffness coefficient	c_{33}^D (Z10^{10} N/m^2)	16.6	14.8	15.2	13.8	16.4	
Mechanical quality factor	Q_m	2000	400	1000	400	1400	250
Temperature stability							
Temp. coeff. of ε_{33}^T [3]	$TK\varepsilon_{33}^T$ (10^{-3}/K)	3	5			2	
Time stability (relative change of parameter per decade of time in %)							
Relative permittivity	C_ε		−4.0				−5.0
Coupling factor	C_k		−2.0				−8.0

[1]Preliminary data, subject to change.

Table 1.5: Properties of lead-free materials (source PI Company).

Property	Symbol/unit	PIC050*	PIC700**
Physical and dielectric properties			
Density	ρ g/cm^3	4.70	5.6
Curie temp.	T_c (°C)	>500	200 [1]
Relative permittivity	$\varepsilon_{33}^T/\varepsilon_0$ [1]	60	700
	$\varepsilon_{11}^T/\varepsilon_0$ [2]	85	1650
Dielectric loss factor	$\tan\delta$ (10^{-3})	<1	30
Electromechanical properties			
Coupling factor	k_p		0.15
	k_t		0.40
	k_{31}		
	k_{33}		
	k_{15}		
Piezoelectric charge coefficient	d_{31} (10^{-12}C/N)		
	d_{33} (10^{-12}C/N)	40	120
	d_{15} (10^{-12}C/N)	80	

*Crystalline material, ** Preliminary data, subject to change, [1] Maximum operating temperature.

As written on their Website, the following properties are valid for all PZT materials from PI:

Specific heat capacity: WK = approx. 350 J/(kg K)

Specific thermal conductivity: WL = approx. 1.1 W/(m K)

Poisson's ratio (lateral contraction): $\nu = 0.34$

Coefficient of thermal expansion:

$\alpha_3 = 4$ to $- 6 \times 10^{-6}$ 1/(K) (approx., in the polarization direction, shorted)

$\alpha_1 = 4$ to 8×10^{-6} 1/(K) (approx. perpendicular to the polarization direction, shorted)

Static compressive strength: >600 MPa

The data was determined using test pieces with geometric dimensions according to En 50324-2 standard and are typical values. All the provided data was determined 24–48 h after the time of polarization at an ambient temperature of 23 ± 2 °C.[30]

Another type of piezoelectric material which is a widely used polymer film is called PVDF. As stated in Ref. [1], PVDF presents a unique combination of properties like:

- flexibility (possibility of application on nonlevel surfaces)
- high mechanical strength
- dimensional stability
- balanced piezoelectric activity in the plane of the film
- high and stable piezoelectric coefficients over time up to approximately 90 °C
- characteristic chemical inertness of PVDF
- continuous polarization for great length spooled onto drums
- thickness between 0.9 and 1 mm
- acoustic impedance close to that of water with a flat response curve

The PVDF (see Figures 1.18 and 1.19) properties are highlighted in Tables 1.6–1.8 and are based on the data presented in [1].

Using the data presented in [2–10] and additional data available in the literature a few tables will be next presented to enable the selection of the suitable type of material and to understand the purpose, actuation, or sensing.

Table 1.9 presents a general comparison between standard PZT and piezoelectric PVDF film. It is clear that while PVDF is light and has a low Young's modulus and low piezoelectric charge coefficients compared to PZT, it presents large voltage coefficients making it suitable for sensing applications.

Table 1.10 shows a comparison of various materials to be used for actuation. Various relevant properties are listed to enable the evaluation of the performances of the different types of actuators.

---

30 To obtain a complete coefficient matrix of the individual materials one should apply to PI Ceramic (info@piceramic.de)

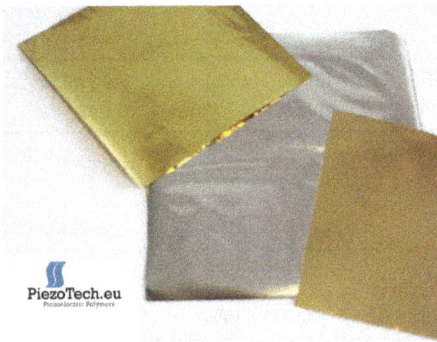

P(VDF-TrFE) Piezoelectric films

Figure 1.18: P(VDF-TrFE) piezoelectric films from PIÉZOTECH S.A.S. (www.piezotech.fr).

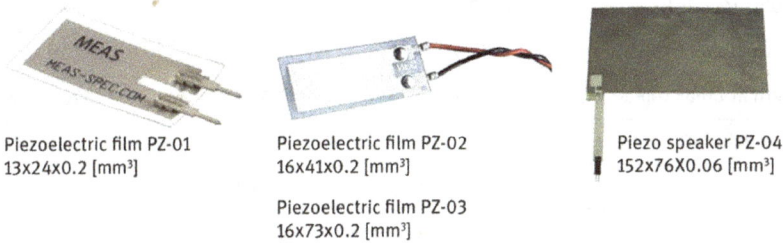

Piezoelectric film PZ-01
13x24x0.2 [mm³]

Piezoelectric film PZ-02
16x41x0.2 [mm³]

Piezoelectric film PZ-03
16x73x0.2 [mm³]

Piezo speaker PZ-04
152x76X0.06 [mm³]

Figure 1.19: Various PVDF piezoelectric films transducers from Images Scientific Instruments (http://www.imagesco.com/).

Table 1.6: Properties of PVDF piezofilm materials bi-oriented film (source PIÉZOTECH S.A.S.[a]).

Property	(9 µm) (±5%) thick	(25 µm) (±5%) thick	(40 µm) (±5%) thick
Piezo/pyroelectric properties at 23 °C			
d_{33} (pC/N)	16 ± 20%	15 ± 20%	15 ± 20%
d_{31} (pC/N)	6 ± 20%	6 ± 20%	6 ± 20%
d_{32} (pC/N)	6 ± 20%	1 ± 20%	6 ± 20%
g_{33} (Vm/N)@ 1 Hz	0.15 ± 20%	0.14 ± 20%	0.14 ± 20%
p_3 (µC/(m²K))[1]	− 20 ± 25%	− 25 ± 25%	− 19 ± 25%
Dielectric properties at 23 °C			
ε_r@ 0.1 kHz	11.5 ± 10%	11.5 ± 10%	11.5 ± 10%

Table 1.6 (continued)

Property	(9 μm) (±5%) thick	(25 μm) (±5%) thick	(40 μm) (±5%) thick
ε_r@ 1 kHz	11.5 ± 10%	11.5 ± 10%	11.5 ± 10%
ε_r@ 10 kHz	11.0 ± 10%	11.0 ± 10%	11.0 ± 10%
tan δ @ 0.1 kHz	0.010 ± 10%	0.010 ± 10%	0.010 ± 10%
tan δ @ 1 kHz	0.015 ± 10%	0.015 ± 10%	0.015 ± 10%
tan δ @ 10 kHz	0.035 ± 10%	0.035 ± 10%	0.035 ± 10%
DC breakdown (V)	750 ± 30%	760 ± 30%	540 ± 30%
Mechanical properties at 23 °C			
Young's modulus (MPa)[2]	2500 ± 20%	3200 ± 20%	2500 ± 20%
Young's modulus (MPa)[3]	2500 ± 20%	3200 ± 20%	2500 ± 20%
Tensile strength at break (MPa)[2]	175 ± 15%	240 ± 15%	170 ± 15%
Tensile strength at break (MPa)[3]	190 ± 15%	60 ± 15%	190 ± 15%
Elongation at break (%)[2]	50 ± 30%	20 ± 30%	50 ± 30%
Elongation at break (%)[2]	50 ± 30%	5 ± 30%	50 ± 30%
Thermal properties at 23 °C			
Melting point (°C)	175 ± 5%	175 ± 5%	175 ± 5%
Transverse direction	90–100	90–100	90–100

[a]www.piezotech.fr
[1] Pyroelectric coefficient, [2] in the machine direction, [3] in the transverse direction.

Table 1.7: Properties of P(VDF-TrFE) copolymer 75/25 film (source PIÉZOTECH S.A.S., www.piezo tech.fr).

Property	12 μm (±5%) thick	25 μm (±5%) thick	50 μm (±5%) thick	110 μm (±5%) thick
Piezo/pyroelectric properties at 23 °C				
d_{33} (pC/N)	16 ± 20%	15 ± 20%	15 ± 20%	15 ± 20%
d_{31} (pC/N)	6 ± 20%	6 ± 20%	6 ± 20%	6 ± 20%
d_{32} (pC/N)	6 ± 20%	6 ± 20%	6 ± 20%	6 ± 20%
g_{33} (Vm/N)@ 1 Hz	0.15 ± 20%	0.18 ± 20%	0.18 ± 20%	0.18 ± 20%
p_3 (μC/(m²K))[1]	− 20 ± 25%	− 19 ± 25%	− 19 ± 25%	− 19 ± 25%

Table 1.7 (continued)

Property	12 µm (±5%) thick	25 µm (±5%) thick	50 µm (±5%) thick	110 µm (±5%) thick
Dielectric properties at 23 °C				
ε_r @ 0.1 kHz	9.4 ± 10%	9.6 ± 10%	9.6 ± 10%	9.6 ± 10%
ε_r @ 1 kHz	9.3 ± 10%	9.4 ± 10%	9.4 ± 10%	9.4 ± 10%
ε_r @ 10 kHz	9.1 ± 10%	9.2 ± 10%	9.2 ± 10%	9.2 ± 10%
tan δ @ 0.1 kHz	0.014 ± 10%	0.015 ± 10%	0.015 ± 10%	0.015 ± 10%
tan δ @ 1 kHz	0.014 ± 10%	0.016 ± 10%	0.016 ± 10%	0.016 ± 10%
tan δ @ 10 kHz	0.028 ± 10%	0.032 ± 10%	0.032 ± 10%	0.032 ± 10%
DC breakdown (V)	575 ± 30%	395 ± 30%		
Mechanical properties at 23 °C				
Young's modulus (MPa)[2]	950 ± 20%	1000 ± 20%	1000 ± 20%	1000 ± 20%
Young's modulus (MPa)[3]	1500 ± 20%	1200 ± 20%	1200 ± 20%	1200 ± 20%
Tensile strength at break (MPa)[2]	90 ± 15%	60 ± 15%	40 ± 15%	40 ± 15%
Tensile strength at break (MPa)[3]	30 ± 15%	20 ± 15%	30 ± 15%	30 ± 15%
Elongation at break (%)[2]	150 ± 30%	300 ± 30%	400 ± 30%	400 ± 30%
Elongation at break (%)[2]	30 ± 30%	300 ± 30%	450 ± 30%	450 ± 30%
Thermal properties at 23 °C				
Melting point (°C)	150 ± 5%	150 ± 5%	150 ± 5%	150 ± 5%
Curie temperature (°C)	135 ± 5%	135 ± 5%	135 ± 5%	135 ± 5%
Transverse direction	90–100	90–100	90–100	90–100

[1] Pyroelectric coefficient, [2] in the machine direction, [3] in the transverse direction

Table 1.8: Properties of P(VDF-TrFE) copolymer 70/30 film (source PIÉZOTECH S.A.S. (www.piezo tech.fr), see Figure 1.18).

Property	20 µm (±5%) thick	25 µm (±5%) thick	40 µm (±5%) thick	6 µm (±5%) thick*
Piezo/pyroelectric properties at 23 °C				
d_{33} (pC/N)	− 20 ± 20%	− 20 ± 20%	− 20 ± 20%	− 19 ± 20%
d_{31} (pC/N)	6 ± 20%	6 ± 20%	6 ± 20%	6 ± 20%

Table 1.8 (continued)

Property	20 μm (±5%) thick	25 μm (±5%) thick	40 μm (±5%) thick	6 μm (±5%) thick*
d_{32} (pC/N)	$6 \pm 20\%$	$6 \pm 20\%$	$6 \pm 20\%$	$6 \pm 20\%$
g_{33} Vm/N@ 1 Hz	$0.15 \pm 20\%$	$0.2 \pm 20\%$	$0.2 \pm 20\%$	$0.2 \pm 20\%$
p_3 (μC/(m^{2}K))[1]				$-20 \pm 25\%$
Dielectric properties at 23 °C				
ε_r from 10 Hz to 1 kHz	$8 \pm 10\%$	$8 \pm 10\%$	$8 \pm 10\%$	$9.4 \pm 10\%$
tan δ @ 1 kHz	$0.016 \pm 10\%$	$0.016 \pm 10\%$	$0.016 \pm 10\%$	$0.016 \pm 10\%$
Mechanical properties at 23 °C				
Young's modulus (MPa)	$1000 \pm 20\%$	$1000 \pm 20\%$	$1000 \pm 20\%$	$950 \pm 20\%$
Tensile strength at break (MPa)	$60 \pm 15\%$	$60 \pm 15\%$	$60 \pm 15\%$	$30 \pm 15\%$
Elongation at break (%)	$60 \pm 30\%$	$60 \pm 30\%$	$60 \pm 30\%$	$30 \pm 30\%$
Thermal properties at 23 °C				
Melting point (°C)	$156 \pm 5\%$	$156 \pm 5\%$	$156 \pm 5\%$	$150 \pm 5\%$
Curie temperature (°C)	$112 \pm 5\%$	$112 \pm 5\%$	$112 \pm 5\%$	$135 \pm 5\%$
Maximal temperature (°C)	95	95	95	120

*P(VDF-TrFE) copolymer 77/23 film; [1] pyroelectric coefficient.

Table 1.9: Comparison between standard PVDF film and standard PZT.

Property	Symbol/Units	PVDF	PZT
Piezoelectric charge coefficients	d_{31} (C/N) or (m/V)	22×10^{-12}	-175×10^{-12}
	d_{33} (C/N) or (m/V)	-30×10^{-12}	400×10^{-12}
Piezoelectric voltage coefficients	g_{31} (Vm/N)	216×10^{-3}	-11×10^{-3}
	g_{33} Vm/N	-330×10^{-3}	25×10^{-3}
Coupling factor	k_{31}	0.14	0.34
Relative dielectric constant	$\varepsilon_r = \varepsilon_{33}^T/\varepsilon_0$	12	1700
Maximal operating temperature	(°C)	80	150
Density	ρ (kg/m^3)	1780	7600
Young's modulus	T (GPa)	2	71
Acoustic impedance	Z (10^6 kg/(m^{2}s))	2.7	30

Table 1.10: Comparison of smart-material-based actuators (based on [2]).

Property	PZT G-1195	PVDF	PMN*	TERFENOL-D	NITINOL
Type of actuation mechanism	Piezoceramic	Piezofilm	Electrostriction	Magnetostriction	Shape Memory Alloy
Maximal strain (10^{-6})	1000	700	1000	2000	20,000
Young's modulus	70	2	~120	$43^{(1)}$ up to $107.5^{(2)}$	$28–41^{(3)}$ & $83^{(4)}$
Bandwidth	High	High	High	Moderate	Low
Response time	µs	µs	µs	µs	s
Strain voltage characteristics	First-order-linear	First-order-linear	Nonlinear	Nonlinear	Nonlinear

[1] At unsaturated magnetic state, [2] at saturated magnetic state [3] for the martensite phase, [4] for the austenite phase, *PMN = lead magnesium niobate

The properties of PZT G-1195 are:

$$\rho = 7.5 \text{g/cm}^3; \quad c_{11}^E = c_{22}^E = 148 \text{GPa}; \quad c_{33}^E = 131 \text{GPa}; \quad c_{12}^E = 76.2 \text{GPa};$$

$$c_{13}^E = c_{23}^E = 74.2 \text{GPa}; \quad c_{44}^E = c_{55}^E = 25.4 \text{GPa}; \quad c_{66}^E = 35.9 \text{GPa}; \quad e_{31} = -2.1 \text{C/m}^2;$$

$$e_{33} = 9.5 \text{C/m}^2; \quad e_{15} = 9.2 \text{C/m}^2.$$

where

$$d_{31} = e_{31} S_{11}^E + e_{31} S_{12}^E + e_{33} S_{13}^E; \quad d_{33} = e_{31} S_{13}^E + e_{31} S_{13}^E + e_{33} S_{33}^E;$$

$$d_{15} = e_{15} S_{44}^E; \quad e_{15} = d_{15} c_{44}^E; \quad e_{33} = d_{31} c_{13}^E + d_{31} c_{13}^E + d_{33} c_{33}^E;$$

$$e_{31} = d_{31} c_{11}^E + d_{31} c_{12}^E + d_{33} c_{13}^E.$$

Other types of actuators are described by Shahinpoor et al. [3] including ion-exchange polymer–noble metal composites (IPMC) and electroactive-ceramics (EAC). Their properties are shown in Table 1.11 in comparison with SMAs actuators.

With the development of giant magnetostrictive materials (GMM) like Terfenol-D and magnetic shape memory materials (MSM) such as NiMnGa alloys new actuators are built based on these two materials and similar ones. Reference [8] presents a study on those promising materials acting as actuators and their properties are summarized in Table 1.12. The comparison includes Terfenol-D GMM, MSN, composite GMM (made of grains of magnetostrictive alloys combined with electrical insulated binder such as a polymer), Galfenol (a magnetostrictive alloy made from iron and

Table 1.11: Comparison of IPMC-, SMA- and EAC-based actuators (from [3]).

Property	IPMC	EAC	SMA
Actuation displacement	> 10%	0.1–0.3%	<8%
Force (MPa)	10–30	30–40	~700
Reaction time	µs to s	µs to s	s to min
Density (g/cm^3)	1–2.5	6–8	5–6
Drive voltage (V)	4–7	50–800	6–10
Power consumption	Watts	Watts	Watts
Fracture toughness	Resilient, elastic	Fragile	Elastic

Table 1.12: Comparison of GMM with other relevant materials for actuation (from [8]).

Property	Terfenol-D GMM	MSM	Composite GMM	Galfenol	Hard PZT PZT-4	Soft PZT MLA
Max. static strain (ppm)$^{(1)}$	1800	50,000	1000	320	600	1250
Coupling coefficient (%)	70	75	35	40	67	65
Young's modulus (GPa)	25	7	20	45	60	40
Max. prestress (MPa)	50	1	30	80	50	40
Max dyn. strain @resonance (ppm)	4000	140*	3000	3500*	1600	2000

$^{(1)}$ ppm = parts per million; *MSN and Galfenol dynamic strains have been calculated with the stress limit.

gallium–FeGa), PZT-4 (a hard piezoceramic material) and a soft PZT multilayer technique for actuators (MLA).

The strains presented in Table 1.12 are peak-to-peak values and the values have been obtained from experiments performed at Cedrat Technologies.[31]

Till now, we have discussed the use of smart materials as actuators. Now, we shall address the second role allocated for the smart structures, the sensors. Due to their ability to be manufactured in any size needed to act as sensors, piezoceramics and piezofilm are good candidates for this role. Their comparison with other sensors is shown in Table 1.13.

---

31 CENELEC normative committee BTTF 63–2 Advanced technical ceramics (WGII: NG13), www.ce drat-technologies.com.

Table 1.13: Comparison of various sensors.

Type of Sensor	Way of activation or property	Sensitivity	Localization (mm)	Bandwidth
Resistance gage	10 V excitation	30 (V/ε)	0.2038	0 Hz–acoustic
Semiconductor gage	10 V excitation	1000 (V/ε)	0.7620	0 Hz–acoustic
Fiber optics	1.016 mm interferometer gage length	10^6 (deg/ε)	1.0160	0 Hz–acoustic
Piezofilm	0.0254 mm thickness	10^4 (V/ε)	<1.0160	0.1 Hz–GHz
Piezoceramics	0.0254 mm thickness	2×10^4 (V/ε)	<1.0160	0.1 Hz–GHz

References

[1] Piezoelectric films-Technical information brochure, issued by PIÉZOTECH S.A.S., www.piezo tech.fr.

[2] Wadley, H. N. G., Characteristics and processing of smart materials, Paper presented at the AGARD SMP lecture Series on "Smart Structures and Materials: Implications for Military Aircraft of New Generation," held in Philadelphia, USA, 30–31 October 1996.

[3] Shahinpoor, M., Bar-Cohen, Y., Xue, T., Simpson, J. O. and Smith, J., Ionic polymer-metal composites (IPMC) as biomimetic sensors and actuators-artificial muscles, Proc. Of the SPIE's 5th Annual International Symposium on Smart Structures and materials, 1–5 March, 1998, San Diego, CA. Paper # 3324–3327.

[4] Giurgiutiu, V., Pomirleanu, R. and Rogers, C. A., Energy-based comparison of solid-state actuators, University of South Carolina, Report # USC-ME-LAMSS-2000-102, March 1, 2000.

[5] Chopra, I., Review of state of art of smart structures and integrated systems, AIAA Journal 40(11), 2002, 2145–2187.

[6] Tzou, H. S., Lee, H.-J. and Arnold, S. M., Smart Materials, Precision Sensors/Actuators, Smart Structures, and Structronic Systems, NASA Publications, Paper 200, 2004. http://digitalcom mons.unl.edu/nasapub/200.

[7] Monner, H. P., Smart materials for active noise and vibration reduction, Keynote paper presented at NOVEM- Noise and vibration: Emerging Methods, Saint Raphaël, France, 18–21 April, 2005.

[8] Claeyssen, F., Hermet, N. L., Barillot, F. and Le Letty, R., Giant dynamic strains in magnetostrictive actuators and transducers, paper presented at ISAGMM 2006 Conference, October 2006, China.

[9] Ramadan, K. S., Sameoto, D. and Evoy, S., A review of piezoelectric polymers as functional materials for electromechanical transducers, Smart materials and Structures 23, 2014, article ID 033001.

[10] Shaikh, A. M., Smart materials and its applications, International Journal of Emerging Technologies in Computational and Applied Sciences (IJETCAS) 12(2), 2015, 193–197.

1.4 Recent Applications of Intelligent Materials and Systems

In what follows, we will try to highlight some of the research being performed and presented in the literature, using intelligent materials and systems. The topics selected represent some of the more active research programs being performed throughout many universities, research institutes and industries.

The first topic deals with morphing aircraft (for typical recent works see [1–5]). This subject has been extensively covered by many researchers and research programs (see the European program SARISTU[32]-Smart Intelligent Aircraft Structure and the American counterpart Morphing wing program at NASA[33]). Though there is no quantitative definition of the "morphing" term, a morphing aircraft would be an adaptable, time-variant airframe, whose changes in geometry would influence its aerodynamic performance (see Figure 1.20). The idea of changing the wing profile according to its mission came from close look at bird's wings, which would change its cross-sectional shape according to its flight phases, like take-off and landing, diving and cruising. The morphing of the wings would be performed using actuators, including SMA, Terfenol and PZT, which connects the topic to smart intelligent materials. The foreseen advantages/benefits of the morphing wing would be: fewer moving parts, weight reduction, improved aerodynamics, reduced fuel consumption of the aircraft and less wiring through the wing. However, some disadvantages/risks are also associated with the morphing approach: how to overcome the weight penalty due to additional actuation systems, high material costs, repair costs, durability of the used materials and what to do in the case of system crash. Mechanisms such as deployable flaps provide the current standard of adaptive airfoil geometry, although this solution places limitations on maneuverability and efficiency, and produces a design that is nonoptimal in many flight regimes. A wide and updated review can be found in Ref. [1]. The paper presents a review of the state oft the art on morphing aircraft and focuses on structural, shape-changing morphing concepts for both fixed and rotary wings, with particular reference to active systems. The authors state that "although many interesting concepts have been synthesized, few have progressed to wing tunnel testing, and even fewer have flown." Reference to inflatable solutions for wing morphing can be found in [32]. Joo et al. [2] discuss in their manuscript composite skin issues and their associated challenges. The need of materials with low in-plane stiffness and relatively high out-of-plane stiffness may be required, leading them to propose a two-stage design process. A 2D skin design based on their proposed design process is demonstrated, manufactured using a rapid prototyping technique and tested. Another group of researchers [3] proposed the use of passive elastomeric matrix composite as a passive morphing skin. The skin includes an elastomer–fiber

---

32 www.saristu.eu
33 www.nasa.gov

(a) Lockheed Martin's morphing wing demonstrator

(b) Morphing trailing edge concept

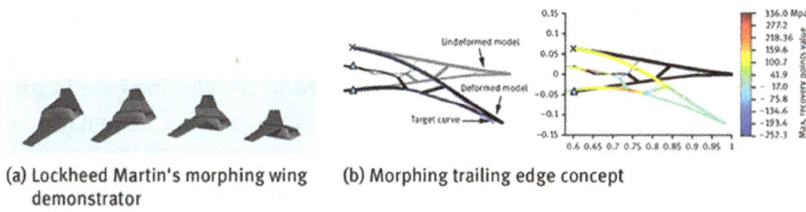

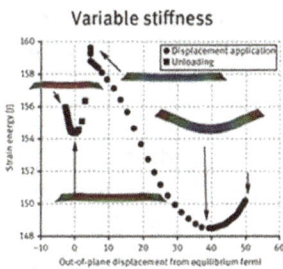

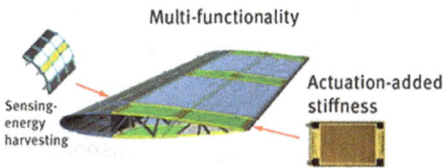

(c) Morphing concepts at ETH

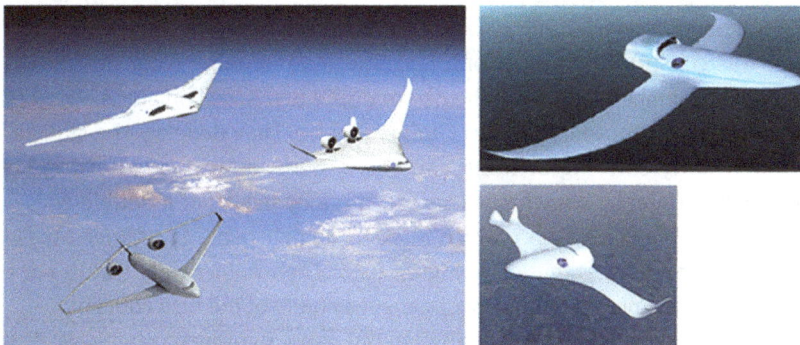

(d) NASA morphing aircraft concepts

Figure 1.20: Various concepts for morphing aircraft: (a) Lockheed Martin, (b) Trailing edge (PoliMI), (c) ETH – DMAVT and (d) NASA. Sources: www.newscientist.com/news/news/jsp?id=as99994484 (a); Prof. Sergio Ricci – DAST PoliMI (b); ETH Zurich – DMAVT – Laboratory of Composite Materials and Adaptive Structures (c); www.nasa.gov (d).

composite surface layer that is supported by a flexible honeycomb structure, each of which exhibit a near zero in-plane Poisson's ratio. The complete prototype morphing skin demonstrated 100% uniaxial extension accompanied by a 100% increase in surface area. They also showed that an out of plane deflection of less than 2.5 mm can be maintained at various levels of area change under pressures of up to 9.58 kPa.

A generic aeroelastic morphing wing analysis combined with an optimized design approach is presented by a group from Delft University of Technology from the Netherlands [4]. The proposed procedure can be used to predict aerodynamic performance, load distribution, aeroelastic deformations, and the required actuation forces and moments and corresponding actuation energy. This procedure was applied theoretically to predict the performances of folding and sweeping wings. Another work performed by Lesieutre et al. [5] presents a 2D compliant cellular truss structure aimed at replacing the fixed internal structure of a given wing. This truss structure had the ability to change span, aspect ratio and the area of the wing. The authors showed that for a 100-lb aircraft, their proposed morphing structure was able to achieve a 74% decrease in span, while maintaining a weight fraction (including actuator weight) similar to that of a conventional aircraft, 12.2%. Scaling of their design showed that sizable decreases in span can also be achieved for larger aircrafts. However, the weight fraction of the structure and actuators needed to deform the structure and limited the feasibility of their 2D cellular truss structure at large scales. Finally, Pecora et al. [15], present their work within a research project funded by the Italian company Alenia Aeronautica S.p.A. A novel morphing architecture for a variable camber trailing edge was investigated. Basing their reference geometry on a full-scale wing of a CS-25 category typical civil regional transportation aircraft, they replaced the conventional flap component with a morphing trailing edge based on compliant ribs. The presented architecture yielded high deformability while keeping good load-sustaining capabilities. Their morphing trailing-edge structure concept was constructed as a set of interconnected morphing wing ribs moved by original actuators embedded within the rib structures and based on SMAs. Another interesting approach is presented by Previtali et al. [26]. They developed a skin capable of concurrently carrying bending and shear loads, as well as allowing for significant levels of in-plane stretching. This was done by means of a double-walled structure called double corrugation which showed a high bending stiffness while achieving a 20% in-plane stretching (see Figure 1.20). Their numerically results were validated against experimental data, showing the feasibility of the new concept.

The second topic deals with harvesting energy using intelligent materials. The aim is to harvest parasitic ambient vibrational energy and transform it to electrical energy to supply power to various devices making them self-sustained. A typical device is presented in Figure 1.21. A cantilever beam equipped with piezoelectric layers is vibrating due to ambient excitation. To reduce the natural frequencies of the beam, a tip mass is added. The aim is to match the natural frequency (usually the first one) to the excitation frequencies. At resonance, the beam would vibrate at its maximal amplitude, yielding a voltage which is harvested and stored on a storage device in the form of a capacitor, super capacitor, battery, etc. The diodes bridge (Figure 1.21b) is needed to prevent the return of the voltage to the piezoelectric layers, as the vibrations are harmonic.

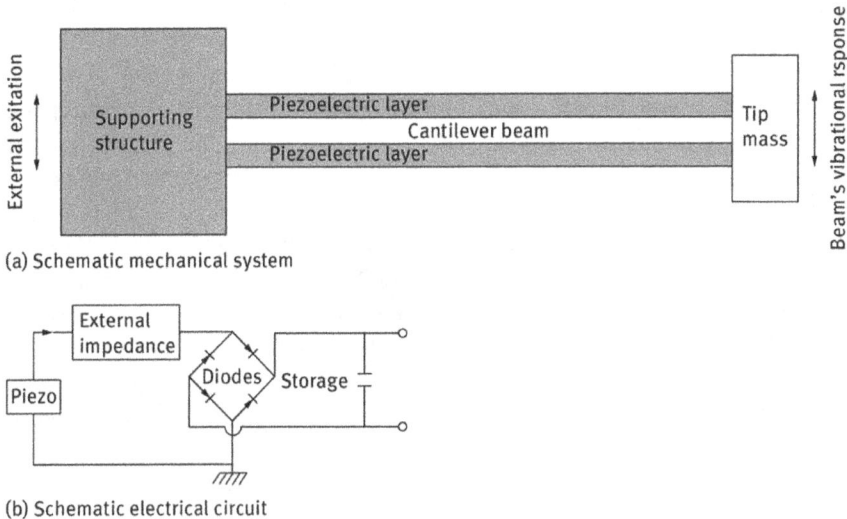

(a) Schematic mechanical system

(b) Schematic electrical circuit

Figure 1.21: A typical harvested based on a cantilever beam equipped with piezoelectric layers: (a) the schematic mechanical system and (b) the associated schematic electrical system.

Huge numbers of papers were devoted to present various aspects of harvesting systems and methods. Typical and recent works are next presented [6–19]. Tadesse et al. [6] present an innovative hybrid mechanism composed of an electromagnetic and piezoelectric harvesters. An aluminum-cantilever-tapered beam having the length of 125 mm was equipped with three pairs of single-crystal piezoelectric plates.[34] A stationary inductive coil was attached to the tip of the cantilever beam such that mass of the tip of cantilever beam was co-centric with the coil. As the beam vibrates, the magnet moves in and out of the coil thus generating voltage according to Faraday's law. The coil attached to the tip of the serves both as additional mass (thus reducing the natural frequency of the cantilever beam) and as electromagnetic harvester. The experimental results showed a harvested power of 0.25 W when using both mechanisms and only 0.25 mW when using the piezoelectric crystals, at an acceleration of 35 g and frequency of 20 Hz (the first natural frequency of the cantilever beam). Increasing the frequency to 100 Hz, a nonresonance frequency resulted in a considerable drop in the

---

34 The piezoelectric plates were made from $Pb(Zn_{1/3}Nb_{2/3})O3$-$PbTiO_3$ (PZN-PT) single crystals (d_{31} mode) – see [33].

harvested, yielding only 0.025 W for the hybrid system. Claiming that there is no accepted definition for the energy-harvesting efficiency, Liao and Sodano [7] defined the efficiency of a harvesting system based on the ratio of the strain energy of the vibrating beam over each cycle to the output power. Their new definition is showed to be analogous to the material loss factor. Their simulations showed that the maximum efficiency would occur normally at the matched impedance; however, for materials with high electromechanical coupling, the maximum power would be generated at the near open- and closed-circuit resonances yielding a lower efficiency. Sun et al. [8] presented a numerical and an experimental study of an aluminum cantilever beam equipped with two single-crystal plates (the same material as the one used in [6] at its clamped boundary and a steel tip mass at its free tip. The reported results showed a maximal power of about 0.6 mW at 80 kΩ external resistive loads. Another approach is shown by Lie et al. [9]. They tackle the problem of harvesting mainly from its electrical side by using switch-mode power electronics to control the voltage and/or charge on a piezoelectric device relative to the mechanical input for optimized energy conversion. Their new approach is demonstrated using a multilayer PVDF polymer device.[35] In these experiments, the active energy harvesting approach increased the harvested energy by a factor of 5 for the same mechanical displacement compared to an optimized diode rectifier-based circuit, yielding about 22 mW at an effective mechanical strain of 1.37%.

Erturk, Renno and Inman, from Virginia Tech., USA [10], present a different beam shape, an alternative to the conventional cantilever beam, for harvesting purposes. Their new beam, an L-shaped beam structure with two concentrated masses, one at the end of horizontal beam and other along the vertical beam and piezoelectric material along the two beams forming the L shape, can be tuned to have the first two natural frequencies relatively close to each other, yielding the possibility of a broader band energy-harvesting system. The numerical study investigated the use of the L-shaped piezoelectric energy harvester configuration as landing gears in unmanned air vehicle applications showing favorably results as compared against the published experimental results of a curved beam configuration used for the same purpose. The team of Guyomar et al. from LGEF, INSA Lyon, France, presents a detailed review of the ways the energy can be harvested and transformed in useful electrical energy using piezoelectric material. They extend the capability of harvesting with piezoelectric material, by adding the pyroelectric effect existing in those materials, thus adding another way of harvesting energy from temperature variations, one of the big common parasitic energy sources.

Farinholt et al. [12] performed a study using PVDF and for comparison an ionically conductive ionic polymer transducer (IPT). Analytical models were compared with experimental results assuming axial loading of the harvesters. They reported a

---

35 From the company Measurement Specialties, Inc. The thickness of the PVDF film was 2 mm.

harvested power of 4.494×10^{-04} mW for PVDF (with a thickness of 64 μm) as compared with 3.15×10^{-06} mW for IPT, where the tip mass for the PVDF was 1208 g (with a first natural frequency of 138 Hz) as compared with 346 g for the second transducer (with a first natural frequency of only 70 Hz). Olivier and Priya [13] presented a harvester based on for-bar electromagnetic device. Their analytical/numerical calculations and experimental results showed a harvested average power of 17.9 mW at about 150 Hz.

An interesting new harvester is advocated by Aldraihem and Baz [14]. In their analytical/numerical study a spring-mass system is placed between the piezoelectric element and the moving base, thus magnifying the strain experienced by the piezoelement leading to the amplification of the electrical power output of the harvester. Their reported results show that the harvested power can be amplified by a factor of 20 as compared to a normal harvester, without a dynamic magnifier.

Zoric et al. [16] tackle the harvesting problem by application of optimal vibration control on a thin-walled composite beam equipped with piezoelectric patches using the fuzzy optimization strategy based on the particle swarm optimization algorithm. Tang et al. [17] present a comprehensive experimental study aimed at investigating the use of magnets for improving the functionality of energy harvesters based on piezoelectric materials under various vibration scenarios. They used the various nonlinearities introduced by magnets to improve the performance of the vibration energy harvester.

Elvin and Elvin [18] studied the effects of large deflections being induced in flexible piezoelectric harvesters. Their numerical predictions were compared with experimental results for static, free vibration case and forced vibrations. They showed that large deflections tend to shift the resonance frequency and increase damping leading to a significant reduction in the output voltage as compared with the common small deflection linear–based harvester. It was shown that when the tip deflection of the vibrating beam exceeds 35% of the beam length these nonlinear dynamic effects become significant.

Mikoshiba et al. [19] tried to increase the harvested energy by applying the resonator lattice system approach. Their basic unit cell is composed of a spring-loaded magnet enclosed in a capped poly(methyl methacrylate) tube equipped with copper coils to create a unit cell that acts both as a resonator and as a linear generator. They report a continuous effective power of 36 mW being generated by a single unit cell across a 1Ω load resistor. Green et al. [22] took the ambient vibration sources as random vibrations often called "white noise." They applied this assumption to two sources of vibration generation: human walking motion and the oscillation of the mid-span of a suspension bridge. They showed that the potential improvements that can be realized through the introduction of these nonlinearities into energy harvesters are sensitive to the type of ambient excitation to which they are subjected. Xiong and Oyadiji [23] used a distributed parameter electromechanical model is used to predict the power output with resistive loads of a piezoelectric-based harvester.

Their results indicate that the convergent and divergent tapered cantilevered and rectangular cantilevered beam designs with partial coverage of piezoelectric layer might be able to generate higher electromechanical coupling coefficient than conventional rectangular cantilevered designs with full coverage. Adding reasonable extra masses with varied locations on vibration energy harvesters might generate larger power density. A different direction had been chosen by Anton et al. [24]. A novel material, a piezoelectric foam, which is a lead-free, polymer-based electret material exhibiting piezoelectric-like properties, is investigated for low-power energy generation. After presenting an overview of the fabrication and operation of piezoelectret foams mechanical tests were performed on various specimens to yield their mechanical properties, yielding Young's moduli between 0.5 and 1 GPa and tensile strengths from 35 to 70 MPa. The d_{33} coefficient was found to be relatively constant at around 175 pC/N from 10 Hz to 1 kHz. Harmonic excitation of a pretensioned $15.2 \times 15.2\,\mathrm{cm}^2$ sample at 60 Hz and displacement of ± 73 (μm) yielded an average power of 6.0 (mW) delivered to a 1 mF storage capacitor. The capacitor was charged to 4.67 V in 30 min, proving the ability of piezoelectret foam to supply power to small electronic components. Abramovich et al. ([29–30]) presented another approach to harvest energy using piezoelectric material. Instead of using the vibration mode, they applied cyclic loads in the thickness direction of disks, thus using the d_{33} coefficient. The influence of the applied compressive load was investigated and a typical power of 0.43 W at a pressure of 40 MPa when using eight PZT disks (one on top of the other) is reported.

Another interesting review was written and presented by Stoppa and Chiolerio [25]. They looked at a new trend in the textile sector dealing with electronic textiles (e-textiles) that are fabrics that feature electronics and interconnections woven into them, presenting physical flexibility and typical size that cannot be achieved with other existing electronic manufacturing techniques. The vision behind wearable computing foresees future electronic systems to be an integral part of our everyday outfits. Such electronic devices have to meet special requirements concerning their wearability. Wearable systems will be characterized by their ability to automatically recognize the activity and the behavioral status of their own user as well as of the situation around her/him, and to use this information to adjust the systems' configuration and functionality. Their review focuses on recent advances in the field of smart textiles and by presenting various available materials and their manufacturing process showing advantages and disadvantages while highlighting possible tradeoffs between flexibility, ergonomics, low power consumption, integration and eventually autonomy.

Other interesting applications using intelligent materials are reported by Yu and Leckey [20]. They applied a sparsely arranged piezoelectric sensor array, quantitative crack detection and imaging approach using a Lamb wave–focusing array algorithm. Also Lamb wave propagation on thin-wall plates and wave interaction with crack damage were studied using three-dimensional elastodynamic finite integration technique. Annamdas and Radhika [21] present an updated review containing the important developments in monitoring and the path forward in wired, wireless- and energy-harvesting methods related to electromechanical impedance-based structural health monitoring for metals and nonmetals. An interesting and applicable research is reported by Spaggiari et al. [27]. They modeled and tested a spring made of SMA named Negator. It is a spiral spring made of a SMA strip and wounded with a given curvature, in such a way that each coil wraps tightly on its inner neighbor. Two features make the Negator springs mechanically attractive: its almost constant force–displacement behavior in the unwinding of the strip, mounted on a rotating drum and it being very long. The authors report the analytical/numerical model of the Negator spring made from SMA. Their experimental results confirmed the applicability of such geometry. Another intelligent material, the MR fluids, is treated by Ghaffari et al. [28]. They present a review on the various models and simulation methods that are currently applied to investigate and study MR fluids. The two general approaches – the continuum model and the discrete phase model – are presented, and various rheological and structural models of MR fluids using the continuum approach are summarized together with computational methods of the discrete approach. Meisel et al. [31] combined the advantages of additive manufacturing approach with SMA to produce 3D printed parts with embedded SMA actuating wires. The manufacturing process using PolyJet 3D printing is presented in detail and preliminary experimental results are highlighted.

References

[1]   Barbarino, S., Bilgen, O., Ajaj, R. M., Friswell, M. I. and Inman, D. J., A review of morphing aircraft, Journal of Intelligent Material Systems and Structures 22(9), 2011, 823–877.
[2]   Joo, J. J., Reich, G. W. and Westfall, J. T, Flexible skin development for morphing aircraft applications via topology optimization, Journal of Intelligent Material Systems and Structures 20(16), 2009, 1969–1985.
[3]   Bubert, E. A., Woods, B. K. S., Lee, K., Kothera, C. S. and Wereley, N. M., Design and fabrication of a passive 1D morphing aircraft skin, Journal of Intelligent Material Systems and Structures 21(17), 2010, 1699–1717.
[4]   De- Breuker, R., Abdalla, M. M. and Gurdal, Z, A generic morphing wing analysis and design framework, Journal of Intelligent Material Systems and Structures 22(10), 2011, 1025–1039.
[5]   Lesieutre, G. A., Browne, J. A. and Frecker, M. I., Scaling of performance, weight, and actuation of a 2-D compliant cellular frame structure for a morphing wing, Journal of Intelligent Material Systems and Structures 22(10), 2011, 979–986.
[6]   Tadesse, Y., Zhang, S. and Priya, S., Multimodal energy harvesting system: piezoelectric and Electromagnetic, Journal of Intelligent Material Systems and Structures 20(5), 2009, 625–632.
[7]   Liao, Y. and Sodano, A., Structural effects and energy conversion efficiency of power harvesting, Journal of Intelligent Material Systems and Structures 25(14), 2014, 505–514.
[8]   Sun, C., Qin, L., Li, F. and Wang, Q-M., Piezoelectric energy harvesting using single crystal Pb (Mg1/3Nb2/3)O3-xPbTiO3 (PMN-PT) device, Journal of Intelligent Material Systems and Structures 20(5), 2009, 559–568.

[9]   Liu, Y., Tian, G., Wang, Y., Lin, J., Zhang, Q. and Hofmann, H. F., Active piezoelectric energy harvesting: general principle and experimental demonstration, Journal of Intelligent Material Systems and Structures 20(5), 2009, 575–585.

[10] Erturk, A., Renno, J. M. and Inman, D. J., Modeling of piezoelectric energy harvesting from an L-shaped beam-mass structure with an application to UAVs, Journal of Intelligent Material Systems and Structures 20(5), 2009, 529–544.

[11] Guyomar, D., Sebald, G., Pruvost, S., Lallart, M., Khodayari, A. and Richard, C., Energy harvesting from ambient vibrations and heat, Journal of Intelligent Material Systems and Structures 20(5), 2009, 609–624.

[12] Farinholt, K. M., Pedrazas, N. A., Schluneker, D. M., Burt, D. W. and Farrar, C. R, An energy harvesting comparison of piezoelectric and ionically conductive polymers, Journal of Intelligent Material Systems and Structures 20(5), 2009, 633–642.

[13] Olivier, J. M. and Priya, S., Design, fabrication, and modeling of a four-bar electromagnetic vibration power generator, Journal of Intelligent Material Systems and Structures 21(16), 2010, 1303–1316.

[14] Aldraihem, O. and Baz, A., Energy harvester with a dynamic magnifier, Journal of Intelligent Material Systems and Structures 22(6), 2011, 521–530.

[15] Pecora, R., Barbarino, S., Concilio, A., Lecce, L. and Russo, S., Design and functional test of a morphing high-lift device for a regional aircraft, Journal of Intelligent Material Systems and Structures 22(10), 2011, 1005–1023.

[16] Zoric, N. D., Simonovic, A. M., Mitrovic, Z. S. and Stupar, S. N, Optimal vibration control of smart composite beams with optimal size and location of piezoelectric sensing and actuation, Journal of Intelligent Material Systems and Structures 24(4), 2012, 499–526.

[17] Tang, L., Yang. Y. and Soh, C.-K., Improving functionality of vibration energy harvesters using magnets, Journal of Intelligent Material Systems and Structures 23(13), 2012, 1433–1449.

[18] Elvin, N. G. and Elvin, A. A, Large deflection effects in flexible energy harvesters, Journal of Intelligent Material Systems and Structures 23(13), 2012, 1475–1484.

[19] Mikoshiba, K., Manimala, J. M. and Sun, C. T., Energy harvesting using an array of multifunctional resonators, Journal of Intelligent Material Systems and Structures 24(2), 2012, 168–179.

[20] Yu, L. and Leckey, C. A. C., Lamb wave–based quantitative crack detection using a focusing array algorithm, Journal of Intelligent Material Systems and Structures 24(9), 2012, 1138–1152.

[21] Annamdas, V. G. M. and Radhika, M. A, Electromechanical impedance of piezoelectric transducers for monitoring metallic and non-metallic structures: a review of wired, wireless and energy-harvesting methods, Journal of Intelligent Material Systems and Structures 24(9), 2013, 1021–1042.

[22] Green, P. L., Papatheou, E. and Sims, N. D., Energy harvesting from human motion and bridge vibrations: an evaluation of current nonlinear energy harvesting solutions, Journal of Intelligent Material Systems and Structures 24(12), 2013, 1494–1505.

[23] Xiong, X. and Oyadiji, S. O., Modal electromechanical optimization of cantilevered piezoelectric vibration energy harvesters by geometric variation, Journal of Intelligent Material Systems and Structures 25(10), 2014, 1177–1195.

[24] Anton, S. R., Farinholt, K. M. and Erturk, A., Piezoelectret foam–based vibration energy harvesting, Journal of Intelligent Material Systems and Structures 25(14), 2014, 1–12.

[25] Stoppa, M. and Chiolerio, A., Wearable electronics and smart textiles: a critical review, Sensors 14, 2014, 11.957–11.992. doi:10.3390/s140711957.

[26] Previtali, F., Arrieta, A. F. and Ermanni, P., Double-walled corrugated structure for bending-stiff anisotropic morphing skins, Journal of Intelligent Material Systems and Structures 26(5), 2015, 599–613.

[27] Spaggiari, A., Dragoni, E. and Tuissi, A., Experimental characterization and modelling validation of shape memory alloy Negator springs, Journal of Intelligent Material Systems and Structures 26(6), 2015, 619–630.

[28] Ghaffari, A., Hashemabadi, H. and Ashtiani, M., A review on the simulation and modeling of magnetorheological fluids, Journal of Intelligent Material Systems and Structures 26(8), 2015, 881–904.

[29] Abramovich, H., Tsikhotsky, E. and Klein, G., An experimental determination of the maximal allowable stresses for high power piezoelectric generators, Journal of Ceramic Science and Technology 4(3), 2013, 131–136.

[30] Abramovich, H., Tsikhotsky, E. and Klein, G., An experimental investigation on PZT behavior under mechanical and cycling loading, Journal of the Mechanical Behavior of Materials 22(3–4), 2013, 129–136.

[31] Meisel, N., Elliott, A. M. and Williams, C. B., A procedure for creating actuated joints via embedding shape memory alloys in PolyJet 3D printing, Journal of Intelligent Material Systems and Structures 26(12), 2015, 1498–1512.

[32] Min, Z., Kien, V. K. and Richard, L. J. Y., Aircraft morphing wing concepts with radical geometry change, The IES Journal Part A: Civil & Structural Engineering 3(3), 2010, 188–195.

[33] Zhang, S., Lebrun, L., Randall, C. A. and Shrout, T. R., Growth and electrical properties of (Mn, F) co-doped $0.92Pb(Zn1/3Nb2/3)O3-0.08PbTiO3$ single crystal, Journal of Crystal Growth 267, 2004, 204–212.

2 Laminated Composite Materials

2.1 Classical Lamination Theory

2.1.1 Introduction

One of the definitions for a composite material, made up of two constituents, one being the fiber (the reinforcement) and the other the glue (the matrix), states that a combination of the two materials would result in better properties than those of the individual components when they are used alone. The main advantages of composite materials over other existing materials like metal or plastics are their high strength and stiffness, combined with low density, allowing for a weight reduction in the finished part. In this chapter, when we are talking about a composite material, we refer to continuous fibers (reinforcements) being embedded into the matrix in the form of adequate glue. Examples of such continuous reinforcements include unidirectional, woven cloth, and helical winding (see Figure 2.1a) and are referred in the literature as a block diagram as shown in Figure 2.1b. Properties of some mostly used continuous-fibers can be found in Appendix A. Continuous-fiber composites are often made into laminates by stacking single sheets of continuous fibers in different orientations to obtain the desired strength and stiffness properties with fiber volumes as high as 60–70%. Fibers produce high-strength composites because of their small diameter; they contain far fewer defects (normally surface defects) compared to the material produced in bulk. On top of it, due to their small diameter, the fibers are flexible and suitable for complicated manufacturing processes, such as small radii or weaving. Materials such as glass, graphite, carbon or aramid are used to produce fibers. The main material for the matrix is a polymer, which has low strength and stiffness. The main functions of the matrix are to keep the fibers in the proper orientation and spacing and to provide protection to the fiber from abrasion and the environment. In polymer–matrix composites, the good and strong bond between the matrix and the reinforcement allows the matrix to transmit the outside loads from the matrix to the fibers through shear loading at the interface.

Two types of polymer matrices are available: thermosets and thermoplastics. A thermoset starts as a low-viscosity resin that reacts and cures during processing, forming a solid. A thermoplastic is a high-viscosity resin that is processed by heating it above its melting temperature. Because a thermoset resin sets up and cures during processing, it cannot be reprocessed by reheating. A thermoplastic can be reheated above its melting temperature for additional processing.

https://doi.org/10.1515/9783110726701-002

(a)

Unidirectional fiber (UD fiber)   Woven cloth (2 directions)   Filament winding

0^0

$0^0 / 90^0$

$\pm 30^0$

$\pm 50^0$

(b)

Composites

Particle-reinforced

Fiber-reinforced

Structural

Large particles

Dispersion strenghtened

Continuous (aligned)

Discontinuous (short)

Laminates

Sandwich

Aligned

Randomly oriented

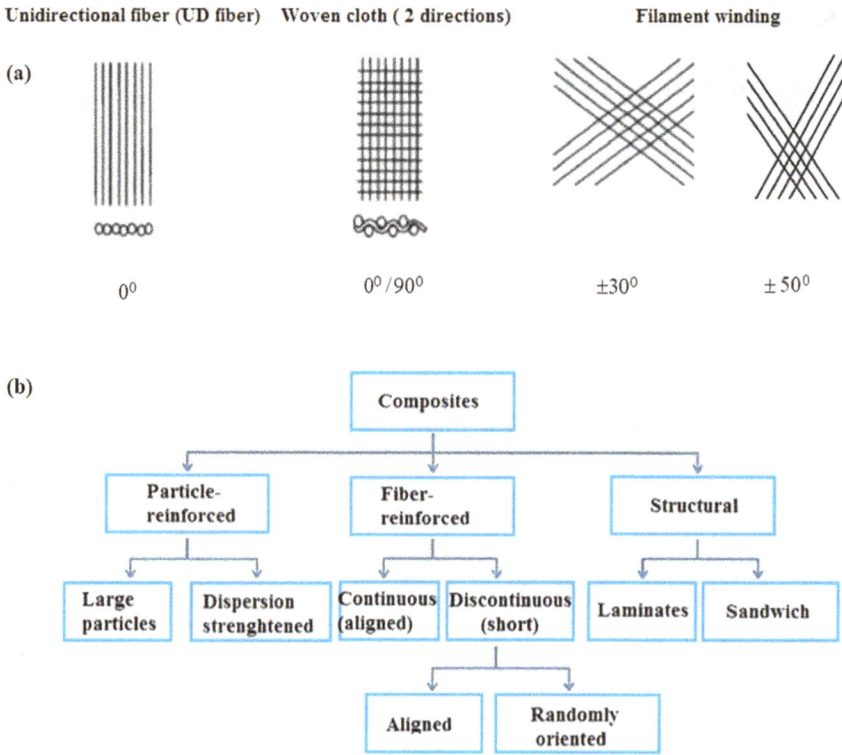

Figure 2.1: Types of composite materials.

2.1.2 Displacements

Under the action of forces, every point in a given material will deform. The displacements in a coordinate system as described in Figure 2.2 will be written as

$$u = \{u_1, u_2, u_3\} \tag{2.1}$$

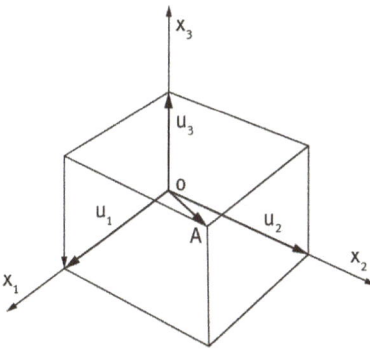

Figure 2.2: The coordinate system and the three components of the displacements, u_1, u_2 and u_3 (note that the vector *OA* is the resultant of the three displacements with the size defined by $OA = \sqrt{x_1^2 + x_2^2 + x_3^2}$).

2.1.3 Strain

The definition of the linear strain tensor ε_{ij} is given by

$$\varepsilon_{ij} = \frac{1}{2}\left(u_{i,j} + u_{j,i}\right), \quad i,j = 1,2,3$$

$$\varepsilon_{ij} = \varepsilon_{ji} \tag{2.2}$$

In the expanded form, we can write strains as

$$\varepsilon_{11} = \frac{\partial u_1}{\partial x_1} \equiv \varepsilon_1; \quad \varepsilon_{22} = \frac{\partial u_2}{\partial x_2} \equiv \varepsilon_2; \quad \varepsilon_{33} = \frac{\partial u_3}{\partial x_3} \equiv \varepsilon_3$$

$$2\varepsilon_{23} = 2\varepsilon_{32} = \gamma_4 = \left(\frac{\partial u_2}{\partial x_3} + \frac{\partial u_3}{\partial x_2}\right) \equiv \varepsilon_4$$

$$2\varepsilon_{13} = 2\varepsilon_{31} = \gamma_5 = \left(\frac{\partial u_1}{\partial x_3} + \frac{\partial u_3}{\partial x_1}\right) \equiv \varepsilon_5 \tag{2.3}$$

$$2\varepsilon_{12} = 2\varepsilon_{21} = \gamma_6 = \left(\frac{\partial u_1}{\partial x_2} + \frac{\partial u_2}{\partial x_1}\right) \equiv \varepsilon_6$$

2.1.4 Stress

The stress tensor can be written using the same notations as for the strains yielding

$$[\sigma] = \begin{bmatrix} \sigma_{11} & \sigma_{12} & \sigma_{13} \\ \sigma_{12} & \sigma_{22} & \sigma_{23} \\ \sigma_{13} & \sigma_{23} & \sigma_{33} \end{bmatrix} = \begin{bmatrix} \sigma_1 & \sigma_6 & \sigma_5 \\ \sigma_6 & \sigma_2 & \sigma_4 \\ \sigma_5 & \sigma_4 & \sigma_3 \end{bmatrix} \tag{2.4}$$

$$\sigma_{ij} = \sigma_{ji}$$

Note that sometimes in the literature, in order to differentiate between normal stresses (σ_{ii}) and shear stresses ($\sigma_{ij}i{\neq}j$), the shear stresses are denoted by τ_{ij}.

2.1.5 Orthotropic Materials

A homogeneous linear elastic material having two perpendicular planes of symmetry is called orthotropic material. The constitutive relations are given as

$$
\begin{Bmatrix} \varepsilon_{11} \\ \varepsilon_{22} \\ \varepsilon_{33} \\ \gamma_{23} \\ \gamma_{13} \\ \gamma_{12} \end{Bmatrix} =
\begin{bmatrix}
\frac{1}{E_1} & -\frac{v_{21}}{E_2} & -\frac{v_{31}}{E_3} & 0 & 0 & 0 \\
-\frac{v_{12}}{E_1} & \frac{1}{E_2} & -\frac{v_{32}}{E_3} & 0 & 0 & 0 \\
-\frac{v_{13}}{E_1} & -\frac{v_{23}}{E_2} & \frac{1}{E_3} & 0 & 0 & 0 \\
0 & 0 & 0 & \frac{1}{G_{23}} & 0 & 0 \\
0 & 0 & 0 & 0 & \frac{1}{G_{13}} & 0 \\
0 & 0 & 0 & 0 & 0 & \frac{1}{G_{12}}
\end{bmatrix}
\begin{Bmatrix} \sigma_{11} \\ \sigma_{22} \\ \sigma_{33} \\ \sigma_{23} \\ \sigma_{13} \\ \sigma_{12} \end{Bmatrix}
\tag{2.5}
$$

where E_1, E_2, E_3 are the longitudinal elastic moduli, G_{23}, G_{13}, G_{12} are the shear elastic moduli and $v_{12}, v_{13}, v_{23}, v_{21}, v_{31}$ and v_{32} are Poisson's ratios. Also due to symmetry of the compliance matrix, the following relationships hold:

$$
\frac{v_{21}}{E_2} = \frac{v_{12}}{E_1}; \quad \frac{v_{31}}{E_3} = \frac{v_{13}}{E_1}; \quad \frac{v_{32}}{E_3} = \frac{v_{23}}{E_2}
\tag{2.6}
$$

The relation between the strain and the stress can be written in a compact form as

$$
\{\varepsilon\} = [s]\{\sigma\}
\tag{2.7}
$$

where the matrix $[s]$ is called the compliance matrix. If we would like to write the stresses as a function of the strains, we have the following expression:

$$
\{\sigma\} = [c]\{\varepsilon\}
\tag{2.8}
$$

where $[c]$ is the stiffness matrix and is equal to $[c] = [s]^{-1}$.

Equation (2.5) written with the stiffness matrix has the following form:

$$
\begin{Bmatrix} \sigma_{11} \\ \sigma_{22} \\ \sigma_{33} \\ \sigma_{23} \\ \sigma_{13} \\ \sigma_{12} \end{Bmatrix} =
\begin{bmatrix}
c_{11} & c_{12} & c_{13} & 0 & 0 & 0 \\
c_{12} & c_{22} & c_{23} & 0 & 0 & 0 \\
c_{13} & c_{23} & c_{33} & 0 & 0 & 0 \\
0 & 0 & 0 & c_{44} & 0 & 0 \\
0 & 0 & 0 & 0 & c_{55} & 0 \\
0 & 0 & 0 & 0 & 0 & c_{66}
\end{bmatrix}
\begin{Bmatrix} \varepsilon_{11} \\ \varepsilon_{22} \\ \varepsilon_{33} \\ \gamma_{23} \\ \gamma_{13} \\ \gamma_{12} \end{Bmatrix}
\tag{2.9}
$$

where the various terms in the stiffness matrix are defined as

$$
c_{11} = \frac{1 - v_{23}v_{32}}{E_2 E_3 \Pi}; \quad c_{22} = \frac{1 - v_{13}v_{31}}{E_1 E_3 \Pi}; \quad c_{33} = \frac{1 - v_{12}v_{21}}{E_1 E_2 \Pi}
$$

$$
c_{12} = \frac{v_{21} + v_{31}v_{23}}{E_2 E_3 \Pi} = \frac{v_{12} + v_{32}v_{13}}{E_1 E_3 \Pi}; \quad c_{13} = \frac{v_{31} + v_{21}v_{32}}{E_2 E_3 \Pi} = \frac{v_{13} + v_{12}v_{23}}{E_1 E_2 \Pi}
$$

$$C_{23} = \frac{v_{32} + v_{12}v_{31}}{E_1 E_3 \Pi} = \frac{v_{23} + v_{21}v_{13}}{E_1 E_2 \Pi}; \quad C_{44} = G_{23}; \quad C_{55} = G_{13}; \quad C_{66} = G_{12} \tag{2.10}$$

where

$$\Pi = \frac{1 - v_{12}v_{21} - v_{23}v_{32} - v_{31}v_{13} - 2v_{21}v_{32}v_{13}}{E_1 E_2 E_3}$$

If the material is defined as transversely isotropic[1] (like for instance PZT), then the following equivalences hold: $G_{12} = G_{13}$; $E_3 = E_2$; and $v_{12} = v_{13}$.

2.1.6 Unidirectional Composites

Unidirectional composites are usually composed of two constituents, the fiber and the matrix (which is the glue holding together the two components). Based on the rule of mixtures, one can calculate the properties of the unidirectional layer based on the properties of the fibers and the matrix and their volume fracture. The assumption to be made when applying the rule of mixtures is that the two constituents are bonded together and they behave like a single body. The longitudinal modulus (or the major modulus), E_{11}, of the layer can be written as

$$E_{11} = E_f V_f + E_m V_m \tag{2.11}$$

where E_f and E_m are the longitudinal moduli for the fibers and the matrix, respectively, and V_f and V_m are their volume fractions.[2]

The major Poisson's coefficient, v_{12}, is given by

$$v_{12} = v_f V_f + v_m V_m \tag{2.12}$$

where v_f and v_m are the longitudinal moduli for the fibers and the matrix, respectively.

One should note that according to Equation (2.6) the minor Poisson's coefficient, v_{21}, would be calculated to be

$$\frac{v_{12}}{E_{11}} = \frac{v_{21}}{E_{22}} \Rightarrow v_{21} = v_{12}\frac{E_{22}}{E_{11}} \tag{2.13}$$

---

1 A transversely isotropic material is a homogenous linear elastic material which has an isotropic plane while in a perpendicular axis to that plane the material is orthotropic.
2 Note that $V_f + V_m = 1$.

The transverse modulus (or the minor modulus), E_{22}, of the layer is given as

$$\frac{1}{E_{22}} = \frac{V_f}{E_f} + \frac{V_m}{E_m} \Rightarrow E_{22} = \frac{E_m}{V_f \frac{E_m}{E_f} + V_m} = \frac{E_m}{V_f \frac{E_m}{E_f} + (1 - V_f)} \tag{2.14}$$

The shear modulus of the layer, G_{12}, is given as

$$\frac{1}{G_{12}} = \frac{V_f}{G_f} + \frac{V_m}{G_m} \Rightarrow G_{12} = \frac{G_m}{V_f \frac{G_m}{G_f} + V_m} = \frac{G_m}{V_f \frac{G_m}{G_f} + (1 - V_f)} \tag{2.15}$$

where G_f and G_m are the shear moduli for the fibers and the matrix, respectively.

To assess the differences between the properties of the fiber and to compare them to those of the matrix, the reader is referred to Table 2.1.

Table 2.1: Typical properties of T300 carbon fibers and 914 epoxy resin matrix.

Property	T300 carbon fibers	914 epoxy resin matrix
Young's modulus, E (GPa)	220	3.3
Shear modulus, G (GPa)	20	1.2
Poisson's ratio, u	0.15	0.37

Table 2.2 presents a comparison between the simplified[3] micromechanics model presented above and experimental values. One can realize that the model well predicts the major Young's modulus (0.998 of the experimental value), E_{11}, under predicts the major Poisson's ratio and minor Young's modulus, E_{22} by 0.735 and 0.813 of the experimental values, respectively. The shear modulus, G_{12}, is highly under-predicted by the simplified micromechanics model to yield only 0.52 of the experimental value.

Table 2.2: Predictions of unidirectional composite properties by simplified micromechanics model and comparison with experimental values.[3]

Equation	Relationship	Predicted values	Experimental values
2.11	$E_{11} = E_f V_f + E_m (1 - V_f)$	124.7	125.0
2.12	$u_{12} = u_f V_f + u_m (1 - V_f)$	0.25	0.34
2.14	$E_{22} = \dfrac{E_m}{V_f (E_m / E_f) + (1 - V_f)}$	7.4	9.1
2.15	$G_{12} = \dfrac{G_m}{V_f (G_m / G_f) + (1 - V_f)}$	2.6	5.0

---

3 Adapted from : Harris, B., *Engineering composite materials*, The Institute of Materials, London, UK, 1999, 193p.

2.1.7 Properties of a Single Ply

A ply has two major dimensions and one, the thickness, very small one compared with the two major ones. Therefore, the 3D presentation of an orthotropic material will be simplified to a 2D presentation (plane stress) by assuming in Equation (2.5) that $\sigma_{33} = 0$. This leads to a reduced compliance matrix for the ply in the form

$$\left\{ \begin{array}{c} \varepsilon_{11} \\ \varepsilon_{22} \\ \gamma_{12} \end{array} \right\} = \left[\begin{array}{ccc} \frac{1}{E_1} & -\frac{v_{21}}{E_2} & 0 \\ -\frac{v_{12}}{E_1} & \frac{1}{E_2} & 0 \\ 0 & 0 & \frac{1}{G_{12}} \end{array} \right] \left\{ \begin{array}{c} \sigma_{11} \\ \sigma_{22} \\ \sigma_{12} \end{array} \right\} \tag{2.16}$$

A third equation, for the strain in the thickness direction, ε_{33}, which has the form

$$\varepsilon_{33} = -\frac{v_{13}}{E_1}\sigma_{11} - \frac{v_{23}}{E_2}\sigma_{22} \tag{2.17}$$

and the remaining two equations for the shear strains that are written as

$$\left\{ \begin{array}{c} v_{23} \\ v_{13} \end{array} \right\} = \left[\begin{array}{cc} \frac{1}{G_{23}} & 0 \\ 0 & \frac{1}{G_{13}} \end{array} \right] \left\{ \begin{array}{c} \sigma_{23} \\ \sigma_{13} \end{array} \right\} \tag{2.18}$$

Calculation of the stresses as a function of strains would yield by the use of Equations (2.16) and (2.18):

$$\left\{ \begin{array}{c} \sigma_{11} \\ \sigma_{22} \\ \sigma_{12} \end{array} \right\} = \left[\begin{array}{ccc} Q_{11} & Q_{12} & 0 \\ Q_{21} & Q_{22} & 0 \\ 0 & 0 & Q_{66} \end{array} \right] \left\{ \begin{array}{c} \varepsilon_{11} \\ \varepsilon_{22} \\ \gamma_{12} \end{array} \right\} = \left[\begin{array}{ccc} \frac{E_1}{(1-v_{12}v_{21})} & \frac{v_{21}E_1}{(1-v_{12}v_{21})} & 0 \\ \frac{v_{12}E_2}{(1-v_{12}v_{21})} & \frac{E_2}{(1-v_{12}v_{21})} & 0 \\ 0 & 0 & G_{12} \end{array} \right] \left\{ \begin{array}{c} \varepsilon_{11} \\ \varepsilon_{22} \\ \gamma_{12} \end{array} \right\} \tag{2.19}$$

with $\quad Q_{12} = Q_{21} \quad$ and $\quad v_{12} \neq v_{21}$

$$\left\{ \begin{array}{c} \sigma_{23} \\ \sigma_{13} \end{array} \right\} = \left[\begin{array}{cc} Q_{23} & 0 \\ 0 & Q_{13} \end{array} \right] \left\{ \begin{array}{c} v_{23} \\ v_{13} \end{array} \right\} = \left[\begin{array}{cc} G_{23} & 0 \\ 0 & G_{13} \end{array} \right] \left\{ \begin{array}{c} v_{23} \\ v_{13} \end{array} \right\} \tag{2.20}$$

2.1.8 Transformation of Stresses and Strains

Consider the two coordinate systems as described in Figure 2.3. The one with indexes 1 and 2 describes the ply orthotropic coordinate system, while the other one (x,y) is an arbitrary one, rotated at a given angle θ relative to the 1, 2 system. The transformation of the stresses and the strains from the 1, 2 coordinate system to the

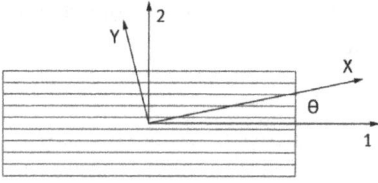

Figure 2.3: Two coordinate systems: 1, 2 the ply orthotropic axis; X,Y arbitrary axis.

x,y coordinate system is done by multiplication of both the stresses and the strains at the ply level by the transformation matrix T as given by[4]

$$\begin{Bmatrix} \sigma_1 \\ \sigma_2 \\ \tau_{12} \end{Bmatrix}^k = [T] \begin{Bmatrix} \sigma_x \\ \sigma_y \\ \tau_{xy} \end{Bmatrix}^k \tag{2.21}$$

$$\begin{Bmatrix} \varepsilon_1 \\ \varepsilon_2 \\ \frac{\gamma_{12}}{2} \end{Bmatrix}^k = [T] \begin{Bmatrix} \varepsilon_x \\ \varepsilon_y \\ \frac{\gamma_{xy}}{2} \end{Bmatrix}^k \tag{2.22}$$

where k is the number of the ply, for which the transformation of strains and stresses is performed.[5] The transformation matrix T is given by

$$[T] = \begin{bmatrix} c^2 & s^2 & 2cs \\ s^2 & c^2 & -2cs \\ -cs & cs & c^2 - s^2 \end{bmatrix} \quad \text{where} \quad \begin{aligned} c &\equiv \cos\theta \\ s &\equiv \sin\theta \end{aligned} \tag{2.23}$$

To obtain the inverse of the matrix T, one needs simply to insert $-\theta$ instead of θ in Equation (2.23) to yield

$$[T]^{-1} = [T(-\theta)] = \begin{bmatrix} c^2 & s^2 & -2cs \\ s^2 & c^2 & 2cs \\ cs & -cs & c^2 - s^2 \end{bmatrix} \tag{2.24}$$

The ply (or lamina) strain–stress relationships transformed to the laminate reference axis (x,y) is written as

$$\begin{Bmatrix} \sigma_1 \\ \sigma_2 \\ \tau_{12} \end{Bmatrix}^k = [T]^{-1}[Q]^k[T] \begin{Bmatrix} \varepsilon_x \\ \varepsilon_y \\ \gamma_{xy} \end{Bmatrix}^k \quad \text{where} \quad [Q]^k = \begin{bmatrix} Q_{11} & Q_{12} & 0 \\ Q_{12} & Q_{22} & 0 \\ 0 & 0 & 2Q_{66} \end{bmatrix}^k \tag{2.25}$$

---

4 See, for example, Ref [1]: *Primer on Composite Materials: Analysis* by J. E. Ashton and J. C. Halpin, TECHNOMIC Publishing Co., Inc., 750 Summer St., Stamford, Conn. 06901, USA, 1969.

5 Note that: $\sigma_{11} \equiv \sigma_1$; $\sigma_{22} \equiv \sigma_2$; $\varepsilon_{11} \equiv \varepsilon_1$; $\varepsilon_{22} \equiv \varepsilon_2$.

where the expressions for Q_{11}, Q_{12}, Q_{22} and Q_{66} are given in Equation (2.19). Performing the matrix multiplication in Equation (2.25) yields

$$\left\{\begin{array}{c} \sigma_1 \\ \sigma_2 \\ \tau_{12} \end{array}\right\}^k = [Q]^k \left\{\begin{array}{c} \varepsilon_x \\ \varepsilon_y \\ \gamma_{xy} \end{array}\right\}^k \quad \text{where} \quad [\bar{Q}]^k = \begin{bmatrix} \bar{Q}_{11} & \bar{Q}_{12} & \bar{Q}_{16} \\ \bar{Q}_{12} & \bar{Q}_{22} & \bar{Q}_{26} \\ \bar{Q}_{16} & \bar{Q}_{26} & \bar{Q}_{66} \end{bmatrix}^k \qquad (2.26)$$

where

$$\bar{Q}_{11} = Q_{11}\cos^4\theta + 2(Q_{12} + 2Q_{66})\sin^2\theta\cos^2\theta + Q_{22}\sin^4\theta$$

$$\bar{Q}_{12} = (Q_{11} + Q_{22} - 4Q_{66})\sin^2\theta\cos^2\theta + Q_{12}(\sin^4\theta + \cos^4\theta)$$

$$\bar{Q}_{22} = Q_{11}\sin^4\theta + 2(Q_{12} + 2Q_{66})\sin^2\theta\cos^2\theta + Q_{22}\cos^4\theta$$

$$\bar{Q}_{16} = (Q_{11} - Q_{12} - 2Q_{66})\sin\theta\cos^3\theta + (Q_{12} - Q_{22} + 2Q_{66})\sin^3\theta\cos\theta$$

$$\bar{Q}_{26} = (Q_{11} - Q_{12} - 2Q_{66})\sin^3\theta\cos\theta + (Q_{12} - Q_{22} + 2Q_{66})\sin\theta\cos^3\theta$$

$$\bar{Q}_{66} = (Q_{11} + Q_{22} - 2Q_{12} - 2Q_{66})\sin^2\theta\cos^2\theta + Q_{66}(\sin^4\theta + \cos^4\theta)$$

(2.27)

Another useful way of presenting the various terms of the matrix $[\bar{Q}]^k$ is the invariant procedure presented by Tsai and Pagano [2]:

$$\bar{Q}_{11} = U_1 + U_2\cos(2\theta) + U_3\cos(4\theta)$$

$$\bar{Q}_{12} = U_4 - U_3\cos(4\theta)$$

$$\bar{Q}_{22} = U_1 - U_2\cos(2\theta) + U_3\cos(4\theta)$$

$$\bar{Q}_{16} = -\frac{1}{2}U_2\sin(2\theta) - U_3\sin(4\theta)$$

$$\bar{Q}_{26} = -\frac{1}{2}U_2\sin(2\theta) + U_3\sin(4\theta)$$

$$\bar{Q}_{66} = U_5 - U_3\cos(4\theta)$$

where

(2.28)

$$U_1 = \frac{1}{8}[3Q_{11} + 3Q_{22} + 2Q_{12} + 4Q_{66}]$$

$$U_2 = \frac{1}{2}[Q_{11} - Q_{22}]$$

$$U_3 = \frac{1}{8}[Q_{11} + Q_{22} - 2Q_{12} - 4Q_{66}]$$

$$U_4 = \frac{1}{8}[Q_{11} + Q_{22} + 6Q_{12} - 4Q_{66}]$$

$$U_5 = \frac{1}{8}[Q_{11} + Q_{22} - 2Q_{12} + 4Q_{66}]$$

Note that the terms U_1, U_4 and U_5 are invariant to a rotation relative to the three axes (perpendicular to the 1–2 plane).

Now, we shall present the addition of the properties of each lamina to form a laminate, which is the structure to be investigated when we apply loads.

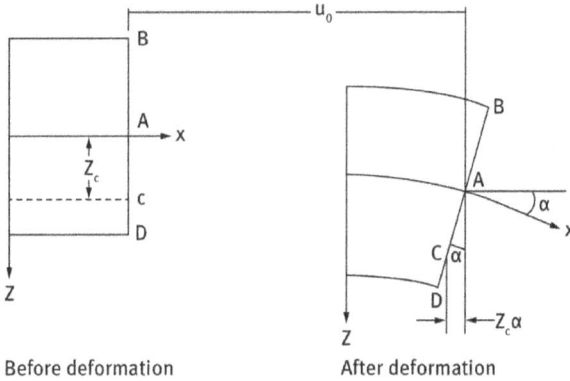

Before deformation   After deformation

Figure 2.4: The plate cross section before and after the deformation.

Referring to Figure 2.4, one can write the displacement in the x-direction of a point at a z distance from the mid-plane as (where w is the displacement in the z-direction)

$$u = u_0 - z\frac{\partial w}{\partial x} \tag{2.29}$$

Similarly, the displacement in the y-direction will be

$$v = v_0 - z\frac{\partial w}{\partial y} \tag{2.30}$$

Then the strains (ε_x, ε_y and γ_{xy}) and the curvatures (κ_x, κ_y and κ_{xy}) can be written as

$$\varepsilon_x \equiv \frac{\partial u}{\partial x} = \frac{\partial u_0}{\partial x} - z\frac{\partial^2 w}{\partial x^2} = \varepsilon_x^0 + z\kappa_x$$

$$\varepsilon_y \equiv \frac{\partial v}{\partial y} = \frac{\partial v_0}{\partial y} - z\frac{\partial^2 w}{\partial y^2} = \varepsilon_y^0 + z\kappa_y \tag{2.31}$$

$$\gamma_{xy} \equiv \frac{\partial u}{\partial y} + \frac{\partial v}{\partial x} = \frac{\partial u_0}{\partial y} + \frac{\partial v_0}{\partial x} - 2z\frac{\partial^2 w}{\partial x \partial y} = \gamma_{xy}^0 + z\kappa_{xy}$$

where $\varepsilon_x^0, \varepsilon_y^0, \gamma_{xy}^0$ are the strains at the neutral plane. In the matrix notation, Equations (2.31) can be given as

$$\left\{\begin{array}{c} \varepsilon_x \\ \varepsilon_y \\ \gamma_{xy} \end{array}\right\} = \left\{\begin{array}{c} \varepsilon_x^0 \\ \varepsilon_y^0 \\ \gamma_{xy}^0 \end{array}\right\} + z \left\{\begin{array}{c} \kappa_x \\ \kappa_y \\ \kappa_{xy} \end{array}\right\} \Rightarrow \quad \{\varepsilon\} = \{\varepsilon^0\} + z\{\kappa\} \tag{2.32}$$

Then the stresses at the lamina level will be given by

$$\{\sigma\}^k = [\bar{Q}]^k\{\varepsilon^0\} + z[\bar{Q}]^k[\kappa] \tag{2.33}$$

Now we shall deal with force (N_x, N_y, N_{xy}) and moment (M_x, M_y, M_{xy}) resultants. Their definitions are given as (h is the total thickness of the laminate)

$$N_x \equiv \int_{-h/2}^{h/2} \sigma_x dz; \quad N_y \equiv \int_{-h/2}^{h/2} \sigma_y dz; \quad N_{xy} \equiv \int_{-h/2}^{h/2} \tau_{xy} dz$$

$$M_x \equiv \int_{-h/2}^{h/2} \sigma_x z dz; \quad M_y \equiv \int_{-h/2}^{h/2} \sigma_y z dz; \quad M_{xy} \equiv \int_{-h/2}^{h/2} \tau_{xy} z dz \tag{2.34}$$

Substituting the expressions of the stresses one obtains expressions for the force and moments resultants as a function of the strain on the mid-plane, ε^0, and the curvature κ (see also Ref. [1]). The short written expressions are

$$\left\{\begin{array}{c} \{N\} \\ \{M\} \end{array}\right\} = \left[\begin{array}{cc} [A] & [B] \\ [B] & [D] \end{array}\right]\left\{\begin{array}{c} \{\varepsilon^0\} \\ \{\kappa\} \end{array}\right\}$$

or

$$\left\{\begin{array}{c} \left\{\begin{array}{c} N_x \\ N_y \\ N_{xy} \end{array}\right\} \\ \left\{\begin{array}{c} M_x \\ M_y \\ M_{xy} \end{array}\right\} \end{array}\right\} = \left[\begin{array}{cc} \left[\begin{array}{ccc} A_{11} & A_{12} & A_{16} \\ A_{12} & A_{22} & A_{26} \\ A_{16} & A_{26} & A_{66} \end{array}\right] & \left[\begin{array}{ccc} B_{11} & B_{12} & B_{16} \\ B_{12} & B_{22} & B_{26} \\ B_{16} & B_{26} & B_{66} \end{array}\right] \\ \left[\begin{array}{ccc} B_{11} & B_{12} & B_{16} \\ B_{12} & B_{22} & B_{26} \\ B_{16} & B_{26} & B_{66} \end{array}\right] & \left[\begin{array}{ccc} D_{11} & D_{12} & D_{16} \\ D_{12} & D_{22} & D_{26} \\ D_{16} & D_{26} & D_{66} \end{array}\right] \end{array}\right]\left\{\begin{array}{c} \left\{\begin{array}{c} \varepsilon_x^0 \\ \varepsilon_y^0 \\ \gamma_{xy}^0 \end{array}\right\} \\ \left\{\begin{array}{c} \kappa_x \\ \kappa_y \\ \kappa_{xy} \end{array}\right\} \end{array}\right\} \tag{2.35}$$

where the various constants are defined as

$$A_{ij} \equiv \int_{-h/2}^{h/2} \overline{Q}_{ij}^k dz = \sum_{k=1}^{n} \overline{Q}_{ij}^k (h_k - h_{k-1})$$

$$B_{ij} \equiv \int_{-h/2}^{h/2} \overline{Q}_{ij}^k z \, dz = \frac{1}{2} \sum_{k=1}^{n} \overline{Q}_{ij}^k (h_k^2 - h_{k-1}^2) \qquad (2.36)$$

$$D_{ij} \equiv \int_{-h/2}^{h/2} \overline{Q}_{ij}^k z^2 dz = \frac{1}{3} \sum_{k=1}^{n} \overline{Q}_{ij}^k (h_k^3 - h_{k-1}^3)$$

where $i, j = 1, 1; \ 1, 2; \ 2, 2; \ 1, 6; \ 2, 6; \ 6, 6$

The way the sum is performed in Equation (2.36) is according to the notations in Figure 2.5. The passage from integral over the thickness of the laminate to the sum over the thickness is dictated by the fact that the individual plies are very thin and the properties within each laminae are assumed constant in the thickness direction.

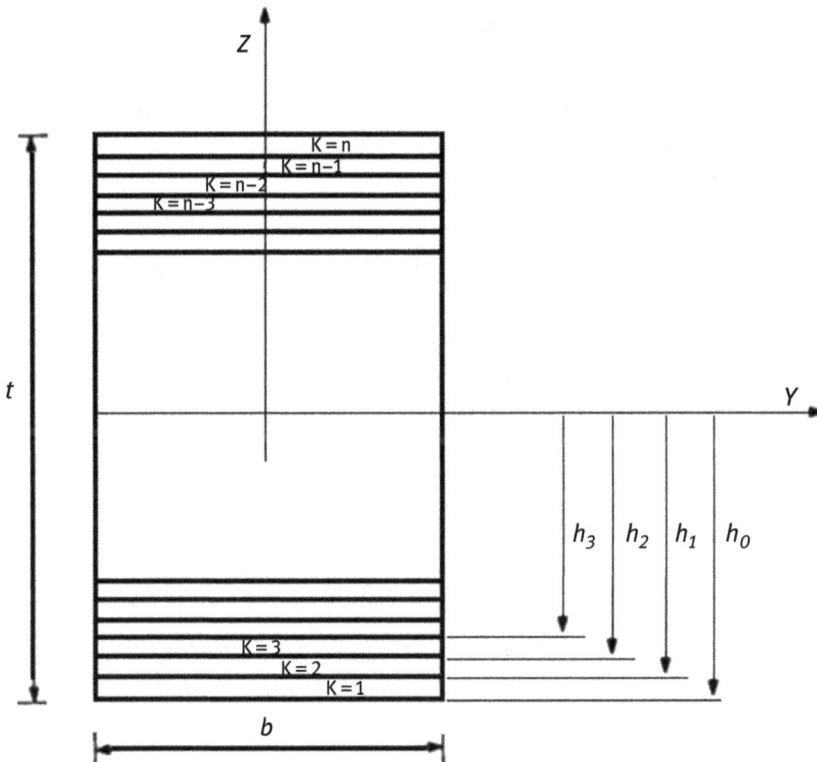

Figure 2.5: Lamina notation within a given laminate.

Finally, the equations of motion for the static case, for a thin plate made up of laminated composite plies, using the classical lamination theory (CLT), are given as (see [1,3]):

$$\frac{\partial N_x}{\partial x} + \frac{\partial N_{xy}}{\partial y} = I_1 \frac{\partial^2 u_0}{\partial t^2} - I_2 \frac{\partial^2}{\partial t^2}\left(\frac{\partial w_0}{\partial x}\right)$$

$$\frac{\partial N_{xy}}{\partial x} + \frac{\partial N_y}{\partial y} = I_1 \frac{\partial^2 v_0}{\partial t^2} - I_2 \frac{\partial^2}{\partial t^2}\left(\frac{\partial w_0}{\partial y}\right)$$

$$\frac{\partial^2 M_x}{\partial x^2} + 2\frac{\partial^2 M_{xy}}{\partial x \partial y} + \frac{\partial^2 M_y}{\partial y^2} + \frac{\partial}{\partial x}\left[N_{xx}\frac{\partial w_0}{\partial x} + N_{xy}\frac{\partial w_0}{\partial y}\right] + \frac{\partial}{\partial y}\left[N_{yy}\frac{\partial w_0}{\partial y} + N_{xy}\frac{\partial w_0}{\partial x}\right] =$$

$$= -p_z + I_1 \frac{\partial^2 w_0}{\partial t^2} - I_3 \frac{\partial^2}{\partial t^2}\left(\frac{\partial^2 w_0}{\partial x^2} + \frac{\partial^2 w_0}{\partial y^2}\right) + I_2 \frac{\partial^2}{\partial t^2}\left(\frac{\partial w_0}{\partial x} + \frac{\partial w_0}{\partial y}\right)$$

(2.37)

where p_z is the load per unit area in the z-direction[6] and the subscript 0 represents the values at the mid-plane of the cross section. N represents the in-plane loads and the various moments of inertia, I_1, I_2 and I_3, are given by (ρ is the mass/unit length)

$$I_j = \int_{-h/2}^{h/2} \rho z^{j-1} dz; \quad j = 1, 2, 3$$

(2.38)

To obtain the equations for a beam, one can use Equation (2.26), while all the derivations with respect to y are identically zero. This yields a one-dimensional equation in the following form:

$$\frac{\partial^2 M_x}{\partial x^2} + \frac{\partial}{\partial x}\left(N_{xx}\frac{\partial w_0}{\partial x}\right) = -p_z + I_1 \frac{\partial^2 w}{\partial t^2} - I_3 \frac{\partial^4 w}{\partial t^2 \partial x^2} + I_2 \frac{\partial^3 w}{\partial t^2 \partial x}$$

(2.39)

where N_x is the axial (in-plane, in the direction of the length of the beam) load. Remembering the relationship between transverse deflection, w, and the bending moment, we can rewrite Equation (2.39) in terms of w only to yield

$$-D_{11}\frac{\partial^2 w}{\partial x^2} = M_x \Rightarrow$$

$$-\frac{\partial^2}{\partial x^2}\left(D_{11}\frac{\partial^2 w}{\partial x^2}\right) + \frac{\partial}{\partial x}\left(N_{xx}\frac{\partial w_0}{\partial x}\right) = -p_z + I_1 \frac{\partial^2 w}{\partial t^2} - I_3 \frac{\partial^4 w}{\partial t^2 \partial x^2} + I_2 \frac{\partial^3 w}{\partial t^2 \partial x}$$

(2.40)

---

6 Note that the coordinate z is normally used for the thickness direction, while x and y coordinates define the plate area.

with its associated boundary conditions

$$\text{Geometric: specify either } w \text{ and } \frac{\partial w}{\partial x}$$
$$\text{Natural: specify either } Q \equiv \frac{\partial M}{\partial x} \text{ and } M. \quad (2.41)$$

Typical boundary conditions, normally used in the literature, can be written as:

$$\text{Simply supported: } w = 0 \text{ and } M = 0$$
$$\text{Clamped: } w = 0 \text{ and } \frac{\partial w}{\partial x} = 0 \quad (2.42)$$
$$\text{Free: } Q \equiv \frac{\partial M}{\partial x} = 0 \text{ and } M = 0$$

The reader should be aware of thermal issues associated with the manufacturing of composite structures due to the differential thermal contraction during the post-curing phase and also as a consequence of any temperature changes during the service life of the structure. This issue is caused by the relatively small axial thermal expansion coefficient of the modern reinforcing fibers (for carbon fibers it is even slightly negative), while the resin matrix has a large thermal coefficient. When cooling from a typical curing temperature, like 140 °C to room temperature, the fibers of the laminate composite will be in compression, while the matrix will show tension stresses [4]. Typical residual stresses due to thermal mismatch between the two components of the laminate are presented in Table 2.3.

Table 2.3: Typical thermal stresses in some common unidirectional composites (from [4]).

Matrix	Fiber	% Fiber volume, V_f	Temperature range ΔT (K)	Fiber residual stress (MPa)	Matrix residual stress (MPa)
Epoxy (high T cure)	T300 carbon	65	120	−19	36
Epoxy (low T cure)	E glass	65	100	−15	28
Epoxy (low T cure)	Kevlar-49	65	100	−16	30
Borosilicate glass	T300 carbon	50	520	−93	93
CAS* glass-ceramic	Nicalon SiC	40	1000	−186	124

*CAS, $CaO - Al_2O_3 - SiO_2$.

Another important data for design is the experimental tension and compression strength as measured during various laboratory tests, as presented in Table 2.4 (see [4]).

Table 2.4: Typical experimental tension and compression strengths for common composite materials (from [4]).

Material	Lay-up	% Fiber volume, V_f	Tensile strength σ_t (GPa)	Compression strengthσ_c (GPa)	Ratio σ_c/σ_t
GRP	ud*	60	1.3	1.1	0.85
CFRP	ud	60	2.0	1.1	0.55
KFRP	ud	60	1.0	0.4	0.40
HTA/913 (CFRP)	$\left[\left(\pm 45^0, 0_2^0\right)_2\right]_s$	65	1.27	0.97	0.77
T800/924 (CFRP)	$\left[\left(\pm 45^0, 0_2^0\right)_2\right]_s$	65	1.42	0.90	0.63
T800/5245(CFRP)	$\left[\left(\pm 45^0, 0_2^0\right)_2\right]_s$	65	1.67	0.88	0.53
SiC/CAS (CMC**)	ud	37	334	1360	4.07
SiC/CAS (CMC)	$\left[0^0, 90^0\right]_{3S}$	37	210	463	2.20

*ud, unidirectional,**CMC, ceramic matrix composites.

References

[1]   Ashton, J. E., Halpin, J. C. and Petit, P. H., Primer on Composite Materials: Analysis, Progress in Material Science Series, Vol. III, Technomic publication, Library of Congress Catalog Card No. 72–81344, 1969.

[2]   Tsai, S. W. and Pagano, N. J., Invariant properties of composite materials. In: Composite Materials Workshop, Tsai, S. W., Halpin, J. C. and Pagano, N. J. (eds.), St. Louis, Missouri, 1967, Technomic Publishing Company, 1968, 233–253.

[3]   Reddy, J. N., Mechanics of Laminated Composite Plates and Shells, 2nd edn., CRC Press LLC, 2004.

[4]   Harris, B., Engineering Composite Materials, London, UK, The Institute of Materials, 1999, 193 p.

2.2 First-Order Shear Deformation Theory (FSDT) Model

The need to remove the somehow restricting assumptions for the CLT which are based on Kirchhoff–Love plate theory [1–3], like the neglecting the influence of the shear strains and the fact that a plane before deformation remains plane after the deformation, led to the derivation of more advanced bending theories for plates, like Mindlin theory of plates [4–6], which includes in-plane shear strains and is an extension of Kirchhoff–Love plate theory incorporating first-order shear effects.

Mindlin's theory assumes that there is a linear variation of displacement across the plate thickness, but the plate thickness does not change during deformation. An additional assumption is that the normal stress through the thickness is ignored; an assumption which is also called the plane stress condition. The Mindlin theory is often called the first-order shear deformation theory of plates and its application to composite materials is next presented. Under the assumptions and restrictions of Mindlin's theory (which has a similarity to Timoshenko's theory for beams [7–10]) the displacement field has five unknowns (u_0, v_0, w_0 – the displacements of the mid-plane in the x-, y- and z-directions, respectively, and – the rotations due to shear about the x- and y-direction, respectively) and is given by(see also Figure 2.6, which is similar to Figure 2.4, but for FSDT approach):

$$u(x, y, z, t) = u_0(x, y, t) + z\phi_x(x, y, t)$$
$$v(x, y, z, t) = v_0(x, y, t) + z\phi_y(x, y, t) \tag{2.43}$$
$$w(x, y, z, t) = w_0(x, y, t)$$

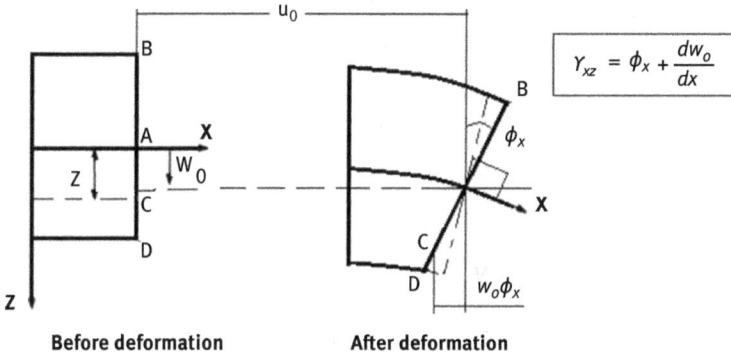

Figure 2.6: The plate cross section before and after the deformation (FSDT approach).

The associated strains [11][7] including nonlinear terms are

$$\varepsilon_x = \frac{\partial u_0}{\partial x} + \frac{1}{2}\left(\frac{\partial w_0}{\partial x}\right)^2 + z\frac{\partial \phi_x}{\partial x}$$

$$\varepsilon_y = \frac{\partial u_0}{\partial x} + \frac{1}{2}\left(\frac{\partial w_0}{\partial x}\right)^2 + z\frac{\partial \phi_x}{\partial x} \tag{2.44}$$

$$\varepsilon_z = 0; \quad \gamma_{xz} = \frac{\partial w_0}{\partial x} + \phi_x; \quad \gamma_{yz} = \frac{\partial w_0}{\partial y} + \phi_x$$

$$\gamma_{xy} = \left(\frac{\partial u_0}{\partial y} + \frac{\partial v_0}{\partial x} + \frac{\partial^2 w_0}{\partial x \partial y}\right) + z\left(\frac{\partial \phi_x}{\partial y} + \frac{\partial \phi_y}{\partial x}\right)$$

---

7 Note that the assumption of constant shear strains across the height of the laminate is a rough approximation of the true strain distribution, which is at least quadratic through the thickness. However, although the rough approximation, the results of the application of Mindlin's plate theory presents very good results compared with experimental ones.

Multiplying Equation (2.44) by the stiffness matrix $[\bar{Q}]$ and integrating through the thickness of the laminate yield the force and moments resultants (as defined in Chapter 1 in Equation (2.34)), with two additional terms, the shear resultants are defined by

$$\left\{ \begin{array}{c} Q_x \\ Q_y \end{array} \right\} = \kappa \int_{-h/2}^{h/2} \left\{ \begin{array}{c} \tau_{xz} \\ \tau_{yz} \end{array} \right\} dz \tag{2.45}$$

where κ is called the shear correction coefficient and is defined by the ratio between the shear strain energies calculated by the actual shear distribution and the constant distribution assumed in the FSDT theory. The value of κ is taken as 5/6 for a rectangular cross section.[8]

The equation of motion will then have the following form:

$$\frac{\partial N_x}{\partial x} + \frac{\partial N_{xy}}{\partial y} = I_1 \frac{\partial^2 u_0}{\partial t^2} + I_2 \frac{\partial^2 \phi_x}{\partial t^2}$$

$$\frac{\partial N_{xy}}{\partial x} + \frac{\partial N_y}{\partial y} = I_1 \frac{\partial^2 v_0}{\partial t^2} + I_2 \frac{\partial^2 \phi_y}{\partial t^2}$$

$$\frac{\partial Q_x}{\partial x} + \frac{\partial Q_y}{\partial y} + \frac{\partial}{\partial x} \left[N_{xx} \frac{\partial w_0}{\partial x} + N_{xy} \frac{\partial w_0}{\partial y} \right]$$

$$+ \frac{\partial}{\partial y} \left[N_{yy} \frac{\partial w_0}{\partial y} + N_{xy} \frac{\partial w_0}{\partial x} \right] = -p_z + I_1 \frac{\partial^2 w_0}{\partial t^2} \tag{2.46}$$

$$\frac{\partial M_x}{\partial x} + \frac{\partial M_{xy}}{\partial y} - Q_x = I_3 \frac{\partial^2 \phi_x}{\partial t^2} + I_1 \frac{\partial^2 u_0}{\partial t^2}$$

$$\frac{\partial M_{xy}}{\partial x} + \frac{\partial M_y}{\partial y} - Q_y = I_3 \frac{\partial^2 \phi_y}{\partial t^2} + I_1 \frac{\partial^2 v_0}{\partial t^2}$$

where p_z is the load per unit area in the z-direction.[9] N represents the in-plane loads and the various moments of inertia, I_1, I_2 and I_3 are given by (ρ is the mass/unit length) Equation (2.38).

In addition to Equations (2.46) that describe the resultants of the force and the moment as a function of the stiffness coefficients A_{ij}, B_{ij} and D_{ij}, the shear resultants Q_x and Q_y are defined as

$$\left\{ \begin{array}{c} Q_y \\ Q_x \end{array} \right\} = \kappa \left[\begin{array}{cc} A_{44} & A_{45} \\ A_{45} & A_{55} \end{array} \right] \left\{ \begin{array}{c} \frac{\partial w_0}{\partial y} + \phi_y \\ \frac{\partial w_0}{\partial x} + \phi_x \end{array} \right\} \tag{2.47}$$

---

8 The accurate value is $\kappa = \frac{10(1+\upsilon)}{12+11\upsilon}$ for a rectangular cross section and $\kappa = \frac{6(1+\upsilon)}{7+6\upsilon}$ for a solid circular cross section.

9 Note that the coordinate z is normally used for the thickness direction, while x and y coordinates define the plate area.

where

$$A_{44} \equiv \kappa \int_{-h/2}^{h/2} \overline{Q}_{44}^{k} dz = \kappa \sum_{k=1}^{n} \overline{Q}_{44}^{k} (h_k - h_{k-1})$$

$$A_{45} \equiv \kappa \int_{-h/2}^{h/2} \overline{Q}_{45}^{k} dz = \kappa \sum_{k=1}^{n} \overline{Q}_{45}^{k} (h_k - h_{k-1}) \tag{2.48}$$

$$A_{55} \equiv \kappa \int_{-h/2}^{h/2} \overline{Q}_{55}^{k} dz = \kappa \sum_{k=1}^{n} \overline{Q}_{55}^{k} (h_k - h_{k-1})$$

and

$$\overline{Q}_{44} = Q_{44} \cos^2\theta + Q_{55} \sin^2\theta$$

$$\overline{Q}_{45} = (Q_{55} - Q_{44}) \cos\theta \sin\theta$$

$$\overline{Q}_{55} = Q_{44} \sin^2\theta + Q_{55} \cos^2\theta \tag{2.49}$$

$$where \quad Q_{44} = G_{23}; \quad and \quad Q_{55} = G_{13} \; .$$

Substituting the resultants defined in terms of the five unknown displacements $(u_0, v_0, w_0, \phi_x, \phi_y)$, we get [11] five differential equations for the five unknown displacements:

$$A_{11} \left[\frac{\partial^2 u_0}{\partial x^2} + \frac{\partial^3 w_0}{\partial x^3} \right] + A_{12} \left[\frac{\partial^2 v_0}{\partial x \partial y} + \frac{\partial^3 w_0}{\partial x \partial y^2} \right] +$$

$$A_{16} \left[2 \frac{\partial^2 u_0}{\partial x \partial y} + \frac{\partial^2 v_0}{\partial x^2} + 3 \frac{\partial^3 w_0}{\partial x^2 \partial y} \right] + A_{26} \left[\frac{\partial^2 v_0}{\partial y^2} + \frac{\partial^3 w_0}{\partial y^3} \right] +$$

$$A_{66} \left[\frac{\partial^2 u_0}{\partial y^2} + \frac{\partial^2 v_0}{\partial x \partial y} + 2 \frac{\partial^3 w_0}{\partial x \partial y^2} \right] + B_{11} \frac{\partial^2 \phi_x}{\partial x^2} + B_{12} \frac{\partial^2 \phi_y}{\partial x \partial y} + \tag{2.50}$$

$$B_{16} \left[2 \frac{\partial^2 \phi_x}{\partial x \partial y} + \frac{\partial^2 \phi_y}{\partial x^2} \right] + B_{26} \frac{\partial^2 \phi_y}{\partial y^2} + B_{66} \left[\frac{\partial^2 \phi_x}{\partial y^2} + \frac{\partial^2 \phi_y}{\partial x \partial y} \right] = I_1 \frac{\partial^2 u_0}{\partial t^2} + I_2 \frac{\partial^2 \phi_x}{\partial t^2}$$

$$A_{22}\left[\frac{\partial^2 v_0}{\partial y^2} + \frac{\partial^3 w_0}{\partial y^3}\right] + A_{12}\left[\frac{\partial^2 u_0}{\partial x \partial y} + \frac{\partial^3 w_0}{\partial x^2 \partial y}\right] +$$

$$A_{16}\left[\frac{\partial^2 u_0}{\partial x^2} + \frac{\partial^3 w_0}{\partial x^3}\right] + A_{26}\left[\frac{\partial^2 u_0}{\partial y^2} + 2\frac{\partial^2 v_0}{\partial x \partial y} + 3\frac{\partial^3 w_0}{\partial x \partial y^2}\right] +$$

$$A_{66}\left[\frac{\partial^2 u_0}{\partial x \partial y} + \frac{\partial^2 v_0}{\partial x^2} + 2\frac{\partial^3 w_0}{\partial x^2 \partial y}\right] + B_{22}\frac{\partial^2 \phi_y}{\partial y^2} + B_{12}\frac{\partial^2 \phi_x}{\partial x \partial y} +$$

$$B_{16}\frac{\partial^2 \phi_x}{\partial x^2} + B_{26}\left[\frac{\partial^2 \phi_x}{\partial y^2} + 2\frac{\partial^2 \phi_y}{\partial x \partial y}\right] + B_{66}\left[\frac{\partial^2 \phi_y}{\partial x^2} + \frac{\partial^2 \phi_x}{\partial x \partial y}\right] = I_1\frac{\partial^2 v_0}{\partial t^2} + I_2\frac{\partial^2 \phi_y}{\partial t^2}$$

$$(2.51)$$

$$\kappa A_{55}\left[\frac{\partial^2 w_0}{\partial x^2} + \frac{\partial \varphi_x}{\partial x}\right] + \kappa A_{45}\left[2\frac{\partial^2 w_0}{\partial x \partial y} + \frac{\partial \varphi_y}{\partial x} + \frac{\partial \varphi_x}{\partial y}\right] + \kappa A_{44}\left[\frac{\partial^2 w_0}{\partial y^2} + \frac{\partial \varphi_y}{\partial y}\right] +$$

$$\frac{\partial}{\partial x}\left[N_{xx}\frac{\partial w_0}{\partial x} + N_{xy}\frac{\partial w_0}{\partial y}\right] + \frac{\partial}{\partial y}\left[N_{yy}\frac{\partial w_0}{\partial y} + N_{xy}\frac{\partial w_0}{\partial x}\right] = -p_z + I_1\frac{\partial^2 w_0}{\partial t^2}$$

$$(2.52)$$

$$B_{11}\left[\frac{\partial^2 u_0}{\partial x^2} + \frac{\partial^3 w_0}{\partial x^3}\right] + B_{12}\left[\frac{\partial^2 v_0}{\partial x \partial y} + \frac{\partial^3 w_0}{\partial x \partial y^2}\right] +$$

$$B_{16}\left[2\frac{\partial^2 u_0}{\partial x \partial y} + \frac{\partial^2 v_0}{\partial x^2} + 3\frac{\partial^3 w_0}{\partial x^2 \partial y}\right] + B_{26}\left[\frac{\partial^2 v_0}{\partial y^2} + \frac{\partial^3 w_0}{\partial y^3}\right] +$$

$$B_{66}\left[\frac{\partial^2 u_0}{\partial y^2} + \frac{\partial^2 v_0}{\partial x \partial y} + 2\frac{\partial^3 w_0}{\partial x \partial y^2}\right] + D_{11}\frac{\partial^2 \phi_x}{\partial x^2} + D_{12}\frac{\partial^2 \phi_y}{\partial x \partial y} +$$

$$D_{16}\left[2\frac{\partial^2 \phi_x}{\partial x \partial y} + \frac{\partial^2 \phi_y}{\partial x^2}\right] + D_{26}\frac{\partial^2 \phi_y}{\partial y^2} + D_{66}\left[\frac{\partial^2 \phi_x}{\partial y^2} + \frac{\partial^2 \phi_y}{\partial x \partial y}\right] -$$

$$\kappa A_{55}\left[\frac{\partial w_0}{\partial x} + \phi_x\right] - \kappa A_{45}\left[\frac{\partial w_0}{\partial y} + \phi_y\right] = I_3\frac{\partial^2 u_0}{\partial t^2} + I_3\frac{\partial^2 \phi_x}{\partial t^2}$$

$$(2.53)$$

$$B_{22}\left[\frac{\partial^2 v_0}{\partial y^2} + \frac{\partial^3 w_0}{\partial y^3}\right] + B_{12}\left[\frac{\partial^2 u_0}{\partial x \partial y} + \frac{\partial^3 w_0}{\partial x^2 \partial y}\right] +$$

$$B_{16}\left[\frac{\partial^2 u_0}{\partial x^2} + \frac{\partial^3 w_0}{\partial x^3}\right] + B_{26}\left[\frac{\partial^2 u_0}{\partial y^2} + 2\frac{\partial^2 v_0}{\partial x \partial y} + 3\frac{\partial^3 w_0}{\partial x \partial y^2}\right] +$$

$$B_{66}\left[\frac{\partial^2 u_0}{\partial x \partial y} + \frac{\partial^2 v_0}{\partial x^2} + 2\frac{\partial^3 w_0}{\partial x^2 \partial y}\right] + D_{22}\frac{\partial^2 \phi_y}{\partial y^2} + D_{12}\frac{\partial^2 \phi_x}{\partial x \partial y} +$$

$$D_{16}\frac{\partial^2 \phi_x}{\partial x^2} + D_{26}\left[\frac{\partial^2 \phi_x}{\partial y^2} + 2\frac{\partial^2 \phi_y}{\partial x \partial y}\right] + D_{66}\left[\frac{\partial^2 \phi_y}{\partial x^2} + \frac{\partial^2 \phi_x}{\partial x \partial y}\right] -$$

$$\kappa A_{44}\left[\frac{\partial w_0}{\partial y} + \phi_y\right] - \kappa A_{45}\left[\frac{\partial w_0}{\partial x} + \phi_x\right] = I_1\frac{\partial^2 v_0}{\partial t^2} + I_2\frac{\partial^2 \phi_y}{\partial t^2}$$

$$(2.54)$$

To solve the five differential equations, 10 boundary conditions should be supplied in the form of the geometric and natural boundary conditions.[10]

To obtain the equations of motion for a beam, using the first-order shear deformation theory, presented before for a plate, one should assume that all the derivations in the y-direction should vanish, and v and ϕ_y should be identically zero. This yields for the general case three coupled equations of motion having the following form (assuming constant properties along the beam):

$$A_{11}\frac{\partial^2 u_0}{\partial x^2} + B_{11}\frac{\partial^2 \phi_x}{\partial x^2} = I_1\frac{\partial^2 u_0}{\partial t^2} + I_2\frac{\partial^2 \phi_x}{\partial t^2}$$

$$B_{11}\frac{\partial^2 u_0}{\partial x^2} + D_{11}\frac{\partial^2 \phi_x}{\partial x^2} - \kappa A_{55}\left[\frac{\partial w_0}{\partial x} + \phi_x\right] = I_3\frac{\partial^2 u_0}{\partial t^2} + I_3\frac{\partial^2 \phi_x}{\partial t^2} \qquad (2.55)$$

$$\kappa A_{55}\left[\frac{\partial^2 w_0}{\partial x^2} + \frac{\partial \varphi_x}{\partial x}\right] + N_{xx}\frac{\partial^2 w_0}{\partial x} = -p_z + I_1\frac{\partial^2 w_0}{\partial t^2}$$

with the following boundary conditions:

$$A_{11}\frac{\partial u_0}{\partial x} + B_{11}\frac{\partial \phi_x}{\partial x} = -N_{xx} \quad \underline{\text{or}} \quad u_0 = 0$$

$$A_{55}\left(\phi_x + \frac{\partial w_0}{\partial x}\right) - N_{xx}\frac{\partial w_0}{\partial x} = 0 \quad \underline{\text{or}} \quad w_0 = 0 \qquad (2.56)$$

$$B_{11}\frac{\partial u_0}{\partial x} + D_{11}\frac{\partial \phi_x}{\partial x} = 0 \quad \underline{\text{or}} \quad \phi_x = 0$$

The reader is now referred to Refs. [12–79], which represent a short list of manuscripts dealing with buckling, vibrations, delaminations, and other aspects of composite beams, plates and shells, including sandwich structures, using classical and FSDT theories. References [80–95] show the introduction of smart materials, like piezoelectric patches, shape memory alloys, optic fibers and advanced control theories to reduce vibrations, increase reliability and buckling loads of composite structures.

References

[1] Love, A. E. H., On the small free vibrations and deformations of elastic shells, Philosophical Transaction of the Royal Society (London) Series A 17, 1888, 491–549.
[2] Reddy, J. N., Theory and Analysis of Elastic Plates and Shells, CRC Press, Taylor and Francis, 2007, 568.
[3] Timoshenko, S. and Woinowsky-Krieger, S., Theory of Plates and Shells, New York, McGraw-Hill, 1959, 580.

---

10 For further discussion about the types of boundary conditions to be imposed the reader is referred to Ref. [11].

[4] Mindlin, R. D., Influence of rotatory inertia and shear on flexural motions of isotropic, elastic plates, ASME Journal of Applied Mechanics 18, 1951, 31–38.

[5] Reddy, J. N., Theory and Analysis of Elastic Plates, Philadelphia, Taylor and Francis, 1999.

[6] Lim, G. T. and Reddy, J. N., On canonical bending relationships for plates, International, Journal of Solids and Structures 40, 2003, 3039–3067.

[7] Timoshenko, S. P., On the correction factor for shear of the differential equation for transverse vibrations of bars of uniform cross-section, Philosophical Magazine, 1921, 744.

[8] Timoshenko, S. P., On the transverse vibrations of bars of uniform cross-section, Philosophical Magazine, 1922, 125.

[9] Rosinger, H. E. and Ritchie, I. G., On Timoshenko's correction for shear in vibrating isotropic beams, Journal of Physics D: Applied Physics 10, 1977, 1461–1466.

[10] Timoshenko, S. P. and Gere, J. M., Mechanics of Materials, Van Nostrand Reinhold Co, 1972.

[11] Reddy, J. N., Mechanics of Laminated Composite Plates and Shells, 2nd edn., CRC Press LLC, 2004, 831.

[12] Chailleux, A., Hans, Y. and Verchery, G., Experimental study of the buckling of laminated composite columns and plates, International Journal of Mechanical Sciences 17(8), August 1975, 489–498.

[13] Wilson, D. W. and Vinson, J. R., Viscoelastic Buckling Analysis of Laminated Composite Columns, Recent Advances in Composites in the United States and Japan, ASTM STP 864, 1985.

[14] Tanigawa, Y., Murakami, H. and Ootao, Y., Transient thermal stress analysis of a laminated composite beam, Journal of Thermal Stresses 12(1), 1989, 25–39.

[15] Ootao, Y., Tanigawa, Y. and Murakami, H., Transient thermal stress and deformation of a laminated composite beam due to partially distributed heat supply, Journal of Thermal Stresses 13(2), 1990, 193–206.

[16] Chandrashekhara, K., Krishnamurthy, K. and Roy, S., Free vibration of composite beams including rotary inertia and shear deformation, Composite Structures 14(4), 1990, 269–279.

[17] Singh, G. and Venkateswara, R., Analysis of the nonlinear vibrations of unsymmetrically laminated composite beams, AAIA Journal 29(10), October 1991, 1727–1735.

[18] Barbero, E. and Tomblinj, J., Buckling testing of composite columns, AIAA Journal 30(11), November 1991, 2798–2800.

[19] Qatu, M. S., In-plane vibration of slightly curved laminated composite beams, Journal of Sound and Vibration 159(2), 8 December 1992, 327–338. doi:10.1016/0022-460X(92)90039-Z.

[20] Barbero, E. J. and Raftoyiannis, I. G., Euler buckling of pultruded composite columns, Composite Structures 24(2), 1993, 139–147.

[21] Abramovich, H. and Livshits, A., Free vibrations of non-symmetric cross-ply laminated composite beams, Journal of Sound and Vibration 176(5), 6 October 1994, 597–612.

[22] Owen, B., Gurdal, Z. and Lee, J., Buckling analysis of laminated composite beams with multiple delaminations, AAIA-94-1573-CP paper, 35th Structures, Structural Dynamics, and Materials Conference Hilton Head, SC, U.S.A., 1994, doi:10.2514/6.1994-1573.

[23] Maiti, D. K. and Sinha, P. K., Bending and free vibration analysis of shear deformable laminated composite beams by finite element method, Composite Structures 29(4), 1994, 421–431.

[24] Abramovich, H., Eisenberger, M. and Shulepov, O., Vibrations and buckling of non-symmetric laminated composite beams via the exact element method, AIAA/ASME/ASCE/AHS/ASC Structures, Structural Dynamics, and Materials Conference, 36th, and AIAA/ASME Adaptive Structures Forum, New Orleans, LA, USA, 1995, doi:http://arc.aiaa.org/doi/abs/10.2514/6.1995-1459.

[25] Kim, Y., Davalos, J. F. and Barbero, E. J., Progressive Failure Analysis of Laminated Composite Beams, Journal of Composite Materials 30(5), March 1996, 536–560.

[26] Abramovich, H., Eisenberger, M. and Shulepov, O., Vibrations and Buckling of Cross-Ply Nonsymmetric Laminated Composite Beams, AIAA Journals 34(5), May 1996, 1064–1069.

[27] Gadelrab, R. M., The effect of delamination on the natural frequencies of a laminated composite beam, Journal of Sound and Vibration 197(3), 31 October 1996, 283–292.

[28] Song, S. J. and Waas, A. M., Effects of shear deformation on buckling and free vibration of laminated composite beams, Composite Structures 37(1), January 1997, 33–43.

[29] Bhattacharya, P., Suhail, H. and Sinha, P. K., Finite Element Free Vibration Analysis of Smart Laminated Composite Beams and Plates, Journal of Intelligent Material Systems and Structures 9(1), January 1998, 20–28.

[30] Kadivar, M. H. and Mohebpour, S. R., Forced vibration of unsymmetric laminated composite beams under the action of moving loads, Composites Science and Technology 58(10), October 1998, 1675–1684.

[31] Yıldırım, V., Governing equations of initially twisted elastic space rods made of laminated composite materials, International Journal of Engineering Science 37(8), June 1999, 1007–1035.

[32] Moradi, S. and Taheri, F., Delamination buckling analysis of general laminated composite beams by differential quadrature method, Composites Part B: Engineering 30(5), July 1999, 503–511.

[33] Lee, J. J. and Choi, S., Thermal buckling and post-buckling analysis of a laminated composite beam with embedded SMA actuator, Composite Structures 47(1–4), December 1999, 695–703.

[34] Rehfield, L. W. and Mueller, U., Design of thin-walled laminated composite to resist buckling, AIAA paper 99-1375, 40th Structures, Structural Dynamics, and Materials Conference and Exhibit St. Louis, MO, U.S.A, 1999, http://arc.aiaa.org/doi/pdf/10.2514/6.1999-1375.

[35] Khdeir, A. A. and Reddy, J. N., Jordan canonical form solution for thermally induced deformations of cross-ply laminated composite beams, Journal of Thermal Stresses 22(3), 1999, 331–346.

[36] Khdeir, A. A., Thermal buckling of cross-ply laminated composite beams, Acta Mechanica 149 (1–4), 2001, 201–213.

[37] Lee, S., Park, T. and Voyiadjis, G. Z., Free vibration analysis of axially compressed laminated composite beam-columns with multiple delaminations, Composites Part B: Engineering 33(8), December 2002, 605–617.

[38] Fares, M. E., Youssif, Y. G. and Hafiz, M. A., Optimization control of composite laminates for maximum thermal buckling and minimum vibrational response, Journal of Thermal Stresses 25(11), 2002, 1047–1064.

[39] Goyal, V. K. and Kapania, R. K., Dynamic Stability of Laminated Composite Beams Subject to Subtangential Loads, 44th AIAA/ASME/ASCE/AHS Structures, Structural Dynamics, and Materials Conference, AAIA Journals, April 2003, doi:10.2514/6.2003-1930.

[40] Anilturk, D. and Chan, W. S., Structural stability of composite laminated column exposed to high temperature or fire, Journal of Composite Materials 37(8), April 2003, 687–700.

[41] Zhang, Z. and Taheri, F., Dynamic pulse buckling and postbuckling of composite laminated beam using higher order shear deformation theory, Composites Part B: Engineering 34(4), 1 June 2003, 391–398.

[42] Zhang, Z. and Taheri, F., Dynamic pulse-buckling behavior of 'quasi-ductile' carbon/epoxy and E-glass/epoxy laminated composite beams, Composite Structures 64(3–4), June 2004, 269–274.

[43] Çallioğlu, H., Tarakcilar, A. R. and Bektaş, N. B., Elastic-plastic stress analysis of laminated composite beams under linear temperature distribution, Journal of Thermal Stresses 27(11), 2004, 1075–1088.

[44] Ganesan, R. and Kowda, V. K., Free-vibration of composite beam-columns with stochastic material and geometric properties subjected to random axial loads, Journal of Reinforced Plastics and Composites 24(1), January 2005, 69–91.

[45] Ganesan, R. and Kowda, V. K., Buckling of composite beam-columns with stochastic properties, Journal of Reinforced Plastics and Composites 24(5), March 2005, 513–543.

[46] Vengallatore, S., Analysis of thermoelastic damping in laminated composite micromechanical beam resonators, Journal of Micromechanics and Microengineering 15(12), 2005, 2398–2404.

[47] Lee, J., Lateral buckling analysis of thin-walled laminated composite beams with monosymmetric sections, Engineering Structures 28(14), December 2006, 1997–2009.

[48] Aydogdu, M., Thermal buckling analysis of cross-ply laminated composite beams with general boundary conditions, Composites Science and Technology 67(10), May 2007, 1096–1104.

[49] Kiral, Z. and Kiral, B. G., Dynamic analysis of a symmetric laminated composite beam subjected to a moving load with constant velocity, Journal of Reinforced Plastics and Composites 27(1), January 2008, 19–32.

[50] Li, J., Hua, H. and Shen, R., Dynamic stiffness analysis for free vibrations of axially loaded laminated composite beams, Composite Structures 84(1), June 2008, 87–89.

[51] Chai, G. B. and Yap, C. W., Coupling effects in bending, buckling and free vibration of generally laminated composite beams, Composites Science and Technology 68(7–8), June 2008, 1664–1670.

[52] Zhen, W. and Wanji, C., An assessment of several displacement-based theories for the vibration and stability analysis of laminated composite and sandwich beams, Composite Structures 84(4), August 2008, 337–349.

[53] Atlihan, G., Çallioğlu, H., Conkur, E. Ş., Topcu, M. and Yücel, U., Free vibration analysis of the laminated composite beams by using DQM, Journal of Reinforced Plastics and Composites 28(7), April 2009, 881–892.

[54] Kiral, B. G. and Kiral, Z., Effect of elastic foundation on the dynamic response of laminated composite beams to moving loads, Journal of Reinforced Plastics and Composites 28(8), April 2009, 913–935.

[55] Kiral, Z., Malgaca, L., Malgaca, M. and Kiral, B. G., Experimental investigation of the dynamic response of a symmetric laminated composite beam via laser vibrometry, Journal of Composite Materials 43(24), November 2009, 2943–2962.

[56] Chai, G. B., Yap, C. W. and Lim, T. M., Bending and buckling of a generally laminated composite beam-column, Proceedings of the Institution of Mechanical Engineers, Part L: Journal of Materials: Design and Applications 224(1), 01 January 2010, 1–7.

[57] Campbell, F. C., Introduction to Composite Materials, ASM International, #05287G documents/10192/1849770/05287G, November 2010, 599 pp.

[58] Baghani, M., Jafari-Talookolaei, R. A. and Salarieh, H., Large amplitudes free vibrations and post-buckling analysis of unsymmetrically laminated composite beams on nonlinear elastic foundation, Applied Mathematical Modelling 35(1), January 2011, 130–138.

[59] Lezgy-Nazargah, M., Shariyat, M. and Beheshti-Aval, S. B., A refined high-order global-local theory for finite element bending and vibration analyses of laminated composite, Acta Mechanica 217(3–4), March 2011, 219–242.

[60] Jafari-Talookolaei, R. A., Salarieh, H. and Kargarnovin, M. H., Analysis of large amplitude free vibrations of unsymmetrically laminated composite beams on a nonlinear elastic foundation, Acta Mechanica 219(1–2), June 2011, 65–75.

[61] Vidal, P. and Polit, O., A sine finite element using a zig-zag function for the analysis of laminated composite beams, Composites Part B: Engineering 42(6), September 2011, 1671–1682.

[62] Jun, L. and Hongxing, H., Free vibration analyses of axially loaded laminated composite beams based on higher-order shear deformation theory, Meccanica 46(6), December 2011, 1299–1317.

[63] Vosoughi, A. R., Malekzadeh, P., Banan, M. R. and Banan, M. R., Thermal buckling and postbuckling of laminated composite beams with temperature-dependent properties, International Journal of Non-Linear Mechanics 47(3), April 2012, 96–102.

[64] Jafari-Talookolaei, R. A., Abedi, M., Kargarnovin, M. H. and Ahmadian, M. T., An analytical approach for the free vibration analysis of generally laminated composite beams with shear effect and rotary inertia, International Journal of Mechanical Sciences 65(1), December 2012, 97–104.

[65] Kargarnovin, M. H., Ahmadian, M. T., Jafari-Talookolaei, R.-A. and Abedi, M., Semi-analytical solution for the free vibration analysis of generally laminated composite Timoshenko beams with single delamination, Composites Part B: Engineering 45(1), February 2013, 587–600.

[66] Kim, N.-I. and Lee, J., Lateral buckling of shear deformable laminated composite I-beams using the finite element method, International Journal of Mechanical Sciences 68, March 2013, 246–257.

[67] Erkliğ, A., Yeter, E. and Bulut, M., The effects of cut-outs on lateral buckling behavior of laminated composite beams, Composite Structures 104, October 2013, 54–59.

[68] Fu, Y., Wang, J. and Hu, S., Analytical solutions of thermal buckling and postbuckling of symmetric laminated composite beams with various boundary conditions, Acta Mechanica 225(1), January 2014, 13–29.

[69] Li, J., Wu, Z., Kong, X., Li, X. and Wu, W., Comparison of various shear deformation theories for free vibration of laminated composite beams with general lay-ups, Composite Structures 108, February 2014, 767–778.

[70] Abadi, M. M. and Daneshmehr, A. R., An investigation of modified couple stress theory in buckling analysis of micro composite laminated Euler–Bernoulli and Timoshenko beams, International Journal of Engineering Science 75, February 2014, 40–53.

[71] Huo, J. L., Li, X., Kong, X. and Wu, W., Vibration analyses of laminated composite beams using refined higher-order shear deformation theory, International Journal of Mechanics and Materials in Design 10(1), March 2014, 43–52.

[72] Lanc, D., Turkalj, G. and Pesic, I., Global buckling analysis model for thin-walled composite laminated beam type structures, Composite Structures 111, May 2014, 371–380.

[73] Mohandes, M. and Ghasemi, A. R., Finite strain analysis of nonlinear vibrations of symmetric laminated composite Timoshenko beams using generalized differential quadrature method, Journal of Vibration and Control, 30 June 2014. doi:10.1177/1077546314538301.

[74] Kuehn, T., Pasternak, H. and Mittelstedt, C., Local buckling of shear-deformable laminated composite beams with arbitrary cross-sections using discrete plate analysis, Composite Structures 113, July 2014, 236–248.

[75] Carrera, E., Filippi, M. and Zappino, E., Free vibration analysis of laminated beam by polynomial, trigonometric, exponential and zig-zag theories, Journal of Composite Materials 48(19), August 2014, 2299–2316.

[76] Panahandeh-Shahraki, D., Mirdamadi, H. R. and Vaseghi, O., Thermoelastic buckling analysis of laminated piezoelectric composite plates, International Journal of Mechanics and Materials in Design, 11, October 2014, 371. doi:10.1007/s10999-014-9284-8.

[77] Sahoo, R. and Singh, B. N., A new trigonometric zigzag theory for buckling and free vibration analysis of laminated composite and sandwich plates, Composite Structures 117, November 2014, 316–332.

[78] Qu, Y., Wu, S., Li, H. and Meng, G., Three-dimensional free and transient vibration analysis of composite laminated and sandwich rectangular parallelepipeds: Beams, plates and solids, Composites Part B: Engineering 73, May 2015, 96–110.

[79] Li, Z.-M. and Qiao, P., Buckling and postbuckling behavior of shear deformable anisotropic laminated beams with initial geometric imperfections subjected to axial compression, Engineering Structures 85(15), February 2015, 277–292.

[80] Raja, S., Rohwer, K. and Rose, M., Piezothermoelastic modeling and active vibration control of laminated composite beams, Journal of Reinforced Plastics and Composites 10(11), November 1999, 890–899.

[81] Choi, S., Lee, J. J., Seo, D. C. and Choi, S. W., The active buckling control of laminated composite beams with embedded shape memory alloy wires, Composite Structures 47(1–4), December 1999, 679–686.

[82] Jeon, B. S., Lee, J. J., Kim, J. K. and Huh, J. S., Low velocity impact and delamination buckling behavior of composite laminates with embedded optical fibers, Smart Materials and Structures 8(1), 1999, 41–48.

[83] Lee, H. J. and Lee, J. J., A numerical analysis of the buckling and post-buckling behavior of laminated composite shells with embedded shape memory alloy wire actuators, Smart Materials and Structures 9(6), 2000, 780–788.

[84] Sun, B. and Huang, D., Vibration suppression of laminated composite beams with a piezo-electric damping layer, Composite Structures 53(4), September 2001, 437–447.

[85] Subramanian, P., Vibration suppression of symmetric laminated composite beams, Smart Materials and Structures 11(6), 2002, 880–886.

[86] Ray, M. C. and Mallik, N., Active control of laminated composite beams using a piezoelectric fiber reinforced composite layer, Smart Materials and Structures 13(1), 2004, 146–153.

[87] Pradhan, S. C. and Reddy, J. N., Vibration control of composite shells using embedded actuating layers, Smart Materials and Structures 13(5), 2004, 1245–1258.

[88] Zabihollah, A., Ganesan, R. and Sedaghati, R., Sensitivity analysis and design optimization of smart laminated beams using layerwise theory, Smart Materials and Structures 15(6), 2006, 1775–1785.

[89] Zhou, H.-M. and Zhou, Y.-H., Vibration suppression of laminated composite beams using actuators of giant magnetostrictive materials, Smart Materials and Structures 16(1), 2007, 198–207.

[90] Zabihollah, A., Sedagahti, R. and Ganesan, R., Active vibration suppression of smart laminated beams using layerwise theory and an optimal control strategy, Smart Materials and Structures 16(6), 2007, 2190–2202.

[91] Foda, M. A., Almajed, A. A. and El Madany, M. M., Vibration suppression of composite laminated beams using distributed piezoelectric patches, Smart Materials and Structures 19(11), 2010, article ID 115018.

[92] Sarangi, S. K. and Ray, M. C., Smart damping of geometrically nonlinear vibrations of laminated composite beams using vertically reinforced 1–3 piezoelectric composites, Smart Materials and Structures 19(7), 2010, article ID 075020.

[93] Asadi, H., Bodaghi, M., Shakeri, M. and Aghdam, M. M., An analytical approach for nonlinear vibration and thermal stability of shape memory alloy hybrid laminated composite beams, European Journal of Mechanics – A/Solids 42, November–December 2013, 454–468.

[94] Gei, M., Springhetti, R. and Bortot, E., Performance of soft dielectric laminated composites, Smart Materials and Structures 22(10), 2013, article ID 104014.

[95] Mareishi, S., Rafiee, M., He, X. Q. and Liew, K. M., Nonlinear free vibration, postbuckling and nonlinear static deflection of piezoelectric fiber-reinforced laminated composite beams, Composites Part B: Engineering 59, March 2014, 123–132.

2.3 Appendix I

2.3.1 Appendix A

Table I-1: Typical properties of mostly used reinforced continuous fibers.[11]

Material	Trade name	Density, ρ (kg/m^2)	Typical fiber diameter (µm)	Young's modulus, E (GPa)	Tensile strength (GPa)
α-Al$_2$O$_3$ (aluminum oxide)	FP (US)	3960	20	385	1.8
Al$_2$O$_3$ + SiO$_2$ + B$_2$O$_3$ (Mullite)	Nextel480 (USA)	3050	11	224	2.3
Al$_2$O$_3$ + SiO$_2$ (alumina-silica)	Altex (Japan)	3300	10–15	210	2.0
Boron (CVD* on tungsten)	VMC (Japan)	2600	140	410	4.0
Carbon (PAN** precursor)	T300 (Japan)	1800	7	230	3.5
Carbon (PAN** precursor)	T800 (Japan)	1800	5.5	295	5.6
Carbon (pitch*** precursor)	Thorne IP755 (USA)	2060	10	517	2.1
SiC (+O) (silicon carbide)	Nicalon (Japan)	2600	15	190	2.5–3.3
SiC (low O) (silicon carbide)	Hi-Nicalon (Japan)	2740	14	270	2.8
SiC (+O + Ti) (silicon carbide)	Tyranno (Japan)	2400	9	200	2.8
SiC (monofilament) (silicon carbide)	Sigma	3100	100	400	3.5
E-glass (silica)		2500	10	70	1.5–2.0

11 The data was taken from: Harris, B., *Engineering composite materials*, The Institute of Materials, London, UK, 1999, 193p, and Jones, R. M., *Mechanics of composite materials*, 2nd edition, Taylor & Francis, Philadelphia, PA 19106, USA, 1999, 519p.

Table I-1 (continued)

Material	Trade name	Density, ρ (kg/m^2)	Typical fiber diameter (µm)	Young's modulus, E (GPa)	Tensile strength (GPa)
E-glass (silica)		2500	10	70	1.5–2.0
Quartz (silica)		2200	3–15	80	3.5
Aromatic polyamide	Kevlar 49 (USA)	1500	12	130	3.6
Polyethylene (UHMW)+	Spectra 100 (USA)	970	38	175	3.0
High carbon steel	E.g., piano wire	7800	250	210	2.8
Aluminum	electrical wire	2680	1670	75	0.27
Titanium	wire	4700	250	115	0.434

*CVD, chemical vapor deposition.
**PAN, polyacrylonitrile. About 90% of the carbon fibers produced are made from PAN.
***pitch represents a viscoelastic material that is composed of aromatic hydrocarbons. Pitch is produced via the distillation of carbon-based materials, such as plants, crude oil and coal,
+ UHMW, ultra-high-molecular-weight *polyethylene* (or polyethene, the most common plastic produced in the world) is a subset of the thermoplastic polyethylene.

3 Piezoelectricity

3.1 Constitutive Equations

Arthur R. von Hippel, in his seminal book, *Dielectric Materials and Applications* (Artech House Publishers, December 19, 1995, 485 pp.), defined a dielectric material as: "Dielectrics . . . are not a narrow class of so-called insulators, but the broad expanse of non-metals considered from the standpoint of their interaction with electric, magnetic, or electromagnetic fields. Thus, we are concerned with gases as well as with liquids and solids, and with the storage of electric and magnetic energy as well as its dissipation."

A dielectric material is an electrical insulator that can be polarized by an applied electric field (Figure 3.1). When a dielectric is placed in an electric field, electric charges do not flow through the material as they do in a conductor, but only slightly shift from their average equilibrium positions causing dielectric polarization. Because of dielectric polarization, positive charges are displaced toward the field and negative charges shift in the opposite direction. This creates an internal electric field that reduces the overall field within the dielectric itself. If a dielectric is composed of weakly bonded molecules, those molecules not only become polarized, but also reorient, so that their symmetry axes align to the field.[1]

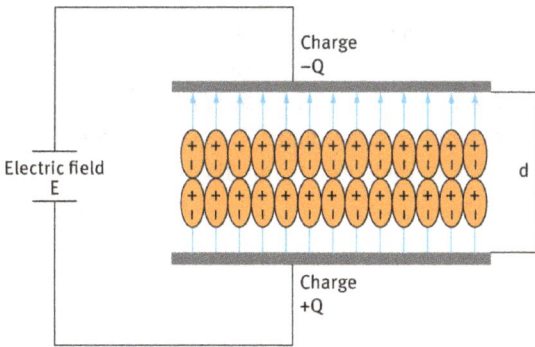

Figure 3.1: A schematic view of a dielectric material.

3.1.1 Classification of Dielectric Materials

All dielectric materials, when subjected to an external electric field, undergo change in dimensions. This is due to the displacements of positive and negative charges within the material. A dielectric crystal lattice may be considered to be made up of

---

1 Taken from Wikipedia, the free encyclopedia, en.wikipedia.org/wiki/Dielectric

https://doi.org/10.1515/9783110726701-003

cations (atoms that have lost an electron to become positively charged) and anions (atoms that have gained electrons to become negatively charged) connected by springs (interionic chemical bonds). When an external electric field is applied to the material, the cations get displaced in the direction of the electric field and the anions get displaced in the opposite direction, resulting in net deformation of the material. The change in dimension may be very small or may be quite significant, depending on the crystal class to which the dielectric belongs.

Among the total of 32 crystal classes, 11 are centrosymmetric (i.e., possess a center of symmetry or inversion center) and 21 are noncentrosymmetric (do not possess a center of symmetry).

When a dielectric material, possessing a center of symmetry, is subjected to an external electric field, due to the symmetry (inversion center), the movements of cations and anions are such that the extension and contraction get cancelled between neighboring springs (chemical bonds) and the net deformation in the crystal is ideally nil. As the chemical bonds are not perfectly harmonic and due to the anharmonicity of the bonds, second-order effects would appear, leading to a small net deformation of the lattice. The deformation in this case is proportional to the square of the electric field. That is, the deformation is independent of the direction of the applied electric field. The effect is called the electrostrictive effect. The anharmonic effect exists in all dielectrics, and so it can be said that all dielectrics are electrostrictive.

When a dielectric material belonging to a noncentrosymmetric class (except the octahedral class) is subjected to an external electric field, there will be asymmetric movement of the neighboring ions, resulting in a significant deformation of the crystal and the deformation is directly proportional to the applied electric field. These materials exhibit an electrostrictive effect due to the anharmonicity of the bonds, but it is masked by the more significant asymmetric displacement. The materials are called piezoelectric materials. The classification of dielectric materials based on their response to external stimuli is shown in Figure 3.2.

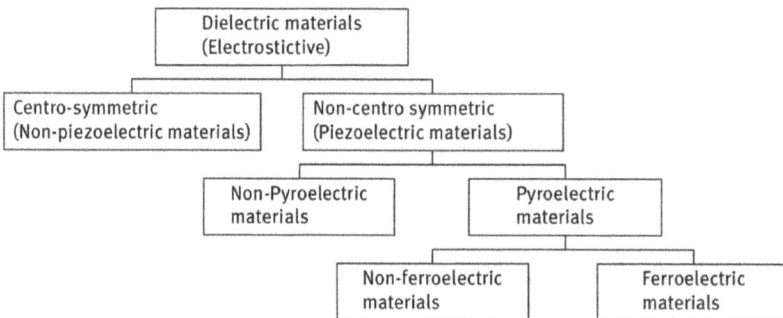

Figure 3.2: Classification of dielectric materials.

Each of these groups of materials exhibits certain special characteristics that make them important engineering materials. The materials belong to the class of smart materials because they exhibit inherent transducer characteristics.

3.1.2 Important Dielectric Parameters

3.1.2.1 Electric Dipole Moment

In an atom or a molecule, when the centers of positive and negative charges are separated by a certain distance d, the atom or the molecule possesses an *electric dipole moment* given by the following equation:

$$\vec{p} = q\vec{d} \tag{3.1}$$

where q is the charge and $\vec{d}$ is the distance between the positive and negative charge centers. $\vec{p}$ is a vector with the direction from the negative charge to the positive one, and its unit is Coulomb meter (C m).

3.1.2.2 Polar and Nonpolar Dielectric Materials

Dielectric materials may be classified as polar and nonpolar. In nonpolar dielectric materials, normally the atoms do not possess an electric dipole moment as the centers of positive and negative charges coincide. Typical examples of nonpolar dielectrics are oxygen, nitrogen, benzene, methane and so on. When these materials are subjected to an external electric field, the centers of positive and negative charges get separated and thus dipole moments are induced. The induced dipole moments disappear once the electric field is removed.

In polar dielectric materials, each atom or molecule possesses a dipole moment as the centers of positive and negative charges do not coincide. Typical examples of polar dielectrics are water, HCl, alcohol and NH_3. Most of the ceramics and polymers fall under this category. When an external electric field is applied to these materials, the electric dipoles tend to orient themselves in the direction of the field.

3.1.2.3 Electric Polarization

A polar dielectric material consists of a large number of atoms or molecules each possessing an electric dipole moment. The total or the net dipole moment of the dielectric material is the vector sum of all the individual dipole moments given by

$$\vec{P}_{total} = \sum_i \vec{p}_i \tag{3.2}$$

The term *electric polarization*, $\vec{P}$, is defined as the total dipole moment per unit volume and is given by

$$\vec{P} = \frac{\sum_i \vec{p}_i}{Vol.}$$ (3.3)

where Vol. is the volume of the material. The unit of $\vec{P}$ is (C/m^2) (Coulomb/meter^2) and it is sometimes called the *surface charge density*. $\vec{P}$ is a vector normal to the surface of the material.

Normally, in a polar dielectric, the individual electric dipoles are all randomly oriented and so the net polarization is zero. When an electric field is applied, the individual dipoles tend to orient themselves in the direction of the electric field and the material develops a finite polarization (see Figure 3.1). The polarization increases with an increase in the electric field and reaches saturation when all the dipole moments are oriented in the direction of the applied field.

3.1.2.4 Dielectric Displacement (Flux Density), Dielectric Constant and Electric Susceptibility

When an electric field E is applied to a dielectric material, the material develops a finite polarization P (induced polarization in nonpolar materials and orientation polarization in polar materials). The electric flux density $\vec{D}$ developed inside the material due to the external field $\vec{E}$ is given by

$$\vec{D} = \varepsilon_0 \vec{E} + \vec{P}$$ (3.4)

where ε_0 is the permittivity of free space.

The term $\vec{D}$ can also be expressed by the relation

$$\vec{D} = \bar{\varepsilon}\vec{E} = \varepsilon_0 \varepsilon_r \vec{E}$$ (3.5)

where ε_r is the *relative permittivity* or *dielectric constant* of the dielectric material. The units of D are (C/m^2) (same as that of P).

The dielectric constant ε_r can also be defined by the ratio

$$\varepsilon_r = \frac{D}{\varepsilon_0 E}$$ (3.6)

Note that the term ε_r is a nondimensional quantity and is always greater than 1.

The polarization vector $\vec{P}$ is directly related to the applied electric field by the relation

$$\vec{P} = \varepsilon_0 \chi \vec{E}$$ (3.7)

where ε_0 is the permittivity of free space (see Equation (3.4)) and χ is called the electric susceptibility of the material.

From the above equations, we can obtain the relation between the dielectric constant ε_r and the electric susceptibility χ as

$$\varepsilon_r = 1 + \chi \tag{3.8}$$

3.1.2.5 The Piezoelectric Effect

Dielectric materials that belong to the class of noncentrosymmetric crystals are classified as piezoelectric materials. When these materials are subjected to an external electric field, there will be asymmetric displacements of anions and cations that cause a considerable net deformation of the crystal. The resulting strain is directly proportional to the applied electric field unlike electrostrictive materials in which the strain is proportional to the square of the electric field (E^2). The strain in a piezoelectric material might be of extensive or compressive nature, depending on the polarity of the applied field. This effect is called the *piezoelectric effect* or, to be more precise, *indirect piezoelectric effect*.

Piezoelectric materials exhibit another unique property; when they are subjected to external strain by applying pressure/stress, the electric dipoles in the crystal get oriented such that the crystal develops positive and negative charges on opposite faces, resulting in an electric field across the crystal. This is exactly the reverse of the above mentioned indirect piezoelectric effect. The brothers, Jacques and Pierre Curie, first observed this effect in quartz crystal in 1880 and called this effect *piezoelectricity*. The source of the word piezoelectricity comes from Greek and means electricity resulting from pressure (*piesi* means pressure in Greek). This effect is called the *direct piezoelectric effect*.

The direct and indirect piezoelectric effects are schematically illustrated in Figures 3.3 and 3.4. In the direct piezoelectric effect, when a poled piezoelectric material is subjected to tensile stress, in the direction parallel to the poling direction, a positive voltage is generated across its faces (Figure 3.3b). When the material is subjected to a compressive stress in the direction of the poling, a negative voltage is generated across its faces (Figure 3.3c). In the indirect piezoelectric effect, when an external voltage is applied to the material, the material gets extended if the polarity of the voltage is the same as that of the field applied during poling (Figure 3.4b) and, when the voltage is applied in the reverse direction, the material is compressed (Figure 3.4c).

Figure 3.5 shows the effect of an alternating field on a poled piezoelectric material. The alternating field makes the material to extend and contract alternately at the same frequency as the applied field. The vibration produces an acoustic field (sound or ultrasonic field) in the vicinity of the material. This effect is used for the generation of acoustic fields.

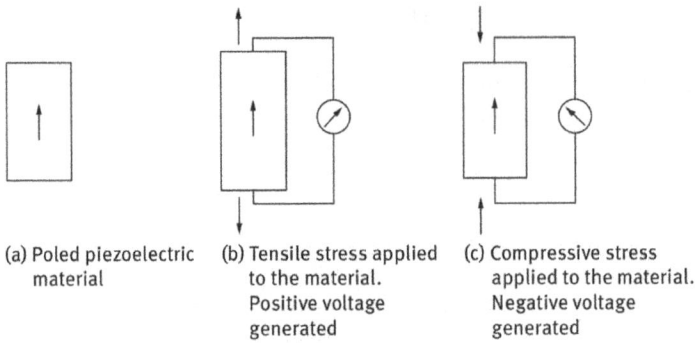

(a) Poled piezoelectric material

(b) Tensile stress applied to the material. Positive voltage generated

(c) Compressive stress applied to the material. Negative voltage generated

Figure 3.3: Direct piezoelectric effect: (a) poled piezoelectric material. (b) When tensile stress is applied to the material, the material develops voltage across its face with the same polarity as the poling voltage. (c) When a compressive stress is applied to the material, the material develops voltage with polarity opposite to that of the poling voltage.

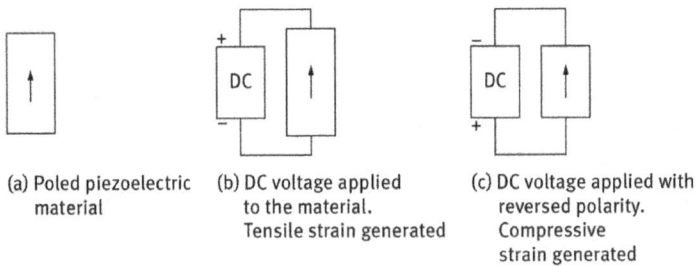

(a) Poled piezoelectric material

(b) DC voltage applied to the material. Tensile strain generated

(c) DC voltage applied with reversed polarity. Compressive strain generated

Figure 3.4: Indirect piezoelectric effect: (a) poled piezoelectric material. (b) When a DC field is applied with the same polarity as the poling field, the material develops tensile strain. (c) When a DC field is applied in the reverse direction, the material develops compressive strain.

The behavior described in Figure 3.4 is for a DC applied voltage. In case an AC voltage is applied (as shown schematically in Figure 3.5), the material would contract and extend alternately according to the frequency of the applied AC voltage and will produce an acoustic field in its vicinity.

The direct and the indirect piezoelectric effects have many applications as the effects involve conversion of mechanical energy into electrical energy and vice versa. The applications include generation and detection of ultrasonic waves, pressure sensors and actuators. Ultrasonic is extensively used both in engineering and medical fields. In engineering, it is used in nondestructive testing of materials (NDT), underwater acoustics (sonar), ultrasonic drilling, energy harvesting and so on; and in medical fields, it is used for diagnosis (sonography), therapy (drug delivery) and surgery. As sensors and actuators, they have a wide variety of applications in both engineering and medical fields.

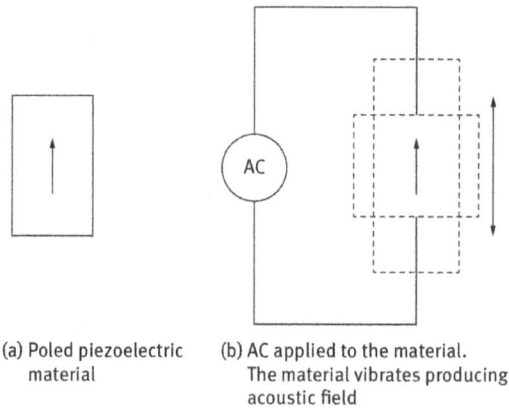

(a) Poled piezoelectric
   material

(b) AC applied to the material.
   The material vibrates producing
   acoustic field

Figure 3.5: Effect of AC field on a piezoelectric material: (a) poled piezoelectric material; (b) AC field is applied to the material. The material is extended and contracted alternately; that is, the material vibrates producing an acoustic field in the vicinity.

3.1.2.6 The Piezoelectric Coefficients

Piezoelectric materials convert mechanical energy to electrical energy (the direct piezoelectric effect) and electrical energy into mechanical energy (the indirect piezoelectric effect).

In the direct piezoelectric effect, the input is mechanical energy and the output is electrical energy. Mechanical input can be in the form of external stress (σ) or strain (ε). The electrical output is in the form of surface charge density (D or P), electric field (E) or voltage (V) (see Figure 3.6).

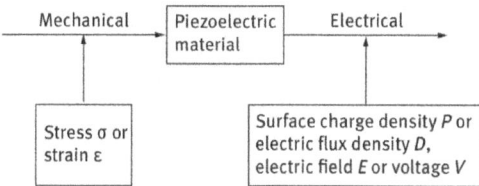

Figure 3.6: The direct piezoelectric effect: input is mechanical and output is electrical.

In the indirect piezoelectric effect, the input is electrical energy and the output is mechanical energy. The electrical input may be in the form of surface charge density (P or D) or electric field (E) or voltage (V), and the mechanical output is in the form of strain (ε) or stress (σ) on the material (see Figure 3.7).

The parameters that describe the sensitivity of a piezoelectric material are the piezoelectric coefficients that relate the input and output parameters. The various piezoelectric coefficients are next defined.

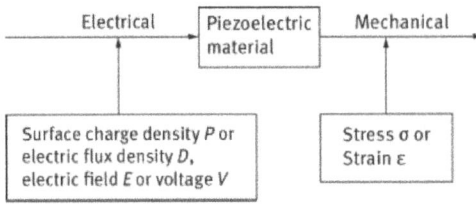

Figure 3.7: The indirect piezoelectric effect: input is electrical and output is mechanical.

In the direct piezoelectric effect, the equations that relate the mechanical input strain ε to the electrical output (in the form of D and E) are

$$D = e\varepsilon \tag{3.9}$$

$$E = h\varepsilon \tag{3.10}$$

and the equations that relate the mechanical input stress σ to the electrical output (in the form of D and E) are

$$D = d\sigma \tag{3.11}$$

$$E = g\sigma \tag{3.12}$$

where the terms e, h, d and g are the *piezoelectric coefficients* that would describe the direct piezoelectric effect.

The piezoelectric coefficients that would describe the indirect piezoelectric effect are denoted by e^*, d^*, h^* and g^*. They are defined by the following relations:

$$\sigma = e^* E \tag{3.13}$$

$$\varepsilon = d^* E \tag{3.14}$$

$$\sigma = h^* D \tag{3.15}$$

$$\varepsilon = g^* D \tag{3.16}$$

Equations (3.13) and (3.14) relate the input electric field (E) to the mechanical output (σ and ε), and Equations (3.15) and (3.16) relate the input charge density (D) to the mechanical output (σ and ε).

Using the relation between D and E (Equation (1.5)), we get the following relations between the various piezoelectric coefficients:

$$h = \frac{e}{\varepsilon}; \quad h^* = \frac{e^*}{\varepsilon}; \quad g = \frac{d}{\varepsilon}; \quad g^* = \frac{d^*}{\varepsilon} \tag{3.17}$$

The piezoelectric effect is a transient effect, which means that the observed parameter is not an absolute value but it is the change in the parameter. In the direct piezoelectric effect, a change in strain $\partial\varepsilon$ (or in the stress $\partial\sigma$) causes a change in the polarization ∂D (or a change in the electric field ∂E), and in the indirect piezoelectric effect, a change in the applied field ∂E (or polarization ∂D) causes a change in the strain $\partial \varepsilon$ (or in the stress $\partial \sigma$). Since D and P are related by Equation (3.4), the variation in D, ∂D, can be replaced by the variation in P, ∂P, for a constant E.

The various piezoelectric coefficients defined earlier are more appropriately defined by the following partial derivatives, which are presented for the direct piezoelectric coefficients as

$$d = \left(\frac{\partial D}{\partial X}\right)_E ; \quad g = -\left(\frac{\partial E}{\partial X}\right)_D ; \quad e = \left(\frac{\partial D}{\partial x}\right)_E ; \quad h = -\left(\frac{\partial E}{\partial x}\right)_D \tag{3.18}$$

and for the indirect piezoelectric coefficients as

$$d^* = \left(\frac{\partial \varepsilon}{\partial E}\right)_\sigma ; \quad g^* = \left(\frac{\partial \varepsilon}{\partial D}\right)_\sigma ; \quad e^* = -\left(\frac{\partial \sigma}{\partial E}\right)_\varepsilon ; \quad h^* = -\left(\frac{\partial \sigma}{\partial D}\right)_\varepsilon \tag{3.19}$$

where the letter written outside the parentheses describes that the derivation was performed for the constant value of the letter (σ = stress; ε = strain; E = electrical field, and D = electric displacement).

From thermodynamics, it can be proved that the following relationships are true:

$$d = d^*, g = g^*, e = e^* \quad \text{and} \quad h = h^* \tag{3.20}$$

The definitions of the various piezoelectric coefficients and their respective units are summarized in Table 3.1. Piezoelectric materials are characterized by the following parameters:
- Piezoelectric coefficients: d, g, e, h
- Electrical parameter: permittivity $\bar{\varepsilon}$
- Elastic parameters: compliance constants defined by the letter s and/or stiffness constants, c

Table 3.1: Piezoelectric coefficients: definitions and units.

Piezoelectric coefficient[1]	Name	Definition	Mathematical definition	Unit
d	Piezoelectric charge or strain constant	$\dfrac{\text{Polarization}}{\text{Stress}}$	$d = \left(\dfrac{\partial D}{\partial X}\right)_E$	$\left[\dfrac{C}{N}\right]$
g	Piezoelectric voltage constant	$\dfrac{\text{Electric field}}{\text{stress}}$	$g = -\left(\dfrac{\partial E}{\partial X}\right)_D$	$\left[\dfrac{V \cdot m}{N}\right]$
e	Piezoelectric constant relating strain to polarization	$\dfrac{\text{Polarization}}{\text{Strain}}$	$e = \left(\dfrac{\partial D}{\partial x}\right)_E$	$\left[\dfrac{C}{m^2}\right]$

Table 3.1 (continued)

Piezoelectric coefficient[1]	Name	Definition	Mathematical definition	Unit
h	Piezoelectric constant relating strain to electric field	Electric field / Strain	$h = -\left(\dfrac{\partial E}{\partial x}\right)_D$	$\left[\dfrac{V}{m}\right]$
d	Piezoelectric charge or strain constant	Strain / Electric field	$d^{*} = \left(\dfrac{\partial \varepsilon}{\partial E}\right)_\sigma$	$\left[\dfrac{m}{V}\right]$
g	Piezoelectric constant relating strain to polarization	Strain / Polarization	$g^{*} = \left(\dfrac{\partial \varepsilon}{\partial D}\right)_\sigma$	$\left[\dfrac{m^2}{C}\right]$
e	Piezoelectric constant relating stress to electric field	Stress / Electric field	$e^{*} = -\left(\dfrac{\partial \sigma}{\partial E}\right)_\varepsilon$	$\left[\dfrac{N}{V \cdot m}\right]$
h	Piezoelectric constant relating stress to polarization	Stress / Polarization	$h^{*} = -\left(\dfrac{\partial \sigma}{\partial D}\right)_\varepsilon$	$\left[\dfrac{N}{C}\right]$

[1]Note: It is easy to verify that the units of the four pairs: d, d^{*}; g, g^{*}; e, e^{*} and h, h^{*} are the same (this can be done using the following identity: $V = N \cdot m/C$).

The governing equation for the direct piezoelectric effect would depend on the mechanical and electrical boundary conditions.

If the material is under "free" conditions where "free" refers to a state when the material is able to change dimensions with the applied field, this will be a normal condition (constant stress condition) with the following governing equation:

$$D = d\sigma + \bar{\varepsilon}^\sigma E \qquad (3.21)$$

where $\bar{\varepsilon}^\sigma$ is the permittivity at constant stress σ.

For the case of a material under clamped condition, "clamped" refers to either a condition where the material is physically clamped or a condition in which it is driven at a sufficiently high frequency at which the device cannot respond to the changing electric field. This is called the constant strain condition with the following governing equation:

$$D = e\sigma + \bar{\varepsilon}^\varepsilon E \qquad (3.22)$$

where $\bar{\varepsilon}^\varepsilon \varepsilon$ is the permittivity at the constant strain ε.

The governing equations for the indirect piezoelectric effect will have the following form (in the case of a short-circuited material, namely for a constant electric field):

$$\varepsilon = s^E \sigma + dE$$

or $\qquad (3.23)$

$$\sigma = c^E \varepsilon - eE$$

where s^E and c^E are the elastic compliance constant and the elastic stiffness constant, respectively, under the constant electric field.

In case the material is under the open-circuit condition (constant charge density), the governing equations for the indirect piezoelectric effect will have the following form:

$$\varepsilon = s^D \sigma + g$$

or
(3.24)

$$\sigma = c^D \varepsilon - hD$$

where s^D and c^D are the elastic compliance constant and the elastic stiffness constant, respectively, under constant charge density (constant electric displacement).

Sometimes, Equations (3.23) and (3.24) are written with thermal expansion contribution to yield

$$\varepsilon = s^E \sigma + dE + \alpha \Delta T$$

or
(3.25)

$$\sigma = c^E \varepsilon - eE + \alpha \Delta T$$

$$\varepsilon = s^D \sigma + gD + \alpha \Delta T$$

or
(3.26)

$$\sigma = c^D \varepsilon - hD + \alpha \Delta T$$

where ΔT is the temperature increase (or decrease) and α is a vector containing the thermal coefficients of expansion in the three directions 1, 2 and 3 of coordinate system (α_1, α_2 and α_3, 0,0,0).

3.1.2.7 The Tensor Form of the Piezoelectric Equations

The piezoelectric coefficients, the elastic constants and the permittivity terms are all tensors as they relate vectors and tensors.

The permittivity matrix, which relates the two vectors D and E, is a second-rank tensor. The elastic compliance constant, which relates the two second-rank tensors σ and ε, is a fourth-rank tensor. The piezoelectric coefficients relate stress/ strain (second-rank tensors) and electric field/ polarization (vectors), and so they are third-rank tensors. The system of coordinates is a right-handed Cartesian coordinate system with X, Y and Z axes being represented by 1, 2 and 3, respectively, and the rotations about the X, Y and Z axes are represented by 4, 5 and 6 as shown in Figure 3.8.

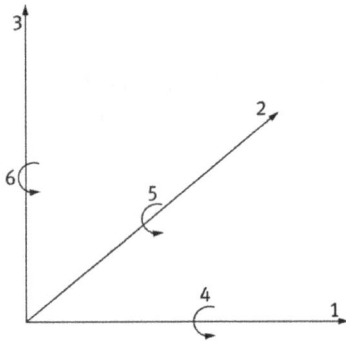

Figure 3.8: Right-handed Cartesian coordinate system. Directions 1, 2 and 3 represent the axes X, Y and Z, respectively, and 4, 5 and 6 represent rotations (anticlockwise) about the three axes X, Y and Z.

3.1.2.8 The Matrix of Piezoelectric Coefficients d

Consider the direct piezoelectric effect that relates the polarization D and the stress σ via the relationship $D = d\sigma$ (see also Equation (3.11)).

The stress $[\sigma]$ is a second-rank tensor with the following components:

$$[\sigma] = \begin{bmatrix} \sigma_{11} & \tau_{12} & \tau_{13} \\ \tau_{21} & \sigma_{22} & \tau_{23} \\ \tau_{31} & \tau_{32} & \sigma_{33} \end{bmatrix} \tag{3.27}$$

Since $\tau_{12} = \tau_{21}$, $\tau_{13} = \tau_{31}$, and $\tau_{23} = \tau_{32}$, $[\sigma]$ is symmetric with only six independent components.

D is a vector with the following components (where 1, 2 and 3 refer to the coordinate system shown in Figure 3.8)

$$\begin{Bmatrix} D_1 \\ D_2 \\ D_3 \end{Bmatrix} \tag{3.28}$$

For simplicity, a *reduced matrix notation* is used for the second-rank tensor σ_{ij}. The two indices i and j, each taking values from 1 to 3, are replaced by a single index that takes values from 1 to 6.

That is,

$$11 \equiv 1;\ 22 \equiv 2;\ 33 \equiv 3;\ 23 = 32 \equiv 4;\ 31 = 13 \equiv 5;\ 12 = 21 \equiv 6 \tag{3.29}$$

The subscripts 1, 2 and 3 indicate normal tensile or compressive stress (or strain) and 4, 5 and 6 indicate shear stress (or shear strain) – rotation about axes 1, 2 and 3, respectively (see also Figure 3.8). Thus, the following relationships hold:

$$\sigma_{11} = \sigma_1;\quad \sigma_{22} = \sigma_2;\quad \sigma_{33} = \sigma_3$$

$$\tau_{23} = \tau_{32} = \sigma_4$$

$$\tau_{31} = \tau_{13} = \sigma_5 \tag{3.30}$$

$$\tau_{12} = \tau_{21} = \sigma_6$$

Therefore, Equation (3.11) will be written in the following matrix form:

$$\begin{Bmatrix} D_1 \\ D_2 \\ D_3 \end{Bmatrix} = \begin{bmatrix} d_{11} & d_{12} & d_{13} & d_{14} & d_{15} & d_{16} \\ d_{21} & d_{22} & d_{23} & d_{24} & d_{25} & d_{26} \\ d_{31} & d_{32} & d_{33} & d_{34} & d_{35} & d_{36} \end{bmatrix} \begin{Bmatrix} \sigma_1 \\ \sigma_2 \\ \sigma_3 \\ \sigma_4 \\ \sigma_5 \\ \sigma_6 \end{Bmatrix} \tag{3.31}$$

Thus, the piezoelectric coefficient d is represented by a 3×6 matrix.

In the indirect piezoelectric effect, the equation that relates the electric field E with the stress σ is written in the matrix form as

$$\begin{Bmatrix} \sigma_1 \\ \sigma_2 \\ \sigma_3 \\ \sigma_4 \\ \sigma_5 \\ \sigma_6 \end{Bmatrix} = \begin{bmatrix} e_{11} & e_{12} & e_{13} \\ e_{21} & e_{22} & e_{23} \\ e_{31} & e_{32} & e_{33} \\ e_{41} & e_{42} & e_{43} \\ e_{51} & e_{52} & e_{53} \\ e_{61} & e_{62} & e_{63} \end{bmatrix} \begin{Bmatrix} E_1 \\ E_2 \\ E_3 \end{Bmatrix} \tag{3.32}$$

Thus, in the indirect piezoelectric effect, the piezoelectric coefficient e is represented by a 6×3 matrix. See Refs. {1–12} for additional information.

3.1.2.9 Mechanical Parameters: Elastic Compliance and Elastic Stiffness Constants

The main mechanical parameters of interest in piezoelectric materials are elastic compliance and elastic stiffness constants. These two constants relate the two second-rank tensors stress and strain, and so they are fourth-rank tensors.

Elastic compliance constant s is defined by the relation

$$\varepsilon_i = s_{ij}\sigma_j \tag{3.33}$$

Elastic stiffness constant c is defined by the relation

$$\sigma_i = c_{ij}\varepsilon_j \tag{3.34}$$

where the subscripts i and j each take values of 1 to 6.

3.1.2.10 Dielectric Parameter: Permittivity

The dielectric parameter of interest in piezoelectric materials is the permittivity $\bar{\varepsilon}$, which relates the vectors D and E and so it is a second-rank tensor:

$$D_i = \bar{\varepsilon}_{ij} E_j \tag{3.35}$$

The subscripts i and j each take values 1, 2 and 3. The tensor form of the equation is

$$
\begin{Bmatrix} D_1 \\ D_2 \\ D_3 \end{Bmatrix} =
\begin{bmatrix} \bar{\varepsilon}_{11} & \bar{\varepsilon}_{12} & \bar{\varepsilon}_{13} \\ \bar{\varepsilon}_{21} & \bar{\varepsilon}_{22} & \bar{\varepsilon}_{23} \\ \bar{\varepsilon}_{31} & \bar{\varepsilon}_{32} & \bar{\varepsilon}_{33} \end{bmatrix}
\begin{Bmatrix} E_1 \\ E_2 \\ E_3 \end{Bmatrix} \tag{3.36}
$$

Thus, the permittivity $\bar{\varepsilon}$ is a 3×3 tensor.

One should note that due to the specific crystal structure of one of the most common piezoelectric materials, PZT, and the assumption that its piezoceramic material is isotropic in a plane perpendicular to the poling direction, several terms in the various matrices, presented above would become equal to zero or equal to each other. The compliance matrix for PZT for the constant electric field E (Equation (3.25)) will be

$$
s^E =
\begin{bmatrix}
\frac{1}{E_1} & -\frac{v_{12}}{E_1} & -\frac{v_{13}}{E_1} & 0 & 0 & 0 \\[6pt]
-\frac{v_{12}}{E_1} & \frac{1}{E_1} & -\frac{v_{13}}{E_1} & 0 & 0 & 0 \\[6pt]
-\frac{v_{13}}{E_1} & -\frac{v_{13}}{E_1} & \frac{1}{E_3} & 0 & 0 & 0 \\[6pt]
0 & 0 & 0 & \frac{2(1+v_{13})}{E_3} & 0 & 0 \\[6pt]
0 & 0 & 0 & 0 & \frac{2(1+v_{13})}{E_3} & 0 \\[6pt]
0 & 0 & 0 & 0 & 0 & \frac{2(1+v_{12})}{E_1}
\end{bmatrix} \tag{3.37}
$$

while its stiffness matrix for the constant electric field E (Equation (3.25)) will have the following form:

$$
c^E =
\begin{bmatrix}
\frac{1-v_{13}^2}{E_1 E_3 \theta} & \frac{v_{12}+v_{13}^2}{E_1 E_3 \theta} & \frac{v_{13}+v_{12}v_{13}}{E_1 E_3 \theta} & 0 & 0 & 0 \\[6pt]
\frac{v_{12}+v_{13}^2}{E_1 E_3 \theta} & \frac{1-v_{13}^2}{E_1 E_3 \theta} & \frac{v_{13}+v_{12}v_{13}}{E_1 E_3 \theta} & 0 & 0 & 0 \\[6pt]
\frac{v_{13}+v_{12}v_{13}}{E_1 E_3 \theta} & \frac{v_{13}+v_{12}v_{13}}{E_1 E_3 \theta} & \frac{1-v_{13}^2}{E_3^2 \theta} & 0 & 0 & 0 \\[6pt]
0 & 0 & 0 & \frac{E_3}{2(1+v_{13})} & 0 & 0 \\[6pt]
0 & 0 & 0 & 0 & \frac{E_3}{2(1+v_{13})} & 0 \\[6pt]
0 & 0 & 0 & 0 & 0 & \frac{E_1}{2(1+v_{12})}
\end{bmatrix} \tag{3.38}
$$

where

$$\theta = \frac{(1+v_{12})(1-v_{12}-2v_{13}^2)}{E_1^2 E_3} \qquad (3.39)$$

The piezo coefficients matrix for PZT will then be :

$$d = \begin{bmatrix} 0 & 0 & d_{31} \\ 0 & 0 & d_{32} \\ 0 & 0 & d_{33} \\ 0 & d_{24} & 0 \\ d_{15} & 0 & 0 \\ 0 & 0 & 0 \end{bmatrix} \qquad (3.40)$$

3.1.2.11 Strain versus Electric Field in Piezoelectric Materials

A typical strain versus electric field curve of a piezoelectric material, which is also ferro-electric (PZT), is shown in Figure 3.9. The strain in both longitudinal and transverse modes both exhibit hysteresis effects. The strain increases with the applied electric field initially linearly and later gets saturated. The strain is of the order of about 10^{-3} for an electric field in the range 1–1.5 kV for PZT, which is the best-known piezoelectric ceramic, used widely for transducer applications. For other piezoelectric materials, it is much less. When the applied electric field is gradually reduced, the strain does not follow the same curve. The change in strain lags behind the change in electric field. When the electric field is zero, a remnant strain is observed in the material. If the electric field is increased in the reverse direction, the strain becomes zero at a particular negative electric field. If the electric field is now increased in the same direction, the strain increases and attains saturation. Thus, the strain versus electric field is a symmetric curve as shown in Figure 3.9. A closed loop is formed when the electric field in the reverse direction is again brought to zero and increased in the positive direction as shown in Figure 3.9.

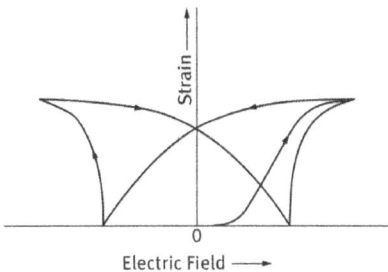

Figure 3.9: Strain versus electric field for a typical piezoelectric material.

3.1.2.12 Piezoelectric Coupling Coefficient k

Piezoelectric coupling coefficient is a measure of the efficiency of a piezoelectric material as a transducer. It quantifies the ability of the piezoelectric material to convert one form of energy (mechanical or electrical) to the other form (electrical or mechanical). It is defined by

$$k^2 = \frac{\text{(Piezoelectric energy density stored in the material)}}{\text{(Electrical energy density)} \cdot \text{(Mechanical energy density)}} \qquad (3.41)$$

If the electrical energy density is W_e, and the mechanical energy density is W_m, then k^2 is written as

$$k^2 = \frac{(W_{em})^2}{W_e W_m} \qquad (3.42)$$

where W_{em} is the piezoelectric energy density.

The expression for k^2 can be obtained in terms of the piezoelectric coefficients (see Appendix A for a detailed derivation):

$$k^2 = \frac{d^2}{\varepsilon^\sigma s^E} \qquad (3.43)$$

The coupling coefficient is the ratio of useable energy delivered by the piezoelectric element to the total energy taken up by the element. The piezoelectric element manufacturers normally specify the theoretical k values that can be in the range of 30–75%. In practice, k values depend on the design of the device and the directions of the applied stimulus and the measured response.

The coupling coefficient is written with subscripts just like piezoelectric constants to denote the direction of the external stimulus and the direction of measurement. The various coupling coefficients are defined in Table 3.2.

Table 3.2: Piezoelectric coupling coefficients: definitions.

Notation	Definition
k_{33}	Coupling coefficient when the electric field is in the direction 3 and the mechanical vibrations are in the same direction, 3.
k_{31}	Coupling coefficient when the electric field is in the direction 3 and the mechanical vibrations are in the direction, 1.
k_t	Coupling coefficient for a thin disk in which the electric field is in the direction 3 (across the thickness of the disk along which the disk is poled) and mechanical vibrations are in the same direction, 3.
k_p	Coupling coefficient for a thin disk in which the electric field is in the direction 3 (across the thickness of the disk along which the disk is poled) and mechanical vibrations are along radial direction.

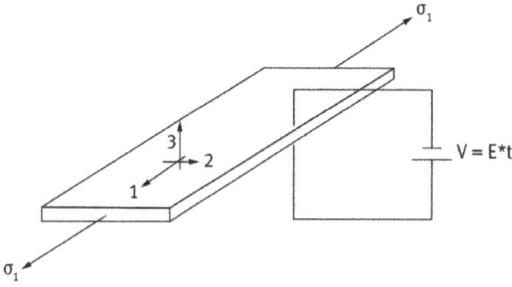

Figure 3.10: A piezoelectric element under uniaxial stress (t = thickness, E = electric field).

To understand the meaning of the coupling coefficients k_{ij}, let us use a simple one-dimensional analysis of the element presented in Figure 3.10. Equations (3.21) and (3.23) are written for a one-dimensional analysis as

$$\varepsilon_1 = s_{11}^E \sigma_1 + d_{31} E_3$$

and $\hfill$ (3.44)

$$D_3 = d_{31} \sigma_1 + \bar{\varepsilon}_{33}^\sigma E_3$$

Excluding from the second equation in Equation (3.44), the term E_3, we obtain

$$\varepsilon_1 = s_{11}^E \left[1 - \frac{d_{31}^2}{s_{11}^E \bar{\varepsilon}_{33}^\sigma} \right] \sigma_1 + \frac{d_{31}}{\bar{\varepsilon}_{33}^\sigma} D_3 \tag{3.45}$$

while rewriting Equation (3.26) for one-dimensional analysis, we get

$$\varepsilon_1 = s_{11}^D \sigma_1 + g_{31} D_3 \tag{3.46}$$

Equating the two equations, we obtain

$$s_{11}^D = s_{11}^E \left[1 - \frac{d_{31}^2}{s_{11}^E \bar{\varepsilon}_{33}^\sigma} \right]$$

and $\hfill$ (3.47)

$$g_{31} = \frac{d_{31}}{\bar{\varepsilon}_{33}^\sigma}$$

The first equation of Equation (3.47) illustrates the importance of boundary conditions for piezoelectric materials. It can be rewritten as a relationship between open circuit (s_{11}^D) and short circuit (s_{11}^E) as

$$s_{11}^D = s_{11}^E \left[1 - k_{31}^2 \right]$$

or

$$k_{31}^2 = \frac{d_{31}^2}{s_{11}^E \bar{\varepsilon}_{33}^\sigma}$$

(3.48)

Accordingly, the extensional coupling coefficient along the 3 direction, when the electrical field is also applied in the 3 direction will be

$$k_{33}^2 = \frac{d_{33}^2}{s_{33}^E \bar{\varepsilon}_{33}^\sigma}$$

(3.49)

For the shear mode in the 1–3 plane, the coefficient will be

$$k_{15}^2 = \frac{d_{15}^2}{s_{55}^E \bar{\varepsilon}_{11}^\sigma}$$

(3.50)

When using the direct piezoelectric effect, it is advised to use the sensor equations, as they are used to sense the piezoelectric output in the form of charge, q. Therefore, Equation (3.21) is rewritten for a PZT material as

$$D_k = d_{ki}\sigma_i + \bar{\varepsilon}_{kl}^\sigma E_l, \quad i = 1, 2, \ldots, 6; \quad k, l = 1, 2, 3$$

(3.51)

In a matrix form, Equation (3.51) will have the following form:

$$
\begin{Bmatrix} D_1 \\ D_2 \\ D_3 \end{Bmatrix} =
\begin{bmatrix} 0 & 0 & 0 & 0 & d_{15} & 0 \\ 0 & 0 & 0 & d_{15} & 0 & 0 \\ d_{31} & d_{31} & d_{33} & 0 & 0 & 0 \end{bmatrix}
\begin{Bmatrix} \sigma_1 \\ \sigma_2 \\ \sigma_3 \\ \sigma_4 \\ \sigma_5 \\ \sigma_6 \end{Bmatrix} +
\begin{bmatrix} \bar{\varepsilon}_{11}^\sigma & 0 & 0 \\ 0 & \bar{\varepsilon}_{22}^\sigma & 0 \\ 0 & 0 & \bar{\varepsilon}_{33}^\sigma \end{bmatrix}
\begin{Bmatrix} E_1 \\ E_2 \\ E_3 \end{Bmatrix}
$$

(3.52)

The output of a piezoelectric sensor is the charge which is related to the electric displacement by the following relationship:

$$q = \iint_A [D_1 \quad D_2 \quad D_3] \begin{Bmatrix} dA_1 \\ dA_2 \\ dA_3 \end{Bmatrix}$$

(3.53)

where dA_1, dA_2 and dA_3 are the components of the electrodes in the 2–3, 1–3 and 1–3 planes (note that for Figure 3.10 only D_3 will appear in the equation). Then, the voltage can be calculated using the following relationship:

$$V_{sensor} = \frac{q}{C_{sensor}} \tag{3.54}$$

where C_{sensor} (measured in Farad) is the capacitance of the piezoelectric sensor. In case the piezoelectric sensor is formed from a thin plate having the length l_s, width w_s and thickness t_s, the capacitance will be given as

$$C_{sensor} = \frac{\bar{\varepsilon}_{33}^\sigma l_s w_s}{t_s} \tag{3.55}$$

The indirect piezoelectric effect, presented by Equation (3.25) (see also Equations (3.37) and (3.40)), would then refer to the actuator equations. When talking about actuator performance two important terms are normally presented: *free displacement* and *blocked force*. The free displacement is defined as the maximum displacement of the piezoelectric effect for a given electric field without the external mechanical load. Its form will be obtained by zeroing the stress vector and the thermal contribution. Its form will then be given by Λ (free strain) which multiplied by the length of the piezoelectric sheet will provide the free displacement (for a PZT material):

$$\Lambda = \begin{Bmatrix} \Lambda_1 \\ \Lambda_2 \\ \Lambda_3 \\ \Lambda_4 \\ \Lambda_5 \\ \Lambda_6 \end{Bmatrix} = \begin{Bmatrix} d_{31}E_3 \\ d_{31}E_3 \\ d_{33}E_3 \\ d_{15}E_2 \\ d_{15}E_1 \\ 0 \end{Bmatrix} \tag{3.56}$$

For a PVDF piezoelectric film, the form of the equation will be (due to the assumption that the surface is not isotropic like for a PZT material)

$$\Lambda = \begin{Bmatrix} \Lambda_1 \\ \Lambda_2 \\ \Lambda_3 \\ \Lambda_4 \\ \Lambda_5 \\ \Lambda_6 \end{Bmatrix} = \begin{Bmatrix} d_{31}E_3 \\ d_{32}E_3 \\ d_{33}E_3 \\ d_{24}E_2 \\ d_{15}E_1 \\ 0 \end{Bmatrix} \tag{3.57}$$

The blocked force is defined as the force required to fully constrain the piezoelectric actuator (no deformation under application of electrical field). Its form will then be obtained by zeroing the strain vector (while the thermal contribution is also zero).

For the piezoelectric sheet in Figure 3.10, the blocked force (F_{bl}) and the free displacement (δ_{free}) will be

$$\varepsilon_1 = s_{11}^E \sigma_1 + d_{31} E_3$$

$$\text{If} \quad \sigma_1 = 0 \quad \varepsilon_1 = \Lambda_1 = d_{31} E_3 \Rightarrow \delta_{free} = d_{31} \left(\frac{l}{t}\right)_p V$$

$$\text{If} \quad \varepsilon_1 = 0 \quad \sigma_1 = -\frac{d_{31} E_3}{s_{11}^E} \Rightarrow F_{bl} = -\frac{d_{31} w_p}{s_{11}^E} V \tag{3.58}$$

where

$$\varepsilon_1 \equiv \Lambda_1 = \frac{\delta_{free}}{l_p} \quad \sigma_1 \equiv \frac{F_{bl}}{w_p t_p} \quad E_3 = \frac{V}{t_p}$$

If we define the stiffness of the actuator as

$$k_{actuator} = \frac{w_p t_p}{s_{11}^E l_p} \equiv \frac{A_p}{s_{11}^E l_p} \tag{3.59}$$

we can write the relationship between the blocked force (taken in its absolute value) and the free displacement as

$$F = F_{bl}\left(1 - \frac{\delta}{\delta_{free}}\right) \quad or \quad \delta = \delta_{free}\left(1 - \frac{F}{F_{bl}}\right) = \delta_{free} - \frac{F}{k_{actuator}} \tag{3.60}$$

$$\Rightarrow F = F_{bl} - \delta \cdot k_{actuator}$$

Equation (3.60) is schematically drawn in Figure 3.11.

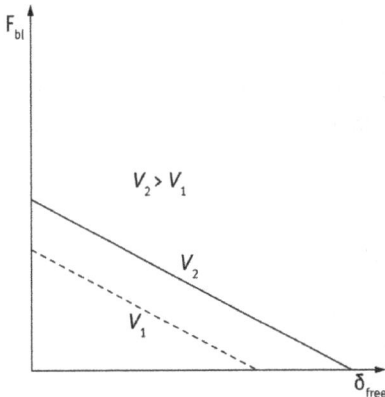

Figure 3.11: A piezoelectric actuator load line (schematic).

Assuming that the piezoelectric actuator can be represented by its stiffness ($k_{actuator}$), when it is acting against an external structure represented by its stiffness as k_{ext}.

The working point on the actuator load line (Figure 3.11) can be found as follows: use of Equation (3.60) leads to the following equations:

$$F_{ext} = F_{bl} - \delta_{ext} k_{actuator}$$

but

(3.61)

$$F_{ext} = \delta_{ext} k_{ext} \quad \Rightarrow \delta_{ext} = \frac{F_{bl}}{k_{ext} + k_{actuator}}$$

The displacement given by Equation (3.61) is the displacement of the two springs connected in parallel. It can be shown that the loading condition at which maximum work is done by the piezoelectric actuator would happen when $k_{ext} = k_{actuator}$, which means that the maximum energy can be extracted from the actuator by matching the external stiffness to its own stiffness.

3.1.2.13 Dynamic Behavior of a Piezoelectric Material

When a piezoelectric material is subjected to an AC electric field, the dimensions of the material change periodically; in other words, the material experiences vibration at the same frequency as that of the applied field. In the direct effect, when a vibrating force is applied to the piezoelectric material, it generates an oscillating electric field at the same frequency.

For the analysis of the dynamic behavior of a vibrating piezoelectric material, an equivalent electrical circuit is used, drawing analogy between mechanical and electrical components. This is illustrated in Figure 3.12. The vibrating force applied to the material is analogous to an alternating voltage. The piezoelectric element behaves as a capacitor of capacitance $C_0 = \varepsilon A/d$, where ε is the permittivity of the material and A and d are the area and thickness of the element, respectively. The mass M (inertia) of the piezoelectric element is equivalent to the inductance L, and the compliance constant is equivalent to the capacitor C. The energy loss due to friction is equivalent to the energy loss due to electrical resistance r in the circuit.

In Figure 3.12, r_m represents the mechanical resistance due to friction that causes energy loss. C_M represents the spring constant of the mechanical system. For the equivalent circuit: V, the applied AC voltage equivalent to the force F. L, inductance equivalent to the mass M. r, electrical resistance equivalent to mechanical friction (energy loss). C, capacitance equivalent to the compliance (related to the spring constant) of the material. C_0, the electrical capacitance of the piezoelectric material.

The impedance of the vibrating system is a function of frequency. The variation of the impedance as a function of frequency for such a system is shown in Figure 3.13. The impedance shows a minimum and a maximum as shown in the figure. The frequency at which the impedance is a minimum is called the *resonance frequency*, and the frequency at which the impedance is a maximum is called the *anti-resonance frequency*.

(a)

(b)

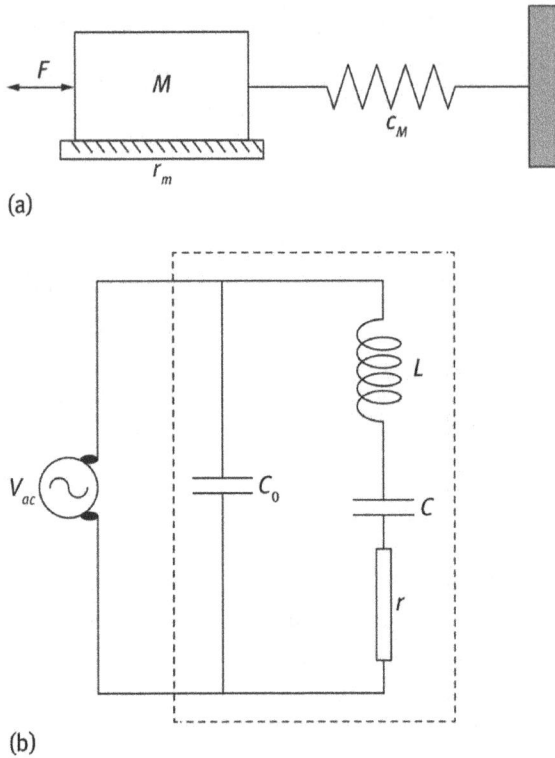

Figure 3.12: Vibrating piezoelectric material can be represented by a mechanical system or by an equivalent electrical circuit. (a) Mechanical system: F is the vibrating force applied on the system of mass M (inertia).

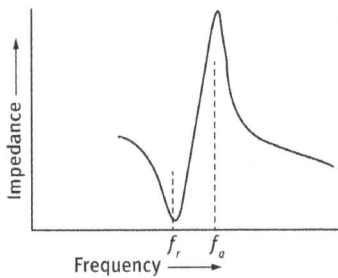

Figure 3.13: Frequency response of a piezoelectric element. The impedance shows a minimum at f_r (the resonance frequency) and a maximum at f_a (antiresonance frequency).

At the resonance frequency f_m, the piezoelectric system vibrates with maximum amplitude. The resonance frequency f_m is equal to the series resonance frequency f_s at which the impedance of the equivalent electrical circuit is zero, assuming that the resistance due to mechanical loss is zero:

$$f_m = f_s = \sqrt{1/L \cdot C}\Big/(2\pi).$$

The antiresonance frequency f_a is equal to the parallel resonance frequency f_p of the equivalent electrical circuit, assuming that the resistance due to mechanical loss is zero:

$$f_a = f_p = \sqrt{(C + C_0)/(L \cdot C \cdot C_0)} \Big/ (2\pi).$$

The resonance and antiresonance frequencies can be experimentally measured for a piezoelectric element. The values of f_r and f_a can be used to evaluate the electromechanical coupling coefficient k. The relation between the coupling coefficient and the frequencies f_r and f_a depends on the shape of the piezoelectric element.

3.2 Pin Force Beam Model

This chapter is devoted to the derivation of the equations of motion for a beam with piezoelectric layers, as depicted in Figure 3.14.

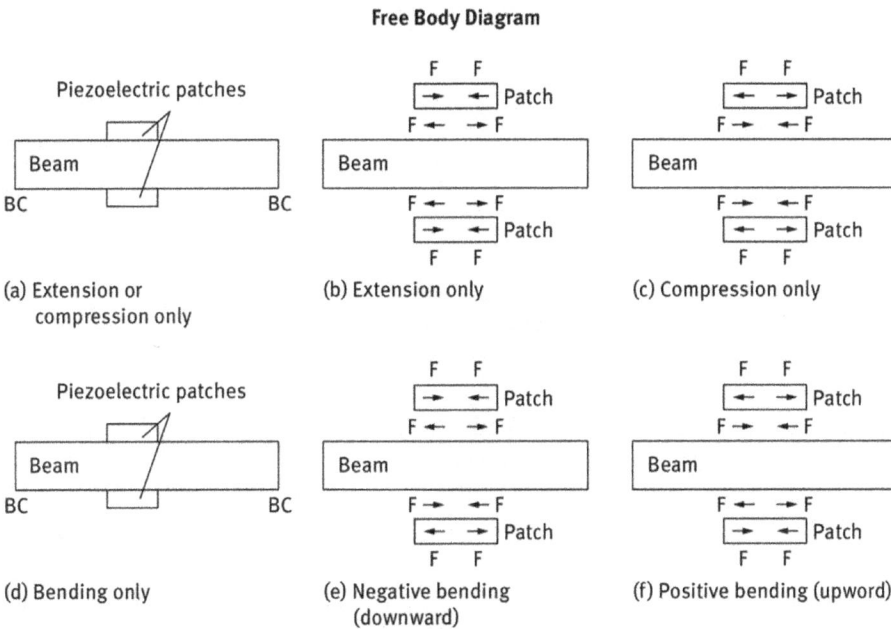

Figure 3.14: Beam with two identical piezoelectric patches acting as actuators.

Let us consider a piezoelectric patch having the following dimensions (see Figure 3.15): length l_p, width w_p, and thickness t_p being attached to a carrying beam, considered to be isotropic (to ease the calculations that will follow) as shown in Figure 3.14.

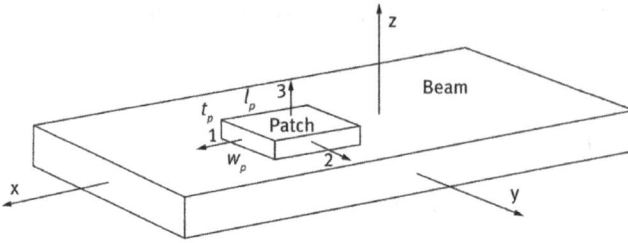

Figure 3.15: A carrying structure with a surface-bonded piezoelectric patch.

An electric voltage V is now applied to the electrodes of the piezoelectric patch, in the thickness direction. As most of the piezoelectric materials have a negative piezoelectric constant d_{31}, the direction of the voltage will be opposite to that of the polarization of the patch, yielding a positive strain in the 1 direction, according to the following relationship:

$$\varepsilon_{\max} = \frac{V}{t_p} d_{31} \equiv \Lambda \tag{3.62}$$

One should note that the maximal strain presented in Equation (3.62) is the strain for a free piezoelectric patch, not attached to any structure.

Multiplying Equation (3.62) by Young's modulus of the piezoelectric patch, E_p, and by the cross section ($w_p \times t_p = A_p$) will give the force acting in the x-direction, namely

$$F_{\max} = V w_p E_p d_{31} = w_p t_p E_p \Lambda = F_{bl} \tag{3.63}$$

Note that the maximal force given in Equation (3.63) is for a case when the ends of the piezoelectric patch are constrained to move in the x-direction, and this is the reason it is usually called in the literature "blocked force," F_{bl}.

When a piezoelectric patch is bonded to a carrying structure, which in our case is a beam (Figure 3.15), the strain and the force being encountered by the patch are influenced by the axial stiffness of the structure. When a voltage is applied across the piezoelectric patch electrodes, due to the fact that the patch is bonded to the carrying structure, the structure will encountered a surface force F, which will cause a reactive force $-F$ on the patch which will cause a strain ε_p. To evaluate this force, F, the strain in the patch, ε_p, will be written as the difference between its free strain, Λ (Equation (3.62)) and the mechanical strain introduced in the patch due to the reaction force, $-F$, yielding the following equation (with the aid of Equation (3.2.1)):

$$\varepsilon_p = \Lambda - \frac{F}{w_p t_p E_p} = \Lambda \left(1 - \frac{F}{F_{bl}}\right)$$

or

$$F = F_{bl}\left(1 - \frac{\varepsilon_p}{\Lambda}\right)$$

(3.64)

The force F acting on the beam can then be calculated by multiplying the strain ε_p by the cross section of the beam (as the same strain will be encountered also by the beam itself) and by its Young's modulus to yield

$$F = w_b h_b E_b \varepsilon_p = (AE)_b \varepsilon_p \tag{3.65}$$

where the subscript b stands for beam's properties (dimensions and Young's modulus) and $A_b = w_b \times h_b$ is the beam's cross section (w_b is the width of the beam and h_b is its height), while $(AE)_b$ is the axial stiffness of the beam.

Equation (3.64) presents the well-known relationship for actuators, namely the nondimensional force F/F_{bl} of the actuator is inversely linear to its strain ratio ε_p/Λ; thus obtaining the maximal force $F = F_{bl}$ for zero strain and for zero force the strain is maximal $\varepsilon_p = \Lambda$.

After presenting the relationship between force and strain, for a beam with a single piezoelectric patch, we shall derive expressions for the axial displacement of a beam sandwiched by two patches (see Figure 3.14).

The first case to be presented will be the model presented in Figure 3.14, dealing with extension or compression of the beam due to equal voltages both in sign and amplitude (in-phase voltages) supplied to the two piezoelectric patches (or actuators), assuming that they are identical in dimensions and properties (both mechanical and electrical ones).

We shall apply the displacement compatibility principle between the axial displacement of the beam and those of the bonded patches. According to Figure 3.14b and c, the axial displacement of the beam can be written as (one should notice that the beam is under traction of $2\,F$):

$$\Delta_b = \frac{2F}{w_b h_b E_b} l_p \tag{3.66}$$

One should note that the length of the beam should be taken as the length of the patch, l_p, for calculating the axial displacement of the beam in Equation (3.66).

The axial displacement of the patch can be found using the first part in Equation (3.64), namely

$$\varepsilon_p = \frac{\Delta l_p}{l_p} = \Lambda - \frac{F}{w_p t_p E_p} \rightarrow \Delta l_p = \left(\frac{V}{t_p} d_{31} - \frac{F}{w_p t_p E_p} \right) l_p \tag{3.67}$$

Equating the two displacements, given in Equations (3.66) and (3.67) gives

$$\Delta L_b = \Delta l_p \rightarrow \frac{2F}{w_b h_b E_b} l_p = \left(\frac{V}{t_p} d_{31} - \frac{F}{w_p t_p E_p}\right) l_p$$

or

$$F = \frac{\frac{V}{t_p} d_{31}}{\frac{2}{(EA)_b} + \frac{2}{(2EA)_p}} = \Lambda \frac{(EA)_b (EA)_p}{(EA)_b + (2EA)_p} = F_{\text{bl}} \frac{(EA)_b}{(EA)_b + (2EA)_p} \tag{3.68}$$

Substituting the expression for F given in Equation (3.68) into Equation (3.66) gives the strain experienced by the beam

$$\varepsilon_b = \frac{\Delta_b}{l_p} = \frac{2F}{(EA)_b} = \Lambda \frac{(2EA)_p}{(EA)_b + (2EA)_p} \tag{3.69}$$

From Equations (3.68) and (3.69), it is clear that for a negligible axial stiffness of the beam relative to the stiffness of the patches $((2EA)_p \gg (EA)_b)$ the force F will tend to zero while the value of the strain experienced by the beam, will be $\varepsilon_b \approx \Lambda$. For the opposite case, namely for a negligible axial stiffness of the patches (usually the more common case) as compared with the axial stiffness of the beam $((2EA)_p \ll (EA)_b)$, the beam's strain will vanish while the axial force will tend to F_{bl}.

The next case to be dealt will be induced bending of the beam depicted in Figure 3.14d–f. Applying out-of-phase voltages (same amplitude, opposite voltage sign) to the two electrodes of the patches, will cause one of the patches to expand, while the other one will contract, leading to a net bending of the beam (Figure 3.14e and f). As before, we would like to calculate the force F, the induced bending moment on the beam, strains, and displacements. Referring to Figure 3.14f, which displays a positive bending moment (the top surface of the beam is under compression, while the bottom is under tension) caused by applying a negative voltage to the top piezoelectric patch (causing it to expand, due to the negative value of d_{31}) and a positive voltage to the bottom patch (causing it to contract), the strain on the upper fiber of the beam can be written as a function of the induced moment M as

$$\varepsilon_b^{\text{top}} = -\frac{M \frac{h_b}{2}}{I_b E_b} = -\frac{(F h_b) \frac{h_b}{2}}{I_b E_b} = -\frac{F h_b^2}{2 I_b E_b} \tag{3.70}$$

where I is the beam's moment of inertia and $(EI)_b$ is its bending stiffness.

The shortening of the beam can be calculated by multiplying the top strain by the length of the patch, l_p, to yield

$$\Delta_b^{\text{top}} = -\frac{F h_b^2}{2 I_b E_b} l_p \tag{3.71}$$

The top patch change in length can be written as in Equation (3.67). Equating the two absolute displacements provides the necessary equation for the induced force F

$$\frac{Fh_b^2}{2I_bE_b}l_p = \left(\frac{V}{t_p}d_{31} - \frac{F}{(AE)_p}\right)l_p \rightarrow F = \frac{\frac{V}{t_p}d_{31}}{\frac{h_b^2}{2(EI)_b} + \frac{1}{(AE)_p}} \tag{3.72}$$

The expression for F obtained before in Equation (3.72) can be written in a compact form to yield

$$F = \frac{\frac{V}{t_p}d_{31}}{\frac{h_b^2}{2(EI)_b} + \frac{1}{(AE)_p}} = F_{bl}\frac{(EI)_b}{(EI)_b + (EI)_p} \tag{3.73}$$

where the bending stiffness of the two patches is defined as $(EI)_p = \frac{h_b^2}{2}(EA)_p$. Remembering that the "blocking moment" is defined by $M_{bl} = F_{bl}h_b$, we can write the expression for the bending moment as

$$M = M_{bl}\frac{(EI)_b}{(EI)_b + (EI)_p} \tag{3.74}$$

Under pure bending, the strain distribution is linear across the beam's cross section and can be written as a function of z coordinate as

$$\varepsilon_b = -\frac{M}{(EI)_b}z = -\frac{M_{bl}}{(EI)_b + (EI)_p}z \tag{3.75}$$

For a positive moment, $z = +h_b/2$ (for the top fiber of the beam) and $z = -h_b/2$ (for the bottom fiber of the beam); we can write the top and bottom beam strains as a function of the free piezoelectric strain Λ as

$$\varepsilon_b^{top} = -\frac{(EI)_p}{(EI)_b + (EI)_p}\Lambda \tag{3.76}$$

$$\varepsilon_b^{bottom} = \frac{(EI)_p}{(EI)_b + (EI)_p}\Lambda$$

As was done before for the case of pure extension or contraction, we would like to evaluate the performance of the beam sandwiched by two piezoelectric patches under induced bending.

From Equations (3.74) and (3.76), it is clear that for a negligible bending stiffness of the beam relative to the stiffness of the patches ($(EI)_p \gg (EI)_b$), the moment M will tend to zero while the values of the strains experienced by the beam, will be

$\varepsilon_b^{top} \approx -\Lambda$ and $\varepsilon_b^{bottom} \approx +\Lambda$. For the opposite case, namely for a negligible bending stiffness of the patches (usually the more common case) as compared with the bending stiffness of the beam ($(EI)_p \ll (EI)_b$), the beam's strain will vanish while the bending moment will tend to M_{block}.

Now, we shall present the induced bending deflection of the beam as a function of the position of the pair of patches and the boundary conditions of the beam. Let us start with the simplest case of a cantilevered beam with the pair of patches being located at its clamped root (see Figure 3.16a).

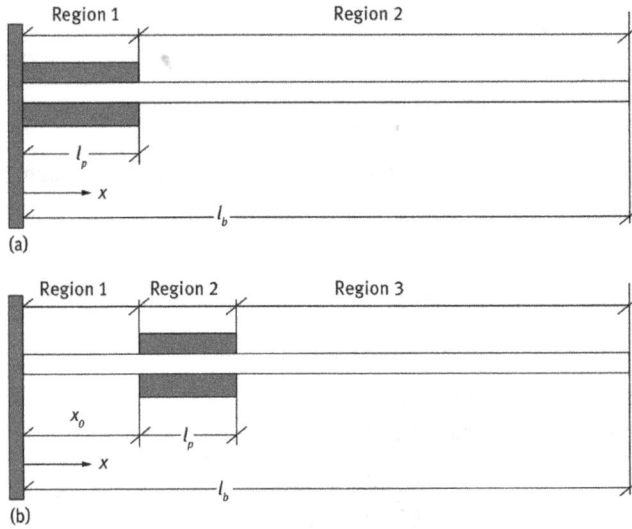

Figure 3.16: A cantilevered beam with a pair of piezoelectric patches located at (a) clamped side and (b) at a distance x_0 from the clamped side.

The relationship between the bending deflection and the applied moment is given according to Bernoulli–Euler beam theory as

$$\frac{d^2w}{dx^2} = \frac{M}{(EI)_b} \tag{3.77}$$

As the moment is not uniform along the beam, the beam is divided into regions, the first region includes the pair of patches near the clamped side of the beam and the second region the beam only, as depicted in Figure 3.16a. For the first region we obtain, after integrating twice the relation in Equation (3.76)

$$w_1(x) = \frac{Mx^2}{2(EI)_b} + A_1x + B_1 \tag{3.78}$$

where A_1 and B_1 are constants to be determined from boundary conditions and continuity requirements between the two regions.

The bending deflection for the second region, while taking into account that the moment is zero along this part, has the following expression:

$$w_2(x) = A_2 x + B_2 \tag{3.79}$$

where A_2 and B_2 are constants to be determined from boundary conditions and continuity requirements between the two regions.

Applying the clamped boundary conditions for the beam (only for the first region), we get

$$w_1(0) = 0 \rightarrow B_1 = 0 \tag{3.80}$$

$$\frac{w_1(0)}{dx} = 0 \rightarrow A_1 = 0$$

$$\rightarrow w_1(x) = \frac{Mx^2}{2(EI)_b} = \frac{M_{bl}x^2}{2\left((EI)_b + (EI)_p\right)}$$

To obtain the shape of the bending deflection for the second region, we shall apply continuity requirements for the deflections and the slopes at the boundaries between the two regions, yielding

$$w_1(l_p) = w_2(l_p) \rightarrow \frac{Ml_p^2}{2(EI)_b} = A_2 l_p + B_2$$

$$\frac{dw_1(l_p)}{dx} = \frac{dw_2(l_p)}{dx} \rightarrow \frac{Ml_p}{(EI)_b} = A_2$$

$$\rightarrow B_2 = -\frac{Ml_p^2}{2(EI)_b} \tag{3.81}$$

$$\rightarrow w_2(x) = \frac{Ml_p}{(EI)_b}\left(x - \frac{l_p}{2}\right) = \frac{M_{bl}l_p}{\left((EI)_b + (EI)_p\right)}\left(x - \frac{l_p}{2}\right)$$

Equations (3.80) and (3.81) describe the shape of the beam's deflection. As one can see from Equation (3.81), the slope of the beam (B_2) is constant in the second region, namely after the region with the pair of patches, the beam rotates only and does not bend (it remains straight).

Now the case depicted in Figure 3.16b will be solved. One should remember that the moment is induced only in Region 2, while the other two regions experience zero bending moments. This implies the following expressions for the deflection:

$$0 < x < x_0 \rightarrow w_1(x) = A_1 x + B_1$$

$$x_0 < x < x_0 + l_p \rightarrow w_2(x) = \frac{Mx^2}{2(EI)_b} + A_2 x + B_2 \tag{3.82}$$

$$x_0 + l_p < x < l_b \rightarrow w_3(x) = A_3 x + B_3$$

where the various constants A_1, B_1, A_2, B_2, A_3 and B_3 are to be determined by applying boundary conditions and continuity requirements at the boundaries between two adjacent regions.

For the first region, we apply the geometric boundary conditions of the beam, yielding

$$w_1(0) = 0 \rightarrow B_1 = 0$$

$$\frac{dw_1(0)}{dx} = 0 \rightarrow A_1 = 0 \tag{3.83}$$

Therefore, in Region 1, both the deflection and the slope are identically zero

$$w_1(x) = 0$$

$$\frac{dw_1(x)}{dx} = 0 \tag{3.84}$$

For Region 2, we can write the following continuity requirements at $x = x_0$:

$$w_1(x_0) = w_2(x_0) \rightarrow 0 = \frac{Mx_0^2}{2(EI)_b} + A_2 x_0 + B_2$$

$$\frac{dw_1(x_0)}{dx} = \frac{dw_2(x_0)}{dx} \rightarrow 0 = \frac{Mx_0}{(EI)_b} + A_2 \tag{3.85}$$

From Equation (3.85), we can solve for the two unknowns, A_2 and B_2 to obtain the deflection and the slope for the second region

$$w_2(x) = \frac{M(x - x_0)^2}{2(EI)_b}$$

$$\frac{dw_2(x)}{dx} = \frac{M(x - x_0)}{(EI)_b} \tag{3.86}$$

For the last part, Region 3, we write the following continuity requirements at $x = x_0 + l_p$:

$$w_2(x_0 + l_p) = w_3(x_0 + l_p) \rightarrow \frac{M l_p^2}{2(EI)_b} = A_3(x_0 + l_p) + B_3$$

$$\frac{dw_2(x_0 + l_p)}{dx} = \frac{dw_3(x_0 + l_p)}{dx} \rightarrow \frac{M l_p}{(EI)_b} = A_3 \qquad (3.87)$$

From Equation (3.87), we can solve for the two unknowns, A_3 and B_3, to obtain the deflection and the slope for Region 3

$$w_3(x) = \frac{M l_p (x - x_0 - 0.5 l_p)}{(EI)_b}$$

$$\frac{dw_3(x)}{dx} = \frac{M l_p}{(EI)_b} = \text{const} \qquad (3.88)$$

Substituting the expression for the moment M, we can write the distribution of the deflection and its accompanied slope along the beam as

$$0 < x < x_0 \rightarrow w_1(x) = 0 \quad \frac{dw_1(x)}{dx} = 0$$

$$x_0 < x < x_0 + l_p \rightarrow w_2(x) = \frac{M_{bl}(x - x_0)^2}{2\left((EI)_b + (EI)_p\right)} \quad \frac{dw_2(x)}{dx} = \frac{M_{bl}(x - x_0)}{\left((EI)_b + (EI)_p\right)} \qquad (3.89)$$

$$x_0 + l_p < x < l_b \rightarrow w_3(x) = \frac{M_{bl} l_p (x - x_0 - 0.5 l_p)}{\left((EI)_b + (EI)_p\right)} \quad \frac{dw_3(x)}{dx} = \frac{M_{bl} l_p}{\left((EI)_b + (EI)_p\right)}$$

A beam on simply supported boundary conditions, equipped with a pair of piezo-electric patches as presented in Figure 3.17, will be solved next to be for induced deflections and slopes along it.

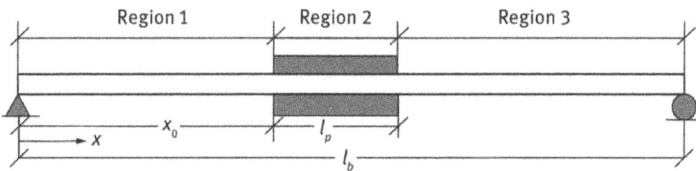

Figure 3.17: A simply supported beam with a pair of piezoelectric patches.

As one can notice, the boundary conditions for the case as presented in Figure 3.17 can be enforced only for Regions 1 and 3, while continuity demands can be applied for the mid-part of the beam, Region 2. Taking into account that the induced moment, M, acts only in Region 1, the equations for the lateral displacements for each part of the beam can be written as (similar to Equation (3.82)):

$$0 < x < x_0 \rightarrow w_1(x) = \overline{A_1}x + \overline{B_1}$$

$$x_0 < x < x_0 + l_p \rightarrow w_2(x) = \frac{Mx^2}{2(EI)_b} + \bar{A}_2 x + \bar{B}_2 \qquad (3.90)$$

$$x_0 + l_p < x < l_b \rightarrow w_3(x) = \overline{A_3}x + \overline{B_3}$$

Treating the first region, we get

$$0 < x < x_0 \quad w_1(x) = \overline{A_1}x + \overline{B_1}$$

$$w_1(0) = 0 \rightarrow \overline{B_1} = 0 \qquad (3.91)$$

$$\rightarrow w_1(x) = \overline{A_1}x \quad \frac{dw_1(x)}{dx} = \overline{A_1}$$

For Region 2, we can write

$$x_0 < x < x_0 + l_p \quad w_2(x) = \frac{Mx^2}{2(EI)_b} + \bar{A}_2 x + \bar{B}_2$$

$$w_1(x_0) = w_2(x_0) \rightarrow \overline{A_1}x_0 = \frac{Mx_0^2}{2(EI)_b} + \bar{A}_2 x_0 + \bar{B}_2 \rightarrow (\overline{A_1} - \bar{A}_2)x_0 = \frac{Mx_0^2}{2(EI)_b} + \bar{B}_2 \frac{dw_1(x_0)}{dx}$$

$$= \frac{dw_2(x_0)}{dx} \rightarrow \overline{A_1} = \frac{Mx_0}{(EI)_b} + \bar{A}_2 \rightarrow (\overline{A_1} - \bar{A}_2) = \frac{Mx_0}{(EI)_b} \qquad (3.92)$$

Applying boundary conditions and continuity of the displacements and slopes at the boundaries between the second and third parts yield for the last region, Region 3:

$$x_0 + l_p < x < l_b \quad w_3(x) = \overline{A_3}x + \overline{B_3}$$

$$w_3(l_b) = 0 \rightarrow \overline{A_3}l_b + \overline{B_3} = 0$$

$$\rightarrow w_3(x) = \overline{A_3}(x - l_b) \quad \frac{dw_3(x)}{dx} = \overline{A_3} \qquad (3.93)$$

$$w_2(x_0 + l_p) = w_3(x_0 + l_p) \rightarrow \frac{M(x_0 + l_p)^2}{2(EI)_b} + \bar{A}_2(x_0 + l_p) + \bar{B}_2 = \overline{A_3}(x_0 + l_p - l_b)$$

$$\frac{dw_2(x_0 + l_p)}{dx} = \frac{dw_3(x_0 + l_p)}{dx} \rightarrow \frac{M(x_0 + l_p)}{(EI)_b} + \bar{A}_2 = \overline{A_3} \rightarrow \frac{M(x_0 + l_p)}{(EI)_b} = (\overline{A_3} - \bar{A}_2)$$

Equations (3.91)–(3.93) provide the necessary six equations to find the six unknowns, $A_1{}^-, B_1{}^-, A_2{}^-, B_2{}^-, A_3{}^-$ and $B_3{}^-$. Their expressions are given as

$$\overline{A_1} = \frac{M}{2(EI)_b} \frac{l_p}{l_b} (2x_0 + 2l_b - l_p) \quad \overline{B_1} = 0$$

$$\overline{A_2} = \frac{M(l_p - l_b) x_0}{(EI)_b} \frac{x_0}{l_b} - \frac{M(l_p - 2l_b) l_p}{2(EI)_b} \frac{l_p}{l_b} \quad \overline{B_2} = \frac{Mx_0^2}{2(EI)_b} \tag{3.94}$$

$$\overline{A_3} = \frac{M(2x_0 + l_p) l_p}{2(EI)_b} \frac{l_p}{l_b} \quad \overline{B_3} = -\frac{M(2x_0 + l_p)}{2(EI)_b} l_p \quad \text{where} \quad \frac{M}{(EI)_b} = \frac{M_{bl}}{(EI)_b + (EI)_p}$$

Using the constants given in Equation (3.94), the deflections and their associated slopes for each of the three regions of the beam can be calculated.

The cases shown earlier present symmetric positions of the piezoelectric patches, leading to either pure bending or pure stretching/shortening of the beam with piezoelectric layers acting as actuators.

The case presented in Figure 3.18 presents one of a series of nonsymmetric possibilities to attach piezoelectric patches to a beam. The common thing for all those configurations is the introduction of both bending and stretching/shortening into the carrying beam.

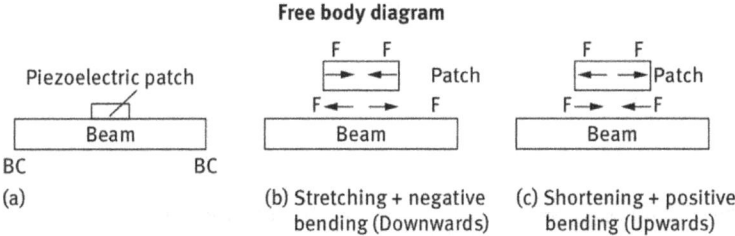

Figure 3.18: A beam with a single piezoelectric patch: (a) model, (b) stretching and negative bending of the beam and (c) shortening and positive bending of the beam.

To simplify the solution, it is assumed that the thickness of the single-piezoelectric patch is negligible, leading to the assumption that the neutral axis of the beam remains in the middle of the beam's cross section. In case this assumption is not valid, one should calculate the new position of the neutral axis using one of the common methods, like modulus weighted neutral axis found in every book dealing with bending stresses in beams.[2]

Referring to Figure 3.18, the top fiber of the beam is under compression, and its strain can be written as the sum of two expressions, the first one due to shortening

---

2 Popov, E. P., Mechanics of Materials (2nd edn.) Hardcover – Prentice Hall; April 7, 1976.

of the fiber and the second one due to its compression because of the positive bending moment, yielding

$$\varepsilon_b^{top} = -\frac{F}{(wh)_b E_b} - \frac{Mh}{2(EI)_b} \tag{3.95}$$

But the moment M and the moment of inertia I can be written as

$$M = F\frac{h_b}{2} \quad I = \frac{w_b h_b^3}{12} \tag{3.96}$$

which makes Equation (3.95) to look as

$$\varepsilon_b^{top} = -\frac{4F}{(wh)_b E_b} \tag{3.97}$$

and the absolute shortening of the top fiber along the length of the patch, l_p, will be (the minus sign is omitted)

$$\Delta l_b^{top} = \varepsilon_b^{top} l_p \tag{3.98}$$

Equating the change in length of the piezoelectric patch (Equation (3.64)) with the change in length of the top fiber, Equation (3.97) leads to the expression for the axial force F,

$$\Delta l_b^{top} = \Delta l_p \quad \rightarrow \quad \frac{4F}{(wh)_b E_b} = \Lambda - \frac{F}{(wh)_p E_p}$$

$$\rightarrow F = \Lambda \frac{(wh)_p E_p (wh)_b E_b}{4(wh)_p E_p + (wh)_b E_b} \tag{3.99}$$

Taking into account that $F_{blocking} = \Lambda E_p (wt)_p$, one can rewrite Equation (3.99) to yield

$$F = F_{bl} \frac{(wh)_b E_b}{4(wh)_p E_p + (wh)_b E_b} \tag{3.100}$$

The expression for the bending moment, while remembering that $M_{bl} = F_{bl}\frac{h_b}{2}$, will be

$$M = M_{bl} \frac{(wh)_b E_b}{4(wh)_p E_p + (wh)_b E_b} \tag{3.101}$$

To solve other nonsymmetric problems, involving two different piezoelectric patches, bonded on the two sides of a carrying beam (see Figure 3.19), where the differences might stem for the application of either not equal voltages, or different thicknesses, or various piezoelectric constants, d_{31} (or all the differences together), the following approach should be used.

Free Body Diagram

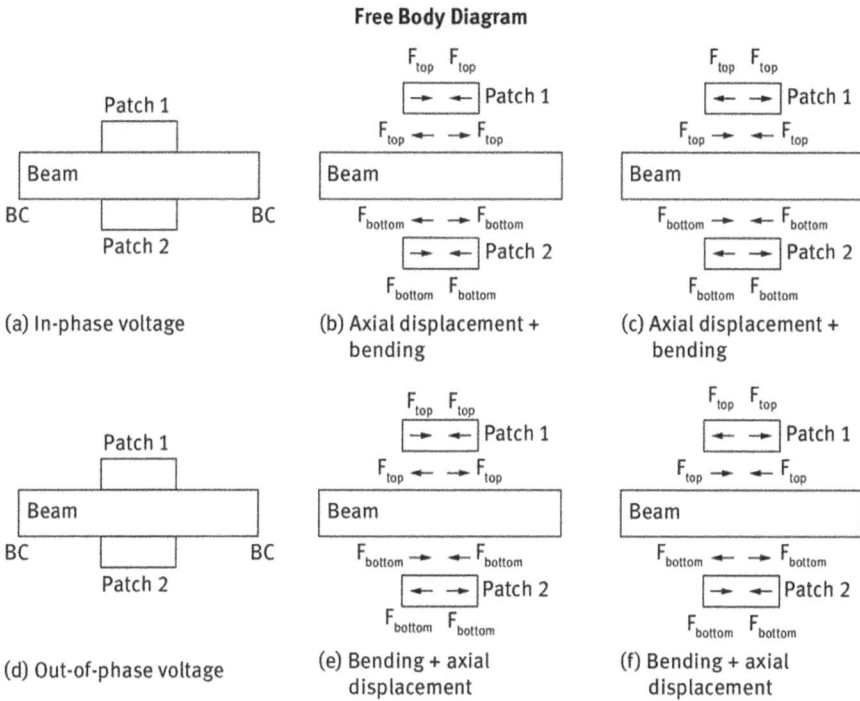

(a) In-phase voltage

(b) Axial displacement + bending

(c) Axial displacement + bending

(d) Out-of-phase voltage

(e) Bending + axial displacement

(f) Bending + axial displacement

Figure 3.19: Beam with two not identical piezoelectric patches acting as actuators.

Assuming the force distribution presented in Figure 3.19, one can define the following two relations:

$$F_1 = \frac{\left(F_{top} + F_{bottom}\right)}{2}$$

$$F_2 = \frac{\left(F_{top} - F_{bottom}\right)}{2}$$

(3.102)

Figure 3.20 presents the beam under the action of the two defined forces, F_1 and F_2. The average between F_{top} and F_{bottom}, the force F_1, leads to axial stretching only provided $F_{top} > F_{bottom}$ and F_{bottom} is positive (as shown in Figure 3.20). In the opposite case, when F_{bottom} is negative and larger than F_{top}, the beam will be shorten. The difference between F_{top} and F_{bottom} divided by 2, the force F_2, leads to a net bending of the beam. The direction of the bending, upward (positive bending) or downward (negative bending) is dependent on the sign of the difference between the two forces.

Free body for abs (F_{top}) > abs (F_{bottom})

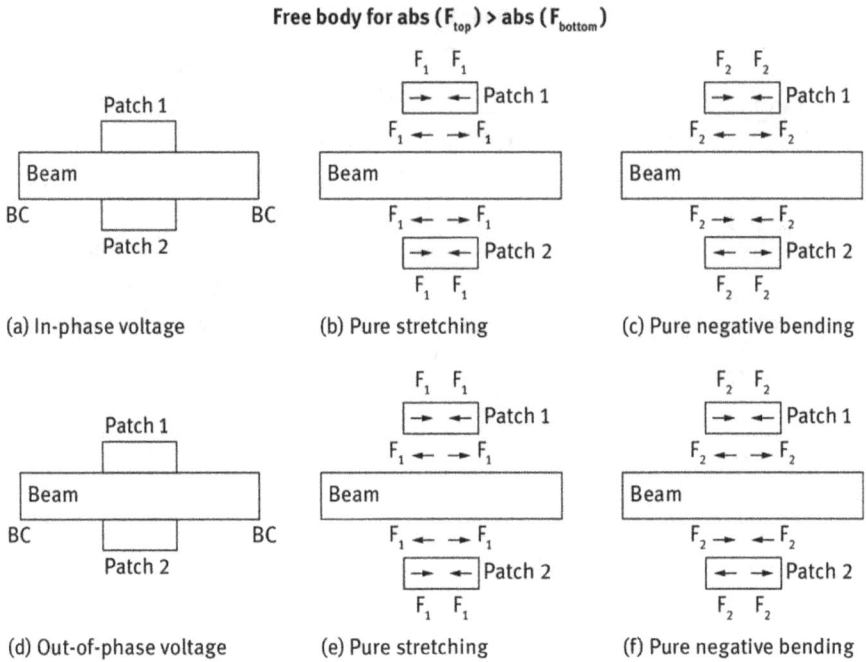

(a) In-phase voltage

(b) Pure stretching

(c) Pure negative bending

(d) Out-of-phase voltage

(e) Pure stretching

(f) Pure negative bending

Figure 3.20: Beam with two not identical piezoelectric patches acting as actuators-equivalent forces F_1 and F_2.

The solution for each case presented Figure 3.20, can now be solved using the equations presented before for the symmetric cases.[3]

3.3 Uniform-Strain Beam Model

The uniform-strain beam model carries its name from the assumption being presented in Figure 3.21a and b, namely, the strain distribution across the cross section of the piezoelectric layer is assumed to be constant, due to its small thickness.

In this chapter, the solution for a carrying structure, in our case a beam, with surface-bonded piezoelectric layers will be presented for the case the bonding layers (BLs) thickness cannot be neglected (see Figure 3.22). To obtain the equations of motion for this problem, a differential element, dx, will be used (Figure 3.22).

The derivation is done following the results presented in [2] (more references on the topic are presented in the references list of this chapter).

---

3 See also Chopra, I. and Sirohi, J., *Smart Structures Theory*, Chapter 4.3, Cambridge University Press, 2014.

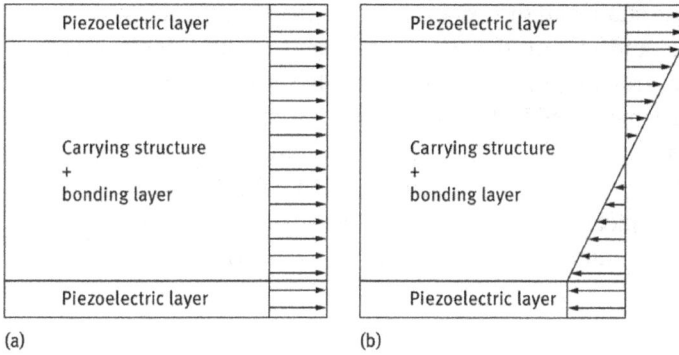

Figure 3.21: Strain distribution across the carrying structure, bonding layers and surface-bonded piezoelectric layers: (a) extension and (b) bending.

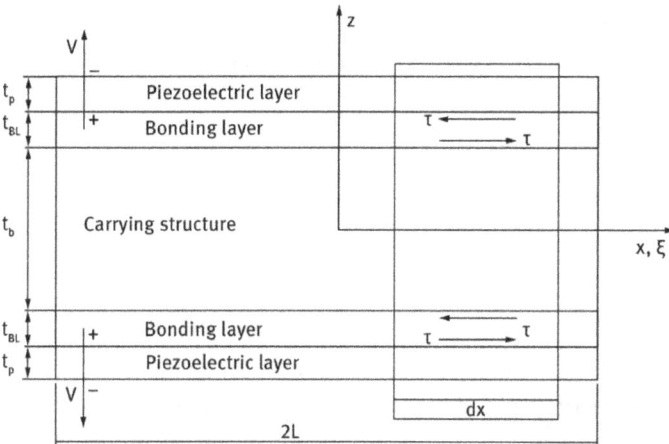

Figure 3.22: Geometry of a carrying structure with surface-bonded piezoelectric layers and finite thickness bonding layers.

Assuming pure one-dimensional shear in the BLs and only extensional strain in the piezoelectric layers and the carrying substructure, the strain–displacement relationships can be written as

$$\varepsilon_p = \frac{du_p}{dx} \tag{3.103}$$

$$\varepsilon_b^{\text{BL}} = \frac{du_b^{\text{BL}}}{dx} \tag{3.104}$$

$$\gamma_{\text{BL}} = \frac{u_p - u_b}{t_{\text{BL}}} \tag{3.105}$$

where ε_p and u_p are the strain and the axial displacement, respectively, in the piezo-electric layer, and ε_b^{BL} and u_b^{BL} are the strain and the axial displacement, respectively, in the carrying structure, the beam at the interface with the bonding BL and y_{BL} is the shear strain experienced by the BL.

Assuming that for the case of bending, being excited by the piezoelectric actuators, the strain distribution is linear (see Figure 3.21b) and for the case of extension (or contraction), as is depicted in Figure 3.21a, the strain distribution would be uniform, the equilibrium equations for the element dx would be written as

$$\frac{d\sigma_p}{dx} - \frac{\tau}{t_p} = 0 \tag{3.106}$$

$$\frac{dM}{dx} + \tau t_b w_p = 0 \tag{3.107}$$

where w_p is the width of the piezoelectric actuator (layer) and t_b is the height of the beam (the carrying structure).

The stress distribution in the beam can be expressed with the aid of the bending moment M [13], as

$$\sigma_b(z) = -\frac{M}{I_b}z \tag{3.108}$$

where

$$I_b = \frac{w_b t_b^3}{12}$$

is the area product of inertia of the beam's cross section about its neutral axis (the contributions of the piezoelectric and BLs are neglected due to their small thickness as compared with the beam's height, t_b. w_b is the beam's width). The stress at the top surface of the beam $(z = t_b/2)$ can be written as

$$\sigma_b^{top} = -\frac{M \cdot (t_b/2)}{\frac{w_b t_b^3}{12}} = -\frac{6M}{w_b t_b^2} \tag{3.109}$$

Defining the moment M as

$$M = -\frac{w_b t_b^2}{6}\sigma_b^{top} \tag{3.108}$$

with the aid of Equation (3.109) and substituting into Equation (3.107) yields

$$\frac{d\sigma_b^{top}}{dx} - 6\frac{w_p}{w_b t_b}\tau = 0$$

Performing the same calculations for the bottom surface of the beam ($z = -t_b/2$), we obtain a similar equation like Equation (3.110), namely

$$\frac{d\sigma_b^{\text{bottom}}}{dx} + 6\frac{w_p}{w_b t_b}\tau = 0 \tag{3.111}$$

Similarly, for the case of pure extension (or contraction), we get for the top and bottom surfaces of the beam the following expressions:

$$\frac{d\sigma_b^{\text{top}}}{dx} + 2\frac{w_p}{w_b t_b}\tau = 0 \tag{3.112}$$

$$\frac{d\sigma_b^{\text{bottom}}}{dx} + 2\frac{w_p}{w_b t_b}\tau = 0 \tag{3.113}$$

A single general expression can be written instead of Equations (3.110)–(3.113) having the following form:

$$\frac{d\sigma_b}{dx} + \alpha\frac{w_p}{w_b t_b}\tau = 0 \quad \text{while}$$

$$\alpha = 2 \quad \text{extension}$$

$$\alpha = 6 \quad \text{bottom bending; } \alpha = -6 \quad \text{top bending} \tag{3.114}$$

The stress–strain expression for the piezoelectric layers is given by

$$\sigma_p = E_p\left(\varepsilon_p - \frac{d_{31}V}{t_p}\right) = E_p(\varepsilon_p - \Lambda) \tag{3.115}$$

where V is the applied voltage across the electrodes of the piezoelectric actuators, E_p is Young's modulus of the piezoelectric material and d_{31} is its constant.

The other two stress–strain equations, for the beam and the BL are given as

$$\sigma_b^{\text{BL}} = E_b u_b^{\text{BL}} \tag{3.116}$$

$$\tau_{\text{BL}} = G_{\text{BL}} \gamma_{\text{BL}} \tag{3.117}$$

where G_{BL} is the shear modulus of the BL.

Equations (3.103)–(3.106) and (3.114)–(3.117) form the eight governing equations for the eight unknowns of the present problem: three stresses ($\sigma_p, \sigma_b^{\text{BL}}, \tau_{\text{BL}}$), two displacements ($u_p, u_b^{\text{BL}}$) and three strains ($\varepsilon_p, \varepsilon_b^{\text{BL}}, \gamma_{\text{BL}}$). Substituting Equations (3.103)–(3.105) into Equations (3.115)–(3.117) and then into Equations (3.106) and (3.114), while assuming the width of the piezoelectric layers matches the width of the beam, namely $w_p = w_b$, yields two coupled equations with the two unknowns, $\varepsilon_b^{\text{BL}}$ and ε_p:

$$\frac{d\varepsilon_p}{dx^2} - \frac{G_{BL}}{E_p} \cdot \frac{\varepsilon_p - \varepsilon_b^{BL}}{t_{BL} t_p} = 0 \tag{3.118}$$

$$\frac{d\varepsilon_b^{BL}}{dx^2} + \frac{G_{BL}}{E_b} \cdot \frac{\varepsilon_p - \varepsilon_b^{BL}}{t_{BL} t_b} \alpha = 0 \tag{3.119}$$

To uncouple the two above equations, some nondimensional quantities are defined as

$$\xi = \frac{x}{L}; \quad \bar{G} = \frac{G}{E_p}; \quad \overline{t_{BL}} = \frac{t_{BL}}{L}; \quad \theta_b = \frac{t_b}{t_p};$$

$$\bar{G} = \frac{G}{E_p} \frac{E_b}{E_p} = \overline{E_b}; \quad \psi = \frac{E_b t_b}{E_p t_p} = \overline{E_b} \theta_b; \quad \theta_{BL} = \frac{t_{BL}}{t_p} \tag{3.120}$$

Uncoupling Equations (3.118) and (3.119) yield a pair of forth-order differential equations in the following form:

$$\varepsilon_p^{iv} - \Gamma^2 \varepsilon_p'' = 0 \tag{3.121}$$

$$\left(\varepsilon_b^{BL}\right)^{iv} - \Gamma^2 \left(\varepsilon_b^{BL}\right)'' = 0 \tag{3.122}$$

where

$$\Gamma^2 = \frac{\bar{G} \theta_{BL}}{\bar{t}_{BL}^2} \left(\frac{\psi + \alpha}{\psi}\right) \tag{3.123}$$

and the differentiation in Equations (3.121) and (3.122) is done with respect to the nondimensional length ξ. The solutions of Equations (3.121) and (3.122) have the following form:

$$\left\{ \begin{array}{c} \varepsilon_p \\ \varepsilon_b^{BL} \end{array} \right\} = \left\{ \begin{array}{c} 1 \\ 1 \end{array} \right\} B_1 + \left\{ \begin{array}{c} 1 \\ 1 \end{array} \right\} B_2 \xi + \left\{ \begin{array}{c} -\psi/\alpha \\ 1 \end{array} \right\} B_3 \sinh \Gamma \xi + \left\{ \begin{array}{c} -\psi/\alpha \\ 1 \end{array} \right\} B_4 \cosh \Gamma \xi \tag{3.124}$$

Application of the four strain boundary conditions would determine the four unknowns, B_1–B_4. The piezoelectric influence, namely the term Λ, does not appear explicitly in Equation (3.124); however, it will be reflected through the boundary conditions of the problem. For the segmented actuators, presented in Figure 3.22, having stress-free ends, the term ε_p must be equal to the piezoelectric strain, Λ, at its both ends. At the ends of the carrying structure, the strains may be equal to arbitrary nonzero strains, arising from nonpiezoelectric action, like mechanical loading and deformations. Therefore, the four boundary conditions are:

$$\begin{array}{llll} \text{at} & \xi = +1: & \varepsilon_p = \Lambda, & \varepsilon_b^{BL} = \varepsilon_b^{BL+} \\ \text{at} & \xi = -1: & \varepsilon_p = \Lambda, & \varepsilon_b^{BL} = \varepsilon_b^{BL-} \end{array} \tag{3.125}$$

where the terms $\varepsilon_b^{\mathrm{BL}+}$ and $\varepsilon_b^{\mathrm{BL}-}$ are carry structure strains at its right (+) and left (−) ends, respectively. Substituting the above four boundary conditions into Equation (3.124) yields

$$B_1 = \frac{\psi}{\alpha + \psi}\left(\frac{\varepsilon_b^{\mathrm{BL}+} + \varepsilon_b^{\mathrm{BL}-}}{2} + \Lambda\frac{\alpha}{\psi}\right) = \frac{\psi}{\alpha + \psi}\left(\varepsilon^+ + \Lambda\frac{\alpha}{\psi}\right) \tag{3.126}$$

$$B_2 = \frac{\psi}{\alpha + \psi}\left(\frac{\varepsilon_b^{\mathrm{BL}+} - \varepsilon_b^{\mathrm{BL}-}}{2}\right) = \frac{\psi}{\alpha + \psi}\varepsilon^- \tag{3.127}$$

$$B_3 = \frac{\alpha}{(\alpha + \psi)\sinh\Gamma}\left(\frac{\varepsilon_b^{\mathrm{BL}+} - \varepsilon_b^{\mathrm{BL}-}}{2}\right) = \frac{\alpha}{(\alpha + \psi)\sinh\Gamma}\varepsilon^- \tag{3.128}$$

$$B_4 = \frac{\alpha}{(\alpha + \psi)\cosh\Gamma}\left(\frac{\varepsilon_b^{\mathrm{BL}+} + \varepsilon_b^{\mathrm{BL}-}}{2} - \Lambda\right) = \frac{\alpha}{(\alpha + \psi)\cosh\Gamma}(\varepsilon^+ - \Lambda) \tag{3.129}$$

To obtain the shear stress τ_{BL}, first we find the displacements u_p and u_b^{BL}, by integrating Equations (3.103) and (3.104) with respect to $d\xi$, and then substituting in Equation (3.105) to yield the shear strain y, in the following form:

$$y = \frac{\tau_{\mathrm{BL}}}{G_{\mathrm{BL}}} = \frac{u_p - u_b^{\mathrm{BL}}}{t_{\mathrm{BL}}} = \frac{-B_3\frac{\cosh\Gamma\xi}{\Gamma}\left(\frac{\psi}{\alpha}+1\right) - B_4\frac{\sinh\Gamma\xi}{\Gamma}\left(\frac{\psi}{\alpha}+1\right)}{t_{\mathrm{BL}}} \tag{3.130}$$

Substituting the values of B_3 and B_4 from Equations (3.128) and (3.129) in Equation (3.130), we get

$$\frac{\tau_{\mathrm{BL}}}{E_b} = \frac{\bar{G}}{\bar{E}_b t_{\mathrm{BL}}\Gamma}\left[\varepsilon^- \frac{\cosh\Gamma\xi}{\sinh\Gamma} + (\varepsilon^+ - \Lambda)\frac{\sinh\Gamma\xi}{\cosh\Gamma}\right] \tag{3.131}$$

Note that the minus (−) sign was removed from Equation (3.131), as negative shear stresses are not meaningful in mechanics.

The explicit expressions for the strains in the piezoelectric layers and the carrying structure can be written for the case of a finite thickness bond as

$$\begin{Bmatrix} \varepsilon_p \\ \varepsilon_b^{\mathrm{BL}} \end{Bmatrix} = \frac{\psi}{\alpha + \psi}\begin{Bmatrix} \varepsilon^+ + \varepsilon^-\xi \\ \varepsilon^+ + \varepsilon^-\xi \end{Bmatrix} - \frac{\psi}{\alpha + \psi}\begin{Bmatrix} \varepsilon^+ \dfrac{\cosh\Gamma\xi}{\cosh\Gamma} + \varepsilon^- \dfrac{\sinh\Gamma\xi}{\sinh\Gamma} \\ -\varepsilon^+ \dfrac{\cosh\Gamma\xi}{\cosh\Gamma} - \varepsilon^- \dfrac{\sinh\Gamma\xi}{\sinh\Gamma} \end{Bmatrix}$$
$$+ \frac{\alpha}{\alpha + \psi}\begin{Bmatrix} 1 + \dfrac{\psi}{\alpha}\dfrac{\cosh\Gamma\xi}{\cosh\Gamma} \\ 1 - \dfrac{\cosh\Gamma\xi}{\cosh\Gamma} \end{Bmatrix}\Lambda \tag{3.132}$$

For the perfect bonding ($\Gamma \to \infty$), we get

$$\begin{Bmatrix} \varepsilon_p \\ \varepsilon_b^{BL} \end{Bmatrix} = \frac{\psi}{\alpha + \psi} \begin{Bmatrix} \varepsilon^+ + \varepsilon^- \xi \\ \varepsilon^+ + \varepsilon^- \xi \end{Bmatrix} + \frac{\alpha}{\alpha + \psi} \begin{Bmatrix} 1 \\ 1 \end{Bmatrix} \Lambda \tag{3.133}$$

yielding $\varepsilon_p = \varepsilon_b^{BL}$, as it should be.

The force transmitted from the piezoelectric layer to the carrying structures can now be calculated for the case of perfect bonding. This force is transferred at the ends of the piezoelectric actuator, at $\xi = \pm 1$, has the following form:

$$F = \frac{E_b t_b^{BL} w_b}{\alpha + \psi} (\varepsilon^+ + \varepsilon^- \xi) + \frac{\alpha E_b t_p w_p}{\alpha + \psi} \Lambda \tag{3.134}$$

If bending is caused by the two piezoelectric layers, the expression for the bending moment is simply

$$M = F t_b^{BL} \tag{3.135}$$

One can see that for the case of finite bonding thickness, the strains (Equation (3.132)) and the stress (Equation (3.131)) are dependent on the voltage applied on the piezo-electric electrodes ($\Lambda = d_{31} V / t_p$) and the strains on the boundaries of the carrying structure (ε_b^{BL+} and ε_b^{BL-}). The terms dependent on ε_b^{BL+} and ε_b^{BL-} are merely additional passive stiffness and, therefore, to show the effect of the finite bonding thickness in the presence of piezoelectric layers, they can be taken as zero, without losing the general description of the problem. This yields

$$\begin{Bmatrix} \dfrac{\varepsilon_p}{\Lambda} \\ \dfrac{\varepsilon_b^{BL}}{\Lambda} \end{Bmatrix} = + \frac{\alpha}{\alpha + \psi} \begin{Bmatrix} 1 + \dfrac{\psi}{\alpha} \dfrac{\cosh \Gamma \xi}{\cosh \Gamma} \\ 1 - \dfrac{\cosh \Gamma \xi}{\cosh \Gamma} \end{Bmatrix} \tag{3.136}$$

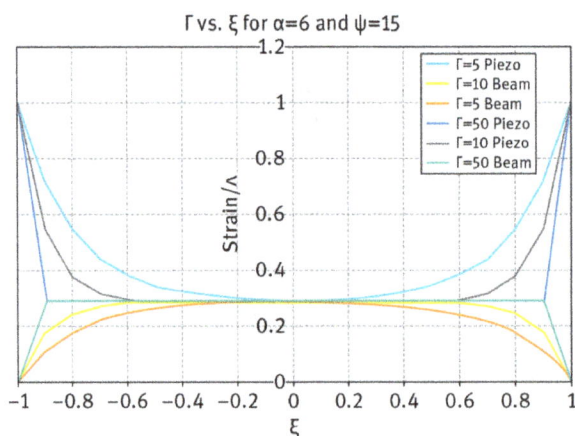

Figure 3.23: The nondimensional strains in the piezoelectric layers and carrying structure along the length of the beam, for $\Gamma = 5, 10, 50$ for $\alpha = 6$ and $\psi = 15$.

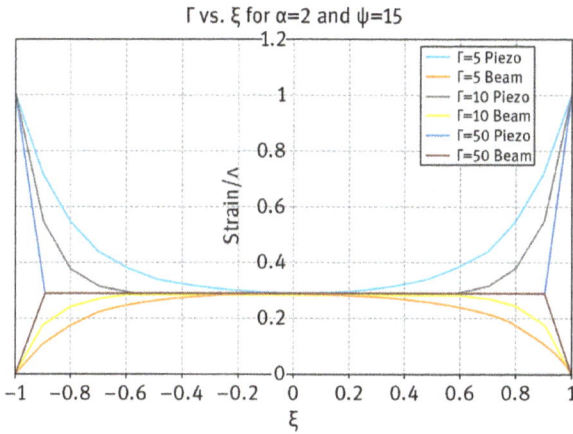

Figure 3.24: The nondimensional strains in the piezoelectric layers and carrying structure along the length of the beam, for $\Gamma = 5, 10, 50$ for $\alpha = 2$ and $\psi = 15$.

The two values of the strains, presented in Equation (3.134) are plotted in Figures 3.23 and 3.24, for various values of Γ, while $\psi = 15$ and $\alpha = 6$ and $\alpha = 2$ (typically for a piezoelectric ceramic bonded on a carrying aluminum beam 10 times its thickness), respectively.

The term Γ defined by Equation (3.123) is called the nondimensional shear lag parameter and represents the effectiveness of the shear–stress transfer between two adjacent bonded layers. As can be seen from Equation (3.123) it is a function of the shear modulus of the BL, GBL, its thickness and the terms ψ and α. Increasing the value of GBL or decreasing the thickness would increase the value of Γ, which will mean that the shear lag is less significant, and the shear stress is effectively transferred at the ends of the piezoelectric layers (see Figure 3.23). The other parameter influencing the nondimensional shear–lag parameter is the term ψ, which is the product of the thickness and moduli ratios between the carrying beam and the piezoelectric layer. If ψ tends to zero, which means that the carrying structure has a very small thickness, it will lead to a perfect bond ($\Gamma \to \infty$) and the strain in the piezoelectric layer will be equal to that in the carrying structure (Equation (3.136)). For the other case, large ψ means that the carrying structure is relatively very thick with a high Young's modulus, as compared with the properties of the piezoelectric layer, which would imply that the strain experienced by the carrying structure is relatively small. Based on what have been written before, a surface-bonded piezoelectric actuator should have a large Γ (namely, almost perfect bonding to the carrying structure) and low ψ (meaning a stiff actuator as compared to the carrying structure stiffness).

One should also note that the two piezoelectric layers in Figure 3.22 have voltages in the same direction, leading to either extension or contraction (depending on the sign of d_{31}) leading to $\alpha = 2$.

If one of the voltages is reversed, the piezoelectric layers will excite bending in the carrying structure and $\alpha = 6$. One should note that due to the high value of α, piezoelectric bonded actuators are more efficient in bending rather than in extension or contraction.

After presenting the case of surface-mounted piezoelectric layers, the case of embedded piezoelectric actuators will be dealt, as is schematically shown in Figure 3.25a and b.

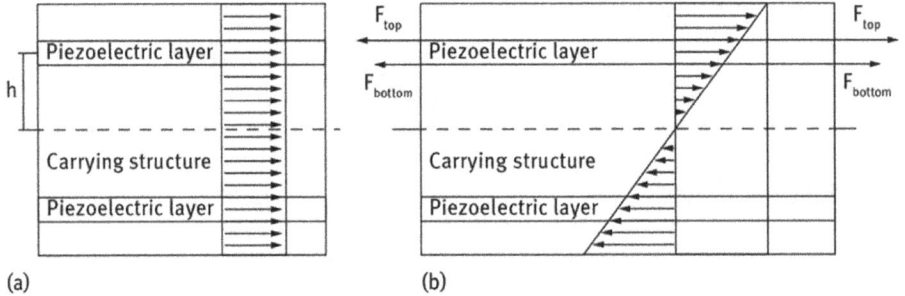

Figure 3.25: Strain distribution across the carrying structure and embedded piezoelectric layers: (a) extension and (b) bending.

Piezoelectric layers can be embedded in a carrying structure made of laminated composite materials. In this case the thickness of the BL is assumed to be negligible, as it is normally five times the diameter of the carbon fiber ($\approx 1\,\mu$m). As stated in [2], the shear–lag parameter has the large value of $\Gamma \approx 1000$, which fully justifies the assumption of perfect bonding.

Unlike the surface-mounted piezoelectric actuators, exciting bending, the embedded ones would experience a linear variation of strain through their thickness, and strain compatibility between the carrying structure and the piezoelectric layers is demanded at both the top and bottom surfaces of the actuators. Taking into account the above assumptions, the equilibrium equations for the stresses in the carrying structure and the piezoelectric layers can be written as

$$\sigma_p = \frac{6F_{\text{DIF}}}{w_p t_p^2}(z - h) + \frac{F_{\text{SUM}}}{w_p t_p} \tag{3.137}$$

$$\sigma_b = \frac{(M_p + M_M)}{I_{\text{eq}}}z \tag{3.138}$$

where M_M is the moment along the carrying structure due to other loadings than the piezoelectric layers; $F_{\text{DIF}} = F_2 - F_1$, $F_{\text{SUM}} = F_1 + F_2$, w_p and t_p are the width and the thickness of the piezoelectric layer, respectively, I_{eq} is the equivalent moment of inertia of the beam's cross section (including the piezoelectric material and the material of the carrying structure, which was assumed to be laminated composite one) and M_p (the piezoelectric-induced bending moment) is defined as

$$M_p = 2\left[\frac{F_{DIF}}{2}t_p + F_{SUM}h\right] \tag{3.139}$$

Using the strain–stress relations for the piezoelectric layer and the carrying structures (Equations (3.115) and (3.116)), and equating the strains in the piezoelectric layer and the carrying structure at their boundaries, namely at $z = (h - t_p/2)$ and $z = (h + t_p/w)$, we obtain two equations for the two unknowns, F_{DIF} and F_{SUM}. Their expressions are given as (see also [2])

$$F_{SUM} = -\frac{12\theta_z}{24\theta_z^2 + \psi\theta_b^2\hat{I} + 2}\frac{M_M}{t_p} - \frac{2E_pt_pw_p + E_bt_bw_p\theta_b^2\hat{I}}{24\theta_z^2 + \psi\theta_b^2\hat{I} + 2}\Lambda \tag{3.140}$$

$$F_{DIF} = -\frac{2_z}{24\theta_z^2 + \psi\theta_b^2\hat{I} + 2}\frac{M_M}{t_p} + \frac{4E_pt_pw_p\theta_b}{24\theta_z^2 + \psi\theta_b^2\hat{I} + 2}\Lambda \tag{3.141}$$

where

$$w_p = w_b; \quad \theta_b = \frac{t_b}{t_p}; \quad \theta_z = \frac{h}{t_p}; \quad \psi = \frac{E_bt_b}{E_pt_p} = \overline{E_b}\theta_b; \quad \hat{I} = \frac{I_{EQ}}{w_pt_p^3/12} \tag{3.142}$$

The term M_M can be expressed as a linear function of the known strains ε_b^{top} and ε_b^{bottom} along the carrying structure, at the ends of piezoelectric layers

$$M_M = -\frac{I_{EQ}E_b}{t_b}\left(\varepsilon_b^{top} + \varepsilon_b^{bottom}\right) - \frac{I_{EQ}E_b}{t_b}\left(\varepsilon_b^{top} - \varepsilon_b^{bottom}\right)\xi \tag{3.143}$$

Using Equations (3.141)–(3.143) and the strain–stress relationship for the beam, the expression for the strain in the actuator and the beam can be written as

$$\varepsilon_b = \varepsilon_p = -\frac{2\psi\theta_b\hat{I}}{\left(24\theta_z^2 + \psi\theta_b^2\hat{I} + 2\right)}\frac{z}{t_p}\left(\frac{\varepsilon_b^{top} + \varepsilon_b^{bottom}}{2} + \frac{\varepsilon_b^{top} - \varepsilon_b^{bottom}}{2}\xi\right)$$
$$+ \frac{24\theta_z}{\left(24\theta_z^2 + \psi\theta_b^2\hat{I} + 2\right)}\frac{z}{t_p}\Lambda \tag{3.144}$$

The normalized moment, applied at the ends of the piezoelectric layers is then given by

$$\frac{M_p}{E_b t_b^2 w_b} = \frac{\left(12\theta_z^2 + 1\right)\hat{I}}{3\left(24\theta_z^2 + \psi\theta_b^2\hat{I} + 2\right)} \left(\frac{\varepsilon_b^{\text{top}} + \varepsilon_b^{\text{bottom}}}{2} + \frac{\varepsilon_b^{\text{top}} - \varepsilon_b^{\text{bottom}}}{2}\xi\right)$$
$$- \frac{2\theta_z\theta_b\hat{I}}{\left(24\theta_z^2 + \psi\theta_b^2\hat{I} + 2\right)}\Lambda \tag{3.145}$$

The analysis for the simpler case of embedded actuators in extension or contraction results in the following expression for the strain in the actuator and the beam:

$$\varepsilon_b = \varepsilon_p = \frac{\bar{E}(\theta_b - 2)}{2 + \bar{E}(\theta_b - 2)} \left(\frac{\varepsilon_b^{\text{top}} + \varepsilon_b^{\text{bottom}}}{2} + \frac{\varepsilon_b^{\text{top}} - \varepsilon_b^{\text{bottom}}}{2}\xi\right) + \frac{2}{2 + \bar{E}(\theta_b - 2)}\Lambda \tag{3.146}$$

where $\bar{E} = \dfrac{E_b}{E_p}$. The force applied at each end of the piezoelectric layer is given by

$$\frac{F}{E_b t_b w_b} = -\frac{(\theta_b - 2)}{2\theta_b + \bar{E}(\theta_b - 2)\theta_b} \left(\frac{\varepsilon_b^{\text{top}} + \varepsilon_b^{\text{bottom}}}{2} + \frac{\varepsilon_b^{\text{top}} - \varepsilon_b^{\text{bottom}}}{2}\xi\right)$$
$$- \frac{2}{2\theta_b + \bar{E}(\theta_b - 2)\theta_b}\Lambda \tag{3.147}$$

Bibliography

[1]   Leo, D. J. , Kothera, C. and Farinholt, K., Constitutive equations for an induced-strain bending actuator with a variable substrate, Journal of Intelligent Material Systems and Structures 14, November 2003, 707–718.

[2]   Crawley, E. F. and De Luis, J., Use of piezoelectric actuators as elements of intelligent structures, AIAA Journal 25(10), October 1987, 1373–1385.

[3]   Crawley, E. F. and Anderson, C.A., Detailed models of piezoceramic actuation of beams, Journal of Intelligent Material Systems and Structures 1(1), January 1990, 4–25.

[4]   Hagood, N., Chung, W. and Von Flotow, A., Modelling of piezoelectric actuator dynamics for active structural control, Journal of Intelligent Material Systems and Structures 1, July 1990, 327–354.

[5]   Wang, K., Modeling of piezoelectric generator on a vibrating beam. For completion of Class Project in ME 5984 Smart Materials, Virginia Polytechnic Institute and State University, April 2001.

[6]   Wang, X., Ehlers, C. and Neitzel, M., An analytical investigation of static models of piezoelectric patches attached to beams and plates, Smart Materials and Structures 6, 1997, 204–213.

[7]   Wang, L., Ruixiang, B. and Cheng, Y., Interfacial debonding behavior of composite beam/ plates with PZT patch, Composite Structures 92(6), 2010, 1410–1415.

[8]   Tylikowski, A., Influence of bonding layer on piezoelectric actuators on a axisymmetric annular plate, Journal of Theoretical and Applied Mechanics 3(38), 2000, 607–621.

[9]   Pietrzakovski, M., Dynamic model of beam-piezoceramic actuator coupling for active vibration control, Journal of Theoretical and Applied Mechanics 1(35), 1997, 3–20.

[10] Galichyan, T. A. and Filippov, D. A., The influence of the adhesive bonding on the magnetoelectric effect in bilayer magnetostrictive-piezoelectric structure, Journal of Physics: Conference Series 572, 2014, 012045, 1–6. doi:10.1088/1742-6596/572/1/012045.

[11] Golub, M. V., Buethe, I., Shpak, A. N., Fritzen, C. P., Jung, H. and Moll, J., Analysis of Lamb wave excitation by the partly de-bonded circular piezoelectric wafer active sensors, 11th European Conference on Non-Destructive Testing (ECNDT 2014), October 6–10, 2014, Prague, Czech Republic.

[12] Pohl, J., Willberg, C., Gabbert, U. and Mook, G., Experimental and theoretical analysis of Lamb wave generation by piezoceramic actuators for structural health monitoring, Experimental Mechanics 52, 429. doi:10.1007/s11340-011-9503-2.

[13] Chopra, I. and Sirohi, J., Smart Structures Theory, Cambridge University Press, 2014, 920.

3.4 Bernoulli–Euler Beam Model

The Bernoulli–Euler beam model (B–EBM) carries its name from the beam theory with the same name, which is usually used for slender beams. The literature shows (Refs. [1–3]) that this model gives more accurate predictions than the uniform-strain model when the bonding layer is assumed to be thin. According to the B–EBM, the strain is assumed to be linear throughout the carrying structure, the beam, and the piezoelectric and bonding layers. The effects of the transverse shear strains are neglected and planes normal to the neutral axis of the beam, before deformation would remain plane after the deformation (these are the usual B–EBM assumptions). The derivation will be done for a laminated composite beam, using the classical lamination theory (see Chapter 2) which leads to the B–EBM in case the layers are isotropic. Figure 3.26 shows a schematic drawing of a beam in bending and its associated coordinate system.

Let us define a coordinate system X, Y, Z, with the X coordinate along the middle plane of the beam, the Z coordinate being perpendicular to the beam's longitudinal axis. As such, the Y coordinate would be in the beam's width direction, thus forming a right-hand coordinate system (see Figure 3.26).

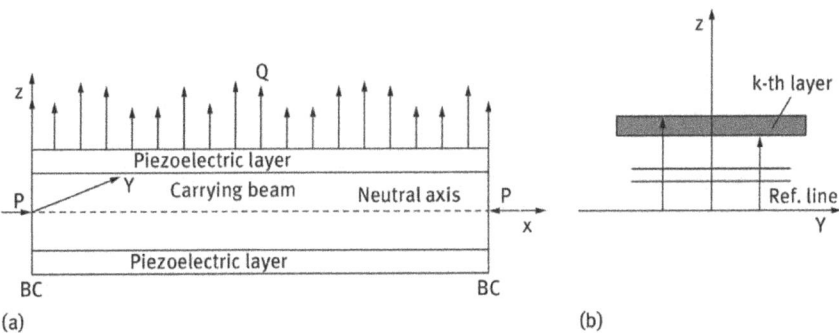

Figure 3.26: Beam with piezoelectric layers.

The only two variables of the model would be an axial displacement (in the x-direction) of a point on the middle plane of the beam, designated $u(x)$ and a lateral displacement (in the z-direction) of a point on the middle plane of the beam, designated to be $w(x)$.

Taking this into account, we can write the axial displacements across the whole cross section of the structure (beam, piezoelectric and bonding layers) as

$$u(x,z) = u_0(x) - z\frac{\partial w(x)}{\partial x} \tag{3.148}$$

$$w(x,z) = w_0(x) \tag{3.149}$$

where $u_0(x)$ and $w_0(x)$ are the displacements of a point on the neutral axis of the carrying beam.

Accordingly, the strain vector would include the mechanical strain, ε^m, and the induced strain caused by the application of voltage across the piezoelectric electrodes, ε^a, both on the same point. One should remember that the induced strain would be zero for nonpiezoelectric layers.

The mechanical strains are obtained by derivation of the displacements presented earlier in Equations (3.148) and (3.149), to yield

$$\varepsilon_x^m(x,z) = \frac{\partial u(x,z)}{\partial x} = \frac{\partial u_0(x)}{\partial x} - z\frac{\partial^2 w(x)}{\partial x^2} \equiv \varepsilon_0(x) - z\kappa(x) \tag{3.150}$$

while all the other mechanical strains are $\varepsilon_y^m(x) = \varepsilon_z^m(x) = \gamma_{xy}^m(x) = \gamma_{yz}^m(x) = \gamma_{zx}^m(x) = 0$, due to the assumptions of the Bernoulli–Euler theory. $\kappa(x)$ is the beam's curvature.

The induced strains $\varepsilon_x^a(x), \varepsilon_z^a(x), \gamma_{xy}^a(x)$ would be (see also [6–7])

$$\left[\varepsilon_x^a(x)\right]_k = \frac{V_k(x)(d_{31})_k}{z_{k+}^a - z_{k-}^a} \tag{3.151}$$

$$\varepsilon_z^a(x) = \gamma_{xy}^a(x) = 0 \tag{3.152}$$

In Equation (3.151), the term $V_k(x)$ is the voltage applied on the kth piezoelectric layer, having a thickness of $(z_{k+}^a - z_{k-}^a)$ (see Figure 3.26b), and $(d_{31})_k$ is the piezoelectric coefficient for the kth layer. As already mentioned above, Equations (3.151) and (3.152) are true only for the piezoelectric layers.

Multiplying the strains by the stiffness coefficients $(\overline{Q_{ij}})_k$, for each layer (see also [8]) yields

$$(\sigma_x)_k = (\overline{Q_{11}})_k (\varepsilon_x^m - \varepsilon_x^a)_k \tag{3.153}$$

where the term $(\overline{Q_{11}})$ (for each single layer) is defined as

$$\overline{Q_{11}} = Q_{11}\cos^4\theta + Q_{22}\sin^4\theta + 2(Q_{12} + 2Q_{66})\sin^2\theta\cos^2\theta \tag{3.154}$$

The θ is the angle between the direction of the fibers on a particular layer and the longitudinal axis, x. The coefficients Q_{11}, Q_{22}, Q_{12} and Q_{66} are defined as a function of the elastic properties of each layer (see also Chapter 2), namely

$$Q_{11} = \frac{\overline{E_{11}}}{1 - v_{12}v_{21}}; \quad Q_{12} = \frac{v_{21}\overline{E_{11}}}{1 - v_{12}v_{21}} = \frac{v_{12}\overline{E_{22}}}{1 - v_{12}v_{21}}; \quad Q_{22} = \frac{\overline{E_{22}}}{1 - v_{12}v_{21}}; \quad Q_{66} = G_{12} \quad (3.155)$$

where $\overline{E_{11}}; \overline{E_{22}}; G_{12}$ are the tension/compression and shear moduli of the single layer and v_{12} is its major Poisson's ratio and the following relationship holds $\frac{\overline{E_{11}}}{\overline{E_{22}}} = \frac{v_{12}}{v_{21}}$ (see Chapter 2):

Integrating along the height of the beam, h, and multiplying by the width, b, yields the force and moment resultants:

$$N_x = b \int_{-h/2}^{+h/2} \sigma_x dz, \quad M_y = b \int_{-h/2}^{+h/2} \sigma_x z dz \quad (3.156)$$

Performing the various integrals in Equation (3.156) yields the following relationship:

$$\left\{ \begin{array}{c} N_x \\ M_y \end{array} \right\} = \begin{bmatrix} A_{11} & B_{11} \\ B_{11} & D_{11} \end{bmatrix} \left\{ \begin{array}{c} \frac{\partial u}{\partial x} \\ -\frac{\partial^2 w}{\partial x^2} \end{array} \right\} - \left\{ \begin{array}{c} E_{11} \\ F_{11} \end{array} \right\} \quad (3.157)$$

The coefficients A_{11}, B_{11}, D_{11} are the well-known expressions for the axial, the coupled bending–stretching, the bending and lateral shear stiffness, respectively, used usually for laminated composite beams. According to the classical lamination theory [8], the integrals are replaced by a summation (see Figure 3.26b), yielding the following expressions:

$$A_{11} = b \sum_{k=1}^{N} (\overline{Q_{11}})_k (z_k - z_{k-1}) \quad (3.158)$$

$$B_{11} = \frac{b}{2} \sum_{k=1}^{N} (\overline{Q_{11}})_k (z_k^2 - z_{k-1}^2) \quad (3.159)$$

$$D_{11} = \frac{b}{3} \sum_{k=1}^{N} (\overline{Q_{11}})_k (z_k^3 - z_{k-1}^3) \quad (3.160)$$

where N is the laminate number of layers, including the piezoelectric ones.

The terms E_{11} and F_{11} in Equation (3.157) represent the induced axial force and the induced moment due to the piezoelectric layers, respectively, and are defined as

$$E_{11} = b \sum_{k=1}^{N_a} (\overline{Q_{11}})_k^a V_k(x,t)(d_{31})_k \quad (3.161)$$

$$F_{11} = \frac{b}{2} \sum_{k=1}^{N_a} (\overline{Q_{11}})_k^a V_k(x,t)(d_{31})_k \left(z_{k+}^a + z_{k-}^a\right) \tag{3.162}$$

where N_a is the laminate number of piezoelectric layers.

To obtain the equations of motion and the associated boundary conditions, we shall use the principle of virtual work for the present problem yielding the variation of the potential as (see [6]):

$$\delta\pi = \int_{t_1}^{t_2} \int_{volume} (\delta E_k - \delta E_P + \delta\overline{W}) \cdot d(volume) dt = 0 \tag{3.163}$$

where E_k stands for the kinetic energy of the beam, E_p is its strain energy, and $\overline{W}$ represents the work of the external forces acting on the beam. Their expressions are

$$E_k = \frac{1}{2} \int_{volume} \rho \cdot \left[\left(\dot{u} - z\frac{\partial \dot{w}}{\partial x} \right)^2 + \dot{w}^2 \right] \cdot d(volume) \tag{3.164}$$

$$E_p = \frac{1}{2} \int_{volume} (\sigma_x \varepsilon_x) \cdot d(volume) \tag{3.165}$$

$$\overline{W} = \int_0^L Q \cdot w \cdot x \cdot dx - P \cdot u(L) + P \cdot u(0) \tag{3.166}$$

As shown in Figure 3.26a, L is the beam's length, P is the compressive loads acting on both ends of the beam, while Q is the distributed lateral load acting along the beam. ($\dot{}$) stands for differentiation with respect to time.

Substituting the expressions presented in Equations (3.164)–(3.166) into Equation (3.163), integrating by parts, while demanding that the variation of the potential, $\delta\pi$, should vanish, which will hold true only if the various expressions in the integral and at the two ends of the beam are identically zero (as the assumed displacements are arbitrary), yields the following equations of motion for a beam with piezoelectric layers, and its associated boundary conditions:

$$\frac{\partial N_x}{\partial x} = \frac{\partial}{\partial t} \left[I_1 \dot{u} + I_2 \frac{\partial \dot{w}}{\partial x} \right] \tag{3.167}$$

$$\frac{\partial}{\partial x} \left[\frac{\partial M_y}{\partial x} + N_x \frac{\partial w}{\partial x} \right] = \frac{\partial}{\partial t} \left[I_1 \dot{w} \right] + Q \tag{3.168}$$

while $(I_1, I_2) = \int \rho \cdot (1, z) \cdot dz$ and ρ being the density of each layer of the beam. The associated possible boundary conditions are

$$N_x + P = 0 \quad \text{or} \quad u = 0 \tag{3.169}$$

$$\frac{\partial M_y}{\partial x} + N_x \frac{\partial w}{\partial x} = 0 \quad \text{or} \quad w = 0 \tag{3.170}$$

$$M_y = 0 \quad \text{or} \quad \frac{\partial w}{\partial x} = 0 \tag{3.171}$$

Note that as is usually presented in the literature, the term N_x is replaced by $-P$ only in the above nonlinear expression.

$$\left[N_x \frac{\partial w}{\partial x} \right]$$

Using the expressions for the forces and moments resultants (Equation (3.157)) and substituting them into Equations (3.167) and (3.168) yields the equations of motion for the beam with piezoelectric layers expressed in the two assumed deflections, u and w:

$$\frac{\partial}{\partial x} \left(A_{11} \frac{\partial u}{\partial x} - B_{11} \frac{\partial^2 w}{\partial x^2} - E_{11} \right) = \frac{\partial}{\partial t} \left[I_1 \dot{u} + I_2 \frac{\partial \dot{w}}{\partial x} \right] \tag{3.172}$$

$$\frac{\partial}{\partial x} \left[\frac{\partial}{\partial x} \left(B_{11} \frac{\partial u}{\partial x} - D_{11} \frac{\partial^2 w}{\partial x^2} - F_{11} \right) - P \frac{\partial w}{\partial x} \right] = \frac{\partial}{\partial t} \left[I_1 \dot{w} \right] + Q \tag{3.173}$$

with their associated possible boundary conditions:

$$A_{11} \frac{\partial u}{\partial x} - B_{11} \frac{\partial^2 w}{\partial x^2} = -P + E_{11} \quad \text{or} \quad u = 0 \tag{3.174}$$

$$\frac{\partial}{\partial x} \left(B_{11} \frac{\partial u}{\partial x} - D_{11} \frac{\partial^2 w}{\partial x^2} - F_{11} \right) - P \frac{\partial w}{\partial x} = 0 \quad \text{or} \quad w = 0 \tag{3.175}$$

$$B_{11} \frac{\partial u}{\partial x} - D_{11} \frac{\partial^2 w}{\partial x^2} = F_{11} \quad \text{or} \quad \frac{\partial w}{\partial x} = 0 \tag{3.176}$$

Assuming constant properties along the beam, Equations (3.172)–(3.176) will have the following form:

$$\left(A_{11} \frac{\partial^2 u}{\partial x^2} - B_{11} \frac{\partial^3 w}{\partial x^3} \right) = I_1 \ddot{u} + I_2 \frac{\partial \ddot{w}}{\partial x} \tag{3.177}$$

$$\left(B_{11} \frac{\partial^3 u}{\partial x^3} - D_{11} \frac{\partial^4 w}{\partial x^4} - P \frac{\partial^2 w}{\partial x^2} \right) = I_1 \ddot{w} + Q \tag{3.178}$$

while two dots above a letter means twice differentiation with respect to time. The associated possible boundary conditions will have the following forms:

$$A_{11} \frac{\partial u}{\partial x} - B_{11} \frac{\partial^2 w}{\partial x^2} = -P + E_{11} \quad \text{or} \quad u = 0 \tag{3.179}$$

$$B_{11}\frac{\partial^2 u}{\partial x^2} - D_{11}\frac{\partial^3 w}{\partial x^3} - P\frac{\partial w}{\partial x} = 0 \quad \underline{\text{or}} \quad w = 0 \tag{3.180}$$

$$B_{11}\frac{\partial u}{\partial x} - D_{11}\frac{\partial^2 w}{\partial x^2} = F_{11} \quad \underline{\text{or}} \quad \frac{\partial w}{\partial x} = 0 \tag{3.181}$$

One should note that the influence of the piezoelectric layers appears only through the boundary conditions. For symmetric laminates, the terms B_{11} and I_2 would vanish yielding the following uncouple equations of motion:

$$A_{11}\frac{\partial^2 u}{\partial x^2} = I_1\ddot{u} \tag{3.182}$$

$$\left(-D_{11}\frac{\partial^4 w}{\partial x^4} - P\frac{\partial^2 w}{\partial x^2} \right) = I_1\ddot{w} + Q \tag{3.183}$$

The boundary conditions will have the following simpler expressions:

$$A_{11}\frac{\partial u}{\partial x} = -P + E_{11} \quad \underline{\text{or}} \quad u = 0 \tag{3.184}$$

$$D_{11}\frac{\partial^3 w}{\partial x^3} + P\frac{\partial w}{\partial x} = 0 \quad \underline{\text{or}} \quad w = 0 \tag{3.185}$$

$$D_{11}\frac{\partial^2 w}{\partial x^2} = -F_{11} \quad \underline{\text{or}} \quad \frac{\partial w}{\partial x} = 0 \tag{3.186}$$

First, we shall solve the static case for two types of induced tractions, due to voltage supplied to the piezoelectric layers. Applying to all the piezoelectric layers identical voltages (both amplitude and sign) will result in a net tension or compression of the whole structure. For that loading case, we will use only Equation (3.182) while $I_1 = 0$ and its associated in plane boundary conditions presented by Equation (3.184). The solution for Equation (3.182) has the general form

$$u(x) = ax + b \tag{3.187}$$

where a and b are the constants to be determined from the boundary conditions of the case.

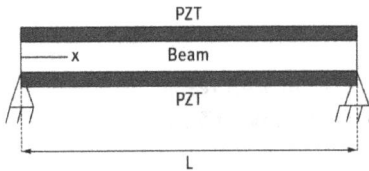

Figure 3.27: A beam with two piezoelectric layers axially restrained.

Let us demand that $u(0) = u(L) = 0$ (no axial displacements at both ends of the beam – Figure 3.27) which would yield a solution in the form of $u(x) = 0$. Substituting this result into the first line of Equation (3.157) gives

$$N_x = -E_{11}^0 \tag{3.188}$$

which after substituted into Equation (3.169) gives

$$P = E_{11}^0 \tag{3.189}$$

where the term E_{11}^0 stands for the static induced axial force. Increasing the voltage being supplied to the piezoelectric layers up to an induced force which is equal to the buckling load of the whole beam, will result in its buckling. The voltage that will cause this buckling, V_{cr}, can be defined to be (with the use of Equation (3.161) for the case the voltage is constant along the beam (thus it is not dependent on both x and time t):

$$V_{cr} = \frac{P_{cr}}{b \sum_{k=1}^{Na} (Q_{11})_k^a (d_{31})_k} ; \quad P_{cr} = \frac{\pi (EI)_{eq.}}{L^2} \tag{3.190}$$

where $(EI)_{eq.}$ is the bending stiffness of the beam with the piezoelectric layers. More on this issue can be found in the next chapter.

When out-of-phase voltages (same amplitude but with different sign) are applied to the upper and lower piezoelectric layers a bending moment will be induced on the beam. Let us solve first the static case of induced bending, namely the term involving the acceleration vanishes:

$$D_{11} \frac{d^4 w}{dx^4} + P \frac{d^2 w}{dx^2} = -Q \tag{3.191}$$

Division by D_{11}, assuming $Q = $ const. and denoting $k^2 = \frac{P}{D_{11}}$, we obtain the general solution for Equation (3.191):

$$w(x) = A_1 \sin(kx) + A_2 \cos(kx) - \frac{Q}{P} x^2 + A_3 x + A_4 \tag{3.192}$$

The four constants A_1–A_4 are to be found by application of four boundary conditions at both ends of the beam with composite layers. When no axial force is presented ($P = 0$), the solution would have the following form (with $Q = $ const.):

$$w(x) = -\frac{Q}{24D_{11}} x^4 + A_1 x^3 + A_2 x^2 + A_3 x + A_4 \tag{3.193}$$

As before, the four constants $A_1 - A_4$ are found from the application of the appropriate boundary conditions of the beam. If the lateral load, $Q(x)$, is a function of the axial coordinate, than integrating four times will yield the out-of-plane deflection of the beam with four constants to be determined by satisfying the boundary conditions of the beam. Let us solve Equation (3.193) for a beam with no lateral load, as presented in Figure 3.28. The voltages supplied to the piezoelectric layer are out-of-phase, namely the same amplitude but with different sign, thus inducing bending on the beam. The boundary conditions of the beam are

$$\text{at}\quad w(0) = 0, D_{11}\frac{d^2 w(0)}{dx^2} = -F_{11}$$

$$\text{at}\quad w(L) = 0, D_{11}\frac{d^2 w(L)}{dx^2} = -F_{11} \tag{3.194}$$

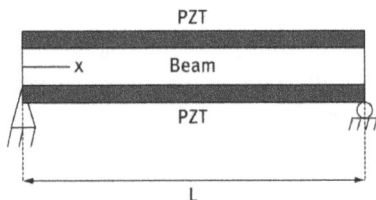

Figure 3.28: A beam on simply supported boundary conditions with two piezoelectric layers.

Application of the boundary conditions yields the following expression for the lateral deflection of a beam with induced bending:

$$w(x) = -\frac{Q}{24D_{11}}x^4 + \frac{QL}{12D_{11}}x^3 - \frac{F_{11}}{2D_{11}}x^2 + \left(\frac{F_{11}L}{2D_{11}} - \frac{QL^3}{24D_{11}}\right)x \tag{3.195}$$

One should note that for the case $Q = 0$, Equation (3.195) simplifies to the following expression:

$$w(x) = -\frac{F_{11}}{2D_{11}}x(x-L) \tag{3.196}$$

which is exactly the deflection equation of a beam on simply supported boundary conditions with two bending moments M_o acting at its both ends (in our case, $M_o = -F_{11}$, $D_{11} = EI$).

Also it is worth mentioning that, as the influence of the piezoelectric actuators in bending enters only through the boundary conditions of the beam, beams having only geometric boundary conditions at its both ends (like a clamped–clamped) will not be influenced by piezoelectric layers stretching from one side to the other side of the carrying beam. For these cases, only patches would induce bending in the beam.

After solving the static case, the dynamic case presented in Equation (3.183), while the beam has no mechanical loads, $Q = 0$ will be dealt. Its form will be

$$D_{11}\frac{\partial^4 w}{\partial x^4} + P\frac{\partial^2 w}{\partial x^2} + I_1 \ddot{w}\partial = 0 \tag{3.197}$$

Division by D_{11} and assuming harmonic vibrations $w(x,t) = W(x)\sin \omega t$, we obtain the following differential equation, with only one variable, x:

$$\frac{d^4 W}{dx^4} + k^2 \frac{d^2 W}{dx^2} - \omega^2 \bar{I}_1 W = 0 \tag{3.198}$$

where

$$k^2 = \frac{P}{D_{11}}; \quad \bar{I}_1 = \frac{I_1}{D_{11}} \tag{3.199}$$

and ω is the natural frequency of the beam.

The solution for Equation (3.198) is obtained by assuming $W = Re^{sx}$ and substituting in it. The result is

$$s^4 + k^2 s^2 - \omega^2 \bar{I}_1 = 0 \tag{3.200}$$

Equation (3.200) has two real and two complex solutions, namely,

$$s_1 = \pm \sqrt{\left[\frac{-k^2 + \sqrt{k^4 + 4\omega^2 \bar{I}_1}}{2} \right]} = \pm \lambda_1$$

$$s_2 = \pm \sqrt{\left[\frac{-k^2 - \sqrt{k^4 + 4\omega^2 \bar{I}_1}}{2} \right]} = \pm i\lambda_2 \tag{3.201}$$

The general solution will then have the following form:

$$W(x) = A_1 \sinh(\lambda_1 x) + A_2 \cosh(\lambda_1 x) + A_3 \sin(\lambda_2 x) + A_4 \cos(\lambda_2 x) \tag{3.202}$$

with the constants A_1–A_4 to be found by the application of the appropriate boundary conditions.

Using the boundary condition of a simply supported beam (Equation (3.194)), one needs to modify the boundary condition for the constant moment $-F_{11}$, to be able to calculate the natural frequencies of a beam under compression and induced bending moments at its two ends due to piezoelectric layers. The expression for the moment can be written as

$$-F_{11} = k_\theta \frac{dw(x)}{dx}; \quad k_\theta = - \frac{F_{11}}{\frac{dw(x)}{dx} \Big|_{x=0,L}} \tag{3.203}$$

Then, the equations for the boundary conditions will be:

$$\text{at} \quad w(0) = 0, D_{11} \frac{d^2 w(0)}{dx^2} = -k_\theta \frac{dw(0)}{dx}$$

$$\text{at} \quad w(L) = 0, D_{11} \frac{d^2 w(L)}{dx^2} = -k_\theta \frac{dw(L)}{dx} \tag{3.204}$$

Inserting the four boundary conditions into Equation (3.202) leads to the following expressions:

$$
\begin{bmatrix}
a_{11} & a_{12} & a_{13} & a_{14} \\
a_{21} & a_{22} & a_{23} & a_{24} \\
a_{31} & a_{32} & a_{33} & a_{34} \\
a_{41} & a_{42} & a_{43} & a_{44}
\end{bmatrix}
\begin{Bmatrix}
A_1 \\
A_2 \\
A_3 \\
A_4
\end{Bmatrix}
=
\begin{Bmatrix}
0 \\
0 \\
0 \\
0
\end{Bmatrix}
\tag{3.205}
$$

where the coefficients of the matrix are given in Appendix B.

Demanding the vanishing of the coefficients' determinant enables the calculation of the natural frequencies of a beam with piezoelectric layers inducing bending and compression loads.

To compare the predictions using either the uniform beam model or the Bernoulli–Euler model, the expressions for the strains in the piezoelectric layer or the carrying

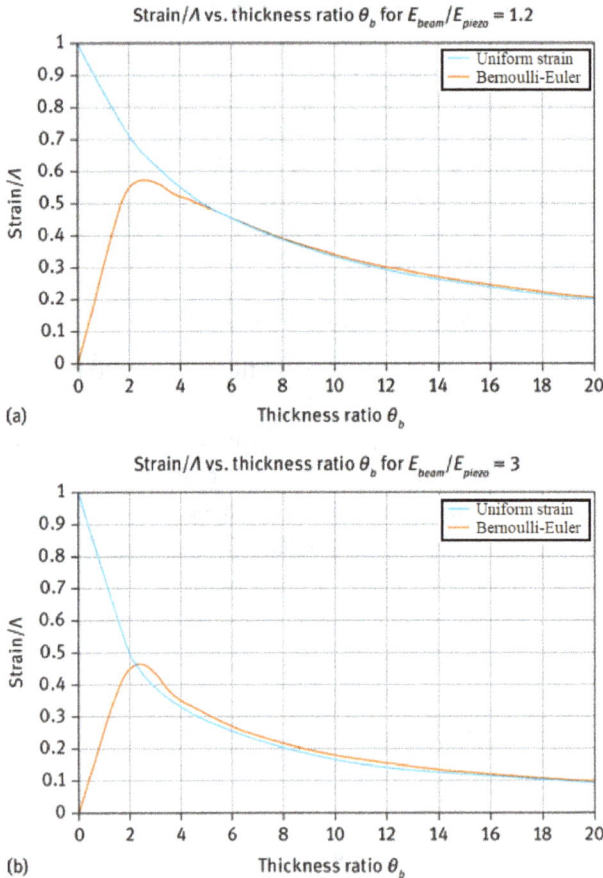

(a)

(b)

Figure 3.29: Bending strain comparison: (a) $E_{beam}/E_{piezo} = 1.2$ and (b) $E_{beam}/E_{piezo} = 3$.

beam are summarized in Table 3.3 (see also [1] and Section 3.3). The beam configuration is given in Figure 3.28.

Figure 3.29 presents the strain/Λ versus the thickness ratio θ_b for $E_{beam}/E_{piezo} = 1.2$ and 3. It is clear from the graphs presented in Figure 3.29 that for large values of the thickness ratio, the prediction of the strains using either the uniform model or the Bernoulli–Euler model yields almost identical results. At low values of θ_b, the uniform model does not predict correctly the strains, while the Bernoulli–Euler model does it well.

Table 3.3: Uniform beam model (without shear lag) versus Bernoulli–Euler beam model[*].

Uniform beam model	Bernoulli–Euler beam model
Extension or compression-induced strains	
$\varepsilon_{piezo} = \dfrac{2\Lambda}{2 + \psi_e}$	$\varepsilon_{piezo} = \dfrac{2\Lambda}{2 + \overline{\psi}_e}$
$\varepsilon_{beam} = \dfrac{2\Lambda}{2 + \psi_e}$	$\varepsilon_{beam} = \dfrac{2\Lambda}{2 + \overline{\psi}_e}$
Bending-induced strains	
$\varepsilon^{*}_{piezo} = -\dfrac{t_b}{2}K = \dfrac{6\Lambda}{6 + \psi_b}$	$\varepsilon^{*}_{piezo} = -\dfrac{t_b}{2}K = \dfrac{6\left(1 + \frac{1}{\theta_b}\right)}{\left(6 + \psi_e\right) + \frac{12}{\theta_b} + \frac{8}{\theta_b^2}}\Lambda$
$\varepsilon^{*}_{beam} = -\dfrac{t_b}{2}K = \dfrac{6\Lambda}{6 + \psi_b}$	$\varepsilon^{*}_{beam} = -\dfrac{t_b}{2}K = \dfrac{6\left(1 + \frac{1}{\theta_b}\right)}{\left(6 + \psi_e\right) + \frac{12}{\theta_b} + \frac{8}{\theta_b^2}}\Lambda$

[*]where $\psi_e = \dfrac{(EA)_{beam}}{(EA)_{piezo}}$; $\quad \psi_b = \dfrac{12(EI)_{beam}}{t^2_{beam}(EA)_{piezo}}$;

for a rectangular cross section we get:

$$\psi_e = \psi_b; \quad \overline{\psi}_e = \dfrac{(EI)_{beam}}{(EA)_{piezo}} = \dfrac{E_{beam}\left[(w \cdot t)_{beam} - 2(w \cdot t)_{piezo}\right]}{(E \cdot w \cdot t)_{piezo}};$$

ε^{*}_{beam} is the strain at the interface between the beam and the piezoelectric layer; $\theta_b = \dfrac{t_{beam}}{t_{piezo}}$;

t_{beam} is the height of the beam only, w is the width of the beam and also of the piezoelectric layer.

References

[1]   Crawley, E. F. and Anderson, E. H., Detailed models of piezoceramic actuation of beams, Journal of Intelligent Material Systems and Structures 1(1), 1990, 4–25.

[2]   Wu, K. and Janocha, H., Optimal thickness and depth for embedded piezoelectric actuators, Proceedings of the Third European Conference on Structural Control, 3ECSC, 12–15 July 2004, Vienna University of Technology, Vienna, Austria.

[3]   Zehetner, C. and Irschik, H., On the static and dynamic stability of beams with an axial piezoelectric actuation, Smart Structures and Systems 4(1), 2008, 67–84.

[4]   Chopra, I. and Sirohi, J., Smart Structures Theory, Cambridge University Press, 2007, 920.

[5]   Leo, D.J., Engineering Analysis of Smart Material Systems, John Wiley & Sons, Inc., 2007, 576.

[6]   Sirohi, J. and Chopra, I., Fundamental understanding of piezoelectric strain sensors, Journal of Intelligent Material Systems and Structures 11(4), April 2000, 246–257.

[7]   Sirohi, J. and Chopra, I., Fundamental behavior of piezoceramic sheet actuators, Journal of Intelligent Material Systems and Structures 11(1), January 2000, 47–61.

[8]   Reddy, J. N., Mechanics of Laminated Composite Plates: Theory and Analysis, CRC Press LLC, 2004, 831.

3.5 First-Order Shear Deformation (Timoshenko-Type) Beam Model

This section is devoted to the derivation of the equations of motion for a laminated beam with piezoelectric layers, as depicted in Figure 3.30.

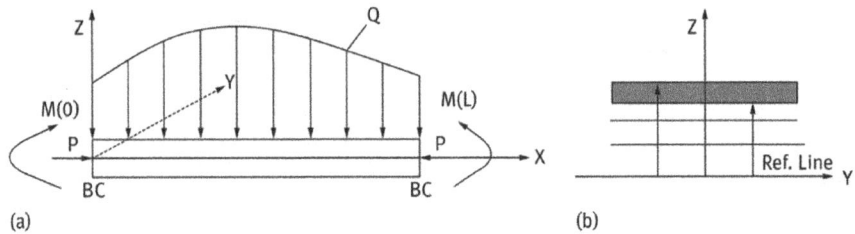

Figure 3.30: (a) Beam's model; (b) cross section of the laminated beam.

First, we shall present the possible displacements of the beam using the first-order shear deformation theory (FSDT), which is identical to the Timoshenko model for isotropic beams.

Let us define a coordinate system X, Y, Z, with the X coordinate along the middle plane of the beam and the Z coordinate being perpendicular to the beam's longitudinal axis. As such, the Y coordinate would be in the beam's width direction, thus forming a right-hand coordinate system (see Figure 3.30a).

The variables of the model would be

(a) An axial displacement (in the x-direction) of a point on the middle plane of the beam, designated to be $U(x,t)$.

(b) A lateral displacement (in the z-direction) of a point on the middle plane of the beam, designated to be $W(x,t)$.

(c) Rotation of the cross section of the beam (around y-axis), Φ (x, t), which is clockwise positive.

The letter t stands for time. Also, it is assumed that the bonding of the layers, including the piezoelectric ones, is perfect and the thickness of the glue can be neglected, and there is no compression in the z-direction.

Based on the above assumptions, the displacement field for any point on the beam's cross section would be as follows:

$$\tilde{U}(x,z,t) = U(x,t) + z \cdot \Phi(x,t) \tag{3.206}$$

$$\tilde{W}(x,z,t) = W(x,t) \tag{3.207}$$

Accordingly, the strain vector would include the mechanical strain, ε^m, and the induced strain caused by the application of voltage across the piezoelectric electrodes, ε^a, both on the same point. One should remember that the induced strain would be zero for nonpiezoelectric layers.

The mechanical strains, $\varepsilon_x^m, \varepsilon_z^m, \gamma_{xz}^m$ (the normal and shear ones) are obtained by derivation of the displacements presented earlier in Equations (3.206) and (3.207):

$$\varepsilon_x^m = \frac{\partial \tilde{U}}{\partial x} + \frac{1}{2}\left(\frac{\partial W}{\partial x}\right)^2 = \frac{\partial U}{\partial x} + z\frac{\partial \Phi}{\partial x} + \frac{1}{2}\left(\frac{\partial W}{\partial x}\right)^2 \tag{3.208}$$

Note that the above expression also includes a nonlinear term, which is important for a beam under axial compression (if exists) due to its interaction

$$\varepsilon_z^m = \frac{\partial \tilde{W}}{\partial z} = 0 \tag{3.209}$$

$$\gamma_{xz}^m = \frac{\partial \tilde{U}}{\partial z} + \frac{\partial \tilde{W}}{\partial x} = \Phi + \frac{\partial W}{\partial x} \tag{3.210}$$

The induced strains $\varepsilon_x^a, \varepsilon_z^a, \gamma_{xz}^a$ would be (see also [4–5])

$$\left(\varepsilon_x^a\right)_k = \frac{V_k(x,t)(d_{31})_k}{z_{k+}^a - z_{k-}^a} \tag{3.211}$$

$$\varepsilon_z^a = \gamma_{xz}^a = 0 \tag{3.212}$$

In Equation (3.211), the term $V_k(x,t)$ is the voltage applied on the kth piezoelectric layer, having a thickness of $z_{k+}^a - z_{k-}^a$ (see Figure 3.30b), and $(d_{31})_k$ is the piezoelectric coefficient for the kth layer. As already mentioned above, Equations (3.211) and (3.212) are true only for the piezoelectric layers.

Multiplying the strains by the stiffness coefficients $(\overline{Q}_{ij})_k$ for each layer (see also [1]) yields

$$(\sigma_x)_k = (\overline{Q}_{11})_k (\varepsilon_x^m - \varepsilon_x^a)_k \quad (\tau_{xz})_k = (\overline{Q}_{55})_k (\varepsilon_{xz})_k \tag{3.213}$$

where the terms $\overline{Q}_{11}$ and $\overline{Q}_{55}$ (for each single layer) are defined as

$$\overline{Q}_{11} = Q_{11}\cos^4\theta + Q_{22}\sin^4\theta + 2(Q_{12} + 2Q_{66})\sin^2\theta\cos^2\theta \tag{3.214}$$

$$\overline{Q}_{55} = G_{13}\cos^2\theta + G_{23}\sin^2\theta \tag{3.215}$$

The θ is the angle between the direction of the fibers on a particular layer and the longitudinal axis x. The coefficients Q_{11}, Q_{22}, Q_{12} and Q_{66} are defined as a function of the elastic properties of the layer (see also Chapter 2), namely,

$$Q_{11} = \frac{\overline{E}_{11}}{1 - v_{12}v_{21}}; \quad Q_{12} = \frac{v_{21}\overline{E}_{11}}{1 - v_{12}v_{21}} = \frac{v_{12}\overline{E}_{22}}{1 - v_{12}v_{21}}$$

$$Q_{22} = \frac{\overline{E}_{22}}{1 - v_{12}v_{21}}; \quad Q_{66} = G_{12} \tag{3.216}$$

where $\overline{E}_{11}, \overline{E}_{22}$ and G_{12} are the tension/compression and shear moduli of the single layer, respectively, and v_{12} is its major Poisson's ratio and the following relationships holds $\overline{E}_{11}/\overline{E}_{22} = v_{12}/v_{21}$ (see Chapter 2).

Integrating along the height of the beam, h, and multiplying by the width, b, yields the force and moment resultants:

$$N_x = b \int_{-h/2}^{+h/2} \sigma_x dz, \quad M_y = b \int_{-h/2}^{+h/2} \sigma_x z dz, \quad Q_{xz} = b\kappa \int_{-h/2}^{+h/2} \tau_{xz} dz. \tag{3.217}$$

Performing the various integrals in Equation (3.217) yields the following relationship:

$$\left\{ \begin{array}{c} N_x \\ M_y \\ Q_{xz} \end{array} \right\} = \left[\begin{array}{ccc} A_{11} & B_{11} & 0 \\ B_{11} & D_{11} & 0 \\ 0 & 0 & A_{55} \end{array} \right] \left\{ \begin{array}{c} \dfrac{\partial U}{\partial x} \\ \dfrac{\partial \Phi}{\partial x} \\ \Phi + \dfrac{\partial W}{\partial x} \end{array} \right\} - \left\{ \begin{array}{c} E_{11} \\ F_{11} \\ 0 \end{array} \right\} \tag{3.218}$$

The coefficients A_{11}, B_{11}, D_{11} and A_{55} are the well-known expressions for the axial, the coupled bending–stretching, the bending and lateral shear stiffnesses, respectively, used usually for laminated composite beams. According to the classical lamination theory ([1]), the integrals are replaced by a summation (see Figure 3.30b), yielding the following expressions:

$$A_{11} = b \sum_{k=1}^{N} (\overline{Q}_{11})_k (z_k - z_{k-1}) \tag{3.219}$$

$$B_{11} = \frac{b}{2} \sum_{k=1}^{N} (\overline{Q}_{11})_k \left(z_k^2 - z_{k-1}^2 \right) \tag{3.220}$$

$$D_{11} = \frac{b}{3} \sum_{k=1}^{N} (\overline{Q}_{11})_k \left(z_k^3 - z_{k-1}^3 \right) \tag{3.221}$$

$$A_{55} = \kappa b \sum_{k=1}^{N} (\overline{Q}_{55})_k (z_k - z_{k-1}) \tag{3.222}$$

where κ is the shear correction factor (taken as 5/6) and N is the number of the layers, including the piezoelectric ones of the laminate.

The terms E_{11} and F_{11} in Equation (3.218) represent the induced axial force and the induced moment due to the piezoelectric layers, respectively, and are defined as

$$E_{11} = b \sum_{k=1}^{N} (\overline{Q}_{11})_k^a V_k(x,t)(d_{31})_k \tag{3.223}$$

$$F_{11} = \frac{b}{2} \sum_{k=1}^{N} (\overline{Q}_{11})_k^a V_k(x,t)(d_{31})_k (z_{k+}^a + z_{k-}^a) \tag{3.224}$$

To obtain the equations of motion and the associated boundary conditions, we shall use the principle of virtual work for the present problem yielding the variation of the potential as (see [6]):

$$\delta \pi = \int_{t_1}^{t_2} \int_{\text{volume}} (\delta E_k - \delta E_p + \delta \overline{W}) \cdot d(\text{volume}) dt = 0 \tag{3.225}$$

where E_k stands for the kinetic energy of the beam, E_p is its strain energy and $\overline{W}$ represents the work of the external forces acting on the beam. Their expressions are

$$E_k = \frac{1}{2} \int_{\text{volume}} \rho \cdot \left[(\dot{U} + z\dot{\Phi})^2 + \dot{W}^2 \right] \cdot d(\text{volume}) \tag{3.226}$$

$$E_p = \frac{1}{2} \int_{\text{volume}} (\sigma_x \varepsilon_x + \tau_{xz} \gamma_{xz}) \cdot d(\text{volume}) \tag{3.227}$$

$$\overline{W} = - \int_0^L Q \cdot W \cdot dx - P \cdot U(L) + P \cdot U(0) - M(L) \cdot \Phi(L) + M(0) \cdot \Phi(0) \tag{3.228}$$

As shown in Figure 3.30a, L is the beam's length, P and M are the compressive loads and the bending moments, respectively, acting on both ends of the beam, while Q is the distributed lateral load acting along the beam. (˙) represents time differentiation.

Substitution of Equations (3.226)–(3.228) into Equation (3.225), and integrating in the z- and y-directions, while $(I_1, I_2, I_3) = \int \rho \cdot (1, z, z^2) \cdot dz$ and ρ being the density of each layer of the beam, yields the following expression:

$$\delta\pi = \int_{t_1}^{t_2} \int_0^L \left(\delta E_k - \delta E_P + \delta \bar{W}\right) dx \, dt = 0 \tag{3.229}$$

where

$$\delta E_k = \left(I_1 \dot{U} + I_2 \dot{\Phi}\right) \delta \dot{U} + I_1 \dot{W} \delta \dot{W} + \left(I_2 \dot{U} + I_3 \dot{\Phi}\right) \delta \dot{\Phi} \tag{3.230}$$

$$\delta \bar{W} = -Q \cdot \delta W + P \cdot \delta U(x = 0) - P \cdot \delta U(x = L) + \\ + M(0) \cdot \delta \Phi(x = 0) - M(L) \cdot \delta \Phi(x = L) \tag{3.231}$$

$$\delta E_P = N_x \frac{\partial(\delta U)}{\partial x} + N_x \frac{\partial W}{\partial x} \frac{\partial(\delta W)}{\partial x} + M_y \frac{\partial(\delta \Phi)}{\partial x} + Q_{xz}\left[\delta \Phi + \frac{\partial(\delta W)}{\partial x}\right] \tag{3.232}$$

Substituting the expressions presented in Equations (3.230)–(3.232) into Equation (3.229), integrating by parts, while demanding that the variation of the potential, $\delta\pi$, should vanish, which will hold true only if the various expressions in the integral and at the two ends of the beam are identically zero (as the assumed displacements are arbitrary), yields the following equations of motion for a beam with piezoelectric layers, and its associated boundary conditions:

$$\frac{\partial N_x}{\partial x} = \frac{\partial\left(I_1 \dot{U} + I_2 \dot{\Phi}\right)}{\partial t} \tag{3.233}$$

$$\frac{\partial}{\partial x}\left(Q_{xz} + N_x \frac{\partial W}{\partial x}\right) = \frac{\partial\left(I_1 \dot{W}\right)}{\partial t} + Q \tag{3.234}$$

$$\frac{\partial M_y}{\partial x} - Q_{xz} = \frac{\partial\left(I_3 \dot{\Phi} + I_2 \dot{U}\right)}{\partial t} \tag{3.235}$$

The associated possible boundary conditions are

$$N_x + P = 0 \quad \underline{\text{or}} \quad U = 0 \tag{3.236}$$

$$Q_{xz} + N_x \frac{\partial W}{\partial x} = 0 \quad \underline{\text{or}} \quad W = 0 \tag{3.237}$$

$$M_y = M \quad \underline{\text{or}} \quad \Phi = 0 \tag{3.238}$$

One should note the nonlinear expression $N_x \frac{\partial W}{\partial x}$ appearing in both Equations (3.234) and (3.237). Assuming small deflections for the beam, and based on Equation (3.236), the term N_x is almost equal to $-P$,

$$N_x \cong -P \tag{3.239}$$

Therefore, as is usually presented in the literature, the term N_x is replaced by $-P$ only in the above nonlinear expression.

Using the expressions for the forces and moments resultants (Equation (3.218)) and substituting them into Equations (3.233)–(3.235) yields the equations of motion for the beam with piezoelectric layers expressed in the three assumed deflections, U, W and Φ:

$$\frac{\partial}{\partial x}\left(A_{11}\frac{\partial U}{\partial x} + B_{11}\frac{\partial \Phi}{\partial x} - E_{11}\right) = \frac{\partial(I_1\dot{U} + I_2\dot{\Phi})}{\partial t} \tag{3.240}$$

$$\frac{\partial}{\partial x}\left[A_{55}\left(\Phi + \frac{\partial W}{\partial x}\right) - P\frac{\partial W}{\partial x}\right] = \frac{\partial(I_1\dot{W})}{\partial t} + Q \tag{3.241}$$

$$\frac{\partial}{\partial x}\left(B_{11}\frac{\partial U}{\partial x} + D_{11}\frac{\partial \Phi}{\partial x} - F_{11}\right) - A_{55}\left(\Phi + \frac{\partial W}{\partial x}\right) = \frac{\partial(I_3\dot{\Phi} + I_2\dot{U})}{\partial t} \tag{3.242}$$

with their associated possible boundary conditions:

$$A_{11}\frac{\partial U}{\partial x} + B_{11}\frac{\partial \Phi}{\partial x} = -P + E_{11} \quad \text{or} \quad U = 0 \tag{3.243}$$

$$A_{55}\left(\Phi + \frac{\partial W}{\partial x}\right) - P\frac{\partial W}{\partial x} = 0 \quad \text{or} \quad W = 0 \tag{3.244}$$

$$B_{11}\frac{\partial U}{\partial x} + D_{11}\frac{\partial \Phi}{\partial x} = F_{11} + M \quad \text{or} \quad \Phi = 0 \tag{3.245}$$

One should note that for the case of uniform properties along the beam, the influence of the induced strains does not appear in the equations of motion, but only at the boundaries of the beam.

To comply with the notations for the pin force model usually presented in the literature (see [20]), the boundary conditions are rewritten as

$$A_{11}\frac{\partial U}{\partial x} + B_{11}\frac{\partial \Phi}{\partial x} = -p_m \quad \text{or} \quad U = 0 \tag{3.246}$$

$$A_{55}\left(\Phi + \frac{\partial W}{\partial x}\right) - P\frac{\partial W}{\partial x} = 0 \quad \text{or} \quad W = 0 \tag{3.247}$$

$$B_{11}\frac{\partial U}{\partial x} + D_{11}\frac{\partial \Phi}{\partial x} = m_m \quad \text{or} \quad \Phi = 0 \tag{3.248}$$

where p_m and m_m, the generalized axial load and moment, including the influence of the piezoelectric layers, are

$$p_m = P - E_{11} \tag{3.249}$$

$$m_m = M + F_{11} \tag{3.250}$$

Note that in Equations (3.241) and (3.244), one should use the nonlinear expression term P rather than p_m:

$$P = p_m + E_{11} \tag{3.251}$$

For uniform properties along the beam, the equations of motion are

$$A_{11} \frac{\partial^2 U}{\partial x^2} + B_{11} \frac{\partial^2 \Phi}{\partial x^2} = I_1 \ddot{U} + I_2 \ddot{\Phi} \tag{3.252}$$

$$A_{55} \left(\frac{\partial \Phi}{\partial x} + \frac{\partial^2 W}{\partial x^2} \right) - P \frac{\partial^2 W}{\partial x^2} = I_1 \ddot{W} + Q \tag{3.253}$$

$$B_{11} \frac{\partial^2 U}{\partial x^2} + D_{11} \frac{\partial^2 \Phi}{\partial x^2} - A_{55} \left(\Phi + \frac{\partial W}{\partial x} \right) = I_3 \ddot{\Phi} + I_2 \ddot{U} \tag{3.254}$$

For a symmetric layup ($B = I_2 = 0$), Equations (3.252)–(3.254) degenerate into an uncoupled equation in the x-direction, and two coupled equations, namely:

$$A_{11} \frac{\partial^2 U}{\partial x^2} = I_1 \ddot{U} \tag{3.255}$$

$$A_{55} \left(\frac{\partial \Phi}{\partial x} + \frac{\partial^2 W}{\partial x^2} \right) - P \frac{\partial^2 W}{\partial x^2} = I_1 \ddot{W} + Q \tag{3.256}$$

$$D_{11} \frac{\partial^2 \Phi}{\partial x^2} - A_{55} \left(\Phi + \frac{\partial W}{\partial x} \right) = I_3 \ddot{\Phi} \tag{3.257}$$

It is worth to notice that similar expressions are presented in [1].[4]
For an isotropic material, Equations (3.255)–(3.257) would be written as

$$EA \frac{\partial^2 U}{\partial x^2} = I_1 \ddot{U} \tag{3.258}$$

$$\kappa GA \left(\frac{\partial \Phi}{\partial x} + \frac{\partial^2 W}{\partial x^2} \right) - P \frac{\partial^2 W}{\partial x^2} = I_1 \ddot{W} + Q \tag{3.259}$$

$$EI \frac{\partial^2 \Phi}{\partial x^2} - \kappa GA \left(\Phi + \frac{\partial W}{\partial x} \right) = I_3 \ddot{\Phi} \tag{3.260}$$

The last two Equations, (3.259) and (3.260) are identical to the ones Equations (5.10) and (5.11) presented on page 175 of [9].

---

4 Ref. [1], page 214, when inserting $A_{16} = B_{16} = 0$ and $V_0 = 0$ in Equations (4.5.2a–d).

3.5.1 Solutions for Static and Dynamic Cases

The influence of the piezoelectric layers, or patches, on small vibrations of the beam can be of two types:

(a) Applying an alternated harmonic voltage to the piezoelectric patches. This will cause the harmonic variation of the term E_{11} or F_{11}. These terms, that appear in the boundary conditions (Equations (3.243)–(3.245)), will cause small harmonic vibrations of the beam. Practically, the contribution of the term F11, would be more interesting, as it causes lateral bending vibrations (see [10–19]).

(b) Applying constant voltage to the piezoelectric layers. For equal voltage for all the piezoelectric layers, while constraining the axial displacement of the beam, at both ends of the beam, then due the term E_{11}, constant axial force P will be induced in the beam, which influences the lateral small vibrations (see [7–8]).

Accordingly, the following relationships can be written for the induced axial force, E_{11}, and the induced bending moment F_{11},

$$E_{11} = E_{11}^0 + E_{11}^{'}(t) \tag{3.261}$$

$$F_{11} = F_{11}^0 + F_{11}^{'}(t) \tag{3.262}$$

where the terms $()^0$ stand for constant contributions and $()'$ describes harmonic, time-dependent ones. As said before, the interesting cases would be

$$F_{11}^0 = E_{11}^{'}(t) = 0; \quad E_{11}^0 \neq 0; \quad F_{11}^{'}(t) \neq 0 \tag{3.263}$$

3.5.2 Calculation of the Axial Induced Force due to E_{11}^0

The first case to be presented is for continuous piezoelectric layers along the beam, as described in Figure 3.31.

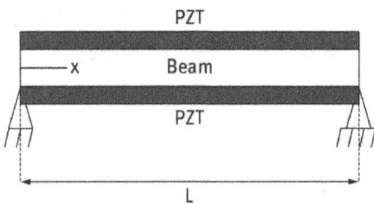

Figure 3.31: A beam sandwiched between two continuous piezoelectric layers.

As shown in Figure 3.31, the axial displacements at both ends of the beam are constrained. Applying identical voltage (both amplitude and sign) would induce compression or tension (depending on the voltage sign) forces in the beam.

For a symmetric time-independent case, Equations (3.255)–(3.257) have the following form:

$$A_{11}\frac{\partial^2 U}{\partial x^2} = 0; \quad W = \Phi = 0 \tag{3.264}$$

The solution for the first equation, Equation (3.255), has the following form:

$$U(x) = ax + b \tag{3.265}$$

where a and b are the constants to be determined from the boundary conditions of the case.

Demanding $U(0) = U(L) = 0$ (no axial displacements at both ends of the beam), yields a solution in the form of $U(x) = 0$. Substituting this result in the first line of Equation (3.218), yields

$$N_x = -E_{11}^0 \tag{3.266}$$

which after substituted in Equation (3.239) gives

$$P = E_{11}^0 \tag{3.267}$$

The second case to be dealt will be a beam with two identical piezoelectric patches as depicted in Figure 3.32a and b.

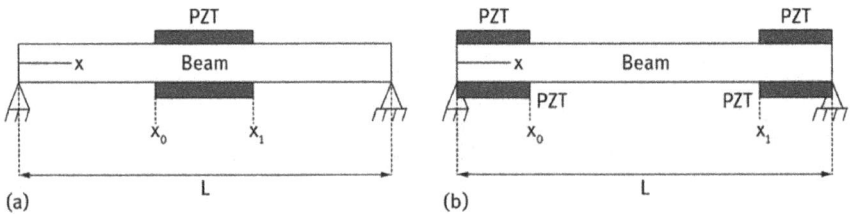

Figure 3.32: A beam equipped with pairs of piezoelectric patches: (a) $x_1 - x_0 = L/3$, (b) $x_1 - x_0 = 2L/3$.

The case presented in Figure 3.32a consists of three parts: the first part, the beam only, extends from $x = 0$ to $x = x_0$. The second part, a beam sandwiched between two identical piezoelectric patches, starts at $x = x_0$ and ends at $x = x_1$. The third part, the beam only, is similar to the first one, and extends from x_1 to $x = L$.

The second case, presented in Figure 3.32b, also consists of three parts: the first part, a beam with a pair of piezoelectric patches, extends from $x = 0$ to $x = x_0$. The second part, from $x = x_0$ to $x = x_1$, is only a beam. The last part, is similar to the first part, and starts at $x = x_1$ and ends at $x = L$. For both cases, the ends of the beam are axially constrained.

Therefore, we can write the distribution of the axial displacement of the beam, presented in Equation (3.265), for each of the three parts of the beam:

$$U_1(x) = a_1 x + b_1, \quad 0 \le x < x_0$$
$$U_2(x) = a_2 x + b_2, \quad x_0 < x < x_1 \tag{3.268}$$
$$U_3(x) = a_3 x + b_3, \quad x_1 < x \le L$$

where the constants, a_1, b_1, a_2, b_2, a_3 and b_3 will be determined using the beam boundary conditions, $U(0) = U(L) = 0$, and continuity requirements at points x_0 and x_1.

The following equations are obtained for the first case (see Figure 3.32a):

$$U_1(0) = a_1 \cdot 0 + b_1 = 0$$

$$U_3(L) = a_3 \cdot L + b_3 = 0$$

$$U_1(x_0) = U_2(x_0) \Rightarrow a_1 \cdot x_0 + b_1 = a_2 \cdot x_0 + b_2$$

$$U_2(x_1) = U_3(x_1) \Rightarrow a_2 \cdot x_1 + b_2 = a_3 \cdot x_1 + b_3 \tag{3.269}$$

$$N_1(x_0) = N_2(x_0) \Rightarrow (A_{11})_I \cdot a_1 = (A_{11})_{II} \cdot a_2 - E_{11}$$

$$N_2(x_1) = N_3(x_1) \Rightarrow (A_{11})_{II} \cdot a_2 - E_{11} = (A_{11})_I \cdot a_3$$

For the second case (see Figure 3.32b), we get

$$U_1(0) = a_1 \cdot 0 + b_1 = 0$$

$$U_3(L) = a_3 \cdot L + b_3 = 0$$

$$U_1(x_0) = U_2(x_0) \Rightarrow a_1 \cdot x_0 + b_1 = a_2 \cdot x_0 + b_2$$

$$U_2(x_1) = U_3(x_1) \Rightarrow a_2 \cdot x_1 + b_2 = a_3 \cdot x_1 + b_3 \tag{3.270}$$

$$N_1(x_0) = N_2(x_0) \Rightarrow (A_{11})_{II} \cdot a_1 - E_{11} = (A_{11})_I \cdot a_2$$

$$N_2(x_1) = N_3(x_1) \Rightarrow (A_{11})_I \cdot a_2 = (A_{11})_{II} \cdot a_3 - E_{11}$$

where $(A_{11})_I$ is the axial stiffness of only the beam, while $(A_{11})_{II}$ is the axial stiffness of the beam and the two piezoelectric patches. The solution of Equation (3.269) or (3.270) and substitution into Equation (3.268) yields the axial displacement distribution for the first case (for $x_1 - x_0 = L/3$, see Figure 3.32a) along the beam:

$$U_1(x) = -\frac{E_{11}}{(A_{11})_I + 2(A_{11})_{II}} x \quad 0 \le x < x_0$$

$$U_2(x) = -\frac{E_{11}}{(A_{11})_I + 2(A_{11})_{II}} (3x_0 - 2x)x_0 \quad x_0 < x < x_1 \tag{3.271}$$

$$U_3(x) = -\frac{E_{11}}{(A_{11})_I + 2(A_{11})_{II}} (x - L) \quad x_1 < x \le L$$

whereas for the second case (for $x_1 - x_0 = 2L/3$, see Figure 3.32b), we get

$$U_1(x) = \frac{2E_{11}}{(A_{11})_I + 2(A_{11})_{II}} x0 \le x < x_0$$

$$U_2(x) = \frac{E_{11}}{(A_{11})_I + 2(A_{11})_{II}} (3x_0 - x)x_0 < x < x_1 \qquad (3.272)$$

$$U_3(x) = \frac{2E_{11}}{(A_{11})_I + 2(A_{11})_{II}} (x - L) \; x_1 < x \le L$$

Similarly, to the continuous case dealt before, substitution of the expression for $U(x)$ into the first line of Equation (3.218) would yield the axial distribution of the force, P, along the beam

$$P = \frac{E_{11} \cdot (A_{11})_I}{(A_{11})_I + 2(A_{11})_{II}} \quad 0 \le x < x_0$$

$$P = \frac{E_{11} \cdot (A_{11})_I}{(A_{11})_I + 2(A_{11})_{II}} \quad x_0 < x < x_1 \qquad (3.273)$$

$$P = \frac{E_{11} \cdot (A_{11})_I}{(A_{11})_I + 2(A_{11})_{II}} \quad x_1 < x \le L$$

As expected, the axial force P is constant along the beam.

3.5.3 Solution of the Equations of Motion for Small Lateral Vibrations

Now the vibrations case of a beam equipped with piezoelectric patches, under constant induced axial force, P, dependent on the voltage supplied to the patches, and dealt above will be derived.

Assuming the following expression for the three variables of the problem:

$$W(x, t) = w(x)e^{i\omega t}; \quad \Phi(x, t) = \phi(x)e^{i\omega t}; \quad U(x, t) = u(x)e^{i\omega t} \qquad (3.274)$$

where $u(x)$, $w(x)$ and $\phi(x)$ are the small amplitudes of vibration with the frequency ω.

Substituting Equation (3.274) into Equations (3.252)–(3.254) yields

$$A_{11}u'' + B_{11}\phi'' = -\omega^2 I_1 u - \omega^2 I_2 \phi \qquad (3.275)$$

$$A_{55}(\phi' + w'') - Pw'' = -\omega^2 I_1 w \qquad (3.276)$$

$$B_{11}u'' + D_{11}\phi'' - A_{55}(\phi + w') = -\omega^2 I_3 \phi - \omega^2 I_2 u \qquad (3.277)$$

where $()'$ is differentiation with respect to x.

For the symmetric case, Equations (3.275)–(3.277) have the form

$$A_{11}u'' = -\omega^2 I_1 u \qquad (3.278)$$

$$A_{55}(\phi' + w'') - Pw'' = -\omega^2 I_1 w \tag{3.279}$$

$$D_{11}\phi'' - A_{55}(\phi + w') = -\omega^2 I_3 \phi \tag{3.280}$$

Equations (3.279) and (3.280) are coupled, while Equation (3.279) has the following solution:

$$u(x) = A \sin(kx) + B \cos(kx); \quad k = \omega \sqrt{\frac{I_1}{A_{11}}} \tag{3.281}$$

The lateral vibrations, with the coupling between w and ϕ are next presented and solved.

To find a solution to the problem, we have to uncouple the two Equations (3.279) and (3.280). This is done using the following procedure:

From Equation (3.279), we get

$$\phi' = \left(\frac{P}{A_{55}} - 1\right)w'' - \frac{I_1}{A_{55}}\omega^2 w \tag{3.282}$$

Derivation of Equation (3.280) with respect to x, and inserting into the result, the second derivation of Equation (3.282) with respect to x, gives the following expression:

$$(D^4 a_2 + D^2 a_1 + a_0)w = 0; \quad D \equiv \frac{d}{dx} \tag{3.283}$$

where

$$a_2 = D_{11}\left(1 - \frac{P}{A_{55}}\right)$$

$$a_1 = D_{11}I_1\frac{\omega^2}{A_{55}} - I_3\omega^2\left(\frac{P}{A_{55}} - 1\right) + P \tag{3.284}$$

$$a_0 = I_3 I_1 \frac{\omega^4}{A_{55}} - I_1\omega^2$$

A solution for Equation (3.283) is obtained assuming $w = \bar{w}e^{sx}$.

Performing the various derivations in Equation (3.283) yields the following characteristic equation for the problem:

$$a_2 s^4 + a_1 s^2 + a_0 = 0 \tag{3.285}$$

Equation (3.285) has two real and two complex solutions, namely

$$\pm \lambda_1; \quad \pm i\lambda_2 \tag{3.286}$$

This leads to the general solution for the lateral displacement of the beam in the form

$$w(x) = C_1 \sinh(\lambda_1 x) + C_2 \cosh(\lambda_1 x) + C_3 \sin(\lambda_2 x) + C_4 \cos(\lambda_2 x) \tag{3.287}$$

To obtain the general solution for the second variable, ϕ, the angle of rotation of the cross section of the beam, the above procedure is again applied for that variable. From Equation (3.280), the variable w' is expressed as a function of ϕ'' and ϕ. This expression is once derived with respect to x, and substituted in Equation (3.279) after being derived once with respect to x. The same expression is now twice derived with respect to x and again substituted in the derivation with respect to x of Equation (3.279). The result of all those derivations is a differential equation for ϕ, similar to Equation (3.283), which has the following solution:

$$\phi(x) = E_1 \sinh(\lambda_1 x) + E_2 \cosh(\lambda_1 x) + E_3 \sin(\lambda_2 x) + E_4 \cos(\lambda_2 x) \tag{3.288}$$

Substitution of Equations (3.287) and (3.288) into the coupled Equations (3.279) and (3.280) would yield the relations between the constants $(C_1–C_4)$ and $(E_1–E_4)$. Defining

$$m_1 = -\frac{1}{\lambda_1}\left(\frac{I_1}{A_{55}}\omega^2 + \lambda_1^2\left(\frac{P}{A_{55}} - 1\right)\right); \quad m_2 = -\frac{1}{\lambda_2}\left(\frac{I_1}{A_{55}}\omega^2 - \lambda_2^2\left(\frac{P}{A_{55}} - 1\right)\right) \tag{3.289}$$

we get

$$E_1 = m_1 C_2; \quad E_2 = m_1 C_1; \quad E_3 = m_2 C_4; \quad E_4 = -m_2 C_3 \tag{3.290}$$

Application of the boundary conditions of simply supported $W(0) = M(0) = W(L) = M(L) = 0$, which can be written in displacements as $W(0) = \phi'(0) = W(L) = \phi'(L) = 0$, for the case of continuous piezoelectric layers, as presented in Figure 3.31, yields the following relationships:

$$\begin{bmatrix} 0 & 1 & 0 & 1 \\ F_{11} & D_{11}m_1\lambda_1 + F_{11} & F_{11} & D_{11}m_2\lambda_2 + F_{11} \\ \sinh(\lambda_1 L) & \cosh(\lambda_1 L) & \sin(\lambda_2 L) & \cos(\lambda_2 L) \\ A & B & C & D \end{bmatrix} \begin{Bmatrix} C_1 \\ C_2 \\ C_3 \\ C_4 \end{Bmatrix} = \begin{Bmatrix} 0 \\ 0 \\ 0 \\ 0 \end{Bmatrix} \tag{3.291}$$

where

$$A = D_{11}m_1\lambda_1 \sinh(\lambda_1 L) + F_{11}; \quad B = D_{11}m_1\lambda_1 \cosh(\lambda_1 L) + F_{11}$$

$$C = D_{11}m_2\lambda_2 \sin(\lambda_2 L) + F_{11}; \quad D = D_{11}m_2\lambda_2 \cos(\lambda_2 L) + F_{11} \tag{3.292}$$

To obtain a unique solution for Equation (3.291), the determinant should vanish, yielding the natural frequencies of the problem, ω_i.

One should note that for the present case, the term, $F_{11,}$ vanishes (as equal voltages were supplied for the upper and lower piezoelectric layers), yielding an induced axial compression force, $-P$, due to the term E_{11}^0, which is constant with time, and the axial constraints at both ends of the beam, as explained before.

For the case of piezoelectric patches, as depicted in Figure 3.32a, and assuming, as before, boundary conditions are simply supported at both ends of the beam, one has to divide the beam into three parts due to different properties along the beam:

(a) The first part, beam only, from $x = 0$ to $x = x_0$, with the following expressions for the displacements:

$$w_1(x) = C_{11} \sinh(\lambda_{11}x) + C_{21} \cosh(\lambda_{11}x) + C_{31} \sin(\lambda_{21}x) + C_{41} \cos(\lambda_{21}x) \qquad (3.293)$$

$$\phi_1(x) = C_{11}m_{11} \cosh(\lambda_{11}x) + C_{21}m_{11} \sinh(\lambda_{11}x) - C_{31}m_{21} \cos(\lambda_{21}x) + C_{41}m_{21} \sin(\lambda_{21}x) \qquad (3.294)$$

(b) The second part, beam and two piezoelectric patches, from $x = x_0$ to $x = x_1$, with the following expressions for the displacements:

$$w_2(x) = C_{12} \sinh(\lambda_{12}x) + C_{22} \cosh(\lambda_{12}x) + C_{32} \sin(\lambda_{22}x) + C_{42} \cos(\lambda_{22}x) \qquad (3.295)$$

$$\phi_2(x) = C_{12}m_{12} \cosh(\lambda_{12}x) + C_{22}m_{12} \sinh(\lambda_{12}x) - C_{32}m_{22} \cos(\lambda_{22}x) + C_{42}m_{22} \sin(\lambda_{22}x) \qquad (3.296)$$

(c) The third part, beam only, from $x = x_0$ to $x = L$, with the following expressions for the displacements:

$$w_3(x) = C_{13} \sinh(\lambda_{13}x) + C_{23} \cosh(\lambda_{13}x) + C_{33} \sin(\lambda_{23}x) + C_{43} \cos(\lambda_{23}x) \qquad (3.297)$$

$$\phi_3(x) = C_{13}m_{13} \cosh(\lambda_{13}x) + C_{23}m_{13} \sinh(\lambda_{13}x) - C_{33}m_{23} \cos(\lambda_{23}x) + C_{43}m_{23} \sin(\lambda_{23}x) \qquad (3.298)$$

Demanding the continuity of the lateral displacements, angles of rotation, shear forces and moments at the two points, x_0 and x_1, namely

$x = x_1$	$x = x_0$
$w_2(x_1) = w_3(x_1)$	$w_1(x_0) = w_2(x_0)$
$\phi_2(x_1) = \phi_3(x_1)$	$\phi_1(x_0) = \phi_2(x_0)$
$Q_2(x_1) = Q_3(x_1)$	$Q_1(x_0) = Q_2(x_0)$
$M_2(x_1) = M_3(x_1)$	$M_1(x_0) = M_2(x_0)$

and applying the boundary conditions, as follows:

$$\text{at} \quad x = 0 \quad w_1(0) = 0 \quad M_1(0) = 0, \quad \frac{d\varphi_1(0)}{dx} = 0$$

$$\text{at} \quad x = L \quad w_3(L) = 0 \quad M_3(L) = 0, \quad \frac{d\varphi_3(L)}{dx} = 0 \qquad (3.299)$$

This results in 12 equations with 12 unknowns written in the matrix notation as

$$[A]\{C\} = \{0\} \tag{3.300}$$

Vanishing of the determinant $[A]$ in Equation (3.300) leads to the characteristic equation of a beam with a pair of piezoelectric, yielding the natural frequencies and their associated mode shapes. The various terms of the A matrix, are given in Appendix E.

3.5.4 Solution of the Static Case

In the previous section, solutions for the cases of equal voltages (in-phase) being supplied to the electrodes of the top and bottom piezoelectric layers, thus causing in plane forces which might cause buckling (for nonmovable axial boundary conditions) of the beam, had been presented. Now, we shall present cases dealing with induced bending of the beam, due to unequal voltages (out of phase Figure 3.33d) being supplied to the electrodes of the piezoelectric layers. The model shown in Figure 3.33a–d is the same as the one presented before in Figure 3.30, with P being mechanical axial force (Figure 3.33a) and both possibilities of continuous and patches of piezoelectric material (Figure 3.33b and c) will be solved for out-of-phase voltages (Figure 3.33d). This topic was discussed in detail in [18], and here only the main results are presented.

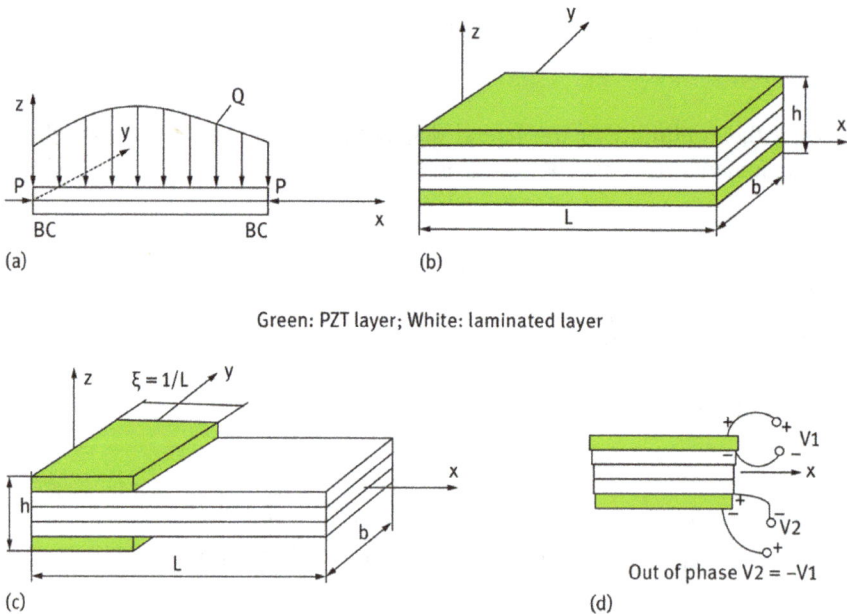

Green: PZT layer; White: laminated layer

Figure 3.33: A piezo-laminated beam: (a) model, (b) continuous piezoelectric patches, (c) piezoelectric patches and (d) electrical connections of the piezoelectric layers.

Equations of motion (3.47)–(3.49), developed earlier for uniform properties along the beam, are now rewritten in the nondimensional form, omitting the inertia terms on right-hand side, yielding

For uniform properties along the beam, the equations of motion have the following form

$$\alpha_0 \bar{U}'' + \beta_0 \bar{\Phi}'' = 0 \tag{3.301}$$

$$c_0 \left(\bar{\Phi}' + \bar{W}'' \right) - \lambda \bar{W}'' = q_0 \tag{3.302}$$

$$\beta_0 \bar{U}'' + \bar{\Phi}'' - c_0 \left(\bar{\Phi} + \bar{W}' \right) = 0 \tag{3.303}$$

with the following nondimensional boundary conditions:

$$\alpha_0 \bar{U}' + \beta_0 \bar{\Phi}' = -\lambda + \hat{E}_{11} \quad \underline{\text{or}} \quad \bar{U} = 0 \tag{3.304}$$

$$c_0 \left(\bar{\Phi} + \bar{W}' \right) - \lambda \bar{W}' = 0 \quad \underline{\text{or}} \quad \bar{W} = 0 \tag{3.305}$$

$$\beta_0 \bar{U}' + \bar{\Phi}' = \hat{F}_{11} \quad \underline{\text{or}} \quad \bar{\Phi} = 0 \tag{3.306}$$

The nondimensional terms are:

$$\begin{gathered} \alpha_0 = \frac{A_{11}L^2}{D_{11}}; \quad \beta_0 = \frac{B_{11}L}{D_{11}}; \quad c_0 = \frac{A_{55}L^2}{D_{11}}; \quad \hat{E}_{11} = \frac{E_{11}L^2}{D_{11}} \\ \hat{F}_{11} = \frac{F_{11}L}{D_{11}}; \quad q_x0 = \frac{QL^3}{D_{11}}; \quad \lambda = \frac{PL^2}{D_{11}}; \quad \bar{U} = \frac{U}{L} \\ \bar{W} = \frac{W}{L}; \quad \bar{\Phi} = \Phi; \quad \xi = \frac{x}{L}; \quad []' = \frac{d[]}{d\xi} \end{gathered} \tag{3.307}$$

Equations (3.301)–(3.303) can be decoupled to yield the following equations:

$$\beta \bar{W}'''' + \lambda \bar{W}'' = -q_0 + \mu q_0'' \tag{3.308}$$

$$\beta \bar{\Phi}''' + \lambda \bar{\Phi}' = q_0 \tag{3.309}$$

$$\beta \bar{U}'''' + \lambda \bar{U}'' = -\chi q_0' \tag{3.310}$$

where

$$\beta = \mu(c_0 - \lambda); \quad \mu = \frac{c_0}{(1 - \beta_0 \chi)}; \quad \chi = \frac{\beta_0}{\alpha_0} \tag{3.311}$$

For a given distribution of the nondimensional transverse distributed loading q_0, Equations (3.308)–(3.310) would be solved by looking for homogeneous and particular solutions. For the case of $q_0 = $ const., the solutions of those equations have the following general form:

$$\bar{W} = A_1 \sin(\gamma\xi) + A_2 \cos(\gamma\xi) + A_3\xi^2 + A_4\xi + A_5 \tag{3.312}$$

$$\bar{\Phi} = B_1 \sin(\gamma\xi) + B_2 \cos(\gamma\xi) + B_3\xi + B_4 \tag{3.313}$$

$$\bar{U} = C_1 \sin(\gamma\xi) + C_2 \cos(\gamma\xi) + C_3\xi^2 + C_4\xi + C_5 \tag{3.314}$$

where

$$\gamma = \sqrt{\frac{\lambda}{\beta}} \tag{3.315}$$

Although one can notice 14 unknowns to be determined (A_1–A_5, B_1–B_4 and C_1–C_5) there are only six available boundary conditions (three at each end of the beam). The eight additional relationships are obtained by substitution of Equations (3.312)–(3.314) into the uncoupled equations of motion, Equations (3.301)–(3.303), while demanding equality on both sides of the equations irrespective of the variable x.

Note that for cases without axial compressive load, P, Equations (3.312)–(3.314) will have the following form:

$$\bar{W} = A_1\xi^4 + A_2\xi^3 + A_3\xi^2 + A_4\xi + A_5 \tag{3.316}$$

$$\bar{\Phi} = B_1\xi^3 + B_2\xi^2 + B_3\xi + B_4 \tag{3.317}$$

$$\bar{U} = C_1\xi^4 + C_2\xi^3 + C_3\xi^2 + C_4\xi + C_5 \tag{3.318}$$

Various solutions for different boundary conditions and loadings can be seen in Table 3.4 (see also [18]). Table 3.5 presents solutions for cases with piezoelectric patches (unlike the continuous piezoelectric layers, presented in Table 3.4).

Table 3.4: Closed-form solutions for $\bar{W}, \bar{\Phi}$ and $\bar{\Phi}$ – continuous piezoelectric layers (lateral load is constant, $q_0 \neq 0$)[a].

Name	Boundary conditions		Closed-form solutions
	$\xi = 0$	$\xi = 1$	
		No axial force $-\lambda = 0$	
S–$S^b(m)$	$\bar{W} = 0$	$\bar{W} = 0$	$\bar{W} = -\dfrac{a}{4}\xi^4 + a\xi^3 + (b-g)\xi^2 - (a+b-g)\xi$
	$\beta_0\bar{U}' + \bar{\Phi}' = \widehat{F}_{11}$	$\beta_0\bar{U}' + \bar{\Phi}' = \widehat{F}_{11}$	$\bar{\Phi} = +a\xi^3 - 3a\xi^2 + 2g\xi + (a-g)$
	$\bar{U} = 0$	$\alpha_0\bar{U} + \beta_0\bar{\Phi} = 0$	$\bar{U} = -\delta a\xi^3 + 3a\delta\xi^2 - 2\delta g\xi$

Table 3.4 (continued)

Name	Boundary conditions $\xi = 0$	$\xi = 1$	Closed-form solutions
$S\text{–}S^b(u)$	$\overline{W} = 0$	$\overline{W} = 0$	$\overline{W} = -\dfrac{a}{4}\xi^4 + a\xi^3 + \left(b - a + \dfrac{\widehat{F}_{11}}{2}\right)\xi^2 - \varepsilon\xi$
	$\beta_0\overline{U}' + \overline{\Phi}' = \widehat{F}_{11}$	$\beta_0\overline{U}' + \overline{\Phi}' = \widehat{F}_{11}$	$\overline{\Phi} = +a\xi^3 - 3a\xi^2 - \left(\dfrac{a}{2} + \widehat{F}_{11}\right)\xi + (\varepsilon - b)$
	$\overline{U} = 0$	$\overline{U} = 0$	$\overline{U} = -\delta a\xi^3 + 3a\delta\xi^2 - \delta a\dfrac{\xi}{2}$
$C\text{–}S^b(m)$	$\overline{W} = 0$	$\overline{W} = 0$	$\overline{W} = -\dfrac{a}{4}\xi^4 + \alpha_0\theta\dfrac{\xi^3}{6} + (b - \theta_1)\xi^2 + \theta_2\xi$
	$\overline{\Phi} = 0$	$\beta_0\overline{U}' + \overline{\Phi}' = \widehat{F}_{11}$	$\overline{\Phi} = +a\xi^3 - \alpha_0\theta\dfrac{\xi^2}{2} + 2\theta_1\xi$
	$\overline{U} = 0$	$\alpha_0\overline{U}' + \beta_0\overline{\Phi}' = 0$	$\overline{U} = -\delta a\xi^3 + \delta\alpha_0\theta\dfrac{\xi^2}{2} + \left(\dfrac{d}{\delta} - \delta\theta\right)\xi$
$C\text{–}S^b(u)$	$\overline{W} = 0$	$\overline{W} = 0$	$\overline{W} = -\dfrac{a}{4}\xi^4 + \eta\dfrac{\xi^3}{3} + \left(b - \dfrac{\mu}{2}\right)\xi^2 + \mu_1\xi$
	$\beta_0\overline{U}' + \overline{\Phi}' = \widehat{F}_{11}$	$\beta_0\overline{U}' + \overline{\Phi}' = \widehat{F}_{11}$	$\overline{\Phi} = +a\xi^3 - \eta\xi^2 + \mu\xi$
	$\overline{U} = 0$	$\overline{U} = 0$	$\overline{U} = -\delta a\xi^3 + \delta\eta\xi^2 + (\eta - \delta a)\xi$
$C\text{–}F$	$\overline{W} = 0$	$c_0(\overline{\Phi} + \overline{W}') = \lambda\overline{W}'$	$\overline{W} = -\dfrac{a}{4}\xi^4 + a\xi^3 + (b - d)\xi^2 - 2b\xi$
	$\overline{\Phi} = 0$	$\beta_0\overline{U}' + \overline{\Phi}' = \widehat{F}_{11}$	$\overline{\Phi} = +a\xi^3 - 3a\xi^2 + 2d\xi$
	$\overline{U} = 0$	$\alpha_0\overline{U}' + \beta_0\overline{\Phi}' = 0$	$\overline{U} = -\delta a\xi^3 + 3a\delta\xi^2 - 2\delta d\xi$
		With axial force $-\lambda \neq 0$	
$C\text{–}F$	$\overline{W} = 0$	$c_0(\overline{\Phi} + \overline{W}') = \lambda\overline{W}'$	$\overline{W} = B_1\sin(\gamma\xi) + B_2\cos(\gamma\xi) + B_0\xi^2 + B_3\xi + B_4$
	$\overline{\Phi} = 0$	$\beta_0\overline{U}' + \overline{\Phi}' = \widehat{F}_{11}$	$\overline{\Phi} = A_1\sin(\gamma\xi) + A_2\cos(\gamma\xi) + A_0\xi + A_3$
	$\overline{U} = 0$	$\alpha_0\overline{U}' + \beta_0\overline{\Phi}' = -\lambda$	$\overline{U} = C_1\sin(\gamma\xi) + C_2\cos(\gamma\xi) + C_3\xi + C_4$
			$\lambda_{cr} = \dfrac{\dfrac{\pi^4}{4}\left(1 - \dfrac{\beta_0^2}{\alpha_0}\right)}{1 + \dfrac{\pi^2}{4c_0}\left(1 - \dfrac{\beta_0^2}{\alpha_0}\right)}$

[a]The expressions for the various terms in the right columns can be found in Appendix C.
[b]The sign of the applied voltage V is opposite to that used for the boundary conditions $C\text{–}F$, where $S\text{–}S(m)$ = simply supported movable; $S\text{–}S(u)$ = simply supported unmovable; $C\text{–}S(m)$ = clamped simply supported movable; $C\text{–}S(u)$ = clamped simply supported unmovable; $C\text{–}F$ = clamped free.

Table 3.5: Closed-form solutions for $\bar{W}, \bar{\Phi}$ and $\bar{\Phi}-$ piezoelectric patches (no lateral load $q_0 = 0$; No axial force $\lambda = 0$)[a].

Boundary conditions	Closed-form solutions	
	Part I $0 \le \xi \le \xi_0$	**Part II** $\xi_0 \le \xi \le 1$
S–S[b](m)	$\bar{W} = -A[\xi^2 + \xi(\xi_0^2 - 2\xi_0)]$	$\bar{W} = -A\xi_0^2(\xi - 1)$
	$\bar{\Phi} = 2A[(\xi - \xi_0) + \frac{\xi_0^2}{2}]$	$\bar{\Phi} = A\xi_0^2$
	$\bar{U} = A_1\xi$	$\bar{U} = A_1\xi_0$
S–S(u)	$\bar{W} = \left(\beta_0 B + \hat{F}_{11}\right)\frac{\xi^2}{2} - F\xi$	$\bar{W} = \frac{\beta_0 B_1(\xi^2 - 1)}{2} - F_1(\xi - 1)$
	$\bar{\Phi} = -\left(\beta_0 B + \hat{F}_{11}\right)\xi + F$	$\bar{\Phi} = -\beta_0 B_1 \xi + F_1$
	$\bar{U} = B\xi$	$\bar{U} = B_1(\xi - 1)$
C–S(m)	$\bar{W} = -D\frac{\xi^3}{3} - E\frac{\xi^2}{2} + I\xi$	$\bar{W} = -D_1\frac{\xi^3}{3} + D_1\frac{\xi^2}{2} + I_1\xi + J$
	$\bar{\Phi} = D\xi^2 + E\xi$	$\bar{\Phi} = D_1\xi^2 - 2D_1\xi + E_1$
	$\bar{U} = -\delta D\xi^2 + C\xi$	$\bar{U} = -\delta D_1\xi^2 + 2\delta_1 D_1\xi + C_1$
C–S[c](u)	$\bar{W} = -d\frac{\xi^3}{3} - e\frac{\xi^2}{2} + i\xi$	$\bar{W} = -d_1\frac{\xi^3}{3} - e_1\frac{\xi^2}{2} + i_1\xi + j$
	$\bar{\Phi} = d\xi^2 + e\xi$	$\bar{\Phi} = d_1\xi^2 + e_1\xi + e_2$
	$\bar{U} = -\delta d\xi^2 + c\xi$	$\bar{U} = -\delta_1 d_1\xi^2 + c_1\xi + c_2$
C–F	$\bar{W} = -A\xi^2$	$\bar{W} = -A\xi_0\left(\xi - \frac{\xi_0}{2}\right)$
	$\bar{\Phi} = 2A\xi$	$\bar{\Phi} = 2A\xi_0$
	$\bar{U} = A_1\xi$	$\bar{U} = A_1\xi_0$
	Part I $0 \le \xi \le 0.5(1 - \xi_0)$	**Part II** $0.5(1 - \xi_0) \le \xi \le (1 + \xi_0)$	**Part III** $0.5(1 + \xi_0) \le \xi \le 1$
S–S(m)	$\bar{W} = A\xi_0\xi$	$\bar{W} = -A[(\xi^2 - \xi) + \frac{(1 - \xi)^2}{4}]$	$\bar{W} = -A\xi_0(\xi - 1)$
	$\bar{\Phi} = -A\xi_0$	$\bar{\Phi} = 2A\left(\xi - \frac{1}{2}\right)$	$\bar{\Phi} = A\xi_0$
	$\bar{U} = 0$	$\bar{U} = A_1[\xi + \frac{(\xi_0 - 1)}{2}]$	$\bar{U} = A_1\xi_0$

[a]The expressions for the various terms in the right columns can be found in Appendix D.
[b]The sign of the applied voltage V is opposite to that used for the boundary conditions C–F, where S–S(m) = simply supported movable; S–S(u) = simply supported unmovable; C–S(m) = clamped simply supported movable; C–S(u) = clamped simply supported unmovable; C–F = clamped free.
[c]Due to the length and cumbersome nature of the constants $c, d, e, i, c_1, c_2, d_1, e_1, e_2, i_2$ and j_2, their explicit expressions are omitted and left for the reader to obtain them; ε_0 is the nondimensional length of the piezoelectric patch.

3.5.5 Experimental Validation for the Induced Axial Force Case

Reference [21] presents an interesting application of altering the natural frequencies of a cantilevered beam, equipped with piezoelectric patches. It is known that for the case of a cantilevered beam, having a free axial displacement at its free end, the induced axial forces cannot change its lateral vibrations as the resultant force on each cross section of a beam is identically zero (as no external axial forces are applied on the beam). However, under certain circumstances, one can alter the natural frequencies of a piezo-laminated beam. This was done by using the following experimental model (more details can be found in [21]) which was mainly made of glass epoxy, having a uniform rectangular hollow cross section. Two glass-epoxy plates were glued at the root and tip of cross sections (see Figure 3.34). The centers of these plates, namely the centers of the root and tip of cross sections, were connected by a 0.6 mm steel wire.

Six pairs of piezoelectric patches were bonded on the wide surfaces of the beam. To be able to apply relatively high voltages (in the range of 1000 V), the piezoelectric patches were macro-fiber composite ones, manufactured by Smart Material Corporation [22]. The piezoelectric patches were electrically connected in parallel, thus identical voltage was applied to all of them. To monitor the behavior of the beam, four pairs of strain gages were bonded along its surfaces (see their location in Figure 3.34).

The mechanism that alters the natural frequencies of the cantilevered beam, is the wire which connects the two sides of the beam. Without the wire, the influence on the lateral vibrations by applying a constant voltage to the collocated piezoelectric patches bonded on the cantilevered beam is negligible, in spite of the internal stresses that are induced inside the beam. Another important feature is the fact that the beam is hollow so the wire is free to move about in it. Preventing the free movement of the wire inside the beam cross section will not enable the altering of the natural frequencies of a cantilevered beam.

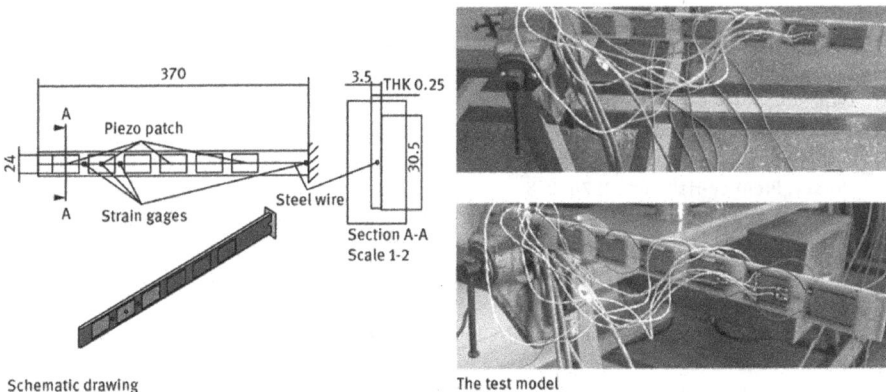

Figure 3.34: The schematic view of the experimental model [21] (all dimensions are in millimeter).

Figure 3.35 presents a comparison between the frequencies that were measured during the test and calculated values. In general, there was a good agreement between both results. Initially the experimental results were lower than the calculated ones. The reasons may be incomplete clamping, and added mass associated with the wires connecting the patches to the voltage source and the strain gages to the strain recorder. At higher voltages, the experimental results exhibit a better agreement with the calculations. The square of the frequency is reduced almost linearly with an increase in voltage. Zero natural frequency indicates buckling of the beam, which can be determined to yield the "buckling" voltage. Its value was experimentally found [21] to be 1168 V, which agreed very well, with the calculated value of 1135 V (a difference of only 2.9%).

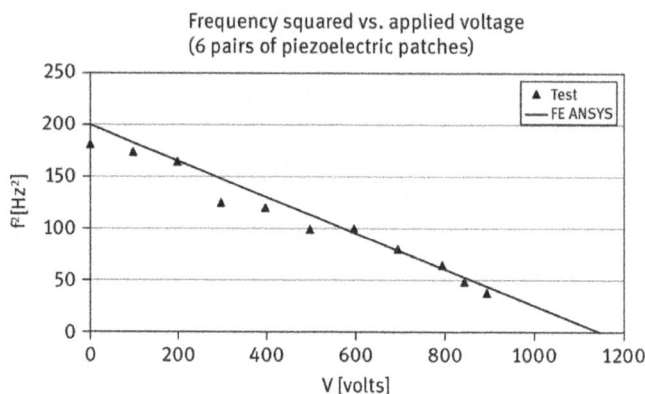

Figure 3.35: Frequency squared versus applied voltage – experimental and numerical (FE-ANSYS) results [21].

References

[1] Reddy, J. N., Mechanics of Laminated Composite Plates: Theory and Analysis, CRC Press LLC, 2004, 831.
[2] Wang, C. M., Reddy, J. N., and Lee, K. H., Shear Deformable Beams and Plates: Relationships with Classical Solutions, Elsevier Science Ltd., June 2000, 296.
[3] Simitses, G. J., An Introduction to the Elastic Stability of Structures, Englewood Cliffs, New Jersey, Prentice-Hall, Inc., 1976, 253.
[4] Sirohi, J. and Chopra, I, Fundamental understanding of piezoelectric strain sensors, Journal of Intelligent Material Systems and Structures 11(4), April 2000, 246–257.
[5] Sirohi, J. and Chopra, I, Fundamental behavior of piezoceramic sheet actuators, Journal of Intelligent Material Systems and Structures 11(1), January 2000, 47–61.
[6] Reddy, J. N., Energy Principles and Variational Methods in Engineering, John Wiley and Sons, August 2002, 608.
[7] Waisman, H. and Abramovich, H, Active stiffening of laminated composite beams using piezoelectric actuators, Composite Structures 58(1), October 2002, 109–120.

[8]   Waisman, H. and Abramovich, H, Variation of natural frequencies of beams using the active stiffening effect, Composites Part B: Engineering 33(6), September 2002, 415–424.

[9]   Magrab, E. B., Vibrations of Elastic Structural Members, Sijthoff & Noordhoff, 1979, 390.

[10] Abramovich, H. and Livshits, A, Dynamic behavior of cross-ply laminated beams with piezoelectric layers, Composite Structures 25(1–4), 1993, 371–379.

[11] Abramovich, H. and Livshits, A, Free vibrations of non-symmetric cross-ply laminated composite beams, Jouranl of Sound and Vibration 176(5), 1994, 597–612.

[12] Miller, S. E. and Abramovich, H., A self-sensing piezolaminated actuator model for shells using a first order shear deformation theory, Journal of Intelligent Material Systems and Structures 6(5), 1995, 624–638.

[13] Pletner, B. and Abramovich, H, Adaptive suspensions of vehicles using piezoelectric sensors, Journal of Intelligent Material Systems and Structures 6(6), November 1995, 744–756.

[14] Abramovich, H. and Pletner, B, Actuation and sensing of piezolaminated sandwich type structures, Composite Structures 38, October 1997, 17–27.

[15] Eisenberger, M. and Abramovich, H, Shape control of non-symmetric piezolaminated composite beams, Composite Structures 38, October 1997, 565–571.

[16] Abramovich, H. and Meyer-Piening, H.-R, Induced vibrations of piezolaminated elastic beams, Composite Structures 43, 1998, 47–55.

[17] Abramovich, H. and Livshits, A, Flexural vibrations of piezolaminated slender beams: a balanced model, Journal of Vibration and Control 8(8), November 2002, 1105–1121.

[18] Abramovich, H, Piezoelectric actuation for smart sandwich structures- closed form solutions, Journal of Sandwich Structures & Materials 5(4), 2003, 377–396.

[19] Edery-Azulay, L. and Abramovich, H, Piezoelectric actuation and sensing mechanisms-closed form solutions, Composite Structures 64(3–4), June 2004, 443–453.

[20] Chopra, I. and Sirohi, J, Smart Structures Theory, Cambridge University Press, 2014, 920.

[21] Abramovich, H, A new insight on vibrations and buckling of a cantilevered beam under a constant piezoelectric actuation, Composite Structures 93, 2011, 1054–1057.

[22] Smart Material Corporation, http://www.smart-material.com.

3.6 Composite Plates with Piezoelectric Patches

In the previous sections, we dealt with the behavior of a beam equipped with piezoelectric patches. In this section, we shall first investigate the behavior of a thin plate with piezoelectric patches, using the CLT of composites.[5] Let us consider a plate in a Cartesian coordinate system were x- and y-axes are parallel to the plate boundaries. For the simplification of the model, let us assume a plate composed of laminated structural materials and piezoelectric layers (piezo-laminated plate), symmetric to its mid-plane, thus the term $B_{ij} = 0$, and having no other coupling terms like $()_{16} = ()_{26} = 0$, and no surface shear stresses. Plane stress conditions are assumed for model [1]. The displacement field is determined according to the

---

5 The first part of this section is based on: Edery-Azulay, L. and Abramovich, H., "Piezolaminated Plates-Highly Accurate Solutions Based on the Extended Kantorovich Method," Composite Structures 84(3), 2008, 241–247.

classical plate theory, where the transverse shear deformations are ignored [1–2] (see also Section 2.1). Based on these assumptions, and integrating the stress components across each lamina including the piezoelectric contribution we get the moment resultants per unit length in the following form:

$$M_x = -D_{11} \frac{\partial^2 w(x,y,t)}{\partial x^2} - D_{12} \frac{\partial^2 w(x,y,t)}{\partial y^2} + F_x(x,y,t)$$

$$M_y = -D_{12} \frac{\partial^2 w(x,y,t)}{\partial x^2} - D_{22} \frac{\partial^2 w(x,y,t)}{\partial y^2} + F_y(x,y,t) \qquad (3.319)$$

$$M_{xy} = -2D_{66} \frac{\partial^2 w(x,y,t)}{\partial x \partial y}$$

where D_{ij} are the various bending stiffnesses of the composite plate and the piezoelectric-induced moments $F_x(x,y,t)$ and $F_y(x,y,t)$ are defined as

$$F_x(x,y,t) = \int_{-h/2}^{h/2} \left(\frac{Q_{13}e_{33}}{Q_{33}} - e_{33} \right) E_3 z dz$$

$$\qquad (3.320)$$

$$F_y(x,y,t) = \int_{-h/2}^{h/2} \left(\frac{Q_{23}e_{33}}{Q_{33}} - e_{23} \right) E_3 z dz$$

where

$$e_{31} = d_{31}Q_{11} + d_{32}Q_{12}$$

$$e_{32} = d_{31}Q_{12} + d_{32}Q_{22} \qquad (3.321)$$

$$e_{33} = d_{33}Q_{33}$$

and d_{31}, d_{32} and d_{33} are the known piezoelectric field-strain constants (in (m/V) or (C/N)), while e_{31}, e_{32} and e_{33} are the known piezoelectric field-stress constants (in (N/mV) or (C/m^2)). E_3 is the electric field, h is the total thickness of the piezoelectric layer, Q_{ij} are the elasticity constants and e_{ij} are the piezoelectric constants [1]. The definition of the induced moments (Equation (3.320)) is based on the assumption that the stress in the thickness direction is assumed to be zero, $\sigma_z = 0$; see [2] for more details.

The well-known equation of motion for thin plates subjected to lateral pressure, $q(x,y)$, is given by

$$\frac{\partial^2 M_x}{\partial x^2} + 2 \frac{\partial^2 M_{xy}}{\partial x \partial y} + \frac{\partial^2 M_y}{\partial y^2} = q(x,y) \qquad (3.322)$$

Substituting Equation (3.319) into (3.322) and assuming static behavior only (then the time dependence is removed) yield

$$D_1 \frac{\partial^4 w(x,y)}{\partial x^4} + D_3 \frac{\partial^4 w(x,y)}{\partial x^2 \partial y^2} + D_2 \frac{\partial^4 w(x,y)}{\partial y^4} = P(x,y)$$

where

$$D_1 \equiv D_{11}, \quad D_2 \equiv D_{22}, \quad D_3 \equiv (2D_{12} + 4D_{66}) \tag{3.323}$$

and

$$P(x,y) = \frac{\partial^2 F_x(x,y)}{\partial x^2} + \frac{\partial^2 F_y(x,y)}{\partial y^2} + q(x,y)$$

A solution to Equation (3.323) cannot be found analytically; therefore, it is sought using the Galerkin method, namely demanding

$$\int_0^a \int_0^b \left[D_1 \frac{\partial^4 w(x,y)}{\partial x^4} + D_3 \frac{\partial^4 w(x,y)}{\partial x^2 \partial y^2} + D_2 \frac{\partial^4 w(x,y)}{\partial y^4} - P(x,y) \right] w(x,y) dx dy = 0 \tag{3.324}$$

where a and b are the length and width of the plate, respectively. As was done by Edery-Azulay and Abramovich [1], the solution of Equation (3.324) is sought by applying the Kantorovich method [3–7], which assumes that the solution $w(x,y)$ is separable as

$$w(x,y) = W(x)W(y) \tag{3.325}$$

One should remember that according to the classical Kantorovich method, the solution of one of the two directions is assumed to be known. Therefore, assuming that $W(y)$ is a priori chosen known function, and substituting it in both Equation (3.325) and the corresponding Galerkin equation (Equation (3.324)) we get

$$\int_0^a \int_0^b \left[D_1 \frac{d^4 W(x)}{dx^4} W(y) + D_3 \frac{d^2 W(x)}{dx^2} \frac{d^2 W(y)}{dy^2} + D_2 W(x) \frac{d^4 W(y)}{dy^4} - P(x,y) \right] W(y) dy W(x) dx = 0 \tag{3.326}$$

To satisfy Equation (3.326), the expression in the square brackets must be zero. This results in an ordinary differential equation for the determination of the unknown $W(x)$ function,[6] and written in the following form:

---

6 In a similar manner, when *W(x)* is a priori known function one obtains an ordinary differential equation for the determination of the unknown *W(y)* function.

$$A_4 \frac{d^4 W(x)}{dx^4} + A_2 \frac{d^2 W(x)}{dx^2} + A_0 W(x) = F(x)$$

where

$$A_4 \equiv D_1 \int_0^b W^2(y)dy, \quad A_2 \equiv D_3 \int_0^b \frac{d^2 W(y)}{dy^2} W(y)dy \qquad (3.327)$$

$$A_0 \equiv D_2 \int_0^b \frac{d^4 W(y)}{dy^4} W(y)dy, \quad F(x) = \int_0^b P(x,y)W(y)dy$$

The integrals presented can be solved numerically, using one of the numerous available commercial mathematical programs like the Maple code [8]. To present a solution for Equation (3.327), let us consider a case, where the piezoelectric-induced bending moments and the mechanical loads are distributed in a double sinusoidal form according to the following expression:[7]

$$\left\{ \begin{array}{c} q(x,y) \\ F_x(x,y) \\ F_y(x,y) \end{array} \right\} = \left\{ \begin{array}{c} q_0 \sin\left(\frac{m\pi}{a}x\right)\sin\left(\frac{n\pi}{b}y\right) \\ f_0 \sin\left(\frac{m\pi}{a}x\right)\sin\left(\frac{n\pi}{b}y\right) \\ f_0 \sin\left(\frac{m\pi}{a}x\right)\sin\left(\frac{n\pi}{b}y\right) \end{array} \right\} \qquad (3.328)$$

Ceramic PZT actuators are normally transversely isotropic where $e_{31} = e_{32}$, therefore, it would yield equal induced bending moments in both x- and y-directions. Substituting Equation (3.328) into (3.327) yields the following differential equation:

$$A_4 \frac{d^4 W(x)}{dx^4} + A_2 \frac{d^2 W(x)}{dx^2} + A_0 W(x) = \left[q_0 - f_0\left(k_x^2 + k_y^2\right) \right] \sin(k_x x)$$

$$(3.329)$$

where

$$k_x = \frac{m\pi}{a}, \quad k_y = \frac{n\pi}{b}$$

with the following solution:

$$W(x) = \sum_{j=1}^4 B_j e^{\lambda_j x} + \frac{f_0\left(k_x^2 + k_y^2\right) - q_0}{-A_4 k_x^4 + A_2 k_x^2 - A_0} \sin(k_x x)$$

where $\qquad (3.330)$

$$\lambda_j = \pm\sqrt{\frac{-A_2 \pm \sqrt{A_2^2 - 4A_4 A_0}}{2A_4}}, j = 1, 2, 3, 4$$

---

7 For piezoelectric layers loaded with a DC electric voltage, the piezoelectric-induced moments are constants all over the plate area.

The four constants B_1, B_2, B_3 and B_4 are determined by enforcing the boundary conditions in the x-direction.

One has to remember that when applying the extended Kantorovich solution after the first Kantorovich solution is obtained, one has to switch the plates' direction to be solved, by using the obtained analytical solution as a specified function and freeing the other direction to be analytical solved. This iterative procedure can be repeated until the result converges to any desired degree of accuracy. A very interesting conclusion reported by Edery-Azulay and Abramovich is [1] that the initial trial functions are neither required to satisfy the geometric nor the natural boundary conditions because the iterative procedure forces the solution to satisfy all the boundary conditions of the plate. This means that for complicated boundary conditions around the plate boundaries, the application of the extended Kantorovich method is a superior tool to obtain accurate results with less calculation effort.

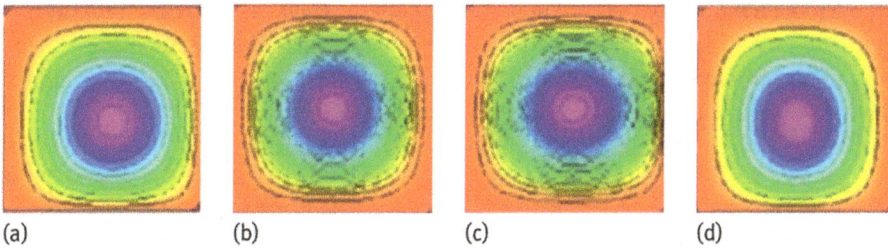

(a)   (b)   (c)   (d)

Figure 3.36: Out-of-plane displacement patterns for a plate actuated by a continuous piezoelectric layer – various boundary conditions: (a) clamped-simply supported-clamped–simply supported; (b) all around clamped; (c) clamped-clamped-clamped-simply supported; (d) clamped-simply supported-clamped-clamped – taken from [1].

Now, we shall deal with another model of the thin plate using an FSDT. Figure 3.37 depicts a rectangular laminated plate in a Cartesian coordinate system. The structure includes elastic layers with planes of symmetry coincident with the coordinate system. The extension-type piezoelectric materials[8] are bonded on the surface of the host structure and poled in the thickness direction; shear-type piezoelectric materials[9] are embedded in the core of the plate and poled in the plane structure direction. The application of an electric field in the thickness direction causes surface actuations to increase or decrease in the plane dimensions, and generates transverse deflection to the embedded actuators; these deformations induce out-of-plane displacements on the host structure.

---

8 In this chapter, extension type piezoelectric material refers to a piezoelectric layer being under an electrical field applied in the thickness direction, causing either extension or compression of the material.
9 As the name implies, shear type piezoelectric material refers to a piezoelectric layer being under an electrical field applied in the thickness direction, inducing shear strains.

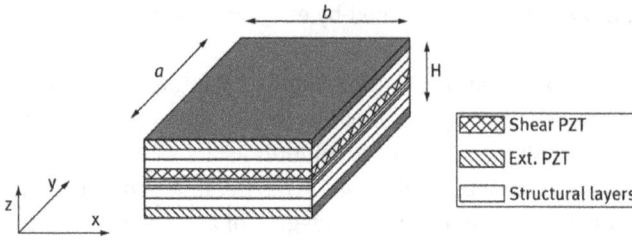

Figure 3.37: Rectangular laminated plate in Cartesian coordinate system.

Applying an electric field in the thickness direction of the shear piezoelectric material will induce shear strains, thus yielding a transverse deflection of the host plate.

The model proposed in the present chapter is useful for rectangular piezo-laminated plate having arbitrary orientations through the thickness. However, for derivations presented here, it is assumed that the principal material coordinates coincide with the coordinates of the problem being analyzed. The constitutive relationship for the converse piezoelectric effect, naming, actuation is given as

$$\{\sigma_{ij}\} = [Q_{ij}]\{\varepsilon_{ij}\} - [e_{ij}]^t\{E_j\}$$ (3.331)

where σ_{ij} and ε_{ij} are the components stresses and strains, respectively, E_j is the electric field, Q_{ij} are elasticity constants and e_{ij} are piezoelectric constants which is related to d_{ij} piezoelectric constants in the following way:

$$[e_{ij}] = [d_{ij}][Q_{ij}]$$ (3.332)

The general relationship in Equation (3.331) has the following detailed form when using orthotropic extension-type piezoelectric materials:

$$\begin{Bmatrix} \sigma_{11} \\ \sigma_{22} \\ \sigma_{33} \\ \sigma_{23} \\ \sigma_{13} \\ \sigma_{12} \end{Bmatrix} = \begin{bmatrix} Q_{11} & Q_{12} & Q_{13} & 0 & 0 & 0 \\ Q_{21} & Q_{22} & Q_{23} & 0 & 0 & 0 \\ Q_{31} & Q_{32} & Q_{33} & 0 & 0 & 0 \\ 0 & 0 & 0 & Q_{44} & 0 & 0 \\ 0 & 0 & 0 & 0 & Q_{55} & 0 \\ 0 & 0 & 0 & 0 & 0 & Q_{66} \end{bmatrix} \begin{Bmatrix} \varepsilon_{11} \\ \varepsilon_{22} \\ \varepsilon_{33} \\ \gamma_{23} \\ \gamma_{13} \\ \gamma_{12} \end{Bmatrix} - \begin{bmatrix} 0 & 0 & e_{31} \\ 0 & 0 & e_{32} \\ 0 & 0 & e_{33} \\ 0 & e_{24} & 0 \\ e_{15} & 0 & 0 \\ 0 & 0 & 0 \end{bmatrix} \begin{Bmatrix} E_1 \\ E_2 \\ E_3 \end{Bmatrix}$$ (3.333)

Shear-type piezoelectric materials are poled in the axial direction (1 or x). The constitutive equations can be obtained from the extension type piezoelectric material, through a 90° rotation around the second direction (2 or y) followed by a 180° rotation around the third direction (3 or z). Successively applying these rotations to

Equation (3.333) yields the following constitutive equations for the shear-type piezoelectric material:

$$
\begin{Bmatrix} \sigma_{11} \\ \sigma_{22} \\ \sigma_{33} \\ \sigma_{23} \\ \sigma_{13} \\ \sigma_{12} \end{Bmatrix} =
\begin{bmatrix}
Q_{11} & Q_{12} & Q_{13} & 0 & 0 & 0 \\
Q_{21} & Q_{22} & Q_{23} & 0 & 0 & 0 \\
Q_{31} & Q_{32} & Q_{33} & 0 & 0 & 0 \\
0 & 0 & 0 & Q_{44} & 0 & 0 \\
0 & 0 & 0 & 0 & Q_{55} & 0 \\
0 & 0 & 0 & 0 & 0 & Q_{66}
\end{bmatrix}
\begin{Bmatrix} \varepsilon_{11} \\ \varepsilon_{22} \\ \varepsilon_{33} \\ \gamma_{23} \\ \gamma_{13} \\ \gamma_{12} \end{Bmatrix} -
\begin{bmatrix}
e_{11} & 0 & 0 \\
e_{12} & 0 & 0 \\
e_{13} & 0 & 0 \\
0 & 0 & 0 \\
0 & 0 & e_{35} \\
0 & e_{26} & 0
\end{bmatrix}
\begin{Bmatrix} E_1 \\ E_2 \\ E_3 \end{Bmatrix}
$$

$$(3.334)$$

The use of piezoelectric patches having thin thickness implies that one can apply only an E_3 electric field. Therefore, one can write the above two constitutive relationships (Equations (3.333) and (3.334)), for the extension and shear-type piezoelectric materials, in the following compact form:[10]

$$
\begin{Bmatrix} \sigma_{11} \\ \sigma_{22} \\ \sigma_{33} \\ \sigma_{23} \\ \sigma_{13} \\ \sigma_{12} \end{Bmatrix} =
\begin{bmatrix}
\hat{Q}_{11} & \hat{Q}_{12} & \hat{Q}_{13} & 0 & 0 & 0 \\
\hat{Q}_{21} & \hat{Q}_{22} & \hat{Q}_{23} & 0 & 0 & 0 \\
\hat{Q}_{31} & \hat{Q}_{32} & \hat{Q}_{33} & 0 & 0 & 0 \\
0 & 0 & 0 & \hat{Q}_{44} & 0 & 0 \\
0 & 0 & 0 & 0 & \hat{Q}_{55} & 0 \\
0 & 0 & 0 & 0 & 0 & \hat{Q}_{66}
\end{bmatrix}
\begin{Bmatrix} \varepsilon_{11} \\ \varepsilon_{22} \\ \varepsilon_{33} \\ \gamma_{23} \\ \gamma_{13} \\ \gamma_{12} \end{Bmatrix} -
\begin{bmatrix}
e_{11} & 0 & e_{31} \\
e_{12} & 0 & e_{32} \\
e_{13} & 0 & e_{33} \\
0 & e_{24} & 0 \\
e_{15} & 0 & e_{35} \\
0 & e_{26} & 0
\end{bmatrix}
\begin{Bmatrix} E_1 \\ E_2 \\ E_3 \end{Bmatrix}
$$

$$(3.335)$$

where $\qquad \hat{Q}_{21} = \hat{Q}_{12}; \quad \hat{Q}_{31} = \hat{Q}_{31}; \quad \hat{Q}_{23} = \hat{Q}_{32}$

Note that when using extension-type piezoelectric actuators the only nonzero components of the piezoelectric tensor would be: e_{31}, e_{32}, e_{33}, e_{24} and e_{15} while for shear-type piezoelectric actuation the only nonzero components of the piezoelectric tensor would be: e_{11}, e_{12}, e_{13}, e_{26} and e_{35}, according to the presented constitutive equations for each piezoelectric mechanism.

It is known that ceramic PZT actuators are transversely isotropic, therefore, $e_{31} = e_{32}$, and $e_{15} = e_{35}$, (see ref. [2]). As we deal with thin plates, we can use the following the assumption $\sigma_{33} \equiv \sigma_z = 0$ for all structural layers, both piezoelectric and elastic. Therefore, we can write

---

10 Q_{ij} is a general symbol. For extension or shear type mechanisms it has the value as in Equation (3.333) or (3.334), respectively.

$$\varepsilon_{33} \equiv \varepsilon_z = -\frac{\widehat{Q}_{13}\varepsilon_{11} + \widehat{Q}_{23}\varepsilon_{22}}{\widehat{Q}_{33}} + \frac{e_{33}E_3}{\widehat{Q}_{33}} \tag{3.336}$$

Substituting Equation (3.336) into (3.335) yields

$$
\left\{
\begin{array}{c}
\sigma_x \\
\sigma_y \\
\tau_{yz} \\
\tau_{xz} \\
\tau_{xy}
\end{array}
\right\}
=
\left[
\begin{array}{ccccc}
\widetilde{Q}_{11} & \widetilde{Q}_{12} & 0 & 0 & 0 \\
\widetilde{Q}_{12} & \widetilde{Q}_{22} & 0 & 0 & 0 \\
0 & 0 & \widetilde{Q}_{44} & 0 & 0 \\
0 & 0 & 0 & \widetilde{Q}_{55} & 0 \\
0 & 0 & 0 & 0 & \widetilde{Q}_{66}
\end{array}
\right]
\left\{
\begin{array}{c}
\varepsilon_x \\
\varepsilon_y \\
\gamma_{yz} \\
\gamma_{xz} \\
\gamma_{xy}
\end{array}
\right\}
-
\left[
\begin{array}{c}
\widetilde{P}_1 \\
\widetilde{P}_2 \\
0 \\
\widetilde{P}_4 \\
0
\end{array}
\right] E_3
$$

where

$$\sigma_{11} \equiv \sigma_x; \quad \sigma_{22} \equiv \sigma_y; \quad \sigma_{23} \equiv \tau_{yz}; \quad \sigma_{13} \equiv \tau_{xz}; \quad \sigma_{11} \equiv \tau_{xy} \tag{3.337}$$

$$\varepsilon_{11} \equiv \varepsilon_x; \quad \varepsilon_{22} \equiv \varepsilon_y; \quad \gamma_{23} \equiv \gamma_{yz}; \quad \gamma_{13} \equiv \gamma_{xz}; \quad \gamma_{11} \equiv \gamma_{xy}$$

and

$$\widetilde{Q}_{11} = \widehat{Q}_{11} - \frac{\widehat{Q}_{13}^2}{\widehat{Q}_{33}}; \quad \widetilde{Q}_{12} = \widehat{Q}_{12} - \frac{\widehat{Q}_{13}\widehat{Q}_{23}}{\widehat{Q}_{33}}; \quad \widetilde{Q}_{22} = \widehat{Q}_{22} - \frac{\widehat{Q}_{23}^2}{\widehat{Q}_{33}}$$

$$\widetilde{Q}_{44} = \widehat{Q}_{44}; \quad \widetilde{Q}_{55} = \widehat{Q}_{55}; \quad \widetilde{Q}_{66} = \widehat{Q}_{66}$$

$$\widetilde{P}_1 = \frac{\widehat{Q}_{13}}{\widehat{Q}_{33}}e_{33} - e_{31}; \quad \widetilde{P}_2 = \frac{\widehat{Q}_{23}}{\widehat{Q}_{33}}e_{33} - e_{32}; \quad \widetilde{P}_4 = e_{35}$$

For the present derivation, where the principal material coordinates would coincide with the coordinates of the problem being analyzed, the extension-type PZT contribution is added to stresses in the x- and y-directions, while the shear-type PZT contributes only to shear stresses in the x-z-direction.

The electric field intensity E_3 is defined as $E_3 = V/h_k$, where V is the applied voltage across the kth layer and h_k is the thickness of the kth PZT layer.

Shear piezoelectric actuation relies on shear deformation. Hence, investigation of a piezo-laminated plate embedded with shear piezoelectric patches or continuous layers must include the transverse shear displacement. In this section, both solution models (exact and approximate) will rely on the displacement field based on the Mindlin theory (FSDT – see also Section 2.2).

The three displacements within the FSDT are assumed to have the following form:

$$u(x, y, z, t) = u^0(x, y, t) + z\phi_x(x, y, t)$$

$$v(x, y, z, t) = v^0(x, y, t) + z\phi_y(x, y, t) \tag{3.338}$$

$$w(x, y, z, t) = w^0(x, y, t)$$

where $u^0(x,y,t)$, $v^0(x,y,t)$ and $w^0(x,y,t)$ are the displacements in the directions x,y and z, respectively, of a point on the mid-surface and $\phi_x(x,y,t)$ and $\phi_y(x,y,t)$ are rotations of the normal to mid-surface about the x- and y-axes, respectively (see also Figure 3.37).

As already discussed in Section 2.2, the FSDT yields five general governing equations of motion:

$$N_{x,x} + N_{xy,y} + p_x = I_1 \ddot{u}_0(x,y,t) + I_2 \ddot{\phi}_x(x,y,t) \tag{3.339a}$$

$$N_{xy,x} + N_{y,y} + p_y = I_1 \ddot{v}_0(x,y,t) + I_2 \ddot{\phi}_y(x,y,t) \tag{3.339b}$$

$$Q_{x,x} + Q_{y,y} + p_z + \left[\overline{N}_{xx}w_{0,x} + \overline{N}_{xy}w_{0,y}\right]_{,x} +$$
$$+ \left[\overline{N}_{yy}w_{0,y} + \overline{N}_{xy}w_{0,x}\right]_{,x} = I_1 \ddot{w}_0(x,y,z,t) \tag{3.339c}$$

$$M_{x,x} + M_{xy,y} - Q_x + m_x = I_1 \ddot{u}_0(x,y,t) + I_3 \ddot{\phi}_x(x,y,t) \tag{3.339d}$$

$$M_{xy,x} + M_{y,y} - Q_y + m_y = I_1 \ddot{v}_0(x,y,t) + I_3 \ddot{\phi}_y(x,y,t) \tag{3.339e}$$

where

$$I_j = \int_{-h/2}^{h/2} \rho z^{j-1} dz; \quad j = 1, 2, 3 \tag{3.340}$$

and N_x, N_y, N_{xy} denote the force resultants; M_x, M_y, M_{xy} denote the moment resultants; Q_x, Q_y represent transverse; p_x, p_y, p_z and m_x, m_y are external loads and moments, respectively and $\overline{N}_{xx}, \overline{N}_{yy}, \overline{N}_{xy}$ are in plane external applied loads. All force and moments resultants are taken per unit length. As widely presented in the literature and to enable comparisons (see, for instance, Refs. [13–22]), we shall restrict the investigation only to the static solution of a plate with piezoelectric-induced forces, with no external mechanical loads and moments. For a plate with a symmetric lay-up, the longitudinal behavior (Equations (3.339a) and (3.339b)) can be solved separately from flexural behavior (Equations (3.339c)–(3.339e)). Based on a FSDT, these three equations of motion can be expressed in terms of the assumed displacements $(u_0, v_0, w_0)^{11}$ and rotations (φ_x, φ_y), as follows:

---

11 Note that, for the sake of convenience, the subscript 0 was omitted from the displacements expressions.

$$\kappa A_{55}\left[\frac{\partial\phi_x(x,y)}{\partial x}+\frac{\partial^2 w(x,y)}{\partial x^2}\right]-\frac{\partial G_{55}(x,y)}{\partial x}+\kappa A_{44}\left[\frac{\partial\phi_y(x,y)}{\partial y}+\frac{\partial^2 w(x,y)}{\partial y^2}\right]=0$$

$$D_{66}\left[\frac{\partial^2\phi_y(x,y)}{\partial x\partial y}+\frac{\partial^2\phi_x(x,y)}{\partial x^2}\right]-\kappa A_{55}\left[\phi_x(x,y)+\frac{\partial w(x,y)}{\partial x}\right]+D_{11}\frac{\partial^2\phi_x(x,y)}{\partial x^2}+$$

$$+D_{12}\frac{\partial^2\phi_y(x,y)}{\partial x\partial y}+G_{55}(x,y)-\frac{\partial F_{11}(x,y)}{\partial x}=0 \tag{3.341}$$

$$(D_{66}+D_{12})\frac{\partial^2\phi_x(x,y)}{\partial x\partial y}-\kappa A_{44}\left[\phi_y(x,y)+\frac{\partial w(x,y)}{\partial y}\right]+D_{66}\frac{\partial^2\phi_y(x,y)}{\partial x^2}+$$

$$+D_{22}\frac{\partial^2\phi_y(x,y)}{\partial y^2}-\frac{\partial F_{22}(x,y)}{\partial x}=0$$

where the terms D_{ij} (i, $j = 1$, 2, 6) and A_{ii} ($i = 4,5$) are the usual bending and transverse shear stiffness coefficients defined according to lamination theory [2].

The piezoelectric-induced forces, $E_{ii}(x,y)$,[12] moments, $F_{ii}(x,y)$ and induced shear piezoelectric forces, $G_{55}(x,y)$ are defined as

$$E_{ii}(x,y)\equiv\int_{-h/2}^{h/2}\tilde{P}_i E_3 dz=\sum_{j=1}^{N}\left(\tilde{P}_i E_3\right)_j;\quad i=1,2$$

$$F_{ii}(x,y)\equiv\int_{-h/2}^{h/2}\tilde{P}_i E_3 z dz=\sum_{j=1}^{N}\left(\tilde{P}_i E_3\right)_j(z_j-z_{j-1});\quad i=1,2 \tag{3.342}$$

$$G_{55}(x,y)\equiv\kappa\int_{-h/2}^{h/2}\tilde{P}_4 E_3 dz=\sum_{j=1}^{N}\left(\tilde{P}_4 E_3\right)_j$$

where κ is a shear factor used in the FSDT (in most of the cases, its value is 5/6) and N is the number of piezoelectric layers in the laminate.

---

12 Note that the terms $E_{ij}(x,y)$ do not appear in Equation (3.341), however they might appear in Equations (3.621a and b), depending on the way the piezoelectric patches are electrically connected (in-phase or out-of-phase)

Based on Navier or Lévy theories for plates [23], one can solve a plate with continuous piezoelectric layers, where at least two opposite boundary conditions are assumed to be simply supported edges.

Let us consider a plate with opposite simply supported edges along the y-direction; one can assume the following form for the plate solution (note that these functions must also satisfy the boundary conditions of the plate):

$$w(x,y) = \overline{w}(x)\sin(\lambda_y y); \quad \phi_x(x,y) = \overline{\phi}_x(x)\sin(\lambda_y y)$$

$$\phi_y(x,y) = \overline{\phi}_y(x)\cos(\lambda_y y); \quad F_{11}(x,y) = \overline{F}_{11}(x)\sin(\lambda_y y)$$

$$F_{22}(x,y) = \overline{F}_{22}(x)\sin(\lambda_y y); \quad G_{55}(x,y) = \overline{G}_{55}(x)\sin(\lambda_y y)$$

(3.343)

where $\quad \lambda_y \equiv \dfrac{n\pi}{b}$

n is the number of half-waves of the plate and b is the plate length along the y-direction.

Assuming constant piezoelectric forces along the x-direction, and substituting the suggested solutions (Equation (3.343)) into the three equilibrium equations, Equation (3.343), one would get a system of three coupled regular differential equations in the following form

$$\kappa A_{55}\left[\frac{d\overline{\phi}_x(x)}{dx} + \frac{d^2\overline{w}(x)}{\partial x^2}\right] - \kappa A_{44}\left[\lambda_y\overline{\phi}_y(x) + \lambda_y^2\overline{w}(x)\right] = 0$$

$$D_{66}\left[\lambda_y\frac{d\overline{\phi}_y(x)}{dx} + \lambda_y^2\overline{\phi}_x(x)\right] - \kappa A_{55}\left[\overline{\phi}_x(x) + \frac{d\overline{w}(x)}{dx}\right] + D_{11}\frac{d^2\overline{\phi}_x(x)}{dx^2} +$$

$$- D_{12}\lambda_y\frac{d\overline{\phi}_y(x)}{dx} + \overline{G}_{55}(x) = 0$$

(3.344)

$$(D_{66} + D_{12})\lambda_y\frac{d\overline{\phi}_x(x)}{dx} - \kappa A_{44}\left[\overline{\phi}_y(x) + \lambda_y\overline{w}(x)\right] + D_{66}\frac{d^2\overline{\phi}_y(x)}{dx^2} +$$

$$- D_{22}\lambda_y^2\overline{\phi}_y(x) = 0$$

The three coupled equations (Equation (3.344)), can be decoupled to yield a single equation for each of the three unknowns $w(x,y)$, $\varphi_x(x,y)$ and $\varphi_y(x,y)$ (see [1]) having the following general pattern:

$$\alpha_1 A^6 + \alpha_2 A^4 + \alpha_3 A^2 + \alpha_4 A = 0$$

(3.345)

where $\qquad A = w, \phi_x \text{ or } \phi_y$

and the coefficients α_i, $i = 1\text{--}4$, differ for each variable. For the variable, $\bar{w}$ Equation (3.345) would be

$$\alpha_1 \frac{d^6 \bar{w}}{dx^6} + \alpha_2 \frac{d^4 \bar{w}}{dx^4} + \alpha_3 \frac{d^2 \bar{w}}{dx^2} + \alpha_4 \bar{w} = 0 \tag{3.346}$$

which leads to the following characteristic equation, if we substitute $(\bar{w}) = \bar{R}e^{\bar{s}x}$

$$\alpha_1 s^3 + \alpha_2 s^2 + \alpha_3 s + \alpha_4 = 0 \tag{3.347}$$

where $\quad\quad\quad\quad\quad s = \bar{s}^2$

The solution of Equation (3.347) yields three real roots, one root being negative and the other two positive. Thus obtaining the roots, we can represent the general solution for the lateral displacement in the thickness direction as follows:

$$\bar{w} = B_1 \sin(r_1 x) + B_2 \cos(r_1 x) + B_3 \sinh(r_2 x) +$$
$$+ B_4 \cosh(r_2 x) + B_5 \sinh(r_3 x) + B_6 \cosh(r_3 x) \tag{3.348}$$

where

$$r_1 = \sqrt{-s_1}; \; r_2 = \sqrt{s_2}; \; r_1 = \sqrt{s_3}.$$

The six constants $B_1\text{--}B_6$ will be found by enforcing the plate's boundary conditions. The expressions for the two bending rotations, $\varphi_x(x,y)$ and $\varphi_y(x,y)$, have the similar forms, however with different constants:

$$\bar{\phi}_x = \bar{B}_1 \sin(r_1 x) + \bar{B}_2 \cos(r_1 x) + \bar{B}_3 \sinh(r_2 x) +$$
$$+ \bar{B}_4 \cosh(r_2 x) + \bar{B}_5 \sinh(r_3 x) + \bar{B}_6 \cosh(r_3 x)$$
$$\bar{\bar{\phi}}_y = \bar{\bar{B}}_1 \sin(r_1 x) + \bar{\bar{B}}_2 \cos(r_1 x) + \hat{B}_3 \sinh(r_2 x) +$$
$$+ \bar{\bar{B}}_4 \cosh(r_2 x) + \bar{\bar{B}}_5 \sinh(r_3 x) + \bar{\bar{B}}_6 \cosh(r_3 x) \tag{3.349}$$

Note that the constants $\bar{B}_1 - \bar{B}_6$ and $\bar{\bar{B}}_1 - \bar{\bar{B}}_6$ can be expressed by the constants, $B_1\text{--}B_6$, by substituting Equation (3.349) into the coupled equations of motion, Equation (3.344).

For other boundary conditions, to obtain solutions for the relevant displacements and the bending rotations of the plate, one should use energy methods, like the Rayleigh–Ritz method.

Next typical results by Edery-Azulay and Abramovich [1] are presented. First, the behavior of a square laminate plate with continuous extension or shear piezoelectric layers is presented. The configuration of the laminated plate is illustrated in Figure 3.37. To simplify the formulations, it is assumed that all layers have a 0° orientation. The length of the squared plate is assumed to be 10 cm. The piezoelectric material is considered to be PZT-5 H with a constant layer thickness of $t = 0.25$ mm, and the structural layers are considered to be made of graphite epoxy, each layer thickness being $t = 0.5$ mm. (The total thickness of the plate is 3 mm.) Table 3.6 summarizes the two material properties.

Table 3.6: Material properties (both mechanical and electrical).

Mechanical property (GPa)	Graphite epoxy	PZT-5 H	Electrical property (Cm^{-2})	PZT-5 H
C_{11}	183.443	99.201	e_{31}	−7.209
C_{22}	11.662	99.201	e_{33}	15.118
C_{33}	11.662	86.856	e_{24}	12.332
C_{12}	4.363	54.016	e_{15}	12.322
C_{13}	4.363	50.778		
C_{23}	3.918	50.778		
C_{44}	2.877	21.100		
C_{66}	7.170	21.100		
C_{66}	7.179	22.593		

Figure 3.38a and b describes the transverse displacement pattern from a top view and the out-of-plane displacements along the x-direction for the line $y = b/2$, of an all edge simply supported plate actuated with extension PZT. These results were found using the present exact mathematical model (Lévy method). For the case of a plate with a shear piezoelectric continuous layer, no lateral displacements were found. Note that a similar behavior was also found for a piezocomposite beam [2].

Figure 3.39a and b shows the transverse displacement pattern from top view and the out-of-plane displacements along the x-direction for the line $y = b/2$, of a plate with two opposite clamped edges (CCSS), actuated by a shear-type PZT layer. For both cases, constant induced piezoelectric moments ($\bar{F}_{11} = \bar{F}_{22} = 1000$ (N mm)) and constant induced shear piezoelectric forces $\bar{G}_{55} = 1000$ N are assumed.

Typical results are now presented for a plate with extension or shear-type piezoelectric patches (also from [1]). Figure 3.40 presents a plate with a piezoelectric patch.

The patch location is determined by two coordinates, ξ and η, which describe the mid location of the patch. The surface and the core layers of the examined plate are made of nonstructural material, for example, foam, while its material properties are taken as zero. There is one shear-type patch and a pair of extension-type patches (see Figure 3.40).

The following results deal with a square plate of 10 cm length and width, where the used patch covers 10% of the plate area. Each extension-type PZT patch has a thickness of $t = 0.5$ mm, while each shear-type PZT patch has a thickness of $t = 0.25$ mm, and for each structural material the thickness is $t = 0.5$ mm. To solve the various cases use was made of the Rayleigh–Ritz method. Table 3.7 summarizes the

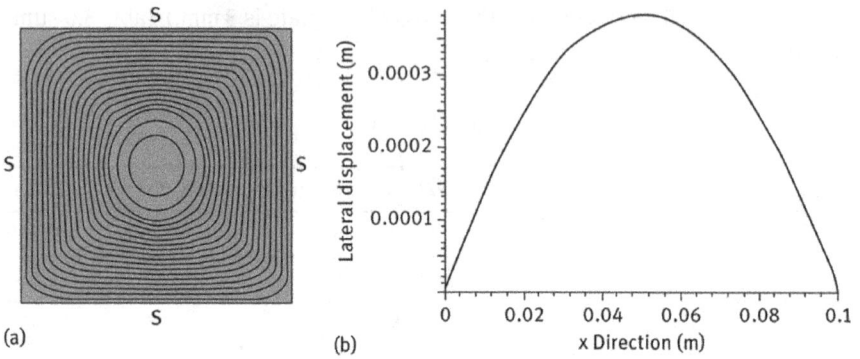

Figure 3.38: An all edge simply supported plate actuated with a continuous extension piezoelectric layer: (a) out-of-plane displacement pattern and (b) out-of-plane displacement.

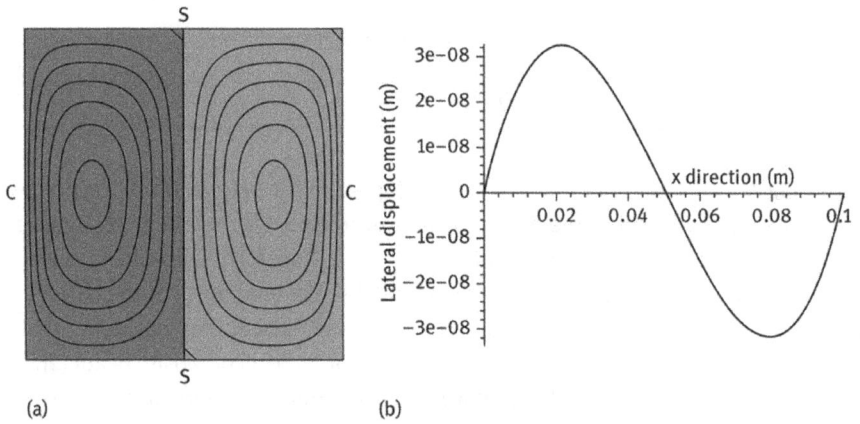

Figure 3.39: A CCSS plate actuated with a continuous shear piezoelectric layer: (a) out-of-plane displacement pattern and (b) out-of-plane displacement.

assumed functions of the rectangular sandwich plates for different boundary conditions, when applying the Rayleigh–Ritz method. One should note that the piezoelectric forces inside the patches are assumed to be constant.

In Figure 3.41a and b, one can see the results for an all edge simply supported plate with a central extension piezoelectric patch. Each line presents the central lateral displacement along the x-direction and the line $y = b/2$, for different number of serial terms (m number of terms in the series in the x-direction) and in Figure 3.41b one can see its transverse displacement pattern for $m = 13$ and $j = 1$. The bold line presents the final calculated displacement curve with $m = 13$.

Figure 3.42a and b describe the lateral displacement along the x-direction, for the line $y = 0.05b$, of an all edge simply supported plate with a side located extension-type piezoelectric patch at $\xi = 0.2a$, $\eta = 0.2b$.

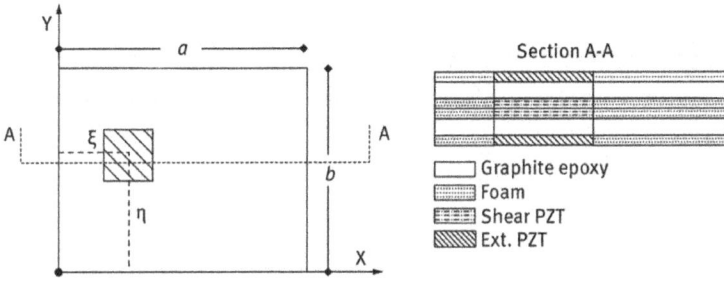

Figure 3.40: Piezoelectric patch geometrical location.

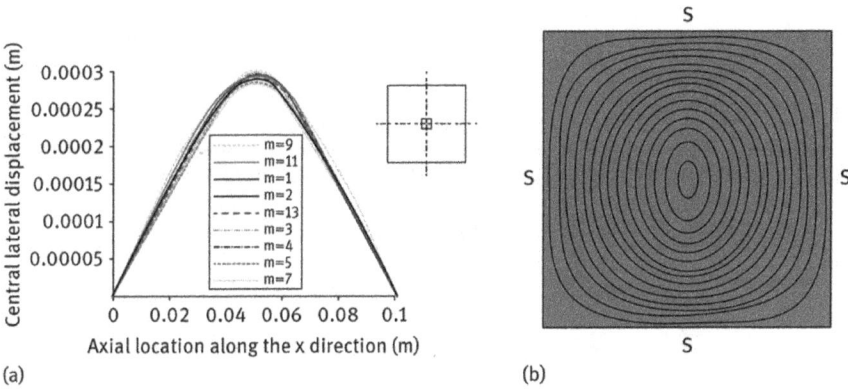

(a)

(b)

Figure 3.41: An all edge simply supported plate with central extension-type PZT patch: (a) central lateral displacement along the x-direction, $y = b/2$; (b) transverse displacement pattern.

Table 3.7: Assumed functions for rectangular piezocomposite plates.

B.C.	$\overline{w}$	$\overline{\phi}_x$	$\overline{\phi}_y$
SSSS	$\sum_{n=1}^{\infty}\sum_{m=1}^{\infty} R_{mn} \cdot SX \cdot SY$	$\sum_{n=1}^{\infty}\sum_{m=1}^{\infty} T_{mn} \cdot CX \cdot SY$	$\sum_{n=1}^{\infty}\sum_{m=1}^{\infty} G_{mn} \cdot SX \cdot CY$
	$\overline{w}(x,0) = 0;\quad \overline{w}(x,b) = 0;$ $\overline{w}(0,y) = 0;\quad \overline{w}(a,y) = 0.$	$\overline{\phi}_x(x,0) = 0;$ $\overline{\phi}_x(x,b) = 0.$	$\overline{\phi}_y(0,y) = 0;$ $\overline{\phi}_y(a,y) = 0.$
CCCC	$\sum_{n=1}^{\infty}\sum_{m=1}^{\infty} R_{mn} \cdot (1 - CX') \cdot (1 - CY')$	$\sum_{n=1}^{\infty}\sum_{m=1}^{\infty} T_{mn} \cdot SX \cdot (1 - CY')$	$\sum_{n=1}^{\infty}\sum_{m=1}^{\infty} G_{mn} \cdot SY \cdot (1 - CX')$
	$\overline{w}(x,0) = 0;\quad \overline{w}(x,b) = 0;$ $\overline{w}(0,y) = 0;\quad \overline{w}(a,y) = 0.$	$\overline{\phi}_x(x,0) = 0;$ $\overline{\phi}_x(x,b) = 0.$	$\overline{\phi}_y(0,y) = 0;$ $\overline{\phi}_y(a,y) = 0.$

Table 3.7 (continued)

B.C.	$\overline{w}$	$\overline{\phi}_x$	$\overline{\phi}_y$
CFCF	$\displaystyle\sum_{\substack{i=1 \\ j=1}}^{\infty} R_i \cdot x^{2i} \cdot y^{2j}$	$\displaystyle\sum_{\substack{i=1 \\ j=1}}^{\infty} T_i \cdot x^{2i-1} \cdot y^{2j}$	$\displaystyle\sum_{\substack{i=1 \\ j=1}}^{\infty} G_i \cdot x^{2i} \cdot y^{2j-1}$
	$\overline{w}(x,0) = 0;$ $\overline{w}(0,y) = 0.$	$\overline{\phi}_x(x,0) = 0.$	$\overline{\phi}_y(0,y) = 0.$

Clamped at $x = 0, y = 0$ and free at $x = a, y = b$, edges

SSCC	$\displaystyle\sum_{n=1}^{\infty}\sum_{m=1}^{\infty} R_{mn} \cdot SX \cdot (1 - CY')$	$\displaystyle\sum_{n=1}^{\infty}\sum_{m=1}^{\infty} T_{mn} \cdot CX \cdot (1 - CY')$	$\displaystyle\sum_{n=1}^{\infty}\sum_{m=1}^{\infty} G_{mn} \cdot SX \cdot SY$
	$\overline{w}(x,0) = 0;$ $\overline{w}(x,b) = 0;$ $\overline{w}(0,y) = 0;$ $\overline{w}(a,y) = 0.$	–	$\overline{\phi}_y(x,0) = 0;$ $\overline{\phi}_y(x,b) = 0.$

Clamped at $y = 0$ and $y = b$, edges

where SS represents simply supported, C represents clamped and F represents free; and

$$SX = \sin\frac{m\pi x}{a}; \quad CX = \cos\frac{m\pi x}{a}; \quad SY = \sin\frac{n\pi y}{b};$$

$$CY = \cos\frac{n\pi y}{b}; \quad CX' = \cos\frac{2m\pi x}{a}; \quad CY' = \cos\frac{2n\pi y}{b}.$$

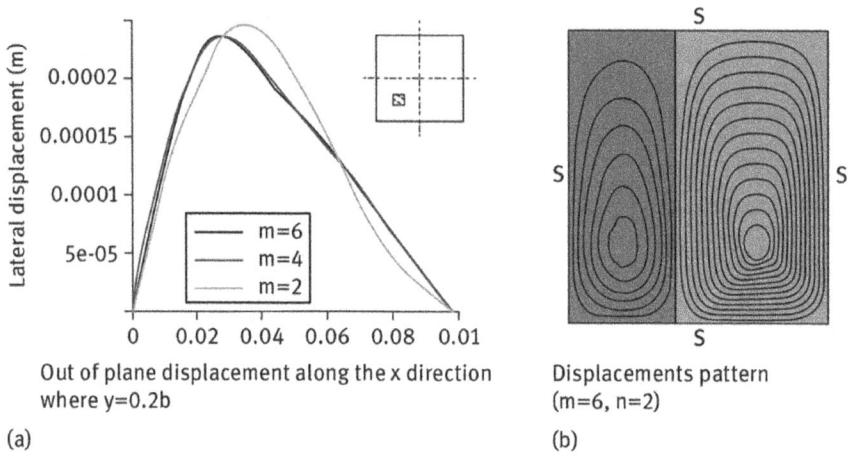

(a) Out of plane displacement along the x direction where y=0.2b

(b) Displacements pattern (m=6, n=2)

Figure 3.42: An all-edge simply supported plate actuated with a side extension piezoelectric patch: (a) out-of-plane displacements and (b) displacement pattern.

The effect of a shear-type PZT patch is depicted in Figure 3.43a and b. A square plate actuated by a central shear patch, located at $\xi = 0.5a$, $\eta = 0.5b$ was used for the results presented. The lateral displacement along the x-direction, for the line $y = b/2$, where the bold line is the final calculated displacement curve with $m = 12$ is shown in Figure 3.43a, while Figure 3.43b shows the transverse displacement pattern of the entire composite plate obtained for $m = 12, n = 1$. One should note that using shear piezoelectric patches one has to use series with even terms only, due to the characteristics of the solution, the first term ($m = 1$) leads to no lateral displacement, and therefore at least two terms are needed to observe the lateral-induced deformation.

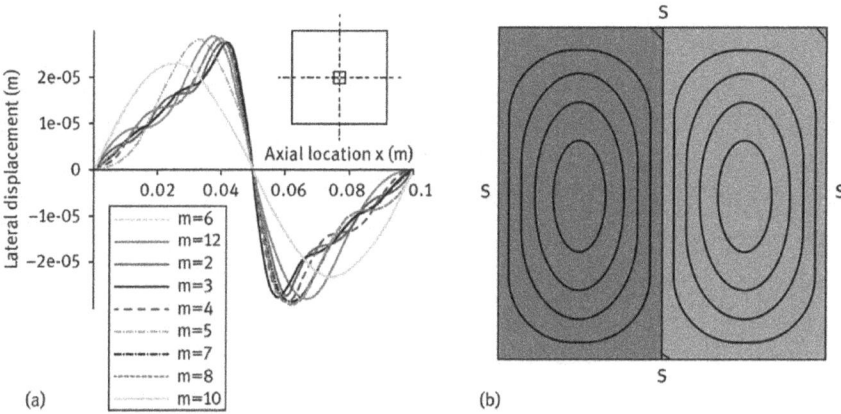

Figure 3.43: An all edge simply supported plate actuated with a central shear-type piezoelectric patch: (a) out-of-plane displacements and (b) displacement pattern.

The point of merit of the present model is its robustness to obtain a solution for a plate with an unlimited number of patches, at arbitrary locations.

In order to check the accuracy of the solution for a plate with more than one patch, one of the cases that have been investigated before is reconsidered here: a simply supported plate with one central extension or shear PZT patch. This single patch is then divided into two patches that together cover the same plate area (with no overlap between them). The flexural behavior of this plate was studied. The plate with the two patches yields identical results as the plate with one patch, using a single piezoelectric type, extension or shear. Figures 3.44 and 3.45 present the results using one or two separate patches, extension or shear type, respectively. One can see the very good agreement between one or two patches having the same area.

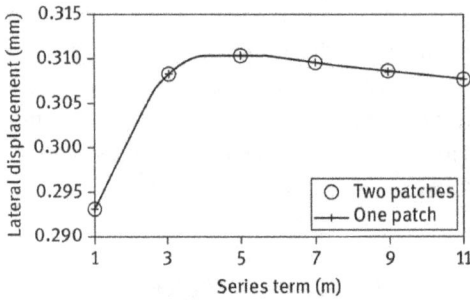

Figure 3.44: Central transverse displacement for an all edge simply supported plate actuated by one or two central extension patches.

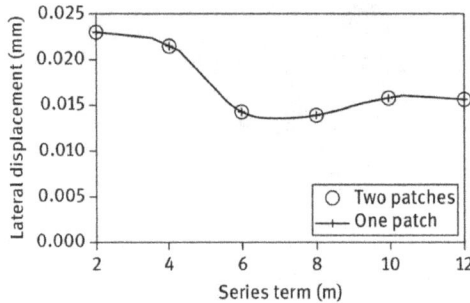

Figure 3.45: Transverse displacement at $x = a/4$, $y = b/2$ for an all edge simply supported plate actuated by one or two centrally located shear-type patches.

References

[1] Edery-Azulay, L. and Abramovich, H., A reliable plain solution for rectangular plates with piezoceramic patches, Journal of Intelligent Material Systems and Structures 18, 2007, 419–433.

[2] Edery-Azulay, L. and Abramovich, H., Actuation and sensing of shear type piezoelectric patches-closed form solutions, Composite Structures 64, 2004, 443–453.

[3] Kerr, A. D., An extension of the Kantorovich method, Quarterly of Applied Mathematics 26, 1968, 219–229.

[4] Kerr, A. D. and Alexander, H., An application of the extended Kantorovich method to the stress analysis of a clamped rectangular plate, Acta Mechanica 6, 1968, 180–196.

[5] Yuan, S. and Jin, Y., Computation of elastic buckling loads of rectangular thin plates using the extended Kantorovich method., Composite Structures 66, 1998, 861–867.

[6] Ungbhakorn, V. and Singhatanadgid, P., Buckling analysis of symmetrically laminated composite plates by the extended Kantorovich method, Composite Structures 73, 2006, 120–128.

[7] Aghdam, M. M. and Falahatgar, S. R., Bending analysis of thick laminated plates using extended Kantorovich method, Composite Structures 62, 2003, 279–283.

[8] Yuan, S., Jin, Y. and Williams, F. W., Bending analysis of Mindlin plates by extended Kantorovich method, Journal of Engineering Mechanics 124, 1998, 1339–1345.

[9]   Maple-9. User's manual. Maple Release VIII; 2002.
[10]   Aldraihem, O. J. and Khdeir, A. A., Smart beams with extension and thickness-shear piezoelectric actuators, Smart Materials and Structures 9, 2000, 1–9.
[11]   Chee, C., Tong, L. and Steven, G., A mixed model for adaptive composite plates with piezoelectric for anisotropic actuation, Computers & Structures 77, 2000, 253–268.
[12]   Chopra, I., Review of state of art of smart structures and integrated systems, AIAA Journal 40, 2002, 2145–2187.
[13]   Edery-Azulay, L. and Abramovich, H., Actuation and sensing of shear type piezoelectric patches-closed form solutions, Journal of Composite Materials 64, 2004, 443–453.
[14]   Lin, C. C., Hsu, C. Y. and Huang, H. N., Finite element analysis on deflection control of plate with piezoelectric actuators, Composite Structures 35(4), 1996, 423–433.
[15]   Mitchell, J. A. and Reddy, J. N., A refined hybrid plate theory for composite laminates with piezoelectric laminae, International Journal of Solids and Structures 32, 1995, 2345–2367.
[16]   Reddy, J. N., Mechanics of Laminated Composite Plate and Shells, Theory and Analysis, 2nd edn., CRC Press LLC, 2004, 831.
[17]   Robaldo, A., Carrera, E. and Benjeddou, A., A unified formulation for finite element analysis of piezoelectric adaptive plates. Seventh International Conference on Computational Structures Technology, Lisbon Portugal, 7–9 September, 2004.
[18]   Shah, D. K., Joshi, S. P. and Chan, W. S., Static structural response of plates with piezoceramic layers, Smart Materials and Structures 2, 1993, 172–180.
[19]   Vel, S. S. and Batra, R. C., Exact solution for the cylindrical bending of laminated plates with embedded piezoelectric shear actuators, Smart Materials and Structures 10, 2000, 240–251.
[20]   Vel, S. S. and Batra, R. C., Exact solution for rectangular sandwich plates with embedded piezoelectric shear actuators, AIAA Journal 39, 2001, 1363–1373.
[21]   Vel, S. S. and Batra, R. C., Analysis of piezoelectric bimorphs and plates with segmented actuators, Thin Walled Structures 39, 2001, 23–44.
[22]   Vinson, J. R. and Sierakowski, R. L., Solid Mechanics and its Applications. The Behavior of Structures Composed of Composite Materials, 2nd edn., Kluwer Academic Publishers, 1986, 435.
[23]   Zhang, X. D. and Sun, C. T., Analysis of a sandwich plate containing a piezoelectric core, Smart Materials and Structures 8, 1999, 31–40.
[24]   Timoshenko, S. and Woinowsky-Krieger, S., Theory of Plates and Shells, New York, McGraw-Hill, 1959, 580.

3.7 Appendix A

In what follows, we shall present and discuss the various derivations of the piezo-electric coupling coefficient or the electromechanical coupling factor denoted as k^2. One of the ways to define the piezoelectric coupling coefficient is as the one presented in Equation (3.41) and presented again as

$$k^2 = \frac{\text{(Piezoelectric energy density stored in the material)}}{\text{(Electrical energy density)} \cdot \text{(Mechanical energy density)}} \quad \text{(a)}$$

The literature presents also other types of definitions using other parameters as are reflected in the typical references [1–6]. Uchino [5] claims that all the conversion rates between mechanical energy and electrical energy and vice versa leading to electromechanical coupling factor or energy transmission coefficient or efficiency are somehow confusing due to their various definitions. Another definition is the one presented by Uchino [5] having the following general form

$$k^2 = \frac{\text{(Stored mechanical energy)}}{\text{(Input electrical energy)}}$$

or (b)

$$k^2 = \frac{\text{(Stored electrical energy)}}{\text{(Input mechanical energy)}}$$

Remembering that the input electrical energy per unit volume can be written as $0.5\varepsilon_0\varepsilon^\sigma E^2 = 0.5\bar{\varepsilon}^\sigma E^2$ while the mechanical stored mechanical energy per unit volume is $0.5d^2E^2/s^E$, and substituting in the first equation of (b) yields

$$k^2 = \frac{0.5d^2E^2/s^E}{0.5\bar{\varepsilon}^\sigma E^2} = \frac{d^2}{\bar{\varepsilon}^\sigma s^E}$$ (c)

The expression in Equation (c) is identical to Equation (3.43). One can obtain the same result as in Equation (c) also for the second expression in Equation (b).[14]

A different approach is presented in [3] where the authors propose to define the piezoelectric coupling coefficient as

$$\kappa^2 = \frac{f_{oc}^2 - f_{oc}^2}{f_{oc}^2}$$ (d)

where f_{oc} is the resonance frequency at open electrodes and f_{sc} is the resonant frequency at short circuit. This definition provides a convenient way to experimentally determine the coupling coefficient.

An alternative definition is presented in [6], namely

$$k^2 = \frac{\text{(Converted energy)}}{\text{(Supplied energy)}}$$ (e)

The authors claim that although the piezoelectric coupling factor is considered as an effective measure to evaluate the effectiveness of piezoelectric materials, the factor k^2, as presented in Equation (e) is not always equal to the value presented in Equation (c).

---

[14] See complete derivation in Chapter 3 Piezoelectricity, Rupitsch, S. J., *Piezoelectric Sensors and Actuators*, Topics in Mining, Metallurgy and Materials Engineering, https://doi.org/10.1007/978-3-662-57534-5_3 © Springer-Verlag GmbH Germany, part of Springer Nature 2019

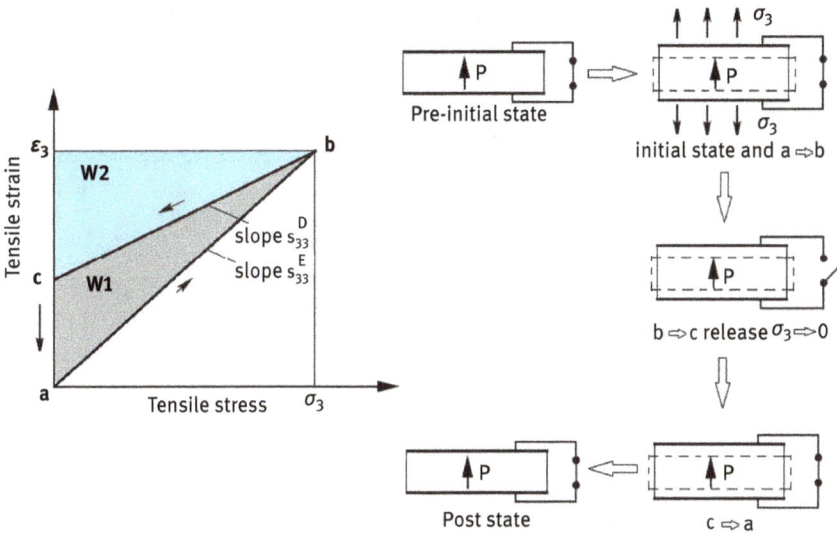

Figure 3.a: A quasistatic stress cycle (adapted from [1]).

For the case of plane stress (see Figure 3.a) they show that Equation (e) leads to

$$k_{33}^2 \equiv \frac{(\text{Converted energy})}{(\text{Supplied energy})} = \frac{0.5 D_3 E_3}{0.5 \varepsilon_3 \sigma_3} = \frac{0.5 \bar{k}_{33}^2 \varepsilon_3^2 / s_{33}^E}{0.5 \varepsilon_3^2 / s_{33}^E} = \bar{k}_{33}^2 \equiv \frac{d_{33}^2}{s_{33}^E \bar{\varepsilon}_{33}^\sigma} \tag{f}$$

This means that the definition present in Equation (c) leads to the expression presented in Equation (b).

For the case of plane strain, the authors [6] use the following constitutive equations (while $\varepsilon_1 = \varepsilon_2 = 0$)

$$\sigma_3 = C_{33}^E \varepsilon_3 - e_{33} E_3$$
$$D_3 = e_{33} \varepsilon_3 + \bar{\varepsilon}_{33}^\varepsilon E_3 \tag{g}$$

leading to the following relationship

$$k_{33}^2 \equiv \frac{(\text{Converted energy})}{(\text{Supplied energy})} = \frac{0.5 D_3 E_3}{0.5 \varepsilon_3 \sigma_3} = \frac{0.5 \left[\kappa_{33}^2 / (1 + \kappa_{33}^2) \right] \varepsilon_3^2 C_{33}^E}{0.5 \varepsilon_3^2 C_{33}^E} = \frac{\kappa_{33}^2}{1 + \kappa_{33}^2} \tag{h}$$

where

$$\kappa_{33}^2 \equiv \frac{e_{33}^2}{C_{33}^E \bar{\varepsilon}_{33}^\sigma}$$

Equation (h) shows that not always the ratio between converted energy to supplied energy leads to Equation (c).

Another issue worth to be outlined is the energy transmission coefficient λ_{max} [5]. Its definition is

$$\lambda_{max} = \left(\frac{Output \quad mechanical \quad energy}{Input \quad electrical \quad energy} \right)_{max}$$

or   (i)

$$\lambda_{max} = \left(\frac{Output \quad electrical \quad energy}{Input \quad mechanical \quad energy} \right)_{max}$$

Note that Equation (i) is similar to Equation (b) with the word *stored* being replaced by the words *output* in Equation (i).

Note also the definition for the efficiency η [5]:

$$\eta = \frac{Output \quad mechanical \quad energy}{Consumed \quad electrical \quad energy}$$

or   (j)

$$\eta = \frac{Output \quad electrical \quad energy}{Consumed \quad mechanical \quad energy}$$

3.7.2 Appendix B

The coefficients appearing in Equation (3.205) have the following form:

$$a_{11} = 0; \, a_{12} = 1; \, a_{13} = 0; \, a_{14} = 1; \tag{a}$$

$$a_{21} = \sinh(\lambda_1 L); \, a_{12} = \cosh(\lambda_1 L); \, a_{13} = \sin(\lambda_2 L); \, a_{14} = \cos(\lambda_2 L); \tag{b}$$

$$a_{31} = k_\theta \lambda_1; \, a_{32} = \lambda_1^2; \, a_{33} = k_\theta \lambda_2; \, a_{34} = -\lambda_2^2 \tag{c}$$

$$
\begin{aligned}
a_{41} &= \lambda_1^2 \sinh(\lambda_1 L) + k_\theta \lambda_1 \cosh(\lambda_1 L); \\
a_{42} &= \lambda_1^2 \cosh(\lambda_1 L) + k_\theta \lambda_1 \sinh(\lambda_1 L); \\
a_{43} &= -\lambda_2^2 \sin(\lambda_2 L) + k_\theta \lambda_2 \cos(\lambda_2 L); \\
a_{44} &= -\lambda_2^2 \cos(\lambda_2 L) - k_\theta \lambda_2 \sin(\lambda_2 L).
\end{aligned}
\tag{d}
$$

3.7.3 Appendix C: Constants Presented in Table 3.4

$$a = \frac{q_0 \alpha_0}{6(\alpha_0 - \beta_0^2)}; b = \frac{q_0}{2c_0}; d = \frac{2\alpha_0 \hat{F}_{11} - \alpha_0 q_0}{4(\beta_0^2 - \alpha_0)}; g = \frac{\alpha_0 \hat{F}_{11}}{2(\beta_0^2 - \alpha_0 \beta_0^2)};$$

$$\delta = \frac{\beta_0}{\alpha_0}; \varepsilon = \frac{q_0}{24} + b + \frac{\hat{F}_{11}}{2}; \theta = \frac{b - d - \frac{a}{4}}{\frac{\alpha_0}{3} - \frac{\beta_0^2 - \alpha_0}{c_0}}; \theta_1 = Z\frac{d + \theta}{2};$$

$$\theta_2 = \frac{(\beta_0^2 - \alpha_0)}{c_0} \theta; \eta_0 = \frac{q_0}{4} + b + \frac{\hat{F}_{11}}{2} + \frac{a}{4} - \frac{a\delta}{2};$$

$$\eta_1 = \beta_0 - \frac{2(\beta_0^2 - \alpha_0)}{\beta_0} - \frac{2}{\beta_0}\left(\frac{\alpha_0}{3} + \frac{2(\beta_0^2 - \alpha_0)}{c_0}\right); \eta = \frac{\eta_0}{\eta_1};$$

$$\mu = a\delta - \hat{F}_{11} - \frac{q_0}{2} + \frac{(2\alpha_0 - \beta_0^2)}{\beta_0}; \mu_1 = 2\frac{(\alpha_0 - \beta_0^2)}{\beta_0^2 c_0}\eta; \mu_1 = 2\frac{(\alpha_0 - \beta_0^2)}{\beta_0^2 c_0}\eta;$$

$$A_0 = \frac{q_0}{\lambda}; B_0 = -\frac{q_0}{2\lambda}; \gamma^2 = \frac{\lambda}{\left(1 - \frac{\beta_0^2}{\alpha_0}\right)\left(1 - \frac{\lambda}{c_0}\right)}; B_1 = \frac{q_0 c_0(c_0 + \gamma c_0 - \gamma\lambda)}{\gamma\lambda(\lambda - c_0)(c_0 - \gamma c_0 + \gamma\lambda)};$$

$$B_3 = -A_3 = -\gamma\left(1 - \frac{\lambda}{c_0}\right)B_1; B_2 = -B_1 \tan\gamma - \frac{1}{\cos\gamma}\left[\frac{\left(\lambda + \alpha_0 \hat{F}_{11}\right)}{\lambda\alpha_0} + \frac{q_0}{\left(1 - \frac{\lambda}{c_0}\right)\gamma\lambda}\right];$$

$$B_4 = -B_2; A_1 = \gamma\left(1 - \frac{\lambda}{c_0}\right)B_2; A_2 = -\gamma\left(1 - \frac{\lambda}{c_0}\right)B_1; C_1 = -\gamma\left(1 - \frac{\lambda}{c_0}\right)\frac{\beta_0}{\alpha_0}B_2;$$

$$C_2 = \gamma\left(\frac{1 - \lambda}{c_0}\right)\frac{\beta_0}{\beta_0}B_1; C_4 = -C_2; C_3 = \frac{\lambda}{\alpha_0\varphi} - \frac{q_0\beta_0}{\alpha_0\lambda}.$$

3.7.4 Appendix D: Constants Presented in Table 3.5

Note that those variables with a bar belong to the part of the beam without a piezo-electric patch. For the case the piezoelectric patch is situated in the middle of the beam, the other two parts have equal properties:

$$A = \frac{-\hat{F}_{11}\alpha_0}{2(\alpha_0 - \beta_0^2)}; A_1 = \frac{\hat{F}_{11}\beta_0}{(\alpha_0 - \beta_0^2)}; B = \frac{\hat{F}_{11}\beta_0(\xi_0 - 1)}{(\xi_0 - 1)(\alpha_0 - \beta_0^2) - (\alpha_0 - \beta_0^2)};$$

$$B_1 = B\frac{\xi_0}{(\xi_0 - 1)}; F = \xi_0\left(\beta_0 B - \bar{\beta}_0 B_1 + \hat{F}_{11}\right) + F_1;$$

$$F_1 = \left(1 + \xi_0^2\right)\frac{\bar{\beta}_0 B_1}{2} - \xi_0^2\frac{\beta_0 B + \hat{F}_{11}}{2}; \delta = \frac{\beta_0}{\alpha_0}; \delta_1 = \frac{\bar{\beta}_0}{\alpha_0}; D_1 = D\left(\frac{\alpha_0 - \beta_0^2}{\bar{\alpha}_0 - \bar{\beta}_0^2}\right)\frac{\bar{\alpha}_0}{\alpha_0};$$

$$C = 2D\frac{\beta_0}{\alpha_0} + A_1; C_1 = A_1\xi_0\frac{\beta_0}{\alpha_0} + A_1\xi_0\left(2 - \xi_0\right)\left[\frac{\beta_0}{\alpha_0} - \frac{\bar{\beta}_0\left(\alpha_0 - \beta_0^2\right)}{\alpha_0\left(\bar{\alpha}_0 - \bar{\beta}_0^2\right)}\right];$$

$$E = 2(A - D); E_1 = 2A\xi_0 + D\xi_0\left(2 - \xi_0\right)\left[1 - \frac{\bar{\alpha}_0\left(\alpha_0 - \beta_0^2\right)}{\alpha_0\left(\bar{\alpha}_0 - \bar{\beta}_0^2\right)}\right]; I = 2D\frac{\left(\alpha_0 - \beta_0^2\right)}{\alpha_0 c_0};$$

$$J = -\frac{2}{3}D_1 - I_1; \ I_1 = 2D_1\frac{\left(\bar{\alpha}_0 - \bar{\beta}_0^2\right)}{\alpha_0 c_0} - E_1$$

3.7.5 Appendix E

The various terms of the matrix A are

$$a_{11} = 0 \quad a_{12} = 1 \quad a_{13} = 0 \quad a_{14} = 1 \quad a_{15} \rightarrow a_{1,12} = 0 \tag{a}$$

$$a_{21} = 0 \quad a_{22} = (D_{11})_1 m_{11}\lambda_{11} \quad a_{23} = 0 \quad a_{24} = (D_{11})_1 m_{21}\lambda_{21}$$
$$a_{25} \rightarrow a_{2,12} = 0 \tag{b}$$

$$a_{31} = \sinh(\lambda_{11}x_0) \quad a_{32} = \cosh(\lambda_{11}x_0) \quad a_{33} = \sin(\lambda_{21}x_0) \quad a_{34} = \cos(\lambda_{21}x_0)$$
$$a_{35} = -\sinh(\lambda_{12}x_0) \quad a_{36} = -\cosh(\lambda_{12}x_0) \quad a_{37} = -\sin(\lambda_{22}x_0) \tag{c}$$
$$a_{38} = -\cos(\lambda_{22}x_0) \quad a_{39} \rightarrow a_{3,12} = 0$$

$$a_{41} = m_{11}\cosh(\lambda_{11}x_0) \quad a_{42} = m_{11}\sinh(\lambda_{11}x_0) \quad a_{43} = -m_{22}\cos(\lambda_{21}x_0)$$
$$a_{44} = m_{22}\sin(\lambda_{21}x_0) \quad a_{45} = -m_{12}\cosh(\lambda_{12}x_0) \quad a_{46} = -m_{12}\sinh(\lambda_{12}x_0) \tag{d}$$
$$a_{47} = +m_{22}\cos(\lambda_{22}x_0) \quad a_{48} = -m_{22}\sin(\lambda_{22}x_0) \quad a_{49} \rightarrow a_{4,12} = 0$$

References

[1] Wolf, K.-D., Electromechanical energy conversion in asymmetric piezoelectric bending actuators, Ph. D. thesis, D17, Technical University of Darmstadt, Darmstadt, Germany, 2000, 65
[2] Kim, M., Kim, J. and Cao, W., Electromechanical coupling coefficient of an ultrasonic array element, Journal of Applied Physics 99, Paper Id: 074102, 20067. doi:10.1063/1.2180487.
[3] Neubauer, M., Schwarzendhal, S.M. and Wallaschek, J., A new solution for the determination of the generalized coupling coefficient for piezoelectric systems, Journal of Vibroengineering 14(1), 2012, 105–110.

[4]   Cheng, S. and Arnold, D. P., Defining the coupling coefficient for electrodynamic transducers, The Journal of the Acoustical Society of America 134(5), 2013, 3561–3572. doi:10.1121/1.4824347.

[5]   Uchino, K., The development of piezoelectric materials and the new perspectives, Chapter 1, In: Advanced Piezoelectric Materials-Science and Technology, 2nd edn., Uchino, K. (ed,), Copyright © 2017 Elsevier Ltd, 93. doi:http://dx.doi.org/10.1016/B978-0-08-102135-4.00001-1.

[6]   Lustig, S. and Elata, D., Ambiguous definitions of the piezoelectric coupling factor, Journal of Intelligent Material Systems and Structures 31(4), 2020, 1689–1696.

4 Shape Memory Alloys

4.1 Basic Behavior of SMA

Shape memory alloys (SMA) are metallic materials that demonstrate the ability to return to some previously defined shape or size when subjected to the appropriate thermal and/or mechanical procedure. Normally, these materials can be plastically deformed at some relatively low temperature, and upon exposure to some higher temperature will return to their shape prior to the deformation. It is custom to denote those materials that would present "shape memory" only due to heating as *one-way shape memory* alloys. Other materials would undergo a change in shape upon recooling having the property of a *two-way shape memory* alloy. Those materials normally present another phenomenon called pseudoelasticity that consists of displaying large elastic strains (6–7%) and their recovery when unloading, without leaving any plastic strains. Both phenomena, the shape memory and pseudoelasticity are enabled due to two metallurgical phases, the parent phase being *austenite* (*A*) (at higher temperature) and at lower temperature the second phase called *martensite* (*M*) (see Figure 4.1).

Historically, in 1932, a Swedish researcher Arne Olander observed the shape and recovery ability of a gold–cadmium alloy (Au–Cd) and noted that it actually created motion [1]. In 1950, Chang and Read at Columbia University observed this unusual motion at the microscopic level by using X-rays to note the changes in crystal structure of Au–Cd [2]. As a result of this study, other similar alloys were discovered including indium–titanium. Then, in 1963, Buehler and co-workers at the US Naval Ordinance Laboratory (NOL) observed the shape memory effect (SME) in a nickel and titanium alloy, today known as (nickel titanium–NOL (Nitinol) [3–4]. Table 4.1 (from [5]) shows some of the SMA materials and their relevant properties.

As already stated above, the SMA's two phases are austenite and martensite, each having different crystal structure yielding different properties. The crystal structure of austenite is generally a cubic one while martensite's one can be tetragonal, orthorhombic or monoclinic. The transformation from parent metallurgical phase (austenite) to the martensite is made due to shear lattice distortion. This transformation is also named as martensitic transformation. Each formed martensitic crystal may have a different orientation direction called a variant [6]. Usually, the assembly of martensitic variants may appear in two forms: the *twinned martensite* (M_t), which is formed by a combination of "self-accommodated" martensitic variants, and *detwinned* (reoriented) *martensite* (M_d), having a dominant specific variant. This reversible phase transformation from austenite (the parent phase) to martensite (the product phase) and vice versa causes the unique behavior of SMAs.

For the simple case of only cooling the SMA material, with no mechanical loads applied, the crystallographic structure would change from austenite to martensite, which is also called the *forward transformation*. This transformation causes the

https://doi.org/10.1515/9783110726701-004

formation of several martensitic variants (e.g., up to 24 variants for Nitinol) and their space arrangement causes a negligible average macroscopic shape change yielding a *twinned martensite*. Heating it from the martensitic phase transforms the crystal back to austenite phase in what it is called the *reverse transformation*, again with no associated shape change.

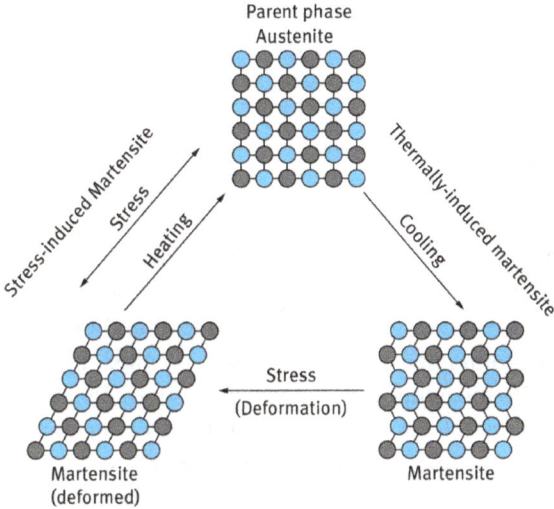

Figure 4.1: Schematic illustration of SME.

Table 4.1: Alloys with SME [5].

Alloy	Composition	Transformation-temperature range		Transformation hysteresis	
		°C	°F	Δ°C	Δ°
AS–Cd	44/49 at.% Cd	−190 to −50	−310 to −60	≈15	≈25
Au–Cd	46.5/50 at.% Cd	30 to 100	85 to 212	≈15	≈25
Cu–Al–Ni	14/14.5 wt.%* Al3/4.5 wt.% Ni	−140 to 100	−220 to 212	≈35	≈65
Cu–Sn	≈15 at.% Sn	−120 to 30	−185 to 85		
Cu–Zn	38.5/41.5 wt.% Zn	−180 to −10	−290 to 15	≈10	≈20
Cu–Zn–X (X = Si, Sn, Al)	a few wt.% of X	−180 to 200	−290 to 390	≈10	≈20
In–Ti	18/23 at.% Ti	60 to 100	140 to 212	≈4	≈7
Ni–Al	36/38 at.% Al	−180 to 100	−290 to 212	≈10	≈20

Table 4.1 (continued)

Alloy	Composition	Transformation-temperature range		Transformation hysteresis	
		°C	°F	Δ°C	Δ°
Ni–Ti	49/51 at.% Ni	−50 to 110	−60 to 230	≈30	≈55
Fe–Pt	≈25 at.% Pt	≈ − 130	≈ − 200	≈4	≈7
Mn–Cu	5/35 at.% Cu	− 250 to 180	− 420 to 355	≈25	≈45
Fe–Mn–Si	32 wt.% Mn, 6 wt.% Si	− 200 to 150	− 330 to 300	≈ 100	≈ 180

*wt.%, percentage by weight.

The phase transformation from the austenite phase to the martensite phase and back has four characteristic temperatures associated with the phase transformation. For the forward transformation, in the absence of a mechanical load and the temperature, the material starts to transform into twinned martensite is called the martensitic start temperature (M_s), while the temperature at which this transformation is completed is named the martensitic finish temperature (M_f). At this temperature and below it, all the material is in the twinned martensitic phase. When performing the reverse transformation, and thus heating, the initiation of the austenite phase will start at the austenitic start temperature (A_s) with a complete transformation at the austenitic finish temperature (A_f). A schematic graph of the four transformation temperatures is presented in Figure 4.2.

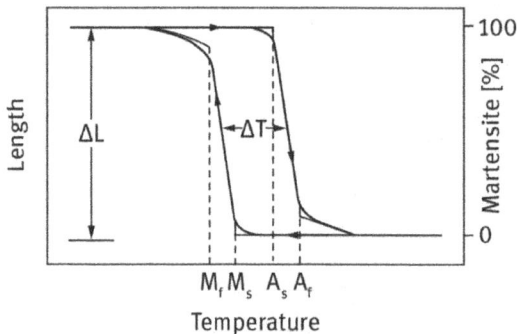

Figure 4.2: Typical transformation versus temperature curve for an SMA specimen.

Till now, we had talked about phase transformations in the absence of mechanical loads being applied on the SMA material. If such a load is being applied on the material while being in its twinned martensitic phase, namely, at a low temperature, it might be possible to induce a detwinning process by reorienting some of the variants. In contrast to the twinning process, which is associated with no change in the shape of the crystal, the detwinning process gives rise to a macroscopic shape change, yielding a deformed configuration with the load being released. Now, without any mechanical load being applied to the deformed material, heating the SMA to a temperature above A_f will bring the detwinned martensite to austenite showing a complete shape recovery. Cooling back to a temperature below M_f causes this time the formation of twinned martensite showing no shape change. This process is sometimes referred to as the SME.

Note that the applied load to start the detwinning process must be above the minimum stress required for detwinning initiation which is named the detwinning start stress (σ_s). The stress at which the complete process of detwinning of the martensite is achieved is termed the detwinning finish stress (σ_f). Also, when the material is cooled in the presence of a stress larger than σ_s which is applied in the austenitic phase, the phase transformation will yield the formation of a detwinned martensite, thus inducing a shape change. Reheating the material will again lead to a shape recovery while the stress is still being applied. One should note that the transformation temperature will increase with the increase of the applied stress (be it tension or compression).

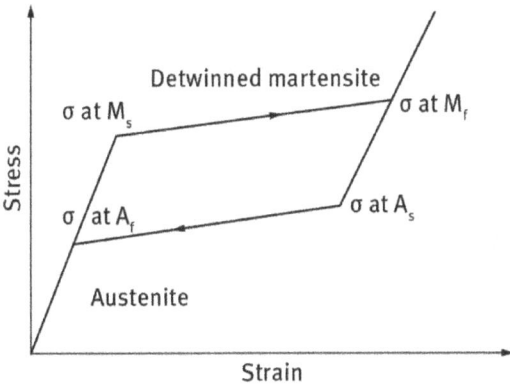

Figure 4.3: Schematic stress–strain curve for the pseudoelasticity behavior.

Another interesting phenomenon is the capability of inducing a phase transformation by only applying a sufficiently high mechanical stress to the material while being in its austenitic phase. This high stress will induce fully detwinned martensite created from austenite. If its temperature is above A_f, a complete shape recovery is obtained when unloading to its austenite phase. This behavior is usually named as pseudoelasticity effect (PE) of SMAs (see Figure 4.3). The loading stress at which the martensite transformation starts is shown in Figure 4.3 as σ at M_s, while σ at M_f denotes the completion of the process. For the unloading process of the SMA, the σ at A_s is the stress at which the austenite phase of the material is initiated, while σ at A_f is the stress at which the material completes its reverse transformation to the

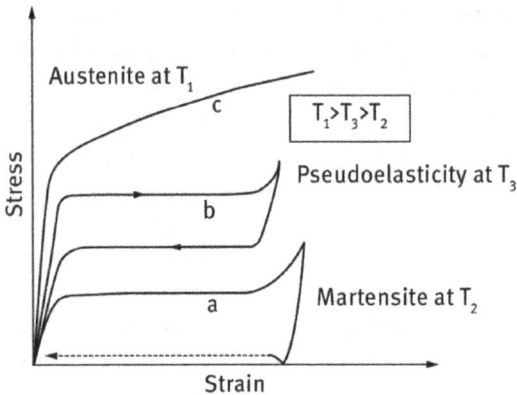

Figure 4.4: Typical stress–strain curves at various temperatures: (a) martensite, (b) pseudoelasticity and (c) austenite.

austenite phase. Note that for the case of a material in its austenite phase tested above its M_s temperature but below the A_f temperature, one can expect only a partial shape recovery. Figure 4.4 presents typical stress–strain curves for three cases, austenite, and the martensite phases and the pseudoelasticity behavior, at three different temperatures T_1, T_2 and T_3, where $T_1 > T_3 > T_2$.

4.1.1 The Shape Memory Effect

As described above, the first effect to be shown by an SMA is its shape memory when it is deformed in its twinned martensitic phase and then unloaded at a temperature below A_s. From this shape, heating the SMA above its A_f, the SMA will return to its original shape by transforming back into the parent austenitic phase. This property of the SMA is presented in Figure 4.5 (from [6]) for a three-dimensional (3D) curve portraying the experimental behavior of Nitinol under uniaxial loading in the form of axial stress σ (see Table 4.2 for the Nitinol properties).

Referring to Figure 4.5, the process starts at point A (the parent phase – austenite). Cooling the material below M_s and M_f in the absence of mechanical causes the formation of twinned martensite resulting in point B on the graph. From this point, the twinned martensite is loaded by a stress exceeding its σ_s (start stress level), causing the starting of the reorientation process – the detwinning process. Note that the stress level is far lower than the permanent plastic yield stress of martensite. The detwinning process will be completed at σ_f (the end of the plateau in the σ–ε graph presented in Figure 4.5). Elastically unloading the SMA material to zero stress, from point C to point D, without heating, keeps its detwinned martensite state. If heating is started, without mechanical stress the back transformation is initiated at A_s (point E) and finishes at A_f (point F). At any temperature above A_f, the material would be in its austenite phase and the original shape of the SMA is recovered (point A). The whole process described above is called *one-way shape memory effect*, as the shape recovery was achieved due to the application of heating after the material had been detwinned due to an applied mechanical stress.

Table 4.2: Nitinol's properties (taken from NiTi Aerospace, Inc., 2235 Polvorosa Ave, San Leandro, CA 94577, www.tiniaerospace.com).

Density	6.45 g/cm^3
Thermal conductivity	10 W/mK
Specific heat	322 J/kgK
Latent heat	24,200 J/kg
Ultimate tensile strength	750 – 900 MPa
Elongation to failure	15.5%
Yield strength (austenite)	560 MPa
Young's modulus (austenite)	75 GPa
Yield strength (martensite)	100 MPa

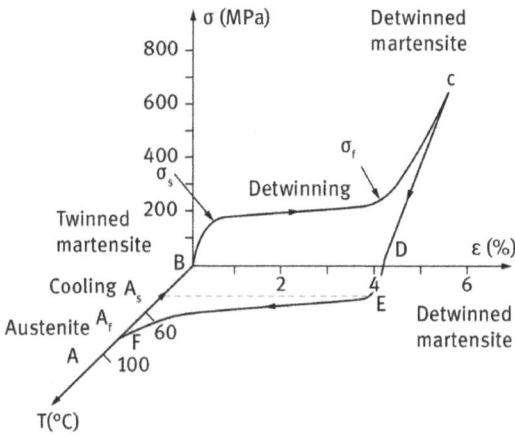

Figure 4.5: A typical experimental Nitinol loading path exhibiting its SME (taken from [6]).

When an SMA is exposed to cyclic thermal loading in the absence of mechanical stresses, repeatable shape changes may occur. This behavior is called *two-way shape memory effect* or training. This process can induce changes in the microstructure, leading to macroscopically permanent changes in the material behavior. The training of an SMA is sometimes needed to obtain a material that its hysteretic response is stable and its inelastic strain saturates.

4.1.2 Pseudoelasticity

The second effect to be exhibited by an SMA is its pseudoelastic behavior which is associated with stress-induced transformation, leading to large strain generation during loading which is reduced to zero upon unloading at temperatures above A_f. As pointed out by Kumar and Lagoudas [6], a pseudoelastic process can be either described by the path $(a \rightarrow b \rightarrow c \rightarrow d \rightarrow e \rightarrow a)$ as presented in Figure 4.6 as path 1, or by the path 2 in Figure 4.6 which is performed at a nominally constant temperature above A_f.

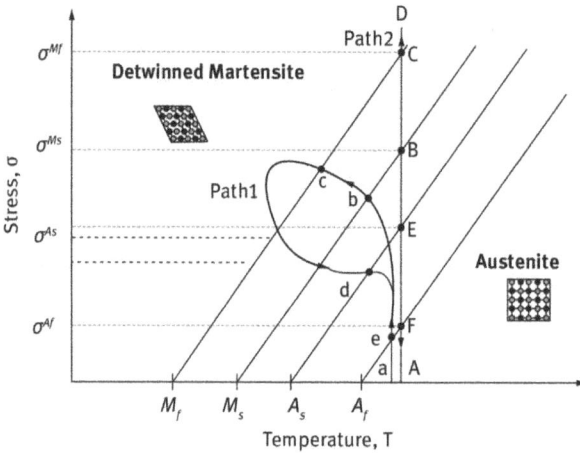

Figure 4.6: A typical phase diagram with two pseudoelastic loading paths (taken from [6]).

Referring to Figure 4.7, the process starts at zero stress at a temperature above A_f at point A. While the material is in its parent (austenite) phase, an elastic loading is applied bringing the curve to point B. The curve intersects the stress σ^{Ms}, namely, the starting of transformation into the martensite phase diagram. This process is accompanied by the generation of large inelastic strains (see Figure 4.7). The curve continues to point C, characterized by the stress σ^{Mf}, at which the whole material is in its martensite phase.

At this point (C), a distinct change in slope on the σ–ε curve associated with the elastic loading of the martensitic phase can be observed. Further increasing of the applied mechanical stress causes only elastic deformation of the martensite in its detwinned phase (the path from C to D). Point D is characterized by the largest stress and associated strain. Starting the unloading process from point D, the martensite

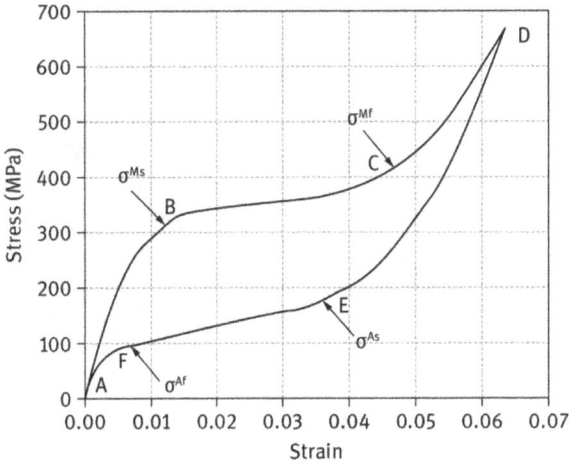

Figure 4.7: A typical SMA pseudoelastic loading cycle (from [6]).

phase is elastically unloaded up to the stress σ^{As} (point E) where it starts to revert back into its austenite phase. From point D to point E, the strain is recovered without permanent plastic strains. Point F, characterized by σ^{Af}, is the point where the curve meets the elastic region of the austenite phase. Further unloading the SMA to zero stress brings the material to the original point A without any residual strains.

One should note that passing the material a complete pseudoelastic cycle will form a hysteresis curve representing the energy being dissipated during this cycle. The size of the hysteresis would depend on the stress levels and the conditions of the testing.

4.1.3 Applications

The outstanding properties of SMAs are attracting technological interest in many fields of sciences and engineering, starting from medical up to aerospace applications. An interesting review of the SMA application in the medical field was written by Petrini and Migliavacca [7]. They claim that the SME and the PE together with good corrosion and bending resistance, biological and magnetic resonance compatibility, might explain the large market penetration, in the last 20 years, of SMA in the production of biomedical devices in dental, orthopedics, vascular, neurological and surgical fields. Typical examples of such applications would be SMA orthodontic wires (Figure 4.8), treatment of orthopedic issues using plates made of Nitinol (Figure 4.9) and various uses of Nitinol devices like stents and filters in the vascular field (Figure 4.10a–c). The reader can find more applications of the SMA in the biomedical sector in Ref. [7].

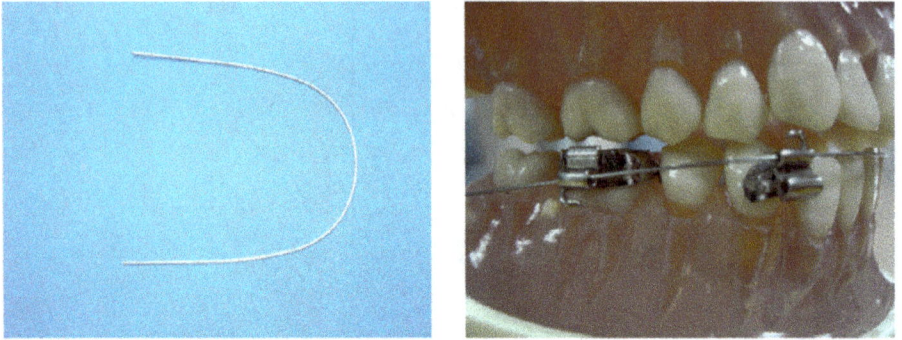

Figure 4.8: Typical SMA orthodontic wires (from [7]).

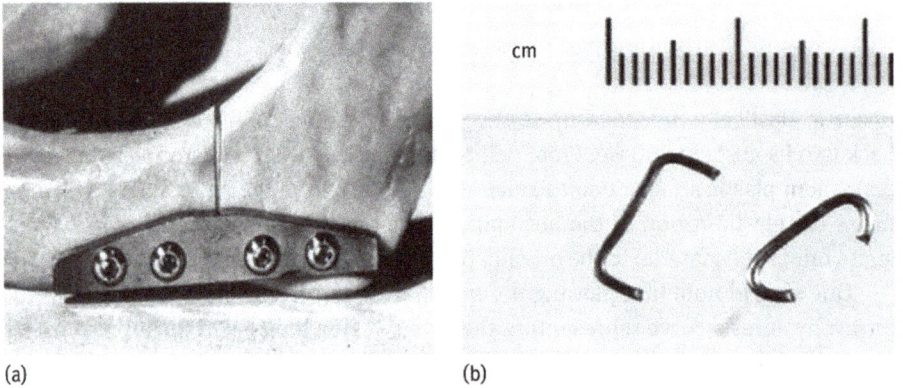

(a)   (b)

Figure 4.9: (a) NiTi plate for mandible fracture [8] and (b) staple before and after heating [9].

(a)   (b)   (c)

Figure 4.10: Use of NiTi in the vascular field: (a1) Venous filter, (a2) Simon filter; (b) device to close ventricular septal defects; (c1) carotid stent, (c2) coronary stent and (c3) femoral stent (from [7]).

Although the biomedical applications of SMA material with an emphasis on Nitinol can be considered as the main usage of the material, one can also find interesting and innovative designs involving SMA in other industrial fields like mechanical, aerospace, and civil engineering. Those applications will include fastening devices,

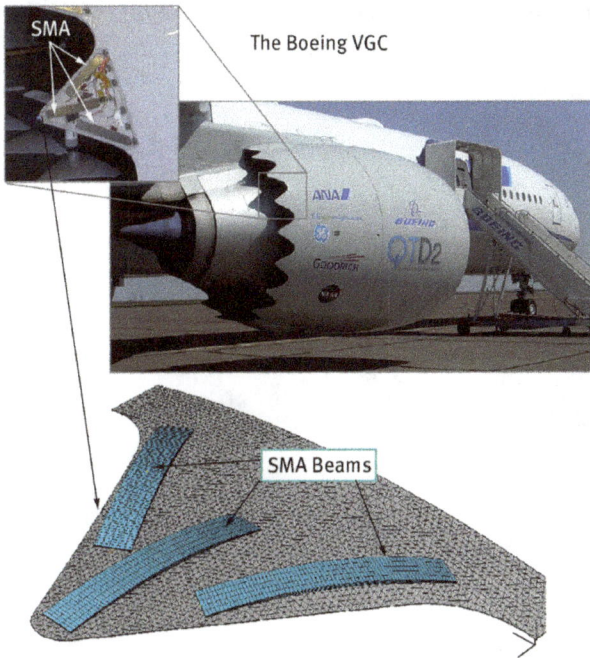

Figure 4.11: Boeing's VGC, a morphing aerospace structure for jet noise reduction [10].

actuators, SMA-actuated valves, springs, damping devices, variable geometry chevrons (VGCs), reconfigurable rotor blades, active hinge pin actuators, variable area nozzles, locking mechanisms, thermal switches, and many other innovative designs to solve practical industrial problems. In Figures 4.11–4.14, some of the above applications are highlighted. Figure 4.11 presents Boeing's VGC, a morphing aerospace structures aimed at the reduction of noise emitted by the aircraft engine. CDI-Aerospace Engineering[1] company presents an interesting application of morphing airfoils using SMA material (see Figure 4.12). A schematic drawing of plane wings with SMA wires which can change shape by applying voltages for heating the wires is presented in Figure 4.13. This mechanics might replace the conventional hydraulic and electromechanical actuators existing today in aerospace industry. Another application of SMA wires was described by Pitt et al. [11] in their overview on the NASA lead SAMPSON smart inlet project. Various uses of SMA wires together with Flexskin or conformal moldline technology (CMT) which is a Boeing-developed technology designed to provide the structural flexibility or compliance needed for gross shape changes, while also providing structural stiffness and smooth surfaces. Flexskin is basically

---

1 www.continuum-dynamics.com

an elastomeric panel with structural rods running through. Some details of the integration of the SMA wires are presented in Figure 4.14.

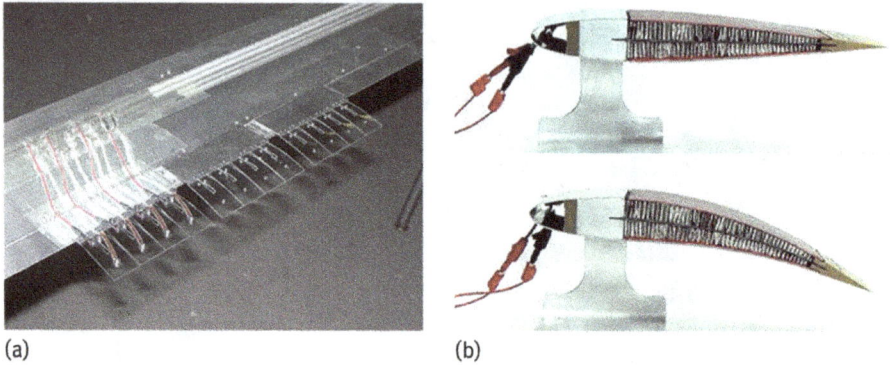

(a)   (b)

Figure 4.12: Morphing airfoils: (a) CDI's SMA-actuated tabs installed on a rotor blade, (b) CDI/Lockheed Martin continuously deformable SMA-controlled airfoil section. Source: CDI-Aerospace Engineering.

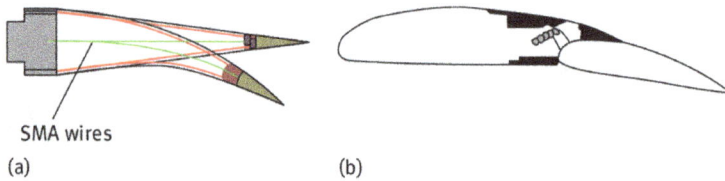

SMA wires
(a)   (b)

Figure 4.13: Changing shape of a wing: (a) SMA-actuated flap, (b) conventional flap.

An actuator with SMA wires providing "latch" control and lockout until electronically released was developed by Autosplice[2] company which claims that the actuator can also provide "burst control" in actual applications (Figure 4.15). S³Lab[3] (Smart Structures and Systems Laboratory) affiliated with INHA University, Korea, presents other interesting application of SMA material in the form of hard disk drive suspension and a robot microgripper as they are depicted in Figure 4.16a and b. The Computational Mechanics & Advanced Materials Group[4] affiliated with the Department of Civil Engineering and Architecture, University of Pavia, Italy, is also active in designing and implementing SMA-based devices presented in Figure 4.17.

---

2 corp.autosplice.com
3 ssslab.com
4 www-2.unipv.it/compmech/mat_const_mod.html

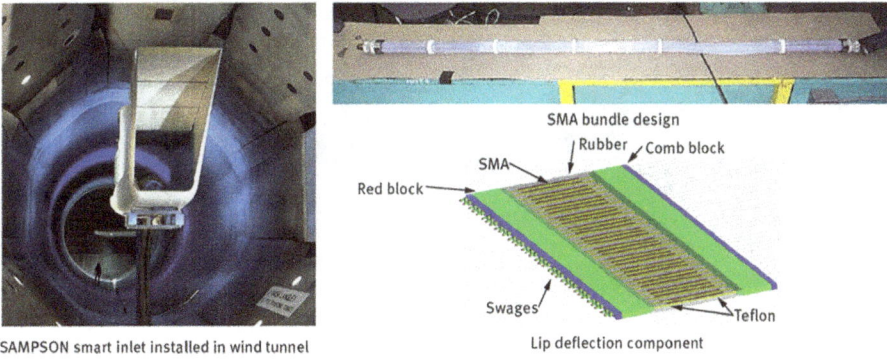

SAMPSON smart inlet installed in wind tunnel

SMA bundle design

Rubber / Comb block

SMA

Red block

Swages

Teflon

Lip deflection component

Figure 4.14: NASA's SAMPSON smart inlet project with integrated CMT/SMA rods.

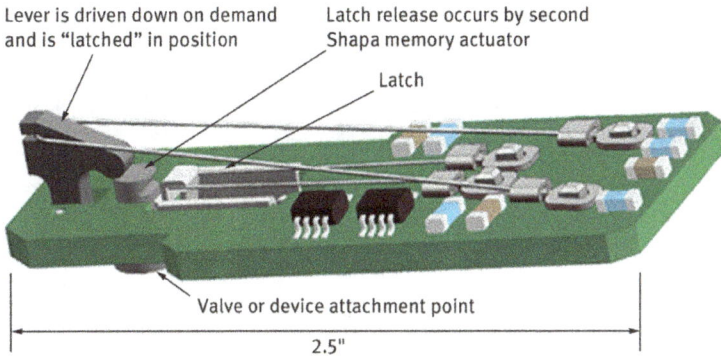

Lever is driven down on demand and is "latched" in position

Latch release occurs by second Shapa memory actuator

Latch

Valve or device attachment point

2.5"

Figure 4.15: SMA-based actuator (from Autosplice company).

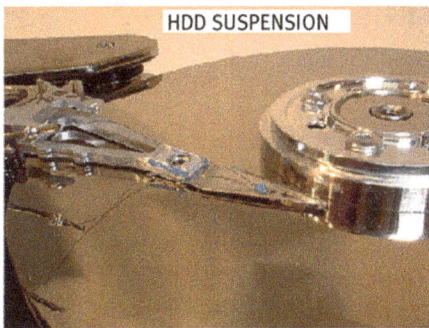

HDD SUSPENSION

Control Input=0.1A
Maximal Displacement=0.1 mm
(a)

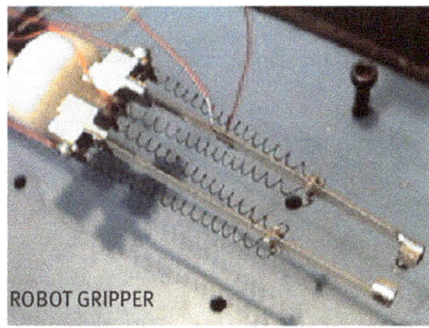

ROBOT GRIPPER

Gripping Force=7.5 mN
Bandwidth=2Hz
(b)

Figure 4.16: S^3 Lab SMA-based devices: (a) HDD suspension, (b) robot gripper.

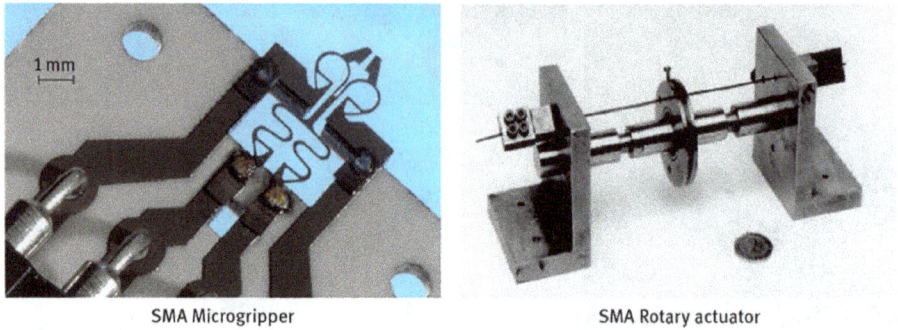

SMA Microgripper   SMA Rotary actuator

Figure 4.17: Computational Mechanics & Advanced Materials Group SMA-based devices. Source: Computational Mechanics & Advanced Materials Group, www-2.unipv.it/compmech/mat_const_mod_html.

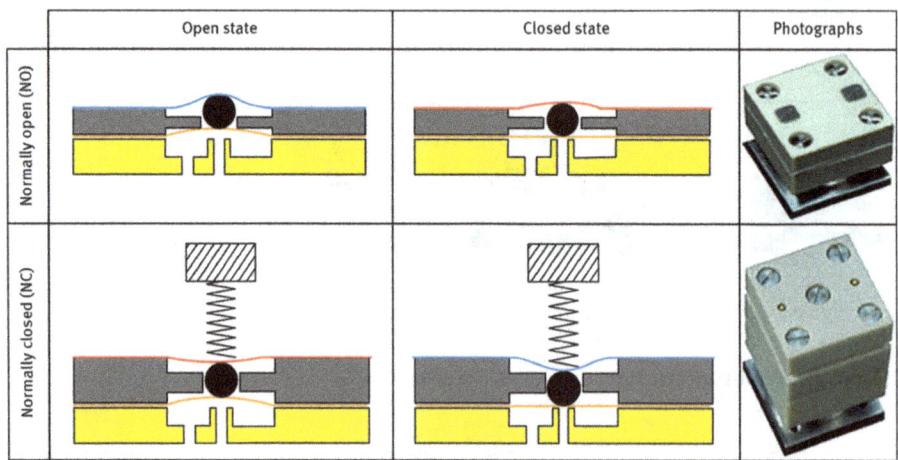

Figure 4.18: The IMT KIT Germany monostable SMA-based microvalve.

The Institute of Microstructure Technology affiliated with the Karlsruhe Institute of Technology, Germany[5] presents a monostable SMA-based microvalve (Figure 4.18) made of a polymer housing, with an integrated fluid chamber, a membrane and an SMA microactuator being deflected by a microball. At zero current, the microvalve is open, allowing the fluid to flow through the valve is a supply pressure is present. Electrically heating the microactuator changes to its planar shape memory state, thus closing the microvalve.

---

5 www.imt.kit.edu/english/1528.php

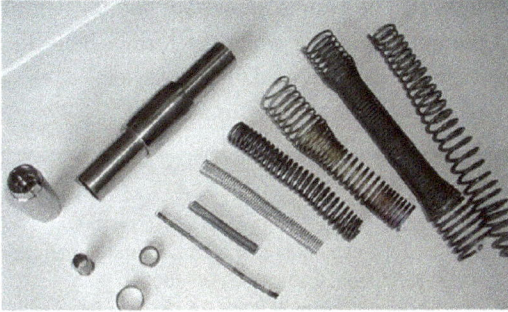

Figure 4.19: The various NiTi SMA springs manufactured by Zuudee Holding Group.

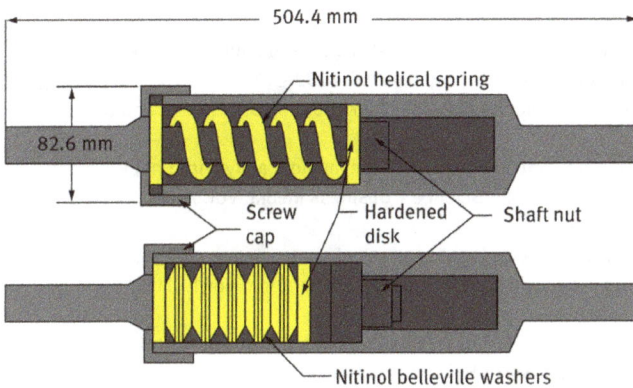

Figure 4.20: The two SMA-based dampers designed and manufactured within NEESR-RC project.

Various SMA springs are manufactured by Zuudee Holding Group,[6] a well-known Chinese manufacturing group. Among the other alloys, they are manufacturing two-ways NiTi SMA springs with a wire diameter between 0.1 and 10 mm for automobile and industrial uses (Figure 4.19). Another interesting application is the use of SMA Nitinol springs and Belleville washers as dampers within the NEESR-RC project led and performed by Georgia Tech[7] faculty, graduate and undergraduate students.

The aim of the project is to retrofit existing houses and buildings with dampers (Figure 4.20) built from SMA to allow a better seismic protection. The SMA-based system is considered to have the ductility and energy dissipation to prevent collapse and the ability to significantly reduce the residual deformations after an earthquake occurs. A tension/compression device was developed for applications as bracing elements in buildings. The device was designed to allow Nitinol helical springs or

---

6 http://zuudee.com/TiNi_SMAAlloys/shape_memory_alloy_NITI_ASTM_F2063-5.html
7 neesrcr.gatech.edu

Belleville washers to be used in compression. The results show that Nitinol helical springs produce good recentering and damping behavior while the Nitinol Belleville washers show a good potential to form the basis for a Nitinol damping device.

References

[1] Ölander, A., An electrochemical investigation of solid cadmium-gold alloys, Journal American Chemical Society 54(10), October, 1932, 3819–3833. doi:10.1021/ja01349a004.

[2] Chang, L. C. and Read, T. A., Plastic deformation and diffusion less phase changes in metals – The gold-cadmium beta-phase, Trans, AIME 191, 1951, 47–52.

[3] Buehler, W. J., Gilfrich, J. W. and Wiley, R. C., Effects of low-temperature phase changes on the mechanical properties of alloys near composition TiNi, Journal of Applied Physics 34(5), 1963, 1475–1477. doi:10.1063/1.1729603.

[4] Wang, F. E., Buehler, W. J. and Pickart, S. J., Crystal structure and a unique martensitic transition of TiNi, Journal of Applied Physics 36(10), 1965, 3232–3239. doi:10.1063/1.1702955.

[5] Shimizu, K. and Tadaki, T., Shape memory alloys, Funakubo, H. (ed.), Gordon and Breach Science Publishers, 1987.

[6] Kumar, P. K. and Lagoudas, D. C., Introduction to shape memory alloys, In: Shape Memory Alloys, Lagoudas, D. C. (ed.), 1, Springer Science + Business Media, Vol. 7, LLC, 2008, 393, doi:10.1007/978-0-387-47685-8.

[7] Petrini, L. and Migliavacca, F., Biomedical applications of shape memory alloys, Journal of Metallurgy 2011, article ID 501483, Hindawi Publishing Corporation, doi:10.1155/2011/501483.

[8] Duerig, T. W., Melton, K. N., Stockel, D. and Wayman, C. M., Engineering Aspects of Shape Memory Alloys, London, UK, Butterworth-Heinemann, 1990.

[9] Laster, Z., MacBean, A. D., Ayliffe, P. R. and Newlands, L. C., Fixation of a frontozygomatic fracture with a shape-memory staple, British Journal of Oral and Maxillofacial Surgery 39(4), 2001, 324–325.

[10] Calkins, F. T., Mabe, J. H. and Butler, G. W., Boeing's variable geometry chevron: morphing aerospace structures for jet noise reduction, SPIE Proceedings Vol. 6171: Smart Structures and Materials 2006: Industrial and Commercial Applications of Smart Structures Technologies, White, E. V. (ed.).

[11] Pitt, D. M., Dunne, J. P. and White, E. V., SAMPSON smart inlet design overview and wind tunnel test Part I – Design overview, In: Smart Structures and Materials 2002: Industrial and Commercial Applications of Smart Structures Technologies, McGowan, A.-M. R. (ed.), Proceedings of SPIE, Vol. 4698, 2002.

4.2 Constitutive Equations

As described in Section 4.1, SMAs possess two distinct properties: the SME and its PE. Since the discovery of the SME property in the Cu–Zn alloy by Greninger and Mooradian [1], and in the Nitinol (NiTi) alloy by Buehler et al. [2–3], Nitinol has been widely used as a smart material due to its large load capacity, high recovery strain (up to 8%), excellent fatigue performance and variable elasticity due to its phase transformation.

To simulate those specific properties, many constitutive laws had been proposed, such as in the form of phenomenological models [4–6], micromechanics models and 3D model for polycrystalline SMA based on a microplane theory [7].

In this section, the one-dimensional (1D) constitutive model of an SMA will be presented and the various adaptations of this model to fit the experimental data will be outlined based on existing empirical methods published in the literature. Finally, a method to heat SMA materials using the electrical current will be described.

4.2.1 One-Dimensional Constitutive Equations for an SMA Material

Tanaka [4–5] showed that a sufficient condition for the Clausius–Duhem inequality[8] to hold is the following equation:

$$\sigma(\varepsilon, T, \xi) = \rho_0 \frac{\partial \Phi}{\partial \varepsilon} \tag{4.1}$$

where the stress σ is a function of the strain ε, the temperature T and the martensite fraction ξ and ρ_0 and Φ are the density of the material and the Helmholtz free energy,[9] respectively.

Using differential calculus, Equation (4.1) can be rewritten as

$$d\sigma = \frac{\partial \sigma}{\partial \varepsilon} d\varepsilon + \frac{\partial \sigma}{\partial T} dT + \frac{\partial \sigma}{\partial \xi} d\xi \equiv E d\varepsilon + \Theta dT + \Omega d\xi \tag{4.2}$$

where $E = \rho_0 \frac{\partial^2 \Phi}{\partial \varepsilon^2}$ is the SMA modulus, $\Theta = \rho_0 \frac{\partial^2 \Phi}{\partial \varepsilon \partial T}$ is related to the thermal coefficient of expansion and $\Omega = \rho_0 \frac{\partial^2 \Phi}{\partial \varepsilon \partial \xi}$ is the transformation coefficient.

---

8 The Clausius–Duhem inequality is a way of expressing the second law of thermodynamics that is used in continuum mechanics. This inequality is particularly useful in determining whether the constitutive relation of a material is thermodynamically allowable.
9 In thermodynamics, the Helmholtz free energy is a thermodynamic potential that measures the "useful" work obtainable from a closed thermodynamic system at a constant temperature.

One should note that according to this model, the only state variables are the mono-axial strain ε, the temperature T and the volume fraction of the martensite phase, ξ.

Simplifying the constitutive model presented in Equation (4.3), one can write

$$(\sigma - \sigma_0) = E(\xi)(\varepsilon - \varepsilon_0) + \Theta(T - T_0) + \Omega(\xi)(\xi - \xi_0) \tag{4.3}$$

where the terms $\sigma_0, \varepsilon_0, T_0$ and ξ_0 stand for the initial state of the material. The modulus of elasticity $E(\xi)$ and the phase transformation coefficient $\Omega(\xi)$ are functions of the martensite volume fraction ξ and can be written as

$$E(\xi) = E_A + \xi(E_M - E_A) \quad \text{and} \quad \Omega(\xi) = -\varepsilon_L E(\xi) \tag{4.4}$$

E_A and E_M represent Young's moduli for the austenite and martensite phases, respectively, and ε_L is the maximum recoverable strain. Tanaka's expressions for the martensite volume fraction were determined by a dissipation potential which depends both on the stress and the temperature and have an exponential form (see [4,5]).

For the transformation from austenite to martensite (namely cooling), the equation is

$$\xi_{A \to M} = 1 - e^{[\alpha_M(M_S - T) + \beta_M \sigma]} \tag{4.5}$$

The equation for the transformation from martensite to austenite (namely heating) is

$$\xi_{M \to A} = e^{[\alpha_A(A_S - T) + \beta_A \sigma]} \tag{4.6}$$

The material constants appearing in Equations (4.5) and (4.6) are defined as

$$\alpha_A = \frac{\ln(0.01)}{(A_S - A_f)}; \quad \beta_A = \frac{\alpha_A}{C_A}$$
$$\alpha_M = \frac{\ln(0.01)}{(M_S - M_f)}; \quad \beta_M = \frac{\alpha_M}{C_M} \tag{4.7}$$

The coefficients E, Θ and Ω together with the other parameters like M_s, M_f, A_s, A_f, C_A and C_M are experimentally determined for each material.

The critical stress–temperature profile used for Tanaka's model is presented in Figure 4.21a.

A further modification of the 1D constitutive equation proposed by Tanaka was introduced by Liang and Rogers [9]. They presented the martensite volume fraction using a cosine formulation. Therefore, their formulation for the transformation from austenite to martensite (namely cooling) has the form

$$\xi_{A \to M} = \frac{1 - \xi_0}{2} \cos\left[a_M(T - M_f) + b_M \sigma\right] + \frac{1 + \xi_0}{2} \tag{4.8}$$

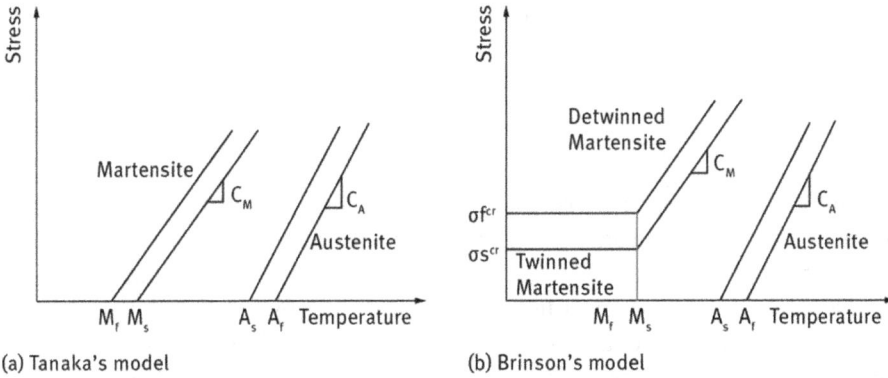

(a) Tanaka's model

(b) Brinson's model

Figure 4.21: Stress–temperature curves: (a) Tanaka's model and (b) Brinson's model.

While for the transformation from martensite to austenite (namely heating) their equation is

$$\xi_{M\to A} = \frac{\xi_0}{2}\cos[a_A(T - A_s) + b_A\sigma] + \frac{\xi_0}{2} \tag{4.9}$$

The material constants appearing in Equations (4.8) and (4.9) are defined as

$$a_M = \frac{\pi}{(M_S - M_f)}; \quad b_M = -\frac{a_M}{C_M}$$
$$a_A = \frac{\pi}{(A_f - A_s)}; \quad b_A = -\frac{a_A}{C_A} \tag{4.10}$$

where ξ_0 is the initial martensite volume fraction. Note that the argument of the cosine function can have a value between 0 and π leading to restrictions for the temperature and the stress, namely, for the transformation from austenite to the martensite phase one would require that

$$M_f \le T \le M_s$$

and

$$C_M(T - M_f) - \frac{\pi}{|b_M|} \le \sigma \le C_M(T - M_f) \tag{4.11}$$

while for the transformation from martensite to austenite (namely heating), the restrictions are

$$A_s \leq T \leq A_f$$

and (4.12)

$$C_A(T - A_s) - \frac{\pi}{|b_A|} \leq \sigma \leq C_A(T - A_s)$$

One should note that both Tanaka and Liang and Rogers models can only correctly represent the phase transformation from martensite to austenite and its reverse transformation, namely, the stress-induced martensitic transformation leading to the PE. Since the SME at lower temperatures is caused by the conversion between stress-induced martensite and temperature-induced martensite, these models cannot be implemented to the detwinning of martensite, which is responsible for the SMA phenomena [10].

To answer this deficiency, a new model was developed by Brinson [6,11] in which the martensite volume fraction was divided into stress-induced fraction, ξ_S, and temperature-induced martensite fraction, ξ_T to yield

$$\xi = \xi_S + \xi_T \tag{4.13}$$

The original constitutive equation of the Brinson model was slightly modified from the Tanaka's one (Equation (4.3)) and was written as

$$(\sigma - \sigma_0) = E(\xi)\varepsilon - E(\xi_0)\varepsilon_0 + \Theta(T - T_0) + \Omega(\xi)\xi - \Omega(\xi_0)\xi_0 \tag{4.14}$$

A simplified constitutive equation [11] leads to the following form:

$$\sigma = E(\xi)(\varepsilon - \varepsilon_L\xi_S) + \Theta(T - T_0) \tag{4.15}$$

To include the SMA at temperatures below M_s, the transformation phase equations of Liang and Roger (see Equations (4.8) and (4.9)) were modified to include two types of martensite volume fractions, the ξ_S and ξ_T [6]. A schematic presentation of the variation of the critical stresses with temperature for the transformation with two types of volume fractions is shown in Figure 4.21 (b). The conversion to detwinned martensite is now written as

$$\xi_S = \frac{1 - \xi_{S_0}}{2} \cos\left[\frac{\pi}{\sigma_s^{cr} - \sigma_f^{cr}}\left[(\sigma - \sigma_f^{cr}) - C_M(T - M_s)\right]\right] + \frac{1 + \xi_{S_0}}{2}$$

(4.16)

$$\xi_T = \xi_{T_0} - \frac{\xi_{T_0}}{1 - \xi_{S_0}}(\xi_S - \xi_{S_0})$$

provided the temperature and the stress are:

$$T > M_s$$

and $\qquad$ (4.17)

$$\sigma_s^{cr} + C_M(T - M_s) < \sigma < \sigma_f^{cr} + C_M(T - M_s)$$

For the case of

$$T < M_s$$

and $\qquad$ (4.18)

$$\sigma_s^{cr} < \sigma < \sigma_f^{cr}$$

we will get the following expressions:

$$\xi_S = \frac{1 - \xi_{S_0}}{2} \cos\left[\frac{\pi}{\sigma_s^{cr} - \sigma_f^{cr}}\left(\sigma - \sigma_f^{cr}\right)\right] + \frac{1 + \xi_{S_0}}{2}$$

(4.19)

$$\xi_T = \xi_{T_0} - \frac{\xi_{T_0}}{1 - \xi_{S_0}}\left(\xi_S - \xi_{S_0}\right) + \Delta T_\xi$$

The term ΔT_ξ will have the following values:

If $\quad M_f < T < M_s \quad$ and $\quad T < T_0$

then: $\quad \Delta T_\xi = \frac{1 - \xi_{T_0}}{2}\left\{\left[\cos\left[a_M(T - M_f)\right] + 1\right]\right\}$ $\qquad$ (4.20)

otherwise, $\quad \Delta T_\xi = 0$

The back conversion to the parent phase, the austenite, is given by

$$\xi = \frac{\xi_0}{2}\left\{\cos\left[a_A\left(T - A_s - \frac{\sigma}{C_A}\right)\right] + 1\right\}$$

(4.21)

$$\xi_S = \xi_{S_0} - \frac{\xi_{S_0}}{\xi_0}(\xi_0 - \xi) \quad \text{and} \quad \xi_T = \xi_{T_0} - \frac{\xi_{T_0}}{\xi_0}(\xi_0 - \xi)$$

One should note that the evolution kinetics as displayed by Brinson model might be incorrect for certain thermomechanical loadings yielding an inadmissible martensite fraction of $\xi > 1$. To overcome this problem, Chung et al. [12] represented the critical stress–temperature curve into eight bands (see Figure 4.22) as follows:

- Band 1: Transformation from austenite or temperature-induced martensite to stress-induced martensite.
- Band 2: Mixture of temperature/stress-induced martensite and austenite (no transformation).
- Band 3: Transformation from martensite to austenite.

- Band 4: Pure austenite (no transformation).
- Band 5: Transformation from austenite or temperature induced martensite to stress-induced martensite and transformation from austenite to temperature-induced martensite.
- Band 6: Transformation from austenite to temperature-induced martensite.
- Band 7: Transformation from temperature-induced martensite to stress-induced martensite.
- Band 8: Mixture of stress/temperature-induced martensite (no transformation).

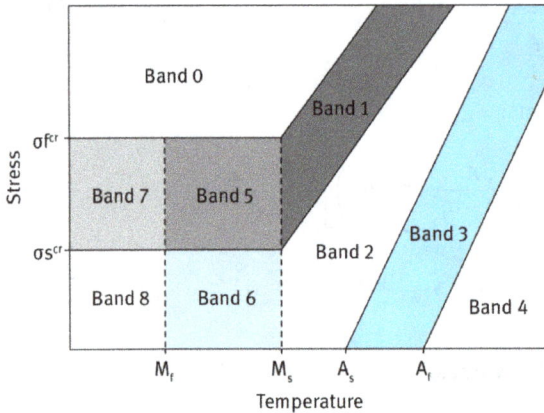

Figure 4.22: The critical stress–temperature curve divided in eight bands (from [12]).

Note that the Brinson model has a "weakness" at band 5. To answer this issue Chung et al. proposed a modified model under the following five constraints:
(a) Under any circumstances, the total martensite fraction should comply with the following constraint: $\xi = \xi_S + \xi_T \leq 1$.
(b) The stress-induced martensite fraction must be 1 when $\sigma = \sigma_f^{cr}(\xi_S = 1)$.
(c) The total martensite fraction must be 1 when $T = M_f(\xi = \xi_S + \xi_T = 1)$.
(d) The transformation kinetics in band 5 must be continuous with band 1 (Figure 4.22).
(e) The functions of the martensite volume fractions are all in cosine forms.

Note that Brinson's model satisfies all the above conditions when only the temperature is decreased or alternatively only the stress is increased. For the case when both the stress and the temperature are simultaneously altered for both the stress, the Brinson model satisfies only conditions (d) and (e). To be consistent with the above five conditions, Equations (4.18)–(4.20) are modified and the formulation of the transformation is given in the following equations [12], provided the critical stresses are constant below M_s:

For the case of

$$T < M_s$$

and $\qquad\qquad\qquad\qquad\qquad\qquad\qquad\qquad\qquad\qquad\qquad$ (4.22)

$$\sigma_s^{cr} < \sigma < \sigma_f^{cr}$$

we will get the following expressions:

$$\xi_S = \frac{1-\xi_{S_0}}{2}\cos\left[\frac{\pi}{\sigma_s^{cr}-\sigma_f^{cr}}\left(\sigma-\sigma_f^{cr}\right)\right] + \frac{1+\xi_{S_0}}{2}$$

$$\xi_T = \Delta T_\xi - \frac{\Delta T_\xi}{1-\xi_{S_0}}\left(\xi_S - \xi_{S_0}\right)$$

$\qquad\qquad\qquad\qquad\qquad\qquad\qquad\qquad\qquad\qquad\qquad$ (4.23)

The term ΔT_ξ will have the following values:

$\qquad$ If $\quad M_f < T < M_s \quad$ and $\quad T < T_0$

$\qquad$ then: $\quad \Delta T_\xi = \dfrac{1-\xi_{S_0}-\xi_{T_0}}{2}\cos\left[\dfrac{\pi}{M_s-M_f}\left(T-M_f\right)\right] + \dfrac{1-\xi_{S_0}+\xi_{T_0}}{2}$ $\qquad$ (4.24)

$\qquad$ otherwise $\quad \Delta T_\xi = \xi_{T_0}$

The modification presented in Equations (4.21)–(4.24) answers the five conditions written above, and it can be shown that the modified Brinson's model will give consistent results for all initial conditions [12]. One should note that already in 1998, Bekker and Brinson [13] developed a consistent mathematical description of the evolution of the martensite fraction during the a thermal thermoelastic phase transformation in an SMA induced by a general thermomechanical loading, yielding a robust kinetics which do not permit volume fractions to exceed unity, however the model additions of Chung et al. to the original Brinson's model capture better the phenomena.

4.2.2 Heating an SMA Material Using Electrical Current

One of the ways to directly heat an SMA material having the shape of a wire is by connecting it to an electrical source. This type of heating is also known as the Joule heating, and its equation is given by [14]

$$(\rho A)c_p\frac{dT(t)}{dt} = I^2R - h_cA_{circ}[T(t) - T_0]$$
$\qquad\qquad\qquad\qquad\qquad\qquad\qquad\qquad\qquad$ (4.25)

where I is denoting the electrical current, R is the wire resistance per unit length, $T(t)$ is the temperature as a function of time, T_0 is the ambient temperature, $\rho A = m$

is the mass per unit length of the wire (ρ being its density and A the cross-sectional area of the wire), c_p is the specific heat, while the terms h_c and A_{circ} are the heat-transfer coefficient and the circumferential area of the unit length of the wire, respectively. Assuming that both the current and the ambient temperature are constant and do not change with time, Equation (4.25) can be solved to yield the following solution for the temperature as a function of time:

$$T(t) - T_0 = \frac{I^2 R}{h_c A_{circ}} \left(1 - e^{-t/t_{ht}}\right) + (T_{start} - T_0)e^{-t/t_{ht}},$$

where $\hspace{8cm}$ (4.26)

$$t_{ht} = \frac{(\rho A)c_p}{h_c A_{circ}}$$

where T_{start} is the temperature at the starting of the heating and t_{ht} is the time constant of the heat-transfer process. For the case $T_{start} = T_0$, we get

$$T(t) - T_0 = \frac{I^2 R}{h_c A_{circ}} \left(1 - e^{-t/t_{ht}}\right) \hspace{3cm} (4.27)$$

Then the steady-state temperature to be measured on the wire will be

$$T_{ss} = \frac{I^2 R}{h_c A_{circ}} \left(1 - e^{-t/t_{ht}}\right) + T_0 \hspace{3cm} (4.28)$$

Analyzing the exponential function presented in Equation (4.27), one can calculate that at $t = 3t_{ht}$ the temperature will be approximately 95% of T_{ss}. Then the time, t_p, required to reach a predefined temperature T_p (thus heating the wire) will be

$$t_p = -t_{ht} \ln \frac{T_{ss} - T_p}{T_{ss} - T_0} \hspace{3cm} (4.29)$$

For the case of cooling the wire, we assume $I = 0$ in Equation (4.26), yielding

$$T(t) - T_0 = (T_{start} - T_0)e^{-t/t_{ht}} \hspace{3cm} (4.30)$$

The steady-state temperature will then be the ambient temperature and the time to reach a predefined temperature will be calculated according to the following equation:

$$t_p = -t_{ht} \ln \frac{T_p - T_0}{T_{start} - T_0} \hspace{3cm} (4.31)$$

Once the time for heating and cooling can be estimated the next question to solve would be what power we need to apply to the Nitinol wire. The power for a given current I (A) and a resistance R (Ω) is given by

$$P = I^2 R \hspace{3cm} (4.32)$$

The resistance R is then calculated using the following equation:

$$R = \rho_r \frac{l}{A}$$ (4.33)

where ρ_r is the wire resistivity (for Nitinol its value is 7.6×10^{-5} Ω/cm) and l and A are its length cross sectional are, respectively. Multiplying the power by time, t will yield the required energy E, to heat a given wire from a starting temperature T_{start} to a predefined temperature T_p. This is given by

$$E = I^2 R \cdot t = c_p \cdot M \cdot (T_p - T_0) + L \cdot M$$ (4.34)

with M being the mass of the wire (in grams) and c_p [J/g*C] and L [J/g] are the specific heat and the latent heat of transformation for the wire, respectively.[10] Substituting the expression for the resistance using Equation (4.33) into Equation (4.34) and solving for the current I, one gets

$$I = \sqrt{\frac{c_p \cdot M \cdot (T_p - T_0) + L \cdot M}{\frac{\rho_r l}{A} t}}$$ (4.35)

Thus, for a given mass of wire, with a known length and cross section, while the material properties c_p, L and ρ_r are known, one can calculate the required current I to raise the temperature from T_0 to the predefined one T_p for a time t.

References

[1] Greninger, A. B. and Mooradian, V. G., Strain transformation in metastable beta copper-zinc and beta copper-tin alloys, Transactions of the Metallurgical Society of AIME 128, 1938, 337–368.

[2] Buehler, W. J., Gilfrich, J. W. and Wiley, R. C., Effects of low-temperature phase changes on the mechanical properties of alloys near composition TiNi, Journal of Applied Physics 34(5), 1963, 1475–1477. doi:10.1063/1.1729603.

[3] Wang, F. E., Buehler, W. J. and Pickart, S. J., Crystal structure and a unique martensitic transition of TiNi, Journal of Applied Physics 36(10), 1965, 3232–3239. doi:10.1063/1.1702955.

[4] Tanaka, K. and Nagaki, S., A thermomechanical description of materials with internal variables in the process of phase transitions, Ingenieur-Archiv 51(5), 1982, 287–299.

[5] Tanaka, K., A thermomechanical sketch of shape memory effect: one-dimensional tensile behavior, Res Mechanica 18(3), 1986, 251–263.

[6] Brinson, L. C., One-dimensional constitutive behavior of shape memory alloys: thermomechanical derivation with non-constant material functions and redefined martensite internal variable, Journal of Intelligent Material Systems and Structures 4(2), 1993, 229–242.

---

10 For NITINOL, c_p = 0.8368 [J/g*C] and L = 20 [J/g]

[7]   Li, L., Li, Q. and Zhang, F., One-dimensional constitutive model of shape memory alloy with an empirical kinetics equation, Journal of Metallurgy 2011, January 2011, 1–14. article ID 563413, doi:10.1155/2011/563413, Hindawi Publishing Corporation.

[8]   Barbarino, S., Saavedra Flores, E. I., Ajaj, R. M., Dayyani, I. and Friswell, M. I., A review on shape memory alloys with applications to morphing aircraft, Smart Materials and Structures 23(6), 2014, article ID 063001.

[9]   Liang, C. and Rogers, C. A., One dimensional thermomechanical constitutive relations for shape memory material, Journal of Intelligent Material Systems and Structures 1, 1990, 207–234.

[10] Prahlad, H. and Chopra, I., Comparative evaluation of shape memory alloy constitutive models with experimental data, Journal of Intelligent Material Systems and Structures 12, 2001, 383–395.

[11] Brinson, L. C. and Huang, M. S., Simplifications and comparisons of shape memory alloy constitutive models, Journal of Intelligent Material Systems and Structures 7, 1996, 108–114.

[12] Chung, J., Heo, J. and Lee, J., Implementation strategy for the dual transformation region in the Brinson SMA constitutive model, Smart Materials and Structure 16(1), 2007, N1–N5.

[13] Bekker, A. and Brinson, L. C., Phase diagram based description of the hysteresis behavior of shape memory alloys, Acta Mater 46, 1998, 3649–3665.

[14] Song, H., Kubica, E. and Gorbet, R., Resistance modelling of SMA wire actuators, International Workshop on SMART MATERIALS, STRUCTURES & NDT in AEROSPACE Conference NDT, 2–4 November 2011, Montreal, Quebec, Canada.

4.3 SMA Models in Literature

In this section, a review will be made on SMA models appearing in the literature besides Tanaka's, Liang and Rogers' and Brinson's models that were presented in Section 4.2. The aim is to emphasize the complexity of the SMA material and the ways researchers try to analyze and form analytical models to be as close as possible to various tests being performed on SMA materials.

A recent review on the various models used to describe the behavior of SMA is given by Cisse et al. [1]. In their review, they present a long and updated list of models currently presented in the literature. Another review is presented by Barbarino et al. [2] in which after presenting the various constitutive models used for describing the SMA material they also present a detailed list of applications in the field of morphing of airfoils. Paiva and Savi present [6] present an overview of the constitutive models of SMA. They divide the models in five classes which capture the general thermomechanical behavior of SMA, including pseudoelesticity, SME and phase transformation because of temperature variation and incomplete phase transformations. The review includes a long list of references on the topic being discussed. Brocca et al. [8] developed a new model based on statically constrained microplane theory, yielding only 1D constitutive law for each plane. Qianhua et al. [9] developed a 3D thermomechanical constitutive model based on experimental results of superelastic NiTi alloy, capturing the transformation hardening, reverse transformation, elastic mismatch between the austenite and martensite phases and temperature dependence of elastic modulus for each phase effects. Arghavani et al. [10] studied the

3D behavior of SMAs and presented a phenomenological constitutive model which completely decouples the pure reorientation mechanism for the pure transformation mechanism. Numerical tests reproduce the main features of the SMA in proportional and nonproportional loadings and a good correlation with experimental results available in the literature is noticed.

Lagoudas et al. [3] developed a unified thermomechanical model based on total specific Gibbs free energy and dissipation potential, and thus being able to include the various phenomenological models under a common framework. The total strain, according to their formulation, has a mechanical part ε_{ij} and a transformation strain tensor $\varepsilon_{ij}^{tr}(\xi)$ being dependable on the martensite volume fraction, ξ. Therefore, the generalized Hook's law would be written as

$$\sigma_{ij} = C_{ijkl}\left[\varepsilon_{kl} - \varepsilon_{kl}^{tr} - \alpha_{kl}(T - T_0)\right] \tag{4.36}$$

where C_{ijkl} is the elastic stiffness matrix, T is the temperature and α_{kl} is the thermal expansion coefficient. Note that detwinning effects have been omitted from their unified model for simplicity [3]. The differences among the different models would then arise from the specific choices of the transformation-hardening function, $f(\xi, \varepsilon_{ij}^{t})$,[11] which physically represents the elastic strain energy due to the interaction between martensitic variants and the surrounding parent phase, and among the martensitic variants themselves. In their work, a linear hardening function was adopted, which convergences in an easy manner in a finite element. As described by Lagoudas et al. [3], the hardening function, f, is assumed to be independent of the transformation strain tensor, ε_{ij}^{t}, which means the absence of kinematic transformation hardening. Then, the following properties of the function are assumed to be kept:

(a) The parent austenitic phase is stress free if no external mechanical loading is applied. Assuming $f(0) = 0$, namely at fully austenitic phase the function would be zero, and thus satisfying the previous condition.

(b) The function must be positive as it represents part of the elastic strain energy stored in the material. To satisfy this condition, appropriate material constants should be chosen.

(c) The function must be continuous during the phase transformation, including return points, for all possible loading paths.

Based on the above conditions, the hardening function $f(\xi)$ is selected as follows:

$$f(\xi) = f^M(\xi) \quad \text{for} \quad \xi > 0$$

and $\tag{4.37}$

$$f(\xi) = f^A(\xi) \quad \text{for} \quad \xi < 0$$

---

[11] ξ is the martensite volume fraction and ε_{ij}^{t} is the transformation strain vector.

where the functions $f^M(\xi)$ and $f^A(\xi)$ are defined as

$$f^M(\xi) = f^{Mo}(\xi) + \frac{1-\xi}{1-\xi^R}\left[f^A\left(\xi^R\right) - f^{Mo}\left(\xi^R\right)\right]$$

$$f^A(\xi) = f^{Ao}(\xi) + \frac{\xi}{\xi^R}\left[f^M\left(\xi^R\right) - f^{Ao}\left(\xi^R\right)\right]$$

(4.38)

and ξ^R is defined as the martensitic volume fraction at the return point, while the functions $f^{Mo}(\xi)$ and $f^{Ao}(\xi)$ have to be chosen for each selected model. The forward phase transformation $(\dot\xi > 0)$ is given by $\xi^R \leq \xi \leq 1$, while the reverse $(\dot\xi < 0)$ one is defined as $0 \leq \xi \leq \xi^R$.

Note that this form of $f(\xi)$ would satisfy all three conditions, (a)–(c), presented above if we choose:

$$f^{Mo}(0) = 0; \quad f^{Ao}(0) = 0; \quad f^{Mo}(\xi) \geq 0; \quad f^{Ao}(\xi) \geq 0; \quad \text{for} \quad 0 \leq \xi \leq 1$$

and

(4.39)

$$f^{Mo}(1) = f^{Ao}(1)$$

For Tanaka's model [4] (the exponential model), the following functions $f^{Mo}(\xi)$ and $f^{Ao}(\xi)$ have to be chosen:

$$f^{Mo}(\xi) = \frac{\rho \Delta s_0}{a_e^M}\left[(1-\xi)lan(1-\xi) + \xi\right] + \left(\mu_1^e + \mu_2^e\right)\xi$$

$$f^{Ao}(\xi) = -\frac{\rho \Delta s_0}{a_e^A}\xi[lan(\xi) - 1] + \left(\mu_1^e - \mu_2^e\right)\xi$$

(4.40)

where ρ is the density and $\Delta s_0 = s_0^M - s_0^A, s_0^M$ and s_0^A being the specific entropy at the martensitic and austenitic phases, respectively. For Liang and Rogers' model [5], the functions would be

$$f^{Mo}(\xi) = -\int_0^\xi \frac{\rho \Delta s_0}{a_c^M}\left[\pi - \cos^{-1}(2\xi - 1)\right]d\xi + \left(\mu_1^c + \mu_2^c\right)\xi$$

$$f^{Ao}(\xi) = -\int_0^\xi \frac{\rho \Delta s_0}{a_c^A}\left[\pi - \cos^{-1}(2\xi - 1)\right]d\xi + \left(\mu_1^c - \mu_2^c\right)\xi$$

(4.41)

where the terms $a_e^M, a_e^A, a_c^M, a_c^A, \mu_1^e, \mu_1^c$ are materials constants, while the parameters μ_1^e, μ_1^c can be determined from the continuity condition $f^{Mo}(1) = f^{Ao}(1)$.

If ones would like to represent the $f(\xi)$ using a polynomial representation up to quadratic terms their form will be

$$f^{Mo}(\xi) = \frac{1}{2}\rho b^M \xi^2 + \left(\mu_1^p + \mu_2^p\right)\xi$$

$$f^{Ao}(\xi) = \frac{1}{2}\rho b^A \xi^2 + \left(\mu_1^p - \mu_2^p\right)\xi$$

(4.42)

Here, b^M and b^A are the linear isotropic hardening moduli for the forward and reverse phase transformation. The parameter μ_2 is introduced to take care of the continuity condition at $\xi = 1$. According to the chosen model, it will have the following expression:

$$\mu_2^e = \frac{\rho \Delta s_0}{2} \left(\frac{1}{a^A} - \frac{1}{a^M} \right): \quad \text{exp.} \quad \text{mod}.$$

$$\mu_2^c = \frac{\pi \rho \Delta s_0}{4} \left(\frac{1}{a_c^M} - \frac{1}{a_c^A} \right): \quad \text{cos.} \quad \text{mod.} \tag{4.43}$$

$$\mu_2^p = \frac{\rho}{2} \left(b^A - b^M \right): \quad \text{poly.} \quad \text{mod.}$$

The various material constants for the various three models are presented in Table 4.3 (from [3]).

Table 4.3: Material constants for the three models, provided M^{os}, M^{of}, A^{os} and A^{of} are known.

Exponential model	Cosine model	Polynomial model
$a_e^A = \frac{lan(0.01)}{A^{0s} - A^{0f}}$	$a_c^A = \frac{\pi}{A^{0f} - A^{0s}}$	$\rho b^A = -\rho \Delta s_0 \left(A^{0f} - A^{0s} \right)$
$a_e^M = \frac{lan(0.01)}{M^{0s} - M^{0f}}$	$a_c^M = \frac{\pi}{M^{0s} - M^{0f}}$	$\rho b^M = -\rho \Delta s_0 \left(M^{0s} - M^{0f} \right)$
$y^e = \frac{\rho \Delta s_0}{2} \left(M^{0s} + 2A^{0f} - A^{0s} \right)$	$y^c = \frac{\rho \Delta s_0}{2} \left(M^{0s} + A^{0f} \right)$	$y^p = \frac{\rho \Delta s_0}{2} \left(M^{0s} + A^{0f} \right)$
$Y_e^* = -\frac{\rho \Delta s_0}{2} \left(A^{0s} - M^{0s} \right) +$ $+ \frac{\rho \Delta s_0}{2 \ln(0.01)} \left(M^{0s} - M^{0f} + A^{0f} - A^{0s} \right)$	$Y_c^* = -\frac{\rho \Delta s_0}{2} \left(A^{0f} - M^{0s} \right) -$ $+ \frac{\rho \Delta s_0}{4} \left(M^{0s} - M^{0f} - A^{0f} + A^{0s} \right)$	$Y_p^* = -\frac{\rho \Delta s_0}{2} \left(A^{0f} - M^{0s} \right) -$ $+ \frac{\rho \Delta s_0}{4} \left(M^{0s} - M^{0f} - A^{0f} + A^{0s} \right)$

It was shown in Ref. [3] that the basic equation for the thermodynamic force has the following form:

$$\bar{\pi} = \sigma_{ij}^{eff} \Lambda_{ij} + \tfrac{1}{2} \Delta S_{ijkl} \sigma_{ij} \sigma_{kl} + \Delta \alpha_{ij} \sigma_{ij} \Delta T + \rho \Delta c \left[\Delta T - T \ln \left(\tfrac{T}{T_0} \right) \right]$$

$$+ \rho \Delta s_0 T - \tfrac{\partial f(\xi)}{\partial \xi} - \rho \Delta u_0 = \pm Y^* \tag{4.44}$$

where $\bar{\pi}$ is the thermodynamic force conjugate to ξ and the various parameters appearing in Equation (4.44) are defined as

$$\Delta S_{ijkl} = S_{ijkl}^M - S_{ijkl}^A, \quad \Delta \alpha_{ij} = \alpha_{ij}^M - \alpha_{ij}^A, \quad \Delta c = c^M - c^A$$

$$\Delta s_0 = s_0^M - s_{i0}^A, \quad \Delta u_0 = u_0^M - u_0^A, \quad \Delta T = T - T_0, \quad Y^* = \sqrt{2Y} \tag{4.45}$$

Note that in Equation (4.44), the plus sign is used for the forward transformation while the minus sign is used for the reverse phase transformation. In addition, the equation can be used to analyze the evolution of the internal state variable ξ. The material constant T^* can be seen as the threshold of the thermodynamic force $\bar{\pi}$ for the starting of the phase transformation. To show that the unified model gives results similar to the various existing models in the literature, Equation (4.44) can be used for Liang and Rogers' cosine model after inserting the appropriate expression for the $f(\xi)$ function, yielding

$$\sigma_{ij}^{\text{eff}} \Lambda_{ij} + \frac{1}{2} \Delta S_{ijkl} \sigma_{ij} \sigma_{kl} + \Delta \alpha_{ij} \sigma_{ij} \Delta T + \rho \Delta c \left[\Delta T - T \ln \left(\frac{T}{T_0} \right) \right] + \rho \Delta s_0 T$$

$$- \frac{\rho \Delta s_0}{\alpha_c^M} \left[\cos^{-1}(2\xi - 1) - \pi \right] - \left(\mu_1^c + \mu_2^c \right) - \frac{f^A \left(\xi^R \right) - f^{Mo} \left(\xi^R \right)}{1 - \xi^R} - \rho \Delta u_0 - Y^* = 0$$

$$(4.46)$$

Then, for a complete cycle (loading and de-loading), we have $\xi^R = 0$ and $f^A \left(\xi^R \right) = f^{Mo} \left(\xi^R \right) = 0$ One can get explicitly the expression of ξ in terms of the applied stress and temperature to yield

$$\xi = \frac{1}{2} \left\{ \cos \left[\alpha_c^M \left(T - M^{Of} \right) - \frac{\alpha_c^M}{C^M H} \left(\sigma_{ij}^{\text{eff}} \Lambda_{ij} + \frac{1}{2} \Delta S_{ijkl} \sigma_{ij} \sigma_{kl} + \Delta \alpha_{ij} \sigma_{ij} \Delta T \right) \right] + 1 \right\} \quad (4.47)$$

One should note that appropriate expressions from Table 4.1 were used to obtain the above equation while the expression for μ_2^c is given in Equation (4.43), and it was assumed that $\Delta c = 0$. The term appearing in Equation (4.47), CM is called martensite stress influence coefficient and has the following form:

$$C_M = - \frac{\rho \Delta s_0}{H} \quad (4.48)$$

If one neglects the terms $\frac{1}{2} \Delta S_{ijkl} \sigma_{ij} \sigma_{kl}$ and $+ \Delta \alpha_{ij} \sigma_{ij} \Delta T$ in Equation (4.47), the form presented by Liang and Rogers [5] is obtained, except the term

$$\frac{\sigma_{ij}^{\text{eff}} \Lambda_{ij}^{12}}{H}$$

Which is used in the present formulation as the effective driving stress instead of the applied stress used by them [5]. The similar procedure can be used to obtain the model proposed by Tanaka [4]. Some years later, Lagoudas et al. [7] updated this unified model which improves the original model [3] by adding the smooth transition in the thermal and mechanical responses when the martensite transformation is initiated and completed. Another characteristic being dealt is the effect of the magnitude of the mechanical applied stress on the generation of favored martensitic variants without explicitly considering martensite reorientation leading to an efficient computational tool. The third improvement deals with the generalization of

---

12 $H = \varepsilon^{tmax}$ being the maximum uniaxial transformation strain.

the critical thermodynamically forces for the transformation concept, which became depended on the direction of the transformation (forward or reverse) and the magnitude of the applied mechanical stress. All three improvements present a high fidelity model with applications over wide SMA material systems.

The model presented above, together with the models of Tanaka, Liang and Rogers and Brinson [15] are characterized in the literature [6] as assumed phase transformation kinetics models that consider preestablished simple mathematical functions to describe the phase transformation kinetics. These models are considered the most popular in the literature, and therefore they have more experimental comparisons, playing an important role within the SMA's behavior modeling context. A comparison between the predictions of the three models of Tanaka, Liang and Rogers and Brinson, based on Table 4.4, are presented in Figure 4.23 for three temperatures.

Table 4.4: Thermomechanical material properties for NiTi (from [15]).

Material properties	Transformation temperature (K)	Model parameters
$E_A = 67 \times 10^3$ MPa	$M_f = 282$	$C_M = 8$ MPa/K
$E_M = 26.3 \times 10^3$ MPa	$M_s = 291.4$	$C_A = 13.8$ MPa/K
$\Omega = 0.55$ MPa/K	$A_s = 307.5$	$\sigma_s^{\text{crit.}} = 100$ MPa
$\varepsilon_R = 0.067$	$A = 322$	$\sigma_f^{\text{crit.}} = 170$ MPa

The literature on SMA presents also other studies, using other hypotheses. Falk and his companions [11–13], present a relatively simple polynomial model based on Devonshire theory [14].[13] According to this model, internal variables or dissipation potential is not necessary to describe both the PE and the shape memory one. Thus, the strain ε and the temperature T are needed for this model. Accordingly, a six-degree polynomial model is presented for the free-energy potential (Λ):

$$\Lambda(\varepsilon, T) = \frac{a}{2}(T - T_M)\varepsilon^2 - \frac{b}{4}\varepsilon^4 + \frac{b^2}{24a(T_A - T_M)}\varepsilon^6 \tag{4.49}$$

where T_A is the temperature above which austenite is stable, T_M is the temperature below which martensite is stable and a and b are positive material constants. The resulting constitutive equation is

---

13 In 1949, Devonshire developed a phenomenological theory to describe the ferroelectric phase transitions and the temperature dependence of dielectric properties for barium titanate, $BaTiO_3$.

(a) $(T = 333K) > A_f$

(b) $M_s < (T = 298K) < A_s$

(c) $(T = 288K) < M_s$

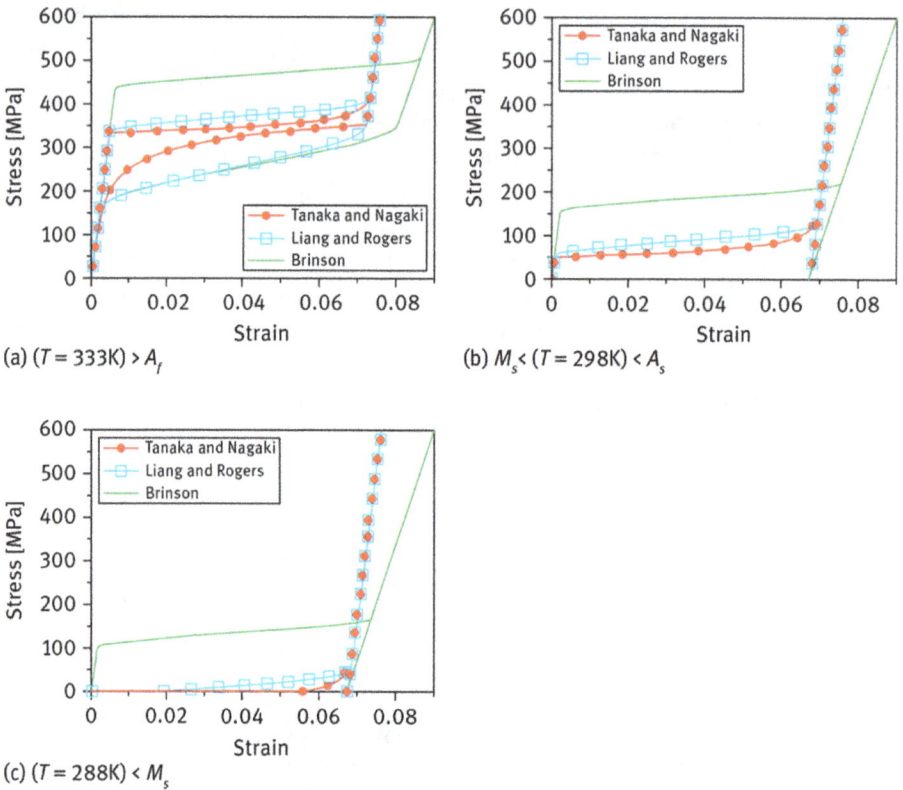

Figure 4.23: Strain–stress curves for three models with assumed transformation kinetics at three representative temperatures.

$$\sigma(\varepsilon, T) = \frac{\partial \Lambda(\varepsilon, T)}{\partial \varepsilon} = a(T - T_M)\varepsilon - b\varepsilon^3 + \frac{b^2}{4a(T_A - T_M)}\varepsilon^5 \qquad (4.50)$$

Typical results for Falk's model are presented in Figure 4.24a–c (from [6]) showing stress–strain curves for three temperatures. Figure 4.24a and b illustrates the martensite detwinning processes, while Figure 4.24c shows the pseudoelastic effect. Although not considering the twining effect, Falk's polynomial represents in a qualitatively coherent way both martensite detwinning process and pseudoelasticity.

Other classes of models are those including internal constraints. The study of Fremond [16–17] is considered such a model. His study considered three volumetric fractions, an austenite fraction and two detwinned variants induced by compression and tension for the 1D case. Three internal variables, the strain, the temperature and the global free energy are included in the 3D model. Paiva et al. [18] used Fremond's model to develop a new model which is able to consider different martensite properties and a new fraction associated with twinned martensite. The

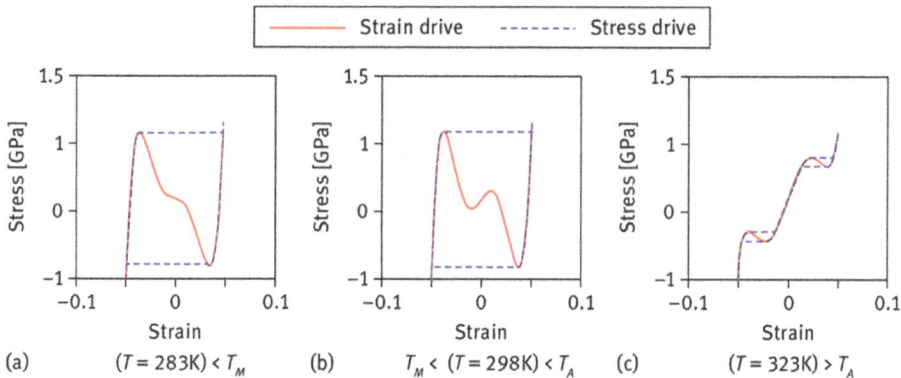

Figure 4.24: Falk's model: strain–stress curves for three representative temperatures.

model also includes plastic strain effects and plastic–phase transformation coupling. The formulation includes elastic strain, temperature and four state variables associated with tensile detwinned martensite, compressive detwinned martensite, austenite and twinned martensite. More details can be found in Ref. [18].

References

[1]   Cisse, C., Zaki, W. and Zineb, T. B., A review of constitutive models and modeling techniques for shape memory alloys, International Journal of Plasticity 76, 2016, 244–284.
[2]   Barbarino, S., Saavedra Flores, E. I., Ajaj, R. M., Dayyani, I. and Friswell, M. I., A review on shape memory alloys with applications to morphing aircraft, Smart Materials and Structures 23(6), 2014, article ID 063001.
[3]   Lagoudas, D. C., Bo, Z. and Qidwai, M. A., A unified thermodynamic constitutive model for SMA and finite element analysis of active metal matrix composites, Mechanics of Composite Materials and Structures 3(2), June, 1996, 153–179.
[4]   Tanaka, K. and Nagaki, S., A thermomechanical description of materials with internal variables in the process of phase transitions, Ingenieur-Archiv 51(5), 1982, 287–299.
[5]   Liang, C. and Rogers, C. A., One dimensional thermomechanical constitutive relations for shape memory material, Journal of Intelligent Material Systems and Structures 1, 1990, 207–234.
[6]   Paiva, A. and Savi, M. A., An overview of constitutive models for shape memory allows, Hindawi Publishing Corporation, mathematical Problems in Engineering 2006, article ID 56876, 1–30. doi:10.1155/MPE/2006,56876.
[7]   Lagoudas, D. C., Hartl, D., Chemisky, Y., Machado, L. and Popov, P., Constitutive model for the numerical analysis of phase transformation in polycrystalline shape memory alloys, International Journal of Plasticity 32–33, 2012, 155–183.
[8]   Brocca, M., Brinson, L. C. and Bazant, Z. P., Three-dimensional constitutive model for shape memory alloys based on microplane model, Journal of the Mechanics and Physics of Solids 50, 2002, 1051–1077.

[9] Qianhua, K., Guozheng, K., Linmao, Q. and Sujuan, G., A temperature-dependent three dimensional super-elastic model considering plasticity for NiTi alloy, International Conference on Experimental Mechanics 2008, Xiaoyuan, H., Xie, H. and Kang, Y. (eds.), Proc. of SPIE Vol. 7375, 73755U, 2009 SPIE, doi:10.1117/12.839357.

[10] Arghavani, J., Auricchio, F., Naghdabadi, R., Reali, A. and Sohrabpour, S., A 3-D phenomenological constitutive model for shape memory alloys under multiaxial loadings, International Journal of Plasticity 26, 2010, 976–991.

[11] Falk, F., Model free-energy, mechanics and thermodynamics of shape memory alloys, ACTA Metallurgica 28(12), 1980, 1773–1780.

[12] Falk, F., One-dimensional model of shape memory alloys, Archives of Mechanics 35(1), 1983, 63–84.

[13] Falk, F. and Konopka, P., Three-dimensional Landau theory describing the martensitic transformation of shape memory alloys, Journal de Physique 2, 1990, 61–77.

[14] Devonshire, A. F., Theory of barium titanate, Philosophical Magazine 40, 1949, 1040–1063.

[15] Brinson, L. C., One dimensional constitutive behavior of shape memory alloys: thermomechanical derivation with non-constant material functions and redefined martensite internal variable, Journal of Intelligent Material Systems and Structures 4, 1993, 229–242.

[16] Fremond, M., Materiaux a Memoire de Forme, Comptes Redus Mathematique, Academie Sciences, Paris 304(7), 1987, 239–244.

[17] Fremond, M., Shape Memory Alloy: A Thermomechanical Macroscopic Theory, CISM Courses and Lectures, Vol. 351, New York, Springer, 1996, 3–68.

[18] Paiva, A., Savi, M. A., Braga, M. B. and Pacheco, C. L., A constitutive model for shape memory alloys considering tensile-compressive asymmetry and plasticity, International Journal of Solids and Structures 42(11–12), 2005, 3439–3457.

5 Electrorheological and Magnetorheological Fluids

5.1 Fundamental Behavior of ER and MR Fluids

Electrorheological (ER) and magnetorheological (MR) materials are both fluids that are considered as intelligent materials. Both fluids belong to a group of non-Newtonian fluids, as they do not obey Newton's law of fluid friction (as their kinetic viscosity coefficient is variable). The ER was patented by Winslow already in 1947 [1], and it consists of suspended polarized particles in a fluid, which under the application of an electrical field will change its apparent viscosity. The response of the ER fluid to an applied electrical field is manifested when the fluids are sheared by the force, thus creating a yield stress that is approximately proportional to the magnitude of the applied field. Their fluidity is manifested by change in their apparent viscosity. This property can be changed as a function of the characteristics of the fluid and the size and density of the dispersed particles [2]. Due to their damping capabilities, the ER fluids have been used as vehicle suspension, absorber and engine mount and have been introduced in clutch, break and valve systems. Typical references can be found in [3–9]. Figure 5.1 presents a test bed for testing ER-fluid-based valves (from [8]) while Figure 5.2 shows an ER-based suspension as manufactured by Fludicon, GmbH, Darmstadt, Germany.

MR fluid is another functional fluid whose yield stress can be changed by applying a magnetic field to obtain an induced yield stress that is 20–50 times larger than that of an ER fluid. MR fluids are also considered as intelligent materials with a reversible and very fast (in a fraction of millisecond) transition from a liquid to a nearly solid state under the presence of external magnetic fields. The reader should note that MR fluids should not be confused with colloidal ferrofluids, in which the particles are about 1×10^3 times smaller than those found in typical MR fluids. The MR fluids can exhibit changes in apparent viscosity of several orders of magnitude for applied magnetic flux densities of order of magnitude 1 T. This unusual property makes them ideal candidates for applications in mechanical systems that would require active control of vibrations or the transmission of torque. Typical examples include shock absorbers, brakes, clutches, seismic vibration dampers, control valves and artificial joints [10–16]. Figure 5.3 presents an MR-based damper for a vehicle suspension.[1] The damper has a built-in MR valve across which the MR fluid is forced. The piston of the MR damper acts like an electromagnet designed to have the correct number of coils to produce the required magnetic field. The damper has a run-through shaft to avoid the accumulation of the MR fluid. An MR clutch is shown in Figure 5.4.

---

[1] From *Intelligent Structures and Systems Lab* (ISSL) at the University of California, Irvine, CA, USA.

https://doi.org/10.1515/9783110726701-005

Although it is known that MR fluids have a good quality of control, due to their restricted efficiency they have not been suitable for applications in vehicle powertrains so far. The figure shows how Magna Powertrain Company[2] could create an efficient MRF clutch by combining clutch design, fluid development and magnetic circuit optimization.

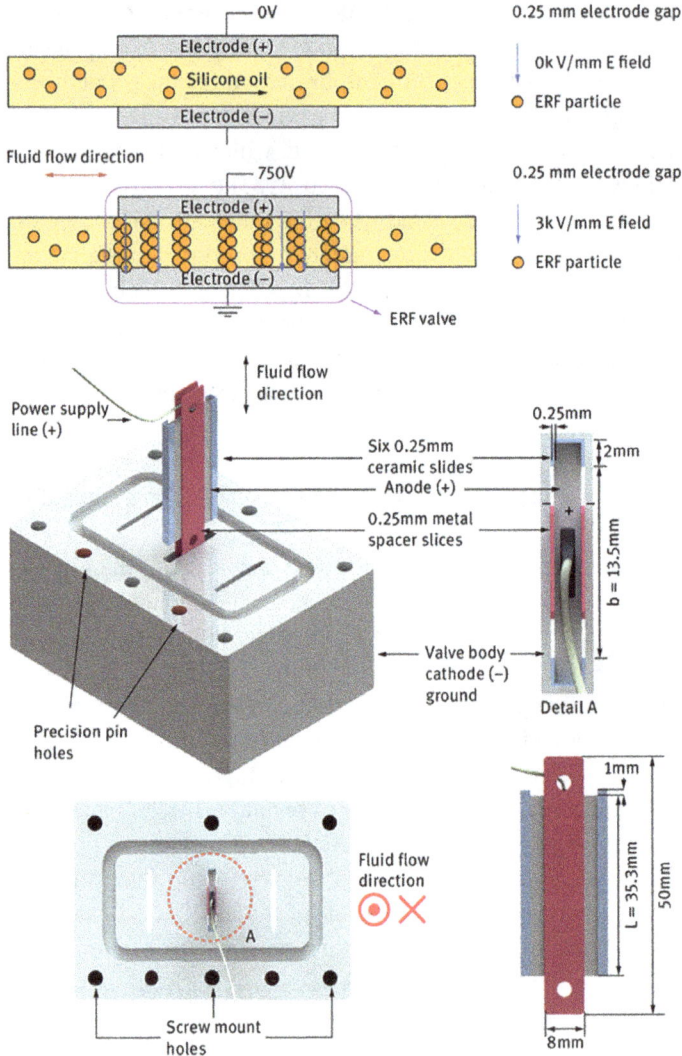

Figure 5.1: ER fluid-based hydraulic rotary actuator in its test bed (from [8]).

2 www.magna.com/capabilities/powertrain-systems

40t Servohydraulic press

8x ER-RheDamp

Figure 5.2: ER machine suspension for a 40-ton servo-hydraulic press (from FLUDICON GmbH, Germany).

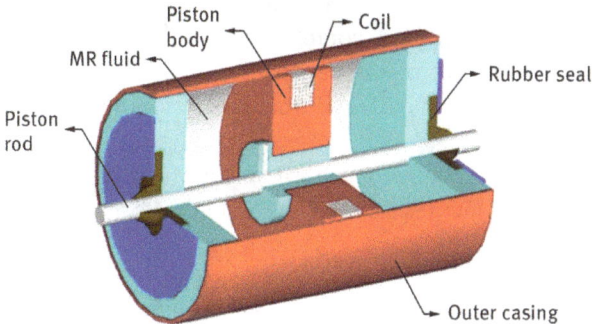

Piston body

Coil

MR fluid

Rubber seal

Piston rod

Outer casing

Figure 5.3: A vehicle suspension MR damper (from ISSL @ University of California, Irvine, USA).

Wen et al. [17] present a heuristic description of the ER effect to enable the understanding of the phenomenon. According to this description, ER fluids have a controllable rheology due to their composition, namely, solid particles suspended in a liquid media. When an electric field is applied, the suspended particles would polarize and align in chains from the plus electrode to the minus electrode. Note that the polarization of the suspended particles is due to the difference in the

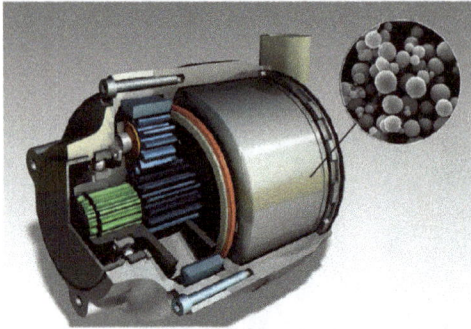

Figure 5.4: A magnetorheological clutch. Source: www.atzonline.com.

dielectric constants between the solid particles and the liquid in the colloid,[3] yielding an effective dipole. These chains are held in place by the applied electric field and they would resist flow (see Figure 5.5). Due to this chain structure, which is able to sustain shear in the direction perpendicular to the applied electric field yielding an increased viscosity, an ER fluid presents the rheology of a gel[4] when the fluid is under an electric field. It was shown [10] that the strength of the gel and its ability to resist flow is directly proportional to the strength of the applied field. This heuristic description can explain the visual effect of the formation of the chains in ER fluids when an electric field is applied; however their actual rheological behavior is much more complicated. As the behavior of ER contains both fluid-like and solid-like phenomenon, traditional rheological instruments must be modified and the data gathered from these modified instruments need a careful interpretation [10].

The properties of the ER fluid are measured using cylindrical rheometers, flow devices or annular pumping tools (see Figures 5.6–5.8), according to its application mode.

Figure 5.6 presents the tool called cylindrical rheometer and some typical experimental results obtained for an ER fluid. The gap between two cylinders is filled with ER fluid. One should note that this gap is very small, 1/100 of the inner cylinder radius. The inner cylinder is turned at varying speeds, and the torque transmitted to the outer cylinder is recorded. This procedure is repeated for different values of the applied electrical field. The torque versus the rotating speed of the inner cylinder is

---

3 A colloid is a homogeneous, noncrystalline substance consisting of large molecules or ultramicroscopic particles of one substance dispersed through a second substance.

4 A gel (its name derived from gelatin, a translucent, colorless, brittle, flavorless food derived from collagen obtained from various animal by-products) is a solid, jelly-like material that can have properties ranging from soft and weak to hard and tough. Gels are defined as a substantially dilute cross-linked system, which exhibits no flow when in steady state.

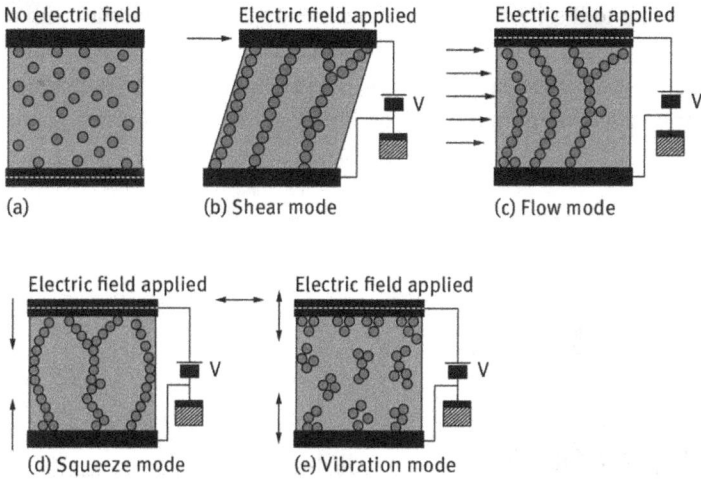

Figure 5.5: ER fluid – various modes of operation (from [2]).

(a)

(b) Shear mode

(c) Flow mode

(d) Squeeze mode

(e) Vibration mode

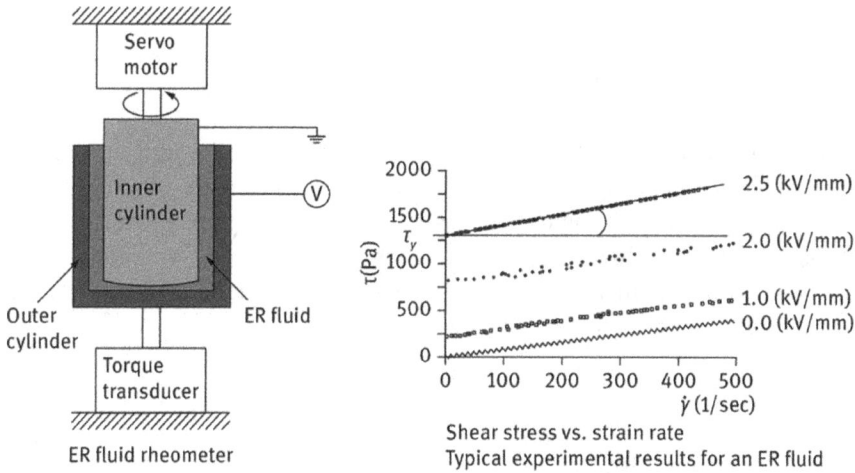

ER fluid rheometer

Shear stress vs. strain rate
Typical experimental results for an ER fluid

Figure 5.6: The ER fluid rheometer and typical experimental results (from [10]).

then converted to shear stress versus shear strain rate can be seen on the graph in Figure 5.6. Then two important properties can be evaluated experimentally (see the same graph in Figure 5.6): τ_y, the yield shear stress at a given electrical field strength and the slope of the experimental curve, η, the plastic viscosity of the ER fluid (see Appendix A at the end of this chapter). The relationship between the shear stress and the shear strain rate can then be written as

$$\tau = \tau_y + \eta\dot{\gamma} \tag{5.1}$$

This equation is similar to the constitutive relationship of a Bingham plastic[5] with a controllable yield stress [10]. One should note that this equation is only an approximation of the actual shear stresses and shear strain rates in an ER fluid; however, it provides useful information for engineering purposes and designing of various structures based on ER fluids.

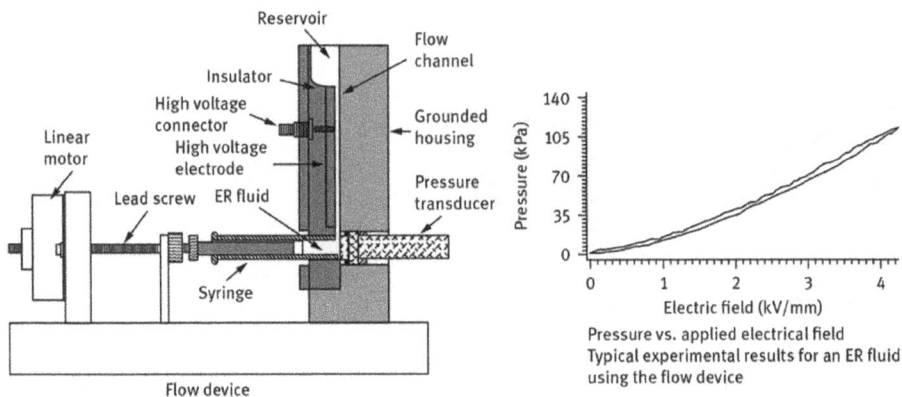

Figure 5.7: The ER fluid flow device and typical experimental results (from [10]).

To be sure that the measured ER fluid obeys the Bingham plastic equation, another tool is employed, called the flow fixture device, shown in Figure 5.7. The device measures the drop through the channel as a function of the applied electric field [10], as is shown in Figure 5.7 (right side). According to Philips,[6] who derived the equations of motion for this device, the pressure drop, Δp, along the length of the flow channel can be expressed as

$$\Delta p = \frac{8\eta QL}{bh^3} + 2\frac{L}{h}\tau_y \tag{5.2}$$

when the parameter $T^*>200$, namely, for a relatively very low flow rate which means a very high yield stress (τ_y). The expression for T^* is given by the following equation:

$$T^* = \frac{bh^2\tau_y}{12Q\eta} \tag{5.3}$$

---

5 A Bingham plastic is a viscoplastic material that behaves like a rigid body at low stresses but flows as a viscous fluid at high stresses. It carries the name of Eugene C. Bingham who proposed its mathematical form in his paper: *Bingham, E. C., An Investigation of the Laws of Plastic Flow, U.S. Bureau of Standards Bulletin, Vol. 13, 1916, 309–353.*

6 Phillips, R. W. Engineering Applications of Fluids with a Variable Yield Stress, Ph.D. Dissertation, University of California, Berkeley, 1969.

where η is the fluid viscosity, τ_y is the yield stress, Q is the fluid flow rate through the channel, L is the length of the flow channel, b is the width of the flow channel and h is the gap between the high voltage electrode and the housing. For $T^* < 0.5$, which means either the yield stress is very low or the flow rate is very high, the relationship between the pressure drop and the yield stress is

$$\Delta p = \frac{12\eta QL}{bh^3} + 3\frac{L}{h}\tau_y \qquad (5.4)$$

Using Equation (5.2) and (5.4), one can obtain the values of the viscosity and yield stress as a function of the applied electrical field. If these results do not match the results using the cylindrical rheometer, then the fluid is not behaving as a Bingham plastic and Equation (5.1) is not valid. Note that in the majority of measurements, these obtained results using two different devices do match and the validity of Equation (5.1) is approved.

The third device, the annular pumping tool [10] (see Figure 5.8), is aimed at given experimental results at stresses below the yield stress, called the pre-yield region. The complex shear modulus of an ER fluid, G', is measured by the device at small strains in the pre-yield region. The device contains an ER fluid at a predetermined electrical field, and the inner cylinder is oscillating with amplitudes below the yield strain (y_y). By varying the frequency of the oscillation, the graph presented in Figure 5.8 is obtained for various applied electrical fields. One should note that G' is changing from zero (when the ER material acts like a pure fluid) to values similar to

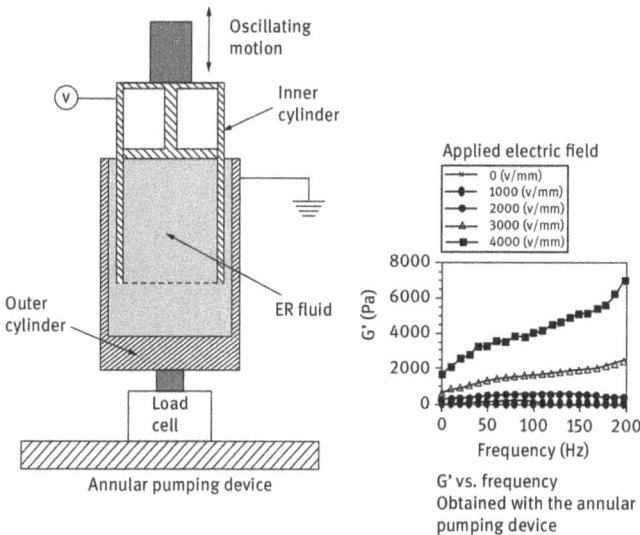

Figure 5.8: The ER fluid annular pumping device and typical experimental results to obtain G' (from [10]).

very soft rubber, while the electrical field is increased, highlighting the transition from fluid to solid for an ER material.

Now we shall address the behavior of MR fluids. As said earlier, the behavior of MR fluids is similar to ER ones, with a great difference in the required power to activate the relevant field and the stability of the rheological fluid (RF) to contaminants (see Table 5.1). The MR fluid is composed of metallic particles in a surrounding liquid. When the magnetic field is off, it will behave like a Newton-type fluid, as shown in Figure 5.9. When a magnetic field is on, the MR fluid would behave like a Bingham-type fluid, as the metallic particles in the liquid tend to form dipoles, leading to chains that oppose the flow of the MR fluid. At zero shear strain rate, a Bingham-type fluid has to overcome the yield stress, τ_y, before allowing the movement of the fluid (Figure 5.9). This yield stress is a function of the strength of the applied magnetic field and the relation can be written as

$$\tau = \tau_y(H) + \eta\dot{\gamma} \tag{5.5}$$

which is identical to Equation (5.1) for ER fluids.

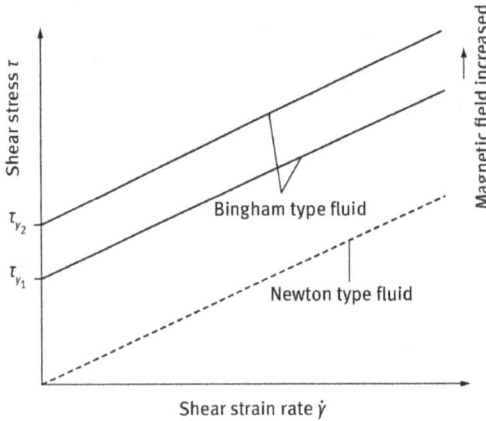

Figure 5.9: Shear stress versus shear strain rate – a schematic view of Newton- and Bingham-type fluids.

To obtain a high yield stress capability, two factors have to be addressed: the size of the particle and its percentage and the applied magnetic field. Increasing the percentage of the particles together with their size will build stronger chain structures (Figure 5.10) leading to higher yield stress. The second factor is expressed as the flux density B and its variation with the magnetic field strength H (see Appendix B). According to Ref. [18], MR fluids based on carbonyl iron have operating yield stresses in the range of 100 kPa.

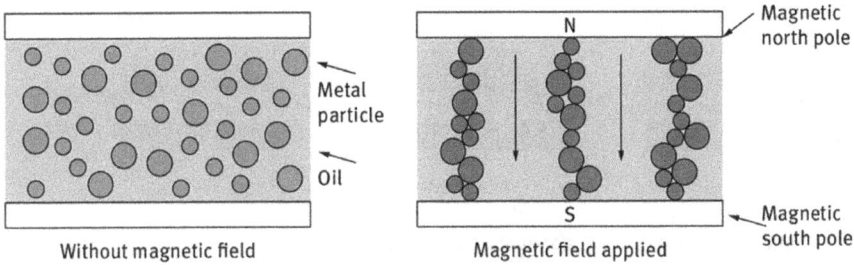

Figure 5.10: MR fluid – schematic behavior without and with the application of a magnetic field. Source: www.intechopen.com.

It is advised to use the linear section of the $H = f(B)$ relationship, presented by Equation (5.i), in Appendix B, thus drastically reducing the hysteresis problem and obtaining a simple design approach. The three modes of operation for MR fluids, the flow (valve) mode, the shear mode and the squeeze mode, are schematically presented in Figure 5.11.

As already described for the ER-type fluids, also the MR fluids are formed from a liquid that provides lubrication (in combination of additives) and carry the metal particles and the suspended particles. The liquid used for MR fluids can be hydrocarbon-, mineral-, or silicon-based oils. Carbonyl iron, iron powder or iron/cobalt alloys are ideal candidates for the suspended particles, having the size in the range of micrometers, as it will provide a high magnetic saturation in the MR fluid [19]. Their volume can reach up to 50% of the entire volume of both the liquid and the particles. The third of the fluid is composed of additives that include surfactants and stabilizers [21]. The additives are needed to control the viscosity of the fluid, the settling rate of the metal particles, the friction among the particles and to prevent the in-use thickening effect after a predefined number of off-duty cycles. As such the additives include suspending agents, friction modifiers, thixotropic[7] liquids and wear (anticorrosive) components. These additives would include grease, various dispersants like ferrous naphthenate ($C_{22}H_{14}O_4Fe$) or ferrous oleate ($C_{36}H_{66}FeO_4$) and various thixotropic additives like lithium or sodium stearates appearing in the form of soap.

The various devices described above for the ER-type fluids are also applicable for their counterpart, the MR fluids (see Figures 5.6 and 5.7). Devices built with MR fluids can use the three modes of operation described above (Figure 5.11) to reach a specific use. Dampers or shock absorbers use the flow (valve) mode of operation (Figure 5.11). Applying current to the coil inside the damper (Figure 5.12) causes the formation of a magnetic field. As the piston is moving, a force is generated leading to a displacement or velocity at its end.

---

7 Thixotropic is a time-dependent shear-thinning property of a fluid.

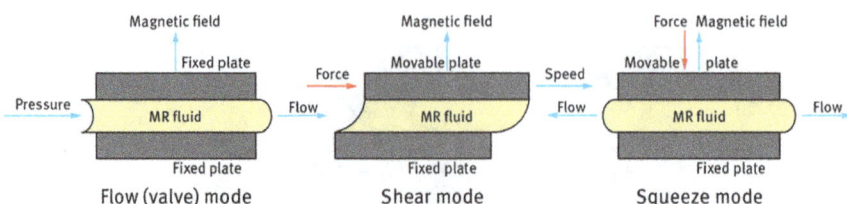

Figure 5.11: MR fluid – various modes of operation (from [18]).

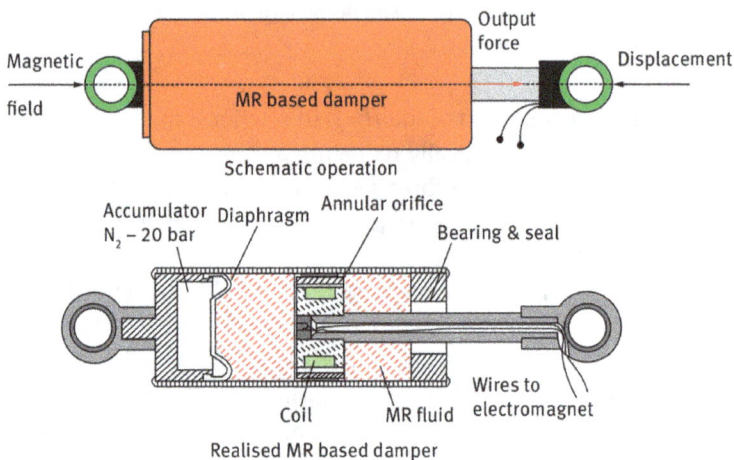

Figure 5.12: A typical MR-based damper. Source: www.intechopen.com.

Brakes and clutches use the shear mode (Figure 5.11) for their operation. Schematic drawings for various popular brakes designs are depicted in Figure 5.13.[8]

The third mode, the squeezing mode (Figure 5.11), is used to damp vibrations for low-motion and high-force applications. Although this use is less common in the literature, some interesting efforts had been done to realize it. Alghamdi and Olabi [22] present in their study a design and construction of a damper for vibration mitigation in vehicles, by using the squeeze mode of MR fluids (Figure 5.14). They used an MR fluid, MRF-140CG produced by Lord Corporation [23], with properties listed in Table 5.1. The low power required to provide a magnetic field (2–24 V @ 1–2 A, leading to 2–48 W) made it ideal for their prototype. By creating tensile and compressive forces on both sides of the piston, high net forces were generated. Their results showed that a very broad changeable damping force can be generated by

8 From: Active Structures Laboratory, ULB (Universite Libre de Brusseles), Belgium, http://scmero. ulb.ac.be

increasing the current supplied to the magnetic coil. This leads to a high range of control over the required damping.

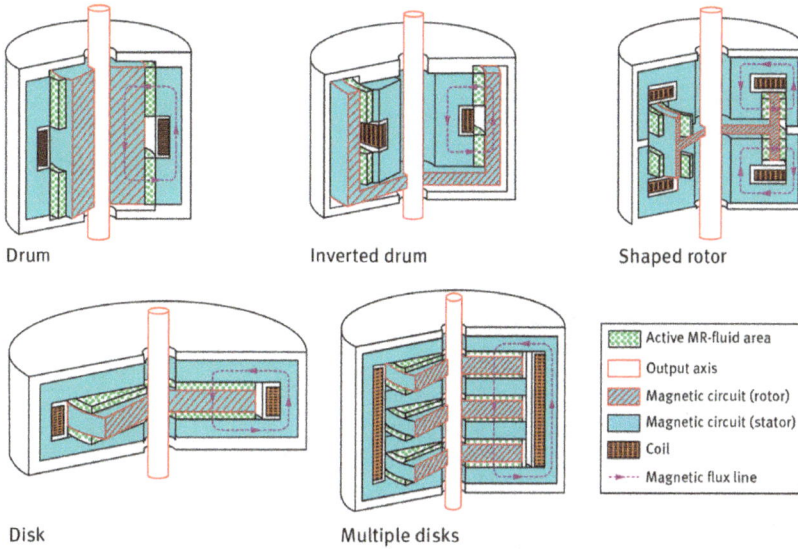

Figure 5.13: Various MR fluid brake designs. Source: www.scmero.ulb.be.

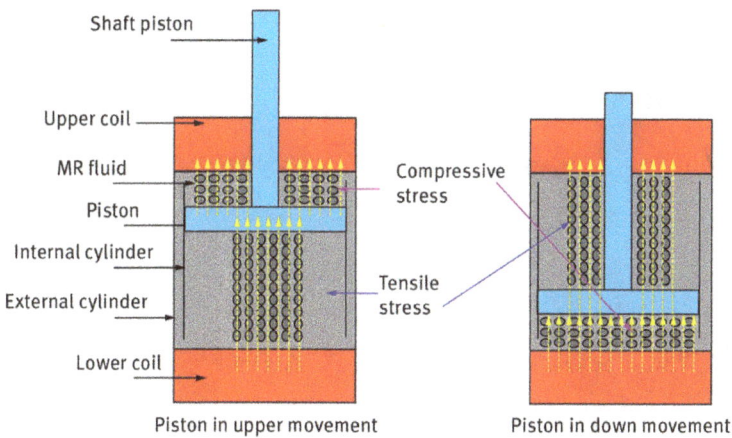

Figure 5.14: The MR-fluid-based damper concept operating in squeeze mode [22].

A recent study [24] presents another concept of using the MR fluid squeeze operation mode to design a damper. This concept is presented schematically in Figure 5.15. The study was focused mainly on the FE investigation on the damper, which revealed a strong influence of the eddy current induced in the conductor part of the damper on the dynamic response of the device. The authors suggested to continue their work accompanied by experiments to understand the phenomenon.

Table 5.1: The MR fluid properties used in [21].

Property	Values or limits
Color	Dark gray
MR fluid code	MRF-140CG
Base fluid	Hydrocarbon
Viscosity (Pa s) @ 40 °C	0.0280 ± 0.070
Density (g/cm^3)	3.54–3.74
Solids content by weight (%)	85.44
Flash point (°C)[1]	>150
Operating temperature (°C)	−40 to 130
Particles type	Carbonyl iron
Size of particles	0.88–4.03 μm

[1]The flash point of a chemical is the lowest temperature where enough fluid can evaporate to form a combustible concentration of gas.

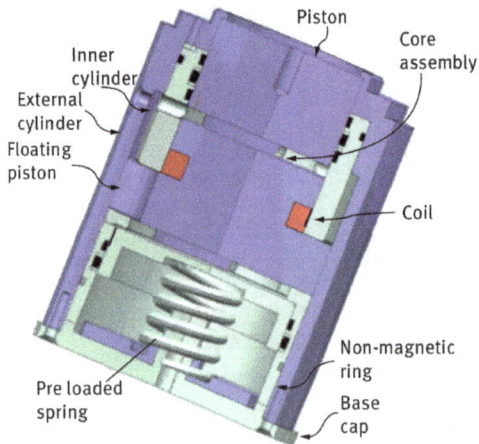

Figure 5.15: A CAD model for an MR-fluid-based damper operating in squeeze mode [24].

For conclusion, the various properties of both the ER and MR fluids are summarized in Table 5.2, displaying the common and various data for the two RFs.

Based on the data displayed in Table 5.2 and the behavior of MR and ER fluids, one can reach the following conclusions:

- The highest level of energy absorption can be reached with MR fluids as their yield stress τ_y is 10 times (approx.) higher than that of ER fluids.
- The activation time of MR fluids (≈ 10 ms) is fast; however, this time is influenced by the geometry of the device.
- MR fluids are insensitive for contaminants and impurities (see Section 5.2) while ER are sensitive to those factors. Use of additives might solve the problem of sedimentation leading to more reliable and durable applications with MR.
- The use of permanent magnets with coil-induced ones makes the MR system more reliable and less sensitive for sedimentation yielding a Fail-Safe operation.
- Although ER and MR fluid devices have similar power requirements of approximately 50 W, their inherent differences in the required voltage and current make the MR devices more attractive as it can be directly run on common low-voltage power sources.
- MR fluids are much more effective than ER fluids. However, one has to account for the peripheral equipment necessary to activate the fluid.
- By matching the densities of the solid and liquid components, or by using nanoparticles, the major problem of the settling time of the ER fluid suspensions might be solved.
- Another problem associated with the ER fluids is the breakdown voltage for air, that is, ~3 kV/mm, which is near the electric field needed for their operation.

Table 5.2: Typical properties of ER and MR fluids (from [20]).

Property	ER fluid	MR fluid
Particulate material	Polymers, zeolites[1], etc.	Ferromagnetic[2], ferrimagnetic[3], etc.
Typical particle size (µm)	0.1–10	0.1–10
Carrier fluid	Oils, dielectric gel and other polymers	Water, synthetic oils, nonpolar and polar liquids, etc.
Density (g/cm^2)	1–2	3–5
Off viscosity (Pa s @ 25 °C)	0.1–0.3	0.1–0.3
Required field	~3 kV/mm	~3 kOe[4]
Yield stress τ_y (kPa)	10	100
Type of activation	High voltage	Electromagnets and/or permanent magnets

[1]Zeolites are microporous, aluminosilicate minerals commonly used as commercial adsorbents and catalysts.

[2]Ferromagnetism is the basic mechanism by which certain materials like iron form permanent magnets, or are attracted to magnets.

[3]Ferrimagnetic materials have populations of atoms with opposing unequal magnetic moments, yielding a spontaneous magnetization.

[4]kOe = kilo Oersted, the units of the auxiliary magnetic field strength (H) in CGI system of units; 1 kOe = 1 Dyne/Maxwell.

References

[1] Winslow, W. M., Method and means for translating electrical impulses into mechanical forces, US Patent Specification 2417850, 1947.

[2] Ahn, Y. K., Yang, B. S. and Morishita, S., Directionally controllable squeeze film damper using electrorheological fluid, Journal of Vibration and Acoustics 124, January 2002, 105–109.

[3] Stanway, R., Sproston, J. L. and EL-Wahed, A. K., Applications of electro-rheological fluids in vibration control: a survey, Smart Materials and Structures 5, 1996, 464–482.

[4] Morishita, S. and Ura, T., ER fluid applications to vibration control devices and an adaptive neural-net controller, Journal of Intelligent Material Systems and Structures 4, 1993, 366–372.

[5] Peel, D. J., Stanway, R. and Bullough, W. A., Dynamic modeling of an ER vibration damper for vehicle suspension applications, Smart Materials and Structures 5, 1996, 591–606.

[6] Duclos, T. G., Carlson, J. D., Chrzan, M. J. and Coulter, J. R., Electrorheological fluids – materials and applications, In: Intelligent Structural Systems, Tzou, H. S. and Anderson, G. L. (eds.), Kluwer Academic Publishers, 1992, 213–241.

[7] Nguyen, Q.-A., Jorgensen, S. J., Ho, J. and Sentis, L., Characterization and testing of an electrorheological fluid valve for control of ERF actuators, Actuators 4, 2015, 135–155.

[8] Dyke, S. J., Spencer, B. F. Jr., Sain, M. K. and Carlson, J. D., Modeling and control of magnetorheological dampers for seismic response reduction, Smart Mater, Struct 5, 1996, 565–575.

[9] Choi, H. J. and Jhonb, M. S., Electrorheology of polymers and nanocomposites, Soft Matter, The Royal Society of Chemistry 5, 2009, 1562–1567. doi:10.1039/b818368f.

[10] Carlson, J. D. and Spencer, B. F. Jr., Magneto-rheological fluid dampers for semi-active seismic control, Proceedings of the 3rd International Conference on Motion and Vibration Control, Vol. 3, 1996, 35–40, Chiba, Japan.

[11] Aslam, M., Liang, Y. X. and Chao, D. Z., Review of magnetorheological (MR) fluids and its applications in vibration control, Journal of Marine Science and Applications 5(3), September 2006, 17–29.

[12] Baranwal, D. and Deshmukh, T. S., MR-fluids technology and its application- a review, International Journal of Emerging Technology and Advanced Engineering 2(12), December 2012, 563–569.

[13] De Vicente, J., Klingenbergb, D. J. and Alvareza, R. H., Magnetorheological fluids: a review, Soft Matter, The Royal Society of Chemistry 7, 2011, 3701–3710. doi:10.1039/c0sm01221a.

[14] Zhu, X., Jing, X. and Cheng, L., Magnetorheological fluid dampers: a review on structure design and analysis, Journal of Intelligent Material Systems and Structures 28(8), 2012, 839–873. doi:10.1177/1045389X12436735.

[15] Kciuk, S., Turczyn, R. and Kciuk, M., Experimental and numerical studies of MR damper with prototype magnetorheological fluid, Journal of Achievements in Materials and Manufacturing Engineering 39(1), 2010, 52–59.

[16] Wen, W., Huang, X. and Sheng, P., Electrorheological fluids: structures and mechanisms, Soft Matter, The Royal Society of Chemistry 4, 2008, 200–210. doi:10.1039/b710948m.

[17] Szary, M. L., The phenomena of electrorheological fluid behavior between two barriers under alternative voltage, Archives of Acoustics 29(2), 2004, 243–258.

[18] Olabi, A. G. and Grunwald, A., Design and application of magneto-rheological fluid, Materials and Design 28, 2007, 2658–2664.

[19] Bin Mazlan, S. A., The behavior of magnetorheological fluids in squeeze mode, Ph.D. Thesis submitted to School of Mechanical and Manufacturing Engineering, Faculty of Engineering and Computing, Dublin City University, North Ireland, August 2008.

[20] Huang, J., Zhang, J. Q., Yang, Y. and Wei, Y. Q., Analysis and design of a cylindrical magnetorheological fluid brake, Journal Material Process Technology 129, 2002, 559–562.

[21] Alghamdi, A. A. and Olabi, A. G., Novel design concept of magneto rheological damper in squeeze mode, 15th International Conference on Experimental Mechanics, ICEM15, Paper No. 2607, Porto, Portugal, 22–27 July 2012.

[22] Lord Corporation company, Lord technical data sheet for MRF-140CG, http://www.lord.com/products-and-solutions/magneto-rheological-(mr)/product.xml.1646/2, 2012.

[23] Sapinski, B. and Goldasz, J., FE Simulation of a Magnetic Circuit in a MR Squeeze-Mode Damper, 22nd International Congress on Sound and Vibration, Florence, Italy, 12–16 July, 2015.

5.2 Modeling ER and MR Fluids

From engineering point of view, the modeling of an ER fluid is based on the assumption that it behaves according to Bingham model of plastics (see Section 5.1), which relates the shear stress developed in the fluid, due to the application of an electrical field in ER fluids to viscosity and yield stress of the fluid (see [1–13]). This model can be written as

$$\tau = \tau_y + \eta_{pl}\dot{y} \qquad (5.6)$$

where τ is the shear stress in the fluid, τ_y is the yield shear stress at a given electrical field strength, η_{pl} is the plastic viscosity of the ER fluid and $\dot{y}$ is the shear strain rate. Another way of presenting the behavior of the ER fluid is to divide the Bingham plastic model (pl) into two regions, pre- and post-yield zones, namely

$$\tau = \eta\dot{y} \quad \text{if} \quad |\tau| < \tau_c$$
$$\tau = \tau_y + \eta_{pl}\dot{y} \quad \text{if} \quad |\tau| > \tau_c \qquad (5.7)$$

where η is the viscosity of the pre-yield and τ_c is a threshold value of the shear stress marking the transition from the pre-yield and to the post-yield region. One should note that in addition to the Bingham model, some researchers [14] would use the Herschel–Bulkley model for ER fluids (see Appendix C).

The models presented in Equations (5.6) and (5.7) show that under shear stress, displacement will not occur until the yield stress has been overcome. Below the yield stress the ER suspension is solid-like and above this value it behaves as of liquid with very high viscosity.

One should note that the Bingham model is valid for uniform, steady flow situations, where transient or starting effects can be neglected. When transient behavior becomes important or when fluids are under dynamic loading, with rapid or impact stresses and the damping issue must be considered, the Bingham model would be inadequate [1].

The capability of ER fluids to withstand normal compressive stresses, namely working in its squeeze working mode (see Figure 5.5d) is another interesting issue for engineers. According to the work done by Monkman [3], ER fluids appear to act as Newtonian fluids when under compressive stress. However, the onset of Bingham plasticity model follows abruptly when the applied electrical field density reaches a certain point. This phenomenon can be achieved either by increasing the applied voltage or by applying force to reduce the gap width (see Figure 5.16 in which an ER fluid layer is compressed between two electrodes constrained to allow only vertical movement) leading to an increase in the electric field intensity between the electrodes. Monkman performed a large number of tests with various types of ER fluids. Figure 5.17 shows typical results for the gap versus the axial compression stress, σ, for four different applied voltages, when using a 30% lithium polymethacrylate suspension in chlorinated paraffin (50LV series) [5]. One should note that the graph shows only the region experiencing compressive stress leading to measurable hardening (after the curve's knee). It can be seen that as the gap closes, the steady-state region represents the completion of the phase change from a Newtonian liquid to an apparent solid state. After the onset of Bingham plasticity, the displacement is plastic and irreversible while the electric field remains. The ER fluid will return to its original liquid-like state the applied electrical field is removed.

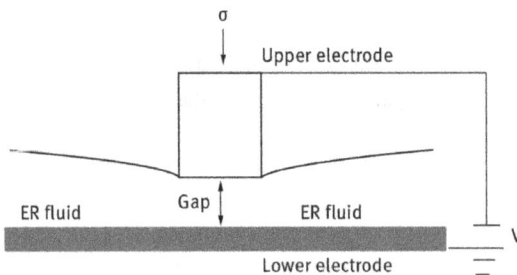

Figure 5.16: An ER fluid subjected to compressive stress – a schematic view (from [3]).

Rearranging the graph presented in Figure 5.17, by dividing the y-axis values by g_0, the initial gap, one can obtain the hardening of the ER fluid for the linear part of the curves yielding the plastic modulus:

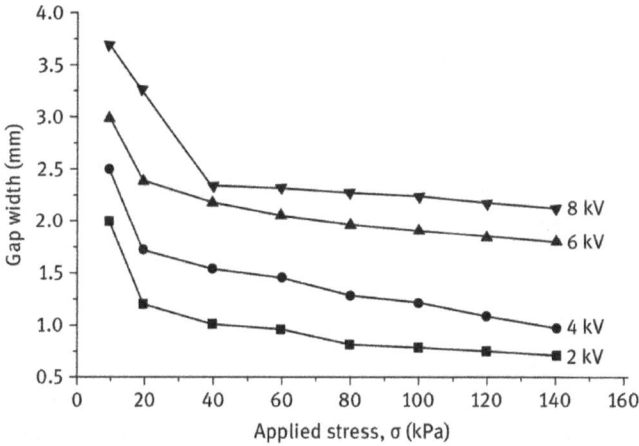

Figure 5.17: Compression characteristics for a 30% lithium polymethacrylate suspension in chlorinated paraffin – an ER fluid (from [3]).

$$E = \frac{\sigma}{\varepsilon} \quad \text{where} \quad \varepsilon = \ln\left(\frac{g}{g_0}\right) \tag{5.8}$$

The strain ε is defined as the true strain due to the relatively large displacements in the width of the gap. The modulus defined in Equation (5.8) is increased at least by an order of magnitude when the ER fluid passes from the liquid state to the plastic solid [3]. It was found that the hardened suspension of a typical ER fluid can withstand a normal stress in excess of 1 GPa [2–4] when a potential of 2000 V is applied across its gap.

It is interesting to note that when applying an electrical field to an RF, its hardened suspension shows an increase in its Young's modulus of elasticity, described by the following relation:

$$E_i = -\frac{d\sigma_i}{d\varepsilon_i} \tag{5.9}$$

where $\qquad\qquad\qquad i = x, y, z$

where σ_i is the normal stress in x-, y- or z-direction and ε_i is the strain in x-, y- or z-direction. It is known [1] that E_i in RF shows strong dependence on direction of the applied electrical field and the Young's modulus of elasticity is valid only for 1 degree-of-freedom systems.

The bulk modulus of elasticity for an RF defines the stiffness of the hardened suspension when its volume is under constant pressure, p (a 3 degree-of-freedom system), and is presented by the following relationship:

$$E_{bulk} = \frac{dp}{d\varepsilon_V}$$

(5.10)

where

$$\varepsilon_V = \frac{dV}{V}$$

p is the applied pressure and V is the initial volume. It is known [1] that the initial value of E_{bulk} for zero applied electrical field is approximately equal to E_{bulk} of fluid phase in an RF.

Besides the squeeze mode dealt above for the compressive stress issue and used in relatively low motion with high force applications, there are two other modes of working for the ER fluid: the first one being called the shear mode (see Figure 5.5b) and the second one is the flow mode or the valve mode (see Figure 5.5c). The shear mode has at least one movable electrical pole, while the valve pole has fixed poles. Typical applications using the shear mode would include clutches, brakes, chucking and locking devices, while valves, dampers and shock absorbers would use the flow mode of operation.

The force developed in a shear design mode is given by [1]

$$F = F_V + F_{rh} = \frac{\eta SA}{h} + \tau_y A$$

(5.11)

where η is the fluid viscosity, τ_y is the yield stress, $A = L*b$ is active shear area, while L and b represent the active length and the active width, respectively, h is the active thickness, namely the gap between the electrodes and S is the relative pole velocity.

The pressure drop developed in an application based on pressure driven flow mode (or the valve mode) is given by the same Equations (5.2)–(5.4) presented in Section 5.1 for the flow fixture device (Figure 5.7), namely

$$\Delta p = \frac{8\eta QL}{bh^3} + c\frac{L}{h}\tau_y; \quad c = 2 \quad \text{when} \quad T^* > 200$$

or

(5.12)

$$\Delta p = \frac{12\eta QL}{bh^3} + c\frac{L}{h}\tau_y; \quad c = 3 \quad \text{when} \quad T^* < 0.5$$

where the expression for T^*, a parameter describing the ratio between the second term and the first term in the right-hand side of Equation (5.27), is given by the following equation:

$$T^* = \frac{bh^2\tau_y}{12Q\eta}$$

(5.13)

and Q is the fluid flow rate.

MR fluids behave similar to their counterpart ER fluids. They would follow the Bingham plastic model in the following form:

$$\tau = \tau_y(H) + \eta \dot{y} \qquad (5.14)$$

where $\tau_y(H)$ is the yield shear stress as a function of the applied magnetic field H, η is the viscosity of the fluid and y' is the shear strain rate. Below the yield stress (showing strains at a range of 10^{-3}), the fluid would behave viscoelastically, namely

$$\tau = Gy \quad \text{for} \quad \tau < \tau_y(H) \qquad (5.15)$$

where G is the complex material modulus. Figure 5.18 schematically shows the behaviour of an MR fluid, including its complexities like hysteresis and sticky behavior.

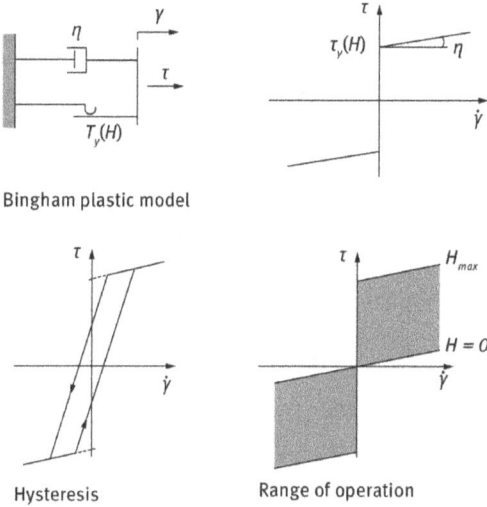

Bingham plastic model

Hysteresis

Range of operation

Figure 5.18: Schematic behavior of an MR fluid. From Active Structures Laboratory, ULB (Universite Libre de Brusseles), Belgium (http://scmero.ulb.ac.be).

A common model to predict the yield stress was developed by Bossis et al. (see [15–17]). According to this model, suitable for the case of $\alpha = \left(\mu_p / \mu_f \right) \geq 1$,[9] which assumes infinite chains of particles aligned in the direction of the applied field, as depicted in Figure 5.19a. Under strain, those chains would deform in line with to the strain, leading to the following relation between the angle θ and the shear strain y:

$$y = \tan(\theta) \qquad (5.16)$$

The field dependence of the yield stress can be written by a power law, $\tau_y \sim H^n$ with $1 < n < 2$. Substituting $n = 2$ provides the case of a linear magnetic material, which is found at low magnetic fields or alternatively for particulates of low permeability [17].

---

9 μ_p is the permeability of the particle and μ_f is the permeability of the fluid.

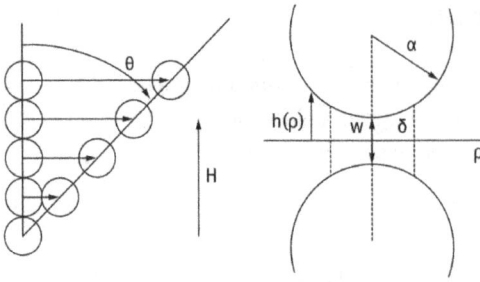

Figure 5.19: Bossis et al. model [17]: (a) deformation of the chain and (b) gap between two particles for the case $\rho < \delta$ (ρ is a polar coordinate while δ is the distance between two spheres as depicted) when $H_g = M_s$.

The effect of saturation on the yield stress is described by a simple model applicable for both ER and MR fluids. This simple model assumes that due to the high permeability of the particles, the magnetic field inside the particles H_i can be neglected.[10] Accordingly, two domains inside the gap between two particles (see Figure 5.19b) can be determined:

- The pole region having $\rho < \delta$ (with ρ being a polar coordinate) and the magnetic field is given by the saturation magnetization, namely, $H_g = M_s$.
- The second domain having $\rho > \delta$ with the magnetic field

$$H_g = \frac{H(a + 0.5w)}{h(\rho)}$$

where

H is the average magnetic field in the suspension, a is the radius of the particles and the term $h(\rho)$ is the distance between the plane of symmetry and the surface of the sphere given by the following equation:

$$h(\rho) = \frac{1}{2}\left(w + \frac{\rho^2}{a}\right) \tag{5.17}$$

with w representing the minimum gap between two spheres. Then demanding $H_g = M_s$ for $\rho = \delta$ yields the distance δ.

For the case

$$\frac{H}{M_s} \ll 1$$

---

10 Note the various symbols for the magnetic field: H_0, the external magnetic field; H_f, the magnetic field inside the MR fluid phase; H_i, the magnetic field inside the MR solid (particles) phase and the units of M_s are kA/m.

the radial force between two spheres can be obtained by integrating the magnetic field on the plane separating the two particles yielding

$$F_r = \frac{\mu_0 a}{2} \int_0 a_0 (H_g - H)^2 \cdot 2\pi\rho d\rho = \pi a^2 \mu_0 M_s^2 \left(\frac{H}{M_s} - \frac{\varepsilon}{w} \right) + \pi a^2 \mu_0 M_s^2 \frac{H}{M_s} \quad (5.18)$$

with $\varepsilon = \dfrac{w}{a}$

One should note that the first term in the right-hand side of Equation (5.18) corresponds to the region $\rho < \delta$, while the second part belongs to $\rho > \delta$.

Approximating the shear strain, namely $\gamma = \tan(\theta) \approx \sin(\theta)$ the shear stress expression is written as [17]

$$\tau(\gamma) = F_r \frac{N}{L^2} \sin(\theta)\cos^2(\theta) = \frac{3}{2} \Phi \frac{F_r}{\pi a^2} \frac{\gamma}{1+\gamma^2} \quad (5.19)$$

where the term N/L^2 is the number of chains per unit surface and Φ is the volume fraction of solid particles.[11] Writing the term ε in terms of the shear strain, namely

$$\varepsilon = 2\left(\sqrt{1+\gamma^2} - 1 \right) \quad (5.20)$$

and using Equations (5.18) and (5.19) for the shear modulus (valid for $\gamma \ll 1$), the yield stress can be written as

$$G = 3\mu_0 \Phi H M_s$$
$$\tau_y = 2.31\Phi\mu_0 M_s^{0.5} H^{1.5} \quad (5.21)$$

Note that the slope of curve of the shear modulus versus the shear strain at $\gamma = 0$ gives the expression for the G, and the maximum for $\gamma = \gamma_c$ would give the yield stress, when

$$\gamma_c = \sqrt{\frac{2H}{1.5M_s + 6H}} \quad (5.22)$$

Lemaire et al. [18] used a similar model presented schematically in Figure 5.20.

According to the model, when a shear strain is developed in the x-direction (perpendicular to the magnetic field) the shear stress can be written as

$$\tau = \frac{F_{shear}}{S} = -h\frac{\delta W}{\delta x} = -h\frac{\delta \mu_{zz}}{\delta x} H^2 \quad (5.23)$$

where

---

[11] The projection of the magnetic field squared on the unit vector joining two adjacent spheres leads to the term $\cos^2(\theta)$, while the term $\sin(\theta)$ comes from the projection of the radial force on the direction of the shear.

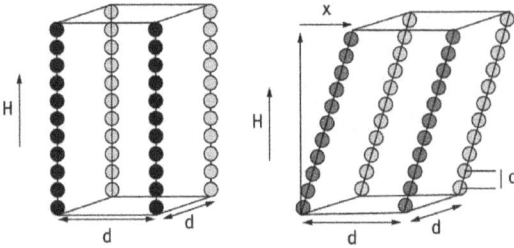

Figure 5.20: A schematic presentation of the chains assumed in Lemaire et al. model [18].

$$W = \frac{1}{2}\mu_{zz}H^2$$

is the magnetostatic energy, H is the average field inside the medium, h is the initial distance along z between two particles inside a chain and μ_{zz} is the diagonal component along the direction of the field of the permeability tensor. Additional details about the model can be obtained from Ref. [18].

Similar models were developed for ER fluids. Rusicka [8] developed a model that captures the behavior of suspensions consisting of solid particles and a carrier oil liquid, which defines the electrorheological fluids. Solutions are provided, and their errors are estimated.

According to the device and mode of operation of the ER fluid, various studies present the relevant constitutive equations. For example, Nilsson and Ohlson [19] derived a constitutive equation which presents the force as a function of displacement and velocity of the two parallel plates sandwiching an ER fluid operating in the squeeze mode (the movement is thus restricted to up and down of the plates). The two circular plates are constrained to move in a perpendicular direction only, causing the ER fluid to flow radially (Figure 5.21). As an electric field is applied, the particles in the suspension (plastic spheres in silicone oil) would form chains between the two parallel plates. They modeled the particle chains as quadratic columns with a side a and distance b between their centers. Their volume fraction was named f. The force between the two parallel plates was found to be

$$F = F_p + F_e = k \cdot f \frac{\pi R^2}{h}\left(3\eta \frac{R^2}{a^2}v_0 + E \cdot \delta\right) \tag{5.24}$$

where F_p and F_e are the vertical forces due to the pore pressure and the force transferred by the columns, respectively, k is a constant $0 < k < 1$, where $k = 0$ means zero electrical field and $k = 1$ means clean liquid phase [19], R is the radius of the circular plates, h is the distance between them, E is Young's modulus of the column material, δ is the axial displacement and η is the fluid viscosity.[12]

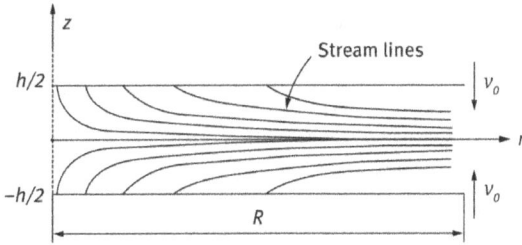

Figure 5.21: A schematic presentation of streamlines of the fluid between two planes moving up and down with a velocity v_0, as assumed in [19].

The realized damper is claimed to provide a force that was shown to be at least 1 order of magnitude higher than the same device operating in shear mode (see also Section 5.1 for the definition of operation modes).

See [10] to use a model based on chain-like aggregates, formed by the interactions between the induced dipoles in the particles forming the ER suspension fluid, similar to the one presented by Bossis [17] for MR fluids, leading to the calculation of the shear stresses for various flow and electric field, showing a good qualitative agreement with experiments.

A 3D constitutive equation for ER fluids was derived by Wineman and Rajagopal [11], based on continuum mechanics approach capturing the experimental behaviour of those fluids. The fluids show a regime of solid-like response when deformed from rest state and a viscoelastic-like manner under sinusoidal shearing stresses exhibiting time-depended response under sudden changes in either the shear strain rate or the electrical field.

For design purposes, additional parameters have to be decided for RF materials. As described in Table 5.3 (from [1]), both the ER and MR fluids would require up to 50 W for operation. However, while ER-fluid-based devices would require a high voltage of up to 5 kV and low current up to 10 mA, a device based on MR fluid would need a low voltage of up to 25 V and a high current of up to 2 A. Reference [1] presents semiempirical equations for the minimum required power as a function of the minimum active volume (MAV) for ER and MR fluids:

---

12 The reported values [19] for the viscosity η are 25×10^{-2} m^2/s at $k = 0$, down to $\tilde{7} \times 10^{-2}$ m^2/s at $k = 1$.

$$P_{min} = \frac{0.001 MAV}{\Delta t} \quad \text{for ER fluid}$$

$$P_{min} = \frac{0.1 MAV}{\Delta t} \quad \text{for MR fluid}$$

(5.25)

where Δt is the required switching time (in seconds).

As summarized in Table 5.3, MR fluids are far less affected by contaminations and/or impurities than the ER ones, which is even highly sensitive to condensed water from the atmosphere. As a result of this, ER fluids have a limited lifetime when working under heavy-duty applications, while MR fluids might show the setting of the solid phase in the suspension. One should be aware that the physical properties of RF fluids are strongly temperature dependent, and this should be taken into account when designing devices based on these fluids.

Table 5.3: Typical properties for rheological fluids (from [1]).

Property	ER fluid	MR fluid
Max yield stress ($\tau_{y max}$) (kPa)	2–5	50–100
Max electrical field	~4 kV/mm (limited by breakdown)	~250 (kA/m) (limited by saturation)
Viscosity (η) (Pa s)	0.1–1.0	0.1–1.0
Temp range of operation (°C)	+10° up to + 90° (ionic, DC) −10° up to + 125° (nonionic, AC)	−40° up to + 150° (limited by carrier fluid)
Stability	Cannot tolerate impurities	Unaffected by most impurities
Response time	<milliseconds	<milliseconds
Density (ρ) (g/cm³)	1–2	3–4
η_p/τ_y^2 (s/Pa)	10^{-7} to 10^{-8}	10^{-10} to 10^{-11}
Max energy density (J/cm³)	0.001	0.1
Typical power supply (watt)	2–50 (2–5 kV @ 1–10 mA)	2–50 (2–25 V @ 1–2 A)
Auxiliary materials	Any conductive surface	Iron or steel
RF volume necessary to operate	High	Low

When comparing the performances of these two RFs, it is noticeable to see that the maximal yield stress of ER is 1 order of magnitude (2–5 kPa) less than MR fluids (50–100 kPa) as presented in Table 5.3. As a result of this fact, for an ER fluid design one would need for the MAV (mega ampere-volt), 2 orders of magnitude greater than an MR-based design, as is shown in the following relationships (from [1]):

$$\mathrm{MAV} \cong \alpha P \left[\tfrac{F_{\mathrm{on}}}{F_{\mathrm{off}}}\right] \times 10^{-2} \quad \text{for ER fluid}$$

and $\hspace{10cm}$ (5.26)

$$\mathrm{MAV} \cong \alpha P \left[\tfrac{F_{\mathrm{on}}}{F_{\mathrm{off}}}\right] \times 10^{-4} \quad \text{for MR fluid}$$

where P is the required power (in watts), F_{on} and F_{off} are minimum "on-state" and "off-state" forces (in N), respectively, and the constant α will have the value of 1 for rotary application (shear mode) or the value of 2 when linear application (valve mode) is pursued.

References

[1]   Szary, M. L., The phenomena of electrorheological fluid behavior between two barriers under alternative voltage, Archives of Acoustics 29(2), 2004, 243–258.
[2]   Klingenberg, D. J., Dierking, D. and Zukoski, C. F., Stress-transfer mechanisms in electrorheological suspensions, Journal of Chemical Society Faraday Transactions 87(3), 1991, 425–430.
[3]   Monkman, G. J., The electrorheological effect under compressive stress, Journal of Physics D: applied Physics 28, 1995, 588–593.
[4]   Davis, L. C. and Ginder, J. M., Electrostatic forces in electrorheological fluids, In: Progress in Electrorheology, Havelka, K. O. and Filisko, F. E. (eds.), New York, Plenum Press, 1995, 107–114.
[5]   Brooks, D. A., A practical High Speed ER actuator, Actuator 1992, 3rd International Conference on New Actuators, Germany, Bremen, 1992, 110–115.
[6]   Hoppe, R. H. W. and Litvinov, W. G., Problems on electrorheological fluid flows, Communications on Pure and Applied Analysis (CPAA) 3, 2004, 809–848.
[7]   Hoppe, R. H. W., Litvinov, W. G. and Rahman, T., Problems of stationary flow of electrorheological fluids in a cylindrical coordinate system, SIAM Journal of Applied Mathematics 65(5), 2005, 1633–1656.
[8]   Ruzicka, M., Modeling, mathematical and numerical analysis of electrorheological fluids, Applications of Mathematics 49(6), 2004, 565–609. http://dml.cz/dmlcz/134585.
[9]   Ursescu, A., Channel flow of electrorheological fluids under an inhomogeneous electric field, Ph.D. Thesis, submitted to Institute of Mechanics, Darmstadt University of Technology (DUT), Darmstadt, Germany, 21st January, 2005.
[10] See, H., Constitutive equation for electrorheological fluids based on the chain model, Journal of Physics D: Applied Physics 32, 2000, 1625–1633.
[11] Wineman, A. S. and Rajagopal, K. R., On constitutive equations for electrorheological materials, Continuum Mechanics Thermodynamics 7, 1995, 1–22.
[12] Gavin, H. P., Hanson, R. D. and Mc-Clamroch, N. H., Control of structures using electrorheological dampers, Paper # 272, 11th World Conference on Earthquake Engineering, Acapulco, Mexico, 23–28 June, 1996, Elsevier Science Ltd.
[13] Prusa, V. and Rajagopal, K. R., Flow of an electrorheological fluid between eccentric rotating cylinders, Preprint No. 2010–025, Necas Center for Mathematical Modeling, Research team 1, Mathematical Institute of the Charles University, Sokolovska 83, 18675, Praha 8, Czech Republic, 23rd of July, 2010, http://ncmm.karlin.mff.cuni.cz/.
[14] Lee, D. Y. and Wereley, N. M., Quasi-steady Herschel–Bulkley analysis of electro- and magneto-rheological flow mode dampers, Journal of Intelligent Material Systems and Structures 10, 1999, 761–769.

[15]   Delivorias, R., The potential of magnetorheological fluid in crashworthiness design, MSc. Thesis, submitted to Automotive Engineering-Vehicle Safety, Department of Mechanical Engineering, Eindhoven University of Technology (EUT), Eindhoven, The Netherlands, December 2005.

[16]   Bossis, G., Lemaire, E., Volkova, O. and Clercx, H., Yield stress in magnetorheological and electrorheological fluids: a comparison between microscopic and macroscopic structural models, Journal of Rheology 41, 1997, 687–704.

[17]   Bossis, G., Volkova, O., Lacis, S. and Meunier, A., Magnetorheology: fluids, structures and rheology, In: Ferrofluids, Odenbach, S. (ed.), Berlin, Springer, 2002.

[18]   Lemaire, E., Meunier, A., Bossis, G., Liu, J., Felt, D., Bashtovoi, P. and Matoussevitch, N., Influence of the particle size on the rheology of magnetorheological fluids, Technical report, Universite de Nice, Laboratoire de Physique Matiere Condensee, France, June 1995.

[19]   Nillson, M. and Ohlson, N. G., An electrorheological fluid in squeeze mode, Journal of Intelligent Material Systems and Structures 11, July 2000, 545–554. doi:10.1106/MB24-94JR-T6LX-648L.

5.3 Damping of ER and MR Fluids

As was described in previous sections, both ER and MR fluids display relatively high damping coefficients, which made them ideal candidates for dampers in various systems. The aim of the present section is to provide the reader with data and formulas to evaluate the damping coefficients for ER and MR fluids for engineering applications. Figure 5.22 presents the large drop in the vertical acceleration due to application of 4 kV electrical field to the ER fluid inserted in their *ER-RheDamp* device (see Figure 5.2). Certainly the reduction of the acceleration by almost seven times from the initial value displays the advanced capabilities of ER fluids to provide effective attenuation of vibrations.

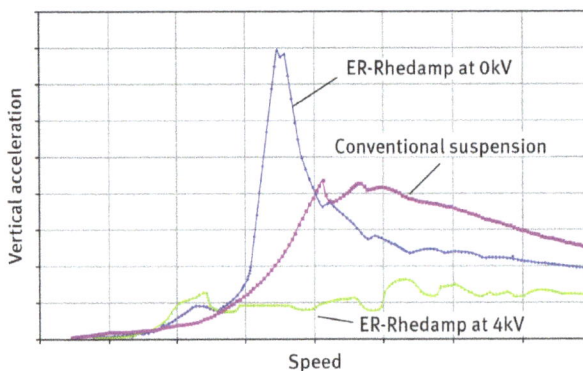

Figure 5.22: The ER damper *ER-RheDamp* performances, designed and manufactured by FLUDICAN GmbH, Darmstadt, Germany.

Kohl and Tichy [1] present useful expressions of the damping coefficients of two types of viscous dampers containing ER fluids: the first one being based on a flow between two stationary parallel plates and the second one has annular flow between two stationary concentric cylinders. The electric field is applied across the gap through which the fluid flows. The various dimensions for the first device are depicted in Figure 5.22. They found an expression of the pressure gradient along the channel as a function of the force F_{plunge} applied to the piston (see Figure 5.23) having the form

$$\frac{dp}{dx} = - \frac{F_{plunge}}{WH_pL} \tag{5.27}$$

where W is the channel's width and L is its length.

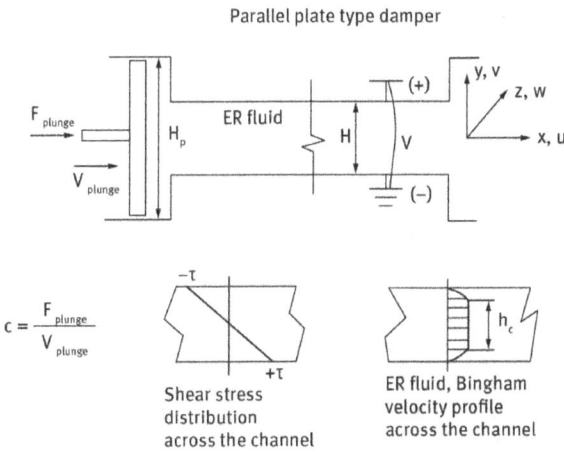

Figure 5.23: Parallel plate-type damper model [1].

Relating the applied force to the velocity of the piston, V_{plunge}, one obtains the damping coefficient, c

$$F_{plunge} = cV_{plunge} = \left(c_N c^*\right) V_{plunge} = \left[\left(\frac{12\eta WL}{H}H^{*2}\right)c^*\right] V_{plunge} \tag{5.28}$$

where c_N is the Newton part of the damping coefficient (when no voltage is applied) and c^* is the correction damping factor defined as

$$c^* = \frac{2}{3}\left(1 + \frac{\tau_y^*}{4H^{*2}}\right)\cos\left(\frac{a}{3}\cos A\right) + \frac{1}{3}\left(1 + \frac{\tau_y^*}{4H^{*2}}\right) \tag{5.29}$$

where

$$A = 1 - \frac{\tau_y^*}{36H^{*6}\left(1 + \frac{\tau_y^*}{4H^{*2}}\right)^3}; \quad \tau_y^* = \left|\frac{\tau_y H_p}{\eta V_{plunge}}\right|; \quad H^* = \frac{H_p}{H} \tag{5.30}$$

and η is the viscosity of the fluid.[13] The correction damping factor, c^*, increases with the electrical field applied to the ER fluid, as the nondimensional yield stress τ_y^* is linearly dependent on the electric field (typical results are presented in [1]).

The second device, treated in [1], was a concentric cylinder type damper, described in Figure 5.24. The authors, main assumption for the thin gap approximation was that the velocity profile in the x-direction (longitudinal direction) is independent on the curvature of the two cylinders. This allowed the use of the velocity distributions and the value of h, found for the previous device. For this type of device, Equations (5.27)–(5.30) are as follows:

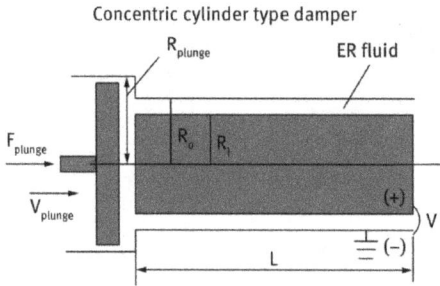

Figure 5.24: Concentric cylinder-type damper model [1].

The expression of the pressure gradient along the concentric cylinder as a function of the force F_{plunge} applied to the piston (see Figure 5.24) is

$$\frac{dp}{dx} = -\frac{F_{\text{plunge}}}{\pi R_{\text{plunge}}^2 L} \tag{5.31}$$

The relation force velocity of the plunge, leading to the damping coefficient c, is

$$F_{\text{plunge}} = cV_{\text{plunge}} = \left(c_N c^*\right)V_{\text{plunge}} = \left[\left(\frac{6\pi\eta R_p^{*4}}{(1-\kappa)^3}\right)c^*\right]V_{\text{plunge}} \tag{5.32}$$

where c_N is the Newton part of the damping coefficient (when no voltage is applied) and c is the correction damping factor defined as

$$c^* = \frac{2}{3}\left(1 + \frac{(1-\kappa)^2\tau_y^*}{2R_p^{*2}}\right)\cos\left(\frac{a}{3}\cos A\right) + \frac{1}{3}\left(1 + \frac{(1-\kappa)^2\tau_y^*}{2R_p^{*2}}\right) \tag{5.33}$$

---

[13] It turns out that the term $a/3$ appearing in Equation (5.29) can be obtained from the requirement that at no application of voltage (which means $\tau_y^* = 0$ or $\tau_y = 0$) the correction factor should be $c^* = 1$ (and also $A = 1$).

where

$$A = 1 - \frac{(1-\kappa)^6 \tau_y^{*3}}{4R_p^{*6}\left(1 + \frac{(1-\kappa)^2 \tau_y^*}{2R_p^{*2}}\right)^3}; \quad \tau_y^* = \left|\frac{\tau_y R_{\text{plunge}}}{\eta V_{\text{plunge}}}\right|; \quad R_p^* = \frac{R_{\text{plunge}}}{R_0}; \quad \kappa = \frac{R_i}{R_0} \quad (5.34)$$

Note that the $a/3$ term in Equation (5.33) is determined as for Equation (5.29) (see footnote 13). As before, the correction-damping factor, c^*, increases with the electrical field applied to the ER fluid, as the nondimensional yield stress τ_y^* is linearly dependent on the electric field (typical results are presented in [1]).

The damping capabilities of MR-type dampers were dealt in details in the literature (see typical Refs. [2–13]). No explicit expressions for damping coefficients, available in MR-fluid-based dampers, are found in those references; however, one can try to deduct them from the various detailed equations available there.

Nishiyama et al. [4] present a study on the behaviour of a flat plate immersed in an MR fluid under low magnetic field. The coefficient of damping for this case can be written as

$$c_{\text{eq}} = \sqrt{\frac{c^2}{2} + \frac{m_A^2 \omega^2}{2}}$$

$$m_A = \frac{k}{\omega^2}\left(1 - \frac{\cos\phi}{a}\right) - m; \quad a \equiv \frac{\Delta z}{\Delta z_g} \quad (5.35)$$

where m_A is the added mass, m is the mass of the flat plate, k is the spring constant connected to the plate, z is the displacement of the plate, z_g is the displacement of the exciter and $\omega = 2\pi f$, where f is the frequency. Note that the amplitude ratio, a, and the phase difference, ϕ, are to be experimentally determined. For more details, one can address Ref. [4].

Another expression for the damping coefficient is given in Ref. [2], where an MR fluid damper was designed and realized for low-frequency application. Carbonyl iron particles were immersed in silicone oil, with a volume fraction of the solid phase being varied from 20% to 35%. The expression for the damping coefficient ζ is given in [2] as

$$\zeta = \frac{1}{\sqrt{2}} \cdot \frac{\sqrt{\left[(a-1^2)\left(a - \sqrt{a^4 - a^2}\right)\right]}}{(a^2 - 1)} \quad (5.36)$$

where a is the amplitude ratio at the peak and a value of $\zeta = 0.2814$ @ $f = 9.24$ Hz is reported. The authors [2] also report that for the dry friction case $\zeta = 0.2139$, whereas when adding the shearing effect, the average damping coefficient was $\zeta_{\text{av}} = 0.277$.

The use of an MR damper as a controllable semiactive device embedded into a control system requires that the selected numerical model is capable of capturing the nonlinear behavior including hysteresis. Reference [6] presents a list of those mathematical models suitable for MR (and sometimes for ER) fluids. That list is presented in various models mentioned in Table 5.4 and next displayed to include their schematic drawings and the equation of motion. Only the most common ones are presented. The first is the idealized mechanical Bingham model proposed by Stanway et al. [32] and includes a viscous damper in parallel with a Coulomb friction element, as shown in Figure 5.25.

Table 5.4: Classification of MR models.

Modeling technique	MR damper models
Bingham models	Original Bingham model [15–32]
	Modified Bingham model [16]
	Gamota and Filisko model [17][1]
	Updated Bingham model by Occhiuzzi et al. [18]
	Three-element model [19][2]
Bi-viscous models	Nonlinear bi-viscous model [32]
	Nonlinear hysteretic bi-viscous model [37]
	Nonlinear hysteretic arctangent model [38]
	Lumped parametric bi-viscous model [39]
Viscoelastic plastic models	General viscoelastic plastic model [40]
	Viscoelastic plastic model by Li et al. [14]
Stiffness viscosity elastoslide model	Stiffness viscosity elastoslide (SVES) model [21]
Hydro mechanical model	Hydromechanical model [41]
Maxwell models	BingMax model by Makris et al. [20][3]
	Maxwell nonlinear slider model [42]
Bouc[4]–Wen[5] models	Simple Bouc–Wen model [34]
	Modified Bouc–Wen model [33]
	Bouc–Wen model for shear mode dampers [43]
	Bouc–Wen model for large-scale mode dampers [44]
	Current-dependent Bouc–Wen model [45]
	Current–frequency–amplitude-dependent Bouc–Wen model [46]
	Nonsymmetrical Bouc–Wen model [35]

Table 5.4 (continued)

Modeling technique	MR damper models
Dahl models[6]	Modified Dahl model [31]
	Viscous Dahl model [25]
LuGre[7] models	Modified LuGre model by Jimenez and Alvarez [22]
	Modified LuGre model by Sakai et al. [23–24]
Hyperbolic tangent models Sigmoid models	Hyperbolic tangent model by Kwok et al. [30] Sigmoid model by Ma et al. and Wang et al. [28–29]
Equivalent models	Equivalent model by Oh [27]
Phase transition models	Phase transition model [26]

[1]The model was initially developed for ER fluids.
[2]The model was initially developed for ER fluids.
[3]The model was initially developed for ER fluids.
[4]Bouc, R., "Modèle mathématique d'hystérésis: application aux systèmes à un degré de liberté". Acustica (in French) Vol. 24, 16–25, 1971.
[5]Wen, Y. K., Method for random vibration of hysteretic systems, Journal of Engineering Mechanics (American Society of Civil Engineers), Vol. 102, No. 2, 249–263, 1976.
[6]Dahl, P. R., A solid friction model. Technical report, The Aerospace Corporation, El Secundo, CA, 1968.
[7]LuGre model resulted from a collaboration between control groups in Lund and Grenoble, France.

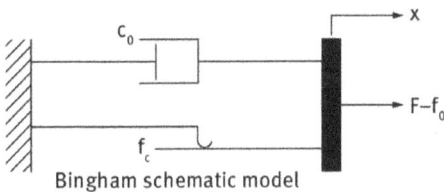

Figure 5.25: Bingham schematic model of an MR fluid damper.

The output force, F, generated by the damper is written as

$$F = f_c \text{sign}(\dot{x}) + c_0 \dot{x} + f_0 \qquad (5.37)$$

where c_0 is the viscous damping coefficient, f_c is the frictional force due to the fluid yield stress, f_0 is an offset force to account for the nonzero mean observed in the measured force due to the fluid accumulator and $\dot{x}$ is the velocity [6]. Another model, named extended Bingham model [17–33], is presented in Figure 5.26. This viscoelastic-plastic model consists of the Bingham model in series with the three-parameter element of a linear solid (Zener element) [12]. The equations of motion for this model is given by

$$F(t) = k_1(x_2 - x_1) + c_1(\dot{x}_2 - \dot{x}_1) + f_0 =$$

$$= f_c sign(\dot{x}) + c_0\dot{x} + f_0 = = k_2(x_3 - x_2) + f_0 \quad \text{for} \quad |F(t)| > f_c \quad (5.38)$$

$$F(t) = k_1(x_2 - x_1) + c_1\dot{x}_2 f_0 = k_2(x_3 - x_2) + f_0 \quad \text{for} \quad |F(t)| < f_c$$

where c_0 and the frictional force f_c for the Bingham model were defined before and the field constants c_1, k_1 and k_2 are associated with the fluid's elastic properties in the pre-yield region [12].

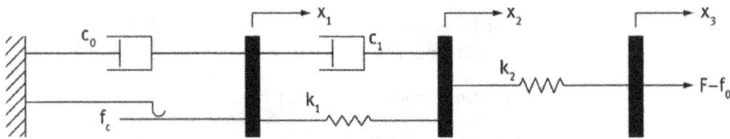

Figure 5.26: The extended Bingham schematic model of an MR fluid damper.

One of the commonly used models is the Bouc–Wen model (Table 5.4) introduced by the Bouc and extended latter by Wen. The model is supposed to mimic the response of hysteretic systems to random excitations [12]. Based on Figure 5.27, the force output can be written as

$$F = c_0\dot{x} + k_0(x - x_0) + \alpha z$$

where $\hspace{8cm}$ (5.39)

$$\dot{z} = -\gamma|\dot{x}|z|z|^{n-1} - \beta\dot{x}|z|^n + \delta\dot{x}$$

and z is the hysteretic component, x_0 is added to model the presence of an accumulator, while $\alpha, \beta, \gamma, \delta$ and n are parameters to control the shape of the force–velocity curve and are functions of the current, amplitude and frequency of excitation.

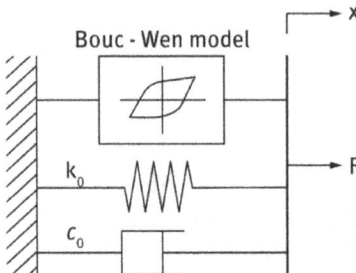

Figure 5.27: The Bouc–Wen schematic model of an MR fluid damper.

As mentioned in [7], to determine the Bouc–Wen characteristic parameters to model the MR fluid damper hysteretic response, Kwok et al. [35] proposed the non-symmetrical Bouc–Wen model (see Table 5.4) having the following modified expressions:

$$F = c_0[\dot{x} - \mu\,\mathrm{sign}(z)] + k_0(x - x_0) + \alpha z$$

with   (5.40)

$$\dot{z} = \{-[\gamma\,\mathrm{sign}(z\dot{x}) + \beta]|z|^n + \delta\}\dot{x}$$

where μ is the scale factor to adjust the velocity.

To better predict the performances of the MR damper in the vicinity of the stress yield pint, Spencer et al. [33] modified the Bouc–Wen model to include two additional mechanical components (a spring and a dashpot) as presented in Figure 5.28. The output force, F, can be written for this model as

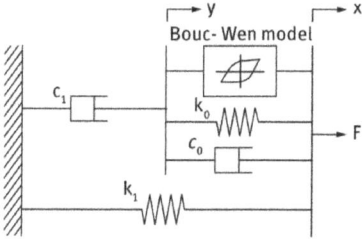

Figure 5.28: The modified Bouc–Wen schematic model of an MR fluid damper.

$$F = \alpha z + c_0(\dot{x} - \dot{y}) + k_0(x - y) + k_1(x - x_0) = c_1\dot{y} + k_1(x - x_0)$$

with

$$\dot{z} = -\gamma|\dot{x} - \dot{y}|z|z|^{n-1} - \beta(\dot{x} - \dot{y})|z|^n + \delta(\dot{x} - \dot{y})$$ (5.41)

and

$$\dot{x} = \frac{[\alpha z + c_0\dot{x} + k_0(x - y)]}{c_0 + c_1}$$

Note that as mentioned in [12], the hysteretic component is again represented by z, the spring k_1 and the initial displacement x_0 allowing the additional stiffness and the presence of an accumulator. The various parameters are assumed to be linearly dependent on the voltage v, applied to the current driver namely

$$\alpha = \alpha(u) = \alpha_a + \alpha_b u$$

$$c_1 = c_1(u) = c_{1a} + c_{1b}u$$

$$c_0 = c_0(u) = c_{0a} + c_{0b}u$$ (5.42)

and

$$\dot{u} = -\eta(u - v)$$

where u is the real signal output and η is a time constant parameter.

Another model similar to Bouc–Wen [7] is the one proposed by Kwok et al. [30], which inserted a hysteretic model to predict the damping force of the MR fluid damper (Figure 5.29). The model can be expressed as

$$F = c\dot{x} + kx + \alpha z + f_0$$

where (5.43)

$$z = \tanh[\beta\dot{x} + \delta\mathrm{sign}(x)]$$

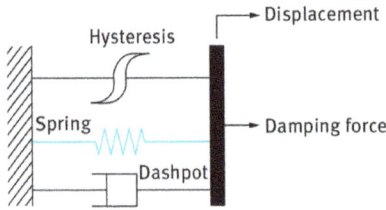

Figure 5.29: The hysteretic schematic model of an MR fluid damper.

where c and k are the viscous and stiffness coefficients, respectively, α is the hysteresis scale factor, z is the hysteretic variable, f_0 is the damper force offset due to the presence of an accumulator and β, δ are the model parameters to be identified. An interesting comparison among the Bingham, Bouc–Wen, hysteretic models and experimental data is supplied by [7] and presented in Figure 5.30. There is a clear deviation between the experimental and numeric models.

Another model, the modified Dahl model, which was reported to reproduce successfully the force–velocity relationship in the low-velocity region [2], is presented next. It consists of simple Dahl's model simulating friction and is able to present the hysteresis and zero slip displacement, but does not capture the Stribeck effect[14] or sticky behavior. Its typical hysteresis is described in Figure 5.31.

The model can then be written as

$$\frac{dF}{dx} = \sigma_0\left[1 - \frac{F}{F_c}\mathrm{sign}(\dot{x})\right]$$

(5.44)

where σ_0 and F_c are the stiffness and the Coulomb friction, respectively. Note that the graph shows different behaviors for increasing and decreasing of the displacement [2].

The force generated by the MR damper based on the modified Dahl model is written as

---

14 The Stribeck effect or the Stribeck curve clearly shows the minimum value of friction as the demarcation between full fluid-film lubrication and some solid asperity interactions.

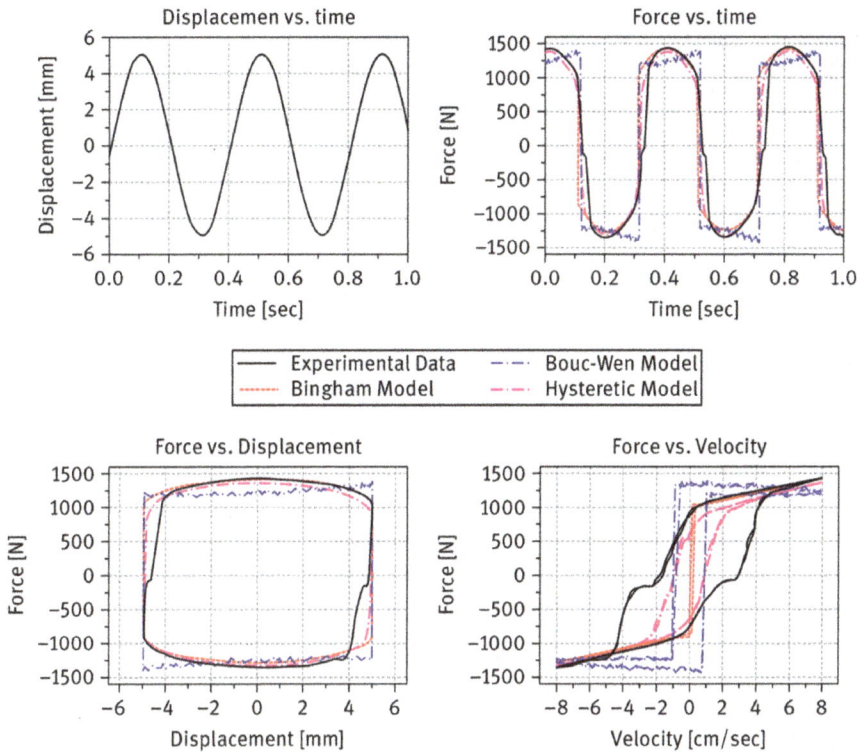

Figure 5.30: MR fluid damper – a comparison between experimental data and predicted values by various models, at 2.5 Hz sinusoidal excitation having an amplitude of 5 mm with a supplied current of 1.5 A (from [7]).

$$F = K_0 x + C_0 \dot{x} + F_d z - f_0$$

with   (5.45)

$$\dot{z} = \sigma \dot{x}[1 - \text{sign}(\dot{x})]$$

and σ is determining the hysteretic loop shape [2]. The relations between the model parameters and the applied magnetic field are obtained by assuming linear relations, similar to the ones used in the modified Bouc–Wen model (5.42), namely

$$C_0 = C_0(u) = C_{0s} + C_{0d}u$$

$$F_d = F_d(u) = F_{ds} + F_{dd}u$$

(5.46)

and

$$\dot{u} = -\eta(u - v)$$

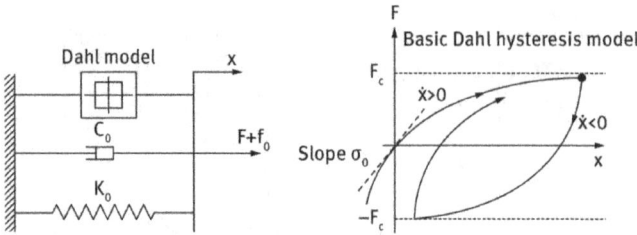

Figure 5.31: The Dahl schematic model and Dahl hysteresis model.

where, as before, v is the applied voltage to the current driver, u is the real signal output and η is a time constant parameter. C_{os} and F_{ds} are the damping coefficient and Coulomb force of the MR damper at 0 V, respectively, and C_{od} and F_{dd} are constants to be determined experimentally.

Wereley et al. [37] proposed a nonlinear bi-viscous model using piecewise linear functions to construct the hysteresis loop (see Figure 5.32) using two damping coefficients, one for the pre-yield condition and the second one for the post-yield [36] one. These functions are written as

$$
f_h = \begin{cases}
c_{po}\dot{x} - f_y & \text{for} \quad \dot{x} \le -\dot{x}_1 \quad \dot{x} > 0 \\
c_{pr}(\dot{x} + v_h) & \text{for} \quad -\dot{x}_1 \le \dot{x} \le \dot{x}_2 \quad \dot{x} > 0 \\
c_{po}\dot{x} + f_y & \text{for} \quad \le \dot{x} \quad \dot{x} > 0\, \dot{x}_2 \\
c_{po}\dot{x} + f_y \quad \dot{x}_1 & \text{for} \quad \le \dot{x} \quad \dot{x} > 0 \\
c_{pr}(\dot{x} + v_h) & \text{for} \quad -\dot{x}_2 \le \dot{x} \le \dot{x}_1 \quad \dot{x} > 0 \\
c_{po}\dot{x} - f_y & \text{for} \quad \dot{x} \le \dot{x}_2 \quad \dot{x} > 0
\end{cases}
$$

$$(5.47)$$

where f_y is a constant derived from a projection of the post-yield branch at zero velocity ($\dot{x} = 0$), v_y is the width of the hysteresis loop and

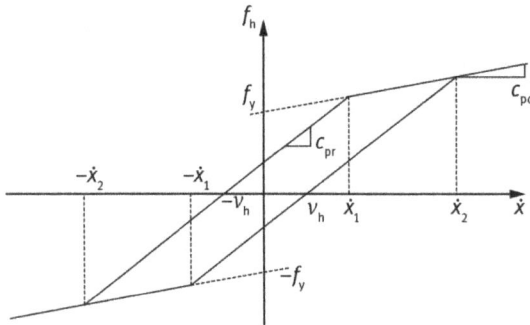

Figure 5.32: The nonlinear hysteresis bi-viscous model of an MR fluid damper.

$$\dot{x}_1 = \frac{f_y - C_{pr}v_h}{C_{pr} - C_{po}}$$

$$\dot{x}_2 = \frac{f_y + C_{pr}v_h}{C_{pr} - C_{po}}$$

(5.48)

where the various parameters in Equation (5.48) are to be found experimentally.

Based on the various models presented above, the force supplied by either the ER of MR fluid can be inserted in the equations of motion of a system, which contains such dampers, to yield the displacement/velocity as a function of time.

References

[1]   Kohl, J. G. and Tichy, J. A., Expressions for coefficients of electrorheological fluid dampers, Lubrication Science 10(2), February 1998, 135–143.

[2]   Prabhu, S. R. B., Harisha, S. R. and Gangadharan, K. V., Design, synthesis and fabrication of magneto-rheological fluid damper for low frequencies application, Journal of Mechanical and Civil Engineering (IOSR-JMCE), e-ISSN: 2278–1684, p-ISSN: 2320-334X, International Conference on Advances in Engineering & Technology, 2014 (ICAET-2014), 60–63.

[3]   Liao, W. H. and Lai, C. Y., Harmonic analysis of a magnetorheological damper for vibration control, Smart Materials and Structures 11, 2002, 288–296.

[4]   Nishiyama, H., Oyama, T. and Fujita, T., Damping characteristics of MR fluids in low magnetic fields, International Journal of Modern Physics B 15(6&7), 2001, 829–836.

[5]   Kelso, S., Denoyer, K., Blankinship, R., Potter, K. and Lindler, J., Experimental validation of a novel stictionless magnetorheological fluid isolator, SPIE conference on Smart Structures and Materials, Paper #5052-24, San Diego, CA, USA, March 2–6, 2003.

[6]   Braz-Cesar, M. T. and Barros, R. C., Experimental behavior and numerical analysis of MR dampers, Proceeding of the 15th World Conference on Earthquake Engineering (15WCEE), Lisbon, Portugal, 24–28 September, 2012.

[7]   Truong, D. Q. and Ahn, K. K., MR fluid damper and its application to force sensorless damping control system, Chapter 15, © 2012 Truong and Ahn, licensee InTech. This is an open access chapter distributed under the terms of the Creative Commons Attribution License (http://crea tivecommons.org/licenses/by/3.0), http://dx.doi.org/10.5772/51391.

[8]   Ambhore, N. H., Hivarale, S. D. and Pangavhane, D. R., A comparative study of parametric models of magnetorheological fluid suspension dampers, International Journal of Mechanical Engineering and Technology (IJMET), ISSN 0976–6340 (Print), ISSN 0976–6359 (Online) 4 (1), January–February 2013, © IAEME, 222–232.

[9]   Dimock, G. A., Yoo, J. H. and Wereley, N. M., Quasi-steady Bingham biplastic analysis of electrorheological and magnetorheological dampers, Journal of Intelligent Material Systems and Structures 13(9), September 2002, 549–559. doi:10.1106/104538902030906.

[10] Yang, G., Spencer, B. F. Jr., Carlson, J. D. and Sain, M. K., Large-scale MR fluid dampers: modeling and dynamic performance considerations, Engineering Structures 24, 2002, 309–323.

[11] Sapinski, B., Linearized characterization of a magnetorheological fluid damper, Mechanics 24 (2), 2005, 144–149.

[12] Butz, T. and Von Stryk, O., Modelling and simulation of electro-and magnetorheological fluid dampers, Journal of Applied Mathematics and Mechanics-ZAMM (Zeitschrift fur Angewandte Mathhematik und Mechanik) 82(1), 2002, 3–20.

[13] Ambhore, N. H., Hivarale, S. D. and Pangavhane, D. R., A study of Bouc–Wen model of magnetorheological fluid damper for vibration control, International Journal of Engineering Research and Technology (IJERT), ISSN 2279–0181, 2(2), February 2013, 1–6.

[14] Li, W. H., Yao, G. Z., Chen, G., Yeo, S. H. and Yap, F. F., Testing and steady state modeling of a linear MR damper under sinusoidal loading, Smart Materials and Structures 9, 2000, 95–102.

[15] Bingham, E. C., An investigation of the laws of plastic flow, U.S. Bureau of Standards Bulletin 13, 1916, 309–353.

[16] Nakamura, M. and Sawada, T., Numerical study on the laminar pulsating flow of slurry, The Journal of Non-Newtonian Fluid Mechanics 22(2), 1987, 191–206.

[17] Gamota, D. R. and Filisko, F. E., Dynamic mechanical studies of electrorheological materials: moderate frequencies, Journal of Rheology 35(3), 1991, 399–425.

[18] Occhiuzzi, A., Spizzuoco, M. and Serino, G., Experimental analysis of magnetorheological dampers for structural control, Smart Materials and Structures 12, 2003, 703–711.

[19] Powel, J. A., Modelling the oscillatory response of an electrorheological fluid, Smart Materials and Structures 3, 1994, 416–438.

[20] Makris, N., Burton, S. A. and Taylor, D. P., Electrorheological damper with annular ducts for seismic protection applications, Smart Materials and Structures 5, 1996, 551–564.

[21] Madhaven, V., Wereley, N. M. and Kamath, G. M., Hysteresis modelling of semi-active magneto-rheological helicopter dampers, Journal of Intelligent Material Systems and Structures 10(8), 1999, 624–633.

[22] Jimenez, R. and Alvarez, L., Real time identification of structures with magnetorheological dampers, Proc. 41st IEEE Conference on Decision and Control 1, 2002, 1017–1022.

[23] Sakai, C., Ohmori, H. and Sano, A., Modeling of MR damper with hysteresis for adaptive vibration control, Proc. 42nd IEEE Conference on Decision and Control 4, 2003, 3840–3845.

[24] Terasawa, T., Sakai, C., Ohmori, H. and Sano, A., Adaptive identification of MR damper for vibration control CDC, Proc. 43rd IEEE Conference on Decision and Control 3, 2004, 2297–2303.

[25] Ikhouane, F. and Dyke, S. J., Modelling and identification of a shear magnetorheological damper, Smart Materials and Structures 16(3), 2007, 605–616.

[26] Wang, L. X. and Kamath, H., Modelling hysteretic behavior in magnetorheological fluids and dampers using phase-transition theory, Smart Materials and Structures 16(6), 2006, 1725–1733.

[27] Oh, H. U., Experimental demonstration of an improved magnetorheological fluid damper for suppression of vibration of a space flexible structure, Smart Materials and Structures 13(5), 2004, 1238–1244.

[28] Ma, X. Q., Rakheja, S. and Su, C. Y., Relative assessments of current dependent models for magnetorheological fluid dampers, ICNSC '06, Proc. IEEE International Conference on Networking, Sensing and Control, 23–26 April 2006, Ft. Lauderdale, FL, USA, 510–515.

[29] Wang, E. R., Ma, X. Q., Rakheja, S. and Su, C. Y., Modelling the hysteretic characteristics of a magnetorheological fluid damper, The Proceedings of the Institution of Mechanical Engineers 217, 2003, 537–550.

[30] Kwok, N. M., Ha, Q. P., Nguyen, T. H., Li, J. and Samali, B., A novel hysteretic model for magnetorheological fluid dampers and parametric identification using particle swarm optimization, Sensors and Actuators 132(2), 2006, 441–451.

[31] Bastien, J., Michon, G., Manin, L. and Dufour, R., An analysis of the modified Dahl and Masing models: application to a belt tensioner, Journal of sound and vibration 302(4–5), 2007, 841–864. ISSN 0022–460X.

[32] Stanway, R., Sproston, J. L. and Stevens, N. G., Non-linear modelling of an electrorheological vibration damper, Journal of Electrostatics 20(2), 1987, 167–184.

[33] Spencer, B. F. Jr., Dyke, S. J., Sain, M. K. and Carlson, J. D., Phenomenological model of a magnetorheological damper, Journal of Engineering Mechanics 123, 1997, 230–238.

[34] Wen, Y. K., Method for random vibration of hysteretic systems, Journal of Engineering Mechanics (American Society of Civil Engineers) 102(2), 1976, 249–263.

[35] Kwok, N. M., Ha, Q. P., Nguyen, M. T., Li, J. and Samali, B., Bouc–Wen model parameter identification for a MR fluid damper using computationally efficient genetic algorithms, ISA Transactions 46(2), 2007, 167–179.

[36] Rakheja, S., Ma, X. Q. and Su, C. Y., Development and relative assessments of model characterizing the current dependent hysteresis properties of magnetorheological fluid dampers, Journal of Intelligent Materials and Structures 18(5), 2007, 487–502.

[37] Wereley, N. M., Kamath, G. M. and Pang, L., Idealized hysteresis modelling of ER and MR dampers, Journal of Intelligent Material Systems and Structures 9(8), 1998, 642–649.

[38] Li, W. H., Zhang, P. Q., Gong, X. L. and Kosasih, P. B., Characterization and Modeling a MR Damper Under Sinusoidal Loading, in Electrorheological Fluids and Magnetorheological Suspensions (ERMR 2004), Lu, K., Shen, R. and Liu, J. (eds.), World Scientific, 14 June 2005, 769–775.

[39] Sims, N. D., Holmes, N. J. and Stanway, R., A unified modelling and model updating procedure for electrorheological and magnetorheological vibration dampers, Smart Materials and Structures 13(1), 15 December 2003, 100–121.

[40] Katona, M., A visco-elastic-plastic constitutive model with a finite element solution methodology, Technical Report, R866, Civil Engineering Laboratory, Naval Construction Battalion Center, Port Hueneme, California, 93043, June 1978, 159.

[41] Hong, S. R., Choi, S. B., Choi, Y. T. and Wereley, N. M., A hydro-mechanical model for hysteretic damping force prediction of ER damper: experimental verification, Journal of Sound and Vibration 285(4–5), August 2005, 1180–1188.

[42] Chae, Y., Ricles, J. M. and Sause, R., Maxwell nonlinear slider model for seismic response prediction of semi-active controlled magnetorheological dampers, COMPDYN 2011, III ECCOMAS Thematic Conference on Computational Methods in Structural Dynamics and Earthquake Engineering, Papadrakis, M. and Fragiadakis, M. and Plevris (eds.), Corfu, Greece, 26–28 May 2011.

[43] Tse, T. and Chang, C., Shear-mode rotary magnetorheological damper for small-scale structural control experiments, Journal of Structural Engineering 130(6), 2004, 904–911.

[44] Rodriguez, A., Iwata, N., Ikhouane, F. and Rodellar, J., Model identification of a large-scale magnetorheological fluid damper, Smart Materials and Structures 18(1), January 2009, 1–12.

[45] Atabay, E. and Ozkol, I., Application of a magnetorheological damper modeled using the current-dependent Bouc–Wen model for shimmy suppression in a torsional nose landing gear with and without freeplay, Journal of Vibration and Control 20, August 2014, 1622–1644.

[46] Dominguez, A., Sedaghati, R. and Stiharu, I., A new dynamic hysteresis model for magnetorheological dampers, Smart Materials and Structures 15(5), 2006, 1179–1189.

5.4 Appendix III

5.4.1 Appendix A

When talking about viscosity of a material, one has to differentiate between dynamic viscosity (η) and kinematic viscosity (v). Their expressions are given by the following relationships:

$$\eta = \frac{\tau}{\dot{\gamma}}\,(\text{Pa s}) \tag{5.a}$$

$$v = \frac{\eta}{\rho}\left(\frac{m^2}{s}\right) \tag{5.b}$$

where τ is the shear stress in N/m^2 or Pa, $\dot{\gamma}$ is the shear strain rate in 1/s and ρ is the density in kg/m^3.

Note that for SI system, the dynamic viscosity, η, has the following units:

$$1\,\text{Poiseuille} = 1\,\text{Pa s} = 1\frac{N\cdot s}{m^2} = 1\frac{kg}{s\cdot m} \tag{5.c}$$

while for CGS (centimeter, gram, second) system one has

$$1\,\text{Poise} = \frac{1}{10}\,\text{Pa s} = \frac{1}{10}\frac{N\cdot s}{m^2} = 1\frac{g}{s\cdot cm} = 1\frac{\text{dyne}\cdot s}{cm^2} \tag{5.d}$$

Division of the Poise by 100 will lead to a smaller unit, the centipoise (cP), where

$$1\,\text{Poise} = 100\ \text{cP}$$
$$1\ \text{cP} = \frac{1}{100}\,\text{Poise} = \frac{1}{1000}\,\text{Pa s} = \frac{1}{1000}\frac{Ns}{m^2} = \frac{1}{100}\frac{g}{s\cdot cm} = \frac{1}{100}\frac{\text{dyne}\cdot s}{cm^2} \tag{5.e}$$

For the SI system, the kinematic viscosity v can also be expressed in stokes (St) where

$$1\frac{m^2}{s} = 10^4\text{St} = 10^4\frac{cm^2}{s} \tag{5.f}$$

A smaller unit will be the centistoke (cSt) obtained by division of the stoke unit by 100, yielding

$$1\,\text{St} = 100\ \text{cSt}$$

$$1\frac{m^2}{s} = 10^6\ \text{cSt} = 10^6\frac{mm^2}{s} \tag{5.g}$$

Typical values of the dynamic (or absolute) viscosity, η, are given in Table 5.a.

In addition, one should note that the viscosity is temperature dependent according to the following equation [19]:

Table 5.a: Typical values of the dynamic viscosity for various fluids.

Fluid	Dynamic viscosity η (Pa s) at room temperature
Water	1×10^{-3}
Olive oil	$\sim 1 \times 10^{-1}$
Glycerol	$\sim 1 \times 10^{0}$
Honey (liquid)	$\sim 1 \times 10^{+1}$
Glass (liquid)	$\sim 1 \times 10^{+40}$

$$\eta(T) = Ae^{\left(\frac{b}{T+273}\right)}$$ (5.h)

where constants A and b are experimentally determined.

5.4.2 Appendix B

The relation between H (the magnetic field strength) and B (the magnetic flux density) fields is given as

$$H \equiv \frac{B}{\mu_0} - M$$ (5.i)

where μ_0 is the vacuum permeability and its value is given as

$$\mu_0 = 4\pi \times 10^{-7} \mathrm{V \cdot s/(A \cdot m)}$$ (5.j)

and M is the magnetization vector field defined as the net magnetic dipole moment per unit volume.

In SI system, B and H terms have the following units:

$$1B(\mathrm{T}) = 1\Phi\left(\frac{\text{Weber}}{\text{m}^2}\right)$$
$$\text{with} \quad \mathrm{T} \equiv \text{Tesla and } \Phi \equiv \text{magnetic flux}$$ (5.k)
$$H\left(\tfrac{\mathrm{A}}{\mathrm{m}}\right)$$

In CGS system, B and H have the following units:

$$B(\mathrm{G})$$
$$\text{with} \quad \mathrm{G} \equiv \text{Gauss and } 1\mathrm{T} = 10,000 \ \mathrm{G}$$ (5.l)
$$H(\mathrm{Oe}) \quad \text{where} \quad \mathrm{Oe} \equiv \text{Oersteds}$$

5.4.3 Appendix C

The Herschel–Bulkley fluid model [1] is a nonlinear, *non-Newtonian fluid* model which tries to characterize the shear-thinning (like ordinary paint) and the shear-thickening (like a suspension of cornstarch in water) effects of the fluid. The constitutive equation has the following form:

$$\tau = \tau_y + k\dot{\gamma}^n \tag{5.m}$$

where τ is the shear stress, τ_y is the yield shear stress, k is the consistency index, $\dot{\gamma}$ is the shear strain rate and n is the flow index. For the case of $\tau < \tau_y$, the Herschel–Bulkley fluid behaves like a solid, while above the τ_y it will behave like a fluid. For $n > 1$ the fluid is shear-thickening, while for $n < 1$ it will be shear-thinning. If $n = 1$ and $\tau_y = 0$, the Herschel–Bulkley fluid reduces to a Newtonian fluid (see Figure 5.a).

The effective viscosity can be written as

$$\eta_{\text{eff}} = \eta_0 \quad \text{for} \quad |\dot{\gamma}| \leq \dot{\gamma}_y$$

or $\hspace{8cm}$ (5.n)

$$\eta_{\text{eff}} = k|\dot{\gamma}|^{n-1} + \tau_y|\dot{\gamma}|^{-1} \quad \text{for} \quad |\dot{\gamma}| \geq \dot{\gamma}_y$$

where the limiting viscosity η_0 is chosen such that the following equation is fulfilled:

$$\eta_0 = k\dot{\gamma}^{n-1} + \tau_y \dot{\gamma}_y^{-1} \tag{5.o}$$

In the post-yield region, the Herschel–Bulkley fluid displays an apparent viscosity described by the following relationship:

$$\frac{d\tau}{d\gamma} \equiv \eta_{\text{app.}} = nk\dot{\gamma}^{n-1} \tag{5.p}$$

One should note that for $n = 1$, one obtains

$$k = \eta_0 - \frac{\tau_y}{\dot{\gamma}_y} \equiv \eta_{pl} \tag{5.q}$$

Substituting the value of k in Equation (5.m), while $n = 1$ leads to the Bingham fluid equation, namely

$$\tau = \tau_y + \eta_{pl}\dot{\gamma} \tag{5.r}$$

which is similar to Equation (5.6) in Section 5.2.

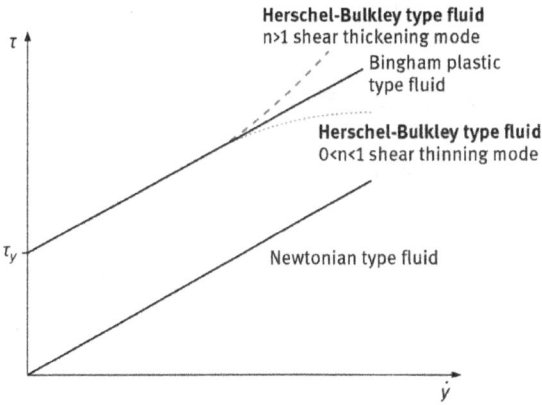

Figure 5.a: Shear stress versus shear strain rate – a schematic representation for the three models of Newtonian, Herschel–Bulkley and Bingham plastic fluids.

Reference

[1]   Herschel, W. H. and Bulkley, R., Konsistenzmessungen von Gummi-Benzollösungen, Kolloid Zeitschrift 39, 1926, 291–300. doi:10.1007/BF01432034.

6 Magnetostrictive and Electrostrictive Materials

6.1 Behavior of Magnetostrictive Materials

It is known that ferromagnetic[1]-type materials change their shape or dimensions (changes in volume of order 10^{-6}) when being magnetized. This property is called magnetostriction. To understand this property, one can picturize the internal structure of a ferromagnetic material as divided into domains, each of which having uniform magnetic polarization. Applying a magnetic field causes the boundaries between the domains to shift and rotate causing a change in the material's dimensions. The capability of those domains to shift and rotate is due to the anisotropy of the crystal structure of the material. This anisotropy leads the preferred directions of magnetization to yield a minimal free energy of the system. A strain is induced in the ferromagnetic material since various crystal directions are associated with different lengths. Magnetostrictive materials are able to convert magnetic energy into mechanical energy, and vice versa leading to manufacturing of both actuators and sensors. In its actuator mode (see a typical actuator in Figure 6.1), a current is supplied to a coil yielding a magnetic field which causes the magnetostrictive material to change its dimensions in the preferred direction, while in the sensor mode, mechanical forces applied to the magnetostrictive material cause changes in its magnetic field, which yield a measurable current in the coil (see a typical sensor in Figure 6.2).

Typical actuator cross section

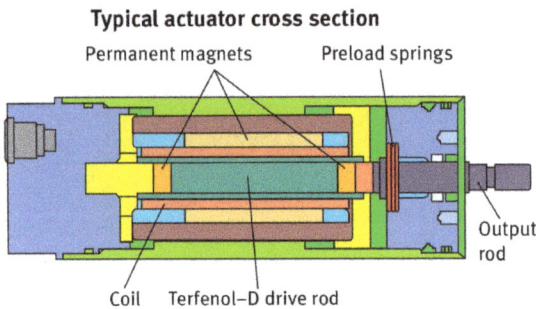

Figure 6.1: A typical magnetostrictive-based actuator (from Olabi and Grunwald [7]).

The magnetostrictive coefficient, λ (may be positive or negative), is defined as

$$\lambda = \frac{\Delta L}{L} @ \text{saturated } H \tag{6.1}$$

---

1 Ferromagnetism – a property of certain materials (such as iron, nickel, cobalt and their alloys) forms permanent magnets, or attracted by magnets.

https://doi.org/10.1515/9783110726701-006

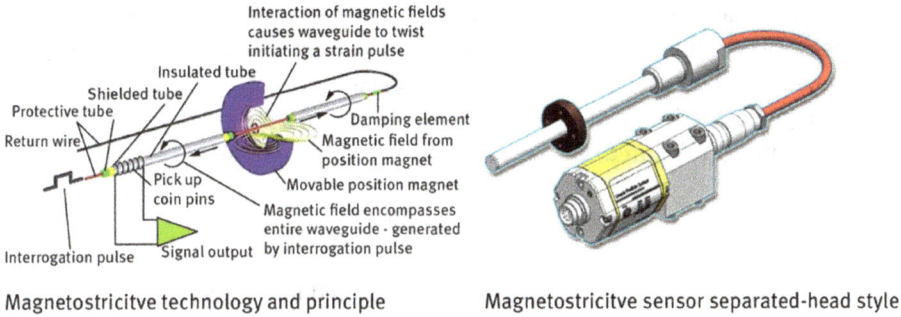

Figure 6.2: A typical magnetostrictive-based sensor (iTarget Sensors Company).

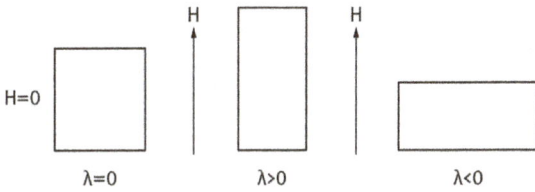

Figure 6.3: Schematic drawing for positive and negative λ.

where L is the original length, ΔL is the change in the length and H is the magnetic field measured in A/m. Note that saturation is the state reached when due to an increase in the applied external magnetic field, H, the magnetization of the material does not increase anymore, leading to "plateau" of the total magnetic flux density B. One should differentiate between the two symbols,[2] H and B, while the first is measured by A/m, the second is measured in Tesla or $N/(A \cdot m)$ (see also Appendix A at the end of this chapter; $H \equiv (B/\mu_0) - M$). Figure 6.3 presents schematically the meaning of positive (expansion) λ and negative (contraction) λ. As the magnetostrictive coefficient λ is known to be anisotropic, Lee [12] derived the simplest equation consistent with anisotropic magnetostriction and containing only two constants, λ_{100} and λ_{111}, representing the total strain when a crystal is magnetized from the ideal demagnetized state to saturation along the [100] and [111] axes. This equation is given as

$$\lambda_s = \frac{3}{2}\lambda_{100}\left(\alpha_1^2\beta_1^2 + \alpha_2^2\beta_2^2 + \alpha_3^2\beta_3^2 - \frac{1}{3}\right) + \\ + 3\lambda_{111}(\alpha_1\alpha_2\beta_1\beta_2 + \alpha_2\alpha_3\beta_2\beta_3 + \alpha_3\alpha_1\beta_3\beta_1)$$

(6.2)

---

2 M is the magnetization vector field and μ_0 is a magnetic constant; see also Appendix 5.1B.

where $\alpha_1, \alpha_2, \alpha_3$ are the direction cosines of the magnetized domain with respect to the reference coordinate system (denoted 1,2,3) and $\beta_1, \beta_2, \beta_3$ are direction cosines of the magnetostriction strain relative to the reference system of coordinates. The saturation magnetostrictive coefficient λ_s can be written, for example, for the case of polycrystalline, un-textured, cubic material as a simple expression

$$\lambda_s = \frac{2}{5}\lambda_{100} + \frac{3}{5}\lambda_{111} \tag{6.3}$$

where 100 and 111 are symbol crystallographic directions. Typical values of λ are presented in Table 6.1 [1] for some common magnetostrictive materials.

Table 6.1: Magnetostrictive coefficients for common ferromagnetic materials.

Material	Magnetostrictive coefficient ($\times 10^{-6}$)			
	λ_{100}	λ_{111}	λ_s (calculated)	λ_s (measured)
Iron	19.5	−18.8	−3.5	−7
Nickel	−45.9	−25.3	−32.9	−34
Cobalt		Polycrystalline		−55
Terfenol-D[1]	90	1640	1020	2000

[1] Terfenol-D ($Tb_xDy_{1-x}Fe_2$), where Ter = terbium, Fe = iron, NOL is Naval Ordnance Laboratory and D = dysprosium.

Terfenol-D is one of the most common magnetostrictive alloys used for actuators, generating strains 100 times greater than traditional magnetostrictives, and 2–5 times greater than traditional piezoceramics. Another interesting magnetostrictive material is Galfenol[3] ($Fe_{100-x}Ga_x$), which is an iron-gallium alloy, discovered in 1999 by the United States Navy researchers [2], where x can be varied to achieve the desired magnetic and mechanical properties. Typical properties are presented in Table 6.2 for the two alloys, while Figure 6.4 presents pieces of Galfenol and Terfenol-D as manufactured by Etrema Products Inc., USA.[4] According to [3], the Galfenol alloys exhibit moderate magnetostriction under very low magnetic fields, have very low hysteresis and demonstrate high tensile strength (in comparison with piezoelectric materials) and limited variation in their magnetomechanical properties for temperatures between −20 and 80 °C [2]. Those alloys are, in general (for Ga content < 20%), machinable, ductile, can be welded, have a high Curie temperature and seem to be corrosion resistant. Accordingly, the Galfenol alloys can also be used for actuation and sensing applications.

---

3 Galfenol ([EQN3117]), where Gal = gallium, Fe = iron, NOL is Naval Ordnance Laboratory.
4 www.etrema.com/

Table 6.2: Typical Terfenol-D and Galfenol physical properties.[1]

Property	Value	
	Terfenol-D	Galfenol
Standard composition	$Tb_{0.3}Dy_{0.7}Fe_{1.92}$	$Fe_{81.6}Ga_{18.4}$
Density (kg/m³)	9200–9300	7800
Hard Young's modulus (GPa)	50–90	60–80
Soft Young's modulus (GPa)	18–55	40–60
Bulk modulus (GPa)	90	125
Speed of sound (m/s)	1395–2444	2265–2775
Tensile strength (MPa)	28–40	350
Compressive strength (MPa)	300–880	–
Fatigue strength (MPa) @ fully reversed ($R = -1$)	–	75
Vickers hardness (HV)[2]	650	227
Minimal laminate thickness (mm)	1.0	0.25
Coefficient of thermal expansion (CTE) (10^{-6}/°C) @ 25 °C	11	11
Specific heat (kJ/(kg K))	0.33	–
Thermal conductivity (w/(m K)) @ 25 °C	13.5	15–20 [3]
Melting point (°C)	1240	1450
Resistivity (Ohm-m)	60×10^{-3}	85×10^{-3}
Curie temperature (°C)	380	670
Strain (estimated linear) (ppm)	800–1200	200–250
Energy density (kJ/m³)	4.9–25	0.3–0.6
Piezomagnetic constant, d_{33} (nm/A)	6–10	20–30
Coupling factor	0.7–0.8	0.6–0.7
Magnetic relative permeability, μ_r	2–10	75–100[4]
Saturation magnetic flux density (Tesla)	1	1.5–1.6

[1] ETREMA Products, Inc., Ames, IA 50010, USA, www.etrema.com.
[2] Vickers hardness is a measure of the hardness of a material, calculated from the size of an impression produced under load by a square-based pyramid diamond indenter having, [HV] = Vickers Pyramid Number.
[3] Estimated values based on carbon low steel.
[4] Highly dependent on Galfenol's stress state. Values ranging from 300 down to 20 have been measured at near-zero stress up to 13 ksi of compression.

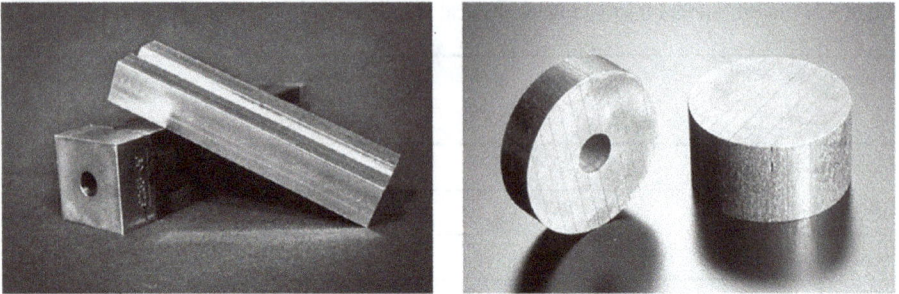

Figure 6.4: Galfenol sticks and Terfenol-D pieces with and without drilled holes (www.etrema.com).

Appendix B presents additional data on both the Terfenol-D and Galfenol giant magnetostrictive materials.

Besides Terfenol-D and Galfenol, another common magnetostrictive composite is the amorphous alloy $Fe_{81}Si_{3.5}B_{13.5}C_2$ with its trade name Metglas 2605SC,[5] having high saturation-magnetostriction constant, λ, of about 20 microstrains and more, coupled with a low magnetic-anisotropy field strength, of less than 1 kA/m (to reach magnetic saturation). Its properties are given in Figure 6.5 and Table 6.3.

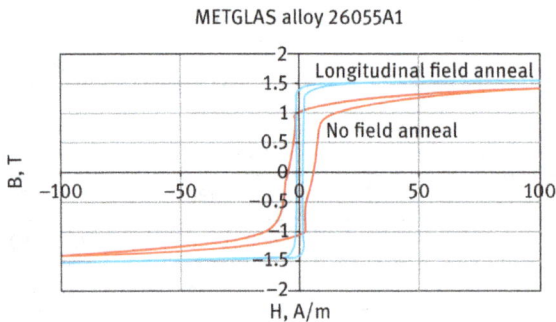

Figure 6.5: Metglas alloy 2605SA1 – a typical hysteresis loop. Source: www.metglas.com/products/magnetic_materials/2605sa1.asp.

One of the most known phenomena for magnetostrictive material is its hysteretic behavior presented in Figure 6.6. The shape of this hysteresis loop is named the "butterfly loop" and can be obtained using the Jiles–Atherton model [4]. Although this model was originally derived for isotropic materials, extensions performed later [5–6] also enables the modeling of anisotropic magnetic materials.

---

5 www.metglas.com/products/magnetic_materials/2605sa1.asp

Table 6.3: Typical Metglas® 2605SA1 & 2605HB1M magnetic alloy properties.

Property	Value
Saturation induction as cast (T)	1.56
Maximum DC permeability (μ) annealed	600,000
Maximum DC permeability (μ) as cast	45,000
Saturation magnetostriction (ppm)	27
Electrical resistivity ($\mu\Omega$cm)	130
Curie temperature (C)	395
Thickness (μm)	23
Standard available widths	
Minimum (mm)	5
Maximum (mm)	213
Density (g/cm^3)	7.18
Vickers hardness (50 g load)	900
Tensile strength (GPa)	1–2
Elastic modulus (GPa)	100–110
Lamination factor (%)	¿84
Thermal expansion (ppm/°C)	7.6
Crystallization temperature (°C)	510
Continuous service temperature (°C)	150

Presently, applications using magnetostrictive devices would include the following: ultrasonic cleaners, high force linear motors, positioners for adaptive optics, active vibration/noise control systems, medical and industrial ultrasonic, pumps and sonar. In addition, magnetostrictive linear motors, reaction mass actuators and tuned vibration absorbers have been designed. Ultrasonic magnetostrictive transducers have been developed for surgical tools, underwater sonar and chemical and material processing [7–11]. The limitations of the magnetostrictive actuators are temperature dependency, exhibit small displacements, when overheated might cause problems, while its advantages are response to relatively low voltages, do not decay over time, less hysteresis, robust to wear-and-tear, high dynamic strains at resonance. Some applications are shown in the following figures. Figure 6.7 presents a concept of using Terfenol-D actuators to change the shape of an airfoil, for morphing purposes.

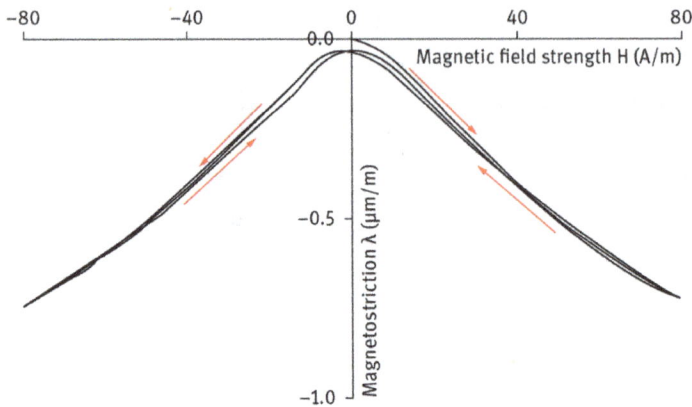

Figure 6.6: Typical magnetostrictive hysteresis loop of Mn–Zn ferrite (from "Magnetostrictive hysteresis loop of Mn–Zn ferrite" by magnetic models – Own work. Licensed under CC BY-SA 4.0 via Commons –https://commons.wikimedia.org/wiki/File:Magnetostrictive_hysteresis _loop_of _Mn-Zn_ferrite.png).

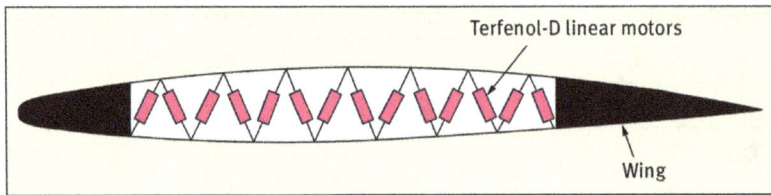

Figure 6.7: Terfenol-D linear motors used as truss-ribs in a two-spar wing of Gulfstream III aircraft. Source: www.machinedesign.com.

Cedrat Technologies[6] manufactures magnetostrictive actuators based on Terfenol-D giant magnetostrictive materials. They can be designed to produce high forces (>20 kN) and large strokes (>200 µm) at low operation voltage (<12 V) in static or dynamic applications (see Figure 6.8).

Magnetostrictive position sensors are essentially sonic-wave-sensing devices. A high-resolution clock measures the time a sonic wave takes to travel the distance between a fixed reference point and a moving magnet. By knowing the speed of the sonic wave, elapsed time is used to calculate the absolute position of the magnet. In addition, the magnet does not touch the waveguide, so there are no parts to wear out. Basic magnetostrictive position sensors have four basic components: the position magnet, a waveguide, a pickup (a sonic-wave converter) and the driver and signal-conditioning electronics. The conductive "waveguide" wire (usually made from

---

6 www.cedrat-technologies.com/en/technologies/actuators/magnetic-actuators-motors.html

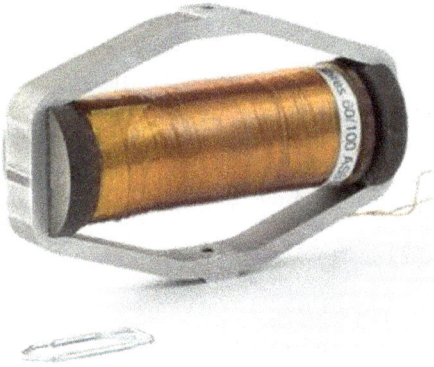

Figure 6.8: CEDRAT magnetostrictive actuators. Source: www.cedrat-technologies.com.

nickel-based magnetostrictive alloy) carries a short burst of electrical current – the interrogation pulse. As it travels along the waveguide, it creates a concentric magnetic field that surrounds the waveguide along it axis. When the waveguide-magnetic field reacts with the permanent magnet field from the position magnet, the magnetostrictive effect results in a strain on the waveguide that creates a pressure wave along the guide traveling at the speed of sound (approx. 2850 m/s) in both directions away from the position magnet. One wave is absorbed at the far end of the waveguide by a damping mechanism. This helps prevent reflections from the end of the waveguide which could cause interference. The other wave travels to the pickup. The complete waveguide, damping module and pickup assembly are commonly referred to as the sensing element (SE). Typical MTS magnetostrictive sensors can be seen in Figure 6.9.

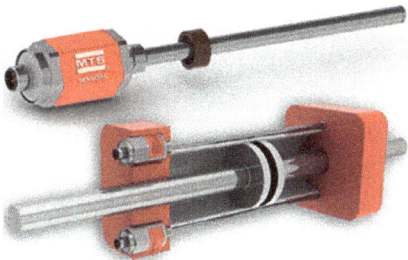

MTS continuous-position feedback sensor (courtesy MTS)

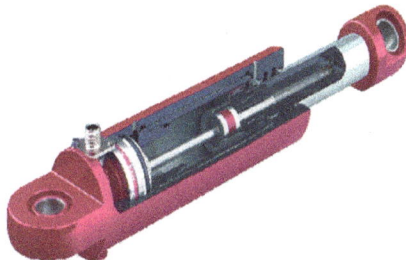

Linear position sensor / contactless / absolute magnetastrictive / with SSI interface - 50 - 2 500 mm, 100 g, 200 V/m, SIL2

Figure 6.9: MTS magnetostrictive sensors.

References

[1]   Gosh, A. K., Introduction to Transducers, Delhi, PHI Learning Private Limited, 2015, 323.

[2]   Clark, A. E., Wun-Fogle, M., Restorff, J. B. and Lograsso, T. A., Magnetic and magnetostrictive properties of Galfenol alloys under large compressive stresses PRICM-4: Int. Symp. on Smart Materials – Fundamentals and System Applications, Pacific Rim Conf. on Advanced Materials and Processing (Honolulu, Hawaii), 2001.

[3]   Atulasimha, J. and Flatau, A. B., A review of magnetostrictive iron–gallium alloys, Smart Materials and Structures 20, 2011, 1–15.

[4]   Jiles, D. C. and Atherton, D. L., Theory of ferromagnetic hysteresis, Journal of Applied Physics 55(5), March 1984, 2115–2120.

[5]   Ramesh, A., Jiles, D. C. and Roderick, J. M., A model of anisotropic anhysteretic magnetization, IEEE Transactions on Magnetics 32(5), 1996, 4234–4236.

[6]   Szewczyk, R., Validation of the anhysteretic magnetization model for soft magnetic materials with perpendicular anisotropy, Materials 7(7), 2014, 5109–5116.

[7]   Olabi, A. G. and Grunwald, A., Design and application of magnetostrictive materials, Materials & Design 29(2), 2008, 469–483.

[8]   Chowdhury, H. A., A finite element approach for the implementation of magnetostrictive material Terfenol –D in automotive Cng Fuel injection actuation, A master thesis in Engineering, submitted to the School of Mechanical and Manufacturing Engineering, Faculty of Engineering and Computing, Ireland, Dublin City University, July 2008, 167.

[9]   Belahcen, A., Magnetoelasticity, magnetic forces and magnetostriction in electrical machines, Ph.D. Thesis, Submitted to Department of Electrical and Communications Engineering, Helsinki University of Technology, Helsinki, Finland, August, 2004, 115.

[10]   Pons, J. L., A comparative analysis of piezoelectric and magnetostrictive actuators in smart structures, Boletin de la Sociedad Espaniola de Ceramica y Vidrio, In English 44(3), 2005, 146–154.

[11]   Poeppelman, C., Characterization of magnetostrictive iron-gallium alloys under dynamic conditions, Undergraduate honors thesis, The Ohio State University, 2010, 62.

[12]   Lee, E. W., Magnetostriction and magnetomechanical effects, Reports on Progress in Physics 18, 1955, 185–229.

6.2 Constitutive Equations of Magnetostrictive Materials

Neglecting thermal effects, the constitutive equations for a magnetostrictive material can be written in tensor expressions (see [5] and [6]) as

$$S_{ij} = s_{ijkl}^{H} T_{kl} + d_{kij} H_k + m_{klij} H_k H_l$$
$$B_j = d_{jkl}^{*} T_{kl} + \mu_{jk}^{T} H_k$$

(6.4)

where S and T are mechanical strain and stress, respectively, d is the magnetostrictive constant (obtained from the slope at the linear part of the S–H curve), B and H are the magnetic flux density and field intensity, respectively. s^H is the elastic compliance under constant magnetic field, μ^T is the permeability under constant stress. m_{klij} is the field magnetostrictive modulus tensor that physically denotes the magnetostrictive strain produced by per unit external magnetic field (its dimension is m^2/A^2). These equations are nonlinear; therefore, for low excitation levels, the linear equations are used [7] to yield a form similar to those used for piezoelectric materials (see Chapter 3):

$$S = s^H T + dH \tag{6.5a}$$

$$B = d^* T + \mu^T H \tag{6.5b}$$

Equation (6.5b) is often referred to as the converse effect, while Equation (6.5a) is known as the direct effect, similar to the piezoelectric equations. These equations are traditionally used for sensing and actuation purposes.

The two magnetostrictive constants d and d^* are defined as

$$d = \frac{\partial S}{\partial H}\bigg|_{@T=\text{const.}} \quad ; \quad d^* = \frac{\partial B}{\partial T}\bigg|_{@H=\text{const.}} \tag{6.6}$$

For small strains one can assume that $d = d^*$. Changing sides in Equations (6.5) yields the following matrix form:

$$\begin{Bmatrix} T \\ B \end{Bmatrix} = \begin{bmatrix} E^H & -e \\ e^* & \mu^S \end{bmatrix} \begin{Bmatrix} S \\ H \end{Bmatrix} \tag{6.7}$$

E^H stands for Young's modulus of the magnetostrictive material at constant magnetic field, μ^S is the permeability at constant strain and the constants e and e^* are defined as

$$e = E^H d; \quad e^* = E^H d^* \tag{6.8}$$

For a polycrystalline ferromagnetic material with the axis x_3 chosen as the direction of the magnetic polarization and under stress, Equations (6.5) have the following matrix notation:

$$\begin{Bmatrix} S_1 \\ S_2 \\ S_3 \\ S_4 \\ S_5 \\ S_6 \end{Bmatrix} = \begin{bmatrix} s_{11}^H & s_{12}^H & s_{13}^H & 0 & 0 & 0 \\ s_{12}^H & s_{11}^H & s_{13}^H & 0 & 0 & 0 \\ s_{13}^H & s_{13}^H & s_{33}^H & 0 & 0 & 0 \\ 0 & 0 & 0 & s_{44}^H & 0 & 0 \\ 0 & 0 & 0 & 0 & s_{44}^H & 0 \\ 0 & 0 & 0 & 0 & 0 & s_{66}^H \end{bmatrix} \begin{Bmatrix} T_1 \\ T_2 \\ T_3 \\ T_4 \\ T_5 \\ T_6 \end{Bmatrix} + \begin{bmatrix} 0 & 0 & d_{31} \\ 0 & 0 & d_{31} \\ 0 & 0 & d_{33} \\ 0 & d_{15} & 0 \\ d_{15} & 0 & 0 \\ 0 & 0 & 0 \end{bmatrix} \begin{Bmatrix} H_1 \\ H_2 \\ H_3 \end{Bmatrix} \tag{6.9}$$

$$\begin{Bmatrix} B_1 \\ B_2 \\ B_3 \end{Bmatrix} = \begin{bmatrix} 0 & 0 & 0 & 0 & d_{15} & 0 \\ 0 & 0 & 0 & d_{15} & 0 & 0 \\ d_{31} & d_{31} & d_{33} & 0 & 0 & 0 \end{bmatrix} \begin{Bmatrix} T_1 \\ T_2 \\ T_3 \\ T_4 \\ T_5 \\ T_6 \end{Bmatrix} + \begin{bmatrix} \mu_{11}^T & 0 & 0 \\ 0 & \mu_{11}^T & 0 \\ 0 & 0 & \mu_{33}^T \end{bmatrix} \begin{Bmatrix} H_1 \\ H_2 \\ H_3 \end{Bmatrix} \quad (6.10)$$

where the strain and stress vectors are defined similarly for piezoelectric materials, namely

$$\begin{Bmatrix} \varepsilon_1 \\ \varepsilon_2 \\ \varepsilon_3 \\ \gamma_{23} \\ \gamma_{13} \\ \gamma_{12} \end{Bmatrix} \equiv \begin{Bmatrix} S_1 \\ S_2 \\ S_3 \\ S_4 \\ S_5 \\ S_6 \end{Bmatrix}; \quad \begin{Bmatrix} \sigma_1 \\ \sigma_2 \\ \sigma_3 \\ \tau_{23} \\ \tau_{13} \\ \tau_{12} \end{Bmatrix} \equiv \begin{Bmatrix} T_1 \\ T_2 \\ T_3 \\ T_4 \\ T_5 \\ T_6 \end{Bmatrix} \quad (6.11)$$

while

ε_i and γ_{ij} are the normal and shear strains, respectively, while σ_i and τ_{ij} are normal and shear stresses, respectively. Assuming a linear relationship between **B** and **H** from the magnetic side and between **S** and **T** from the elastic one, one obtains the internal energy [5] as

$$U = \frac{S_i T_i}{2} + \frac{H_m B_m}{2} = \frac{T_i s_{ij} T_j}{2} + \frac{T_i d_{im} H_m}{2} + \frac{H_m d_{mi} T_i}{2} + \frac{H_m \mu_{mk} H_k}{2} \quad (6.12)$$

$$\equiv U_e + U_{em} + U_{me} + U_m = U_e + 2U_{em} + U_m$$

where U_e and U_m are the pure elastic and magnetic energies of the system, respectively, and $U_{em} = U_{me}$ is the mutual magnetoelastic energy.

An important figure of merit is the coupling factor k defined as

$$k = \frac{U_{me}}{\sqrt{U_e \cdot U_m}} \quad \text{or} \quad k^2 = \frac{U_{me}^2}{U_e \cdot U_m} \quad (6.13)$$

For example, assuming the case of both the magnetic field and the stress is only in the direction of x_3, namely:

$$H_1 = H_2 = 0, \ H_3 \neq 0$$

and $\quad (6.14)$

$$T_1 = T_2 = T_4 = T_5 = T_6 = 0, \ T_3 \neq 0$$

then we obtain from Equations (6.9) and (6.10):

$$B_1 = B_2 = 0, \qquad B_3 \neq 0$$

and (6.15)

$$S_4 = S_5 = S_6 = 0, \qquad S_1 = S_2 \neq 0, \qquad S_3 \neq 0$$

However, since $T_1 = T_2 = 0$, S_1 and S_2 do not contribute to the elastic energy of the system, then the coupling factor k^2 (Equation (6.13)) can be written for this special case as

$$k_{33}^2 = \frac{d_{33}^2}{\mu_{33}^T \cdot s_{33}^H} \qquad (6.16)$$

As given in [8], for the case of longitudinal coupling, assumed to be in the x_3 direction, all the unknowns, strains, stresses and magnetic field are parallel to it. For this case, we can omit the subscripts yielding

$$S = s^H T + d \cdot H \qquad (6.17a)$$

$$B = d \cdot T + \mu^T H \qquad (6.17b)$$

Writing Equation (6.17a) as a function of T and B while Equation (6.17b) as a function of S and H gives

$$S = s^B T + \frac{d}{\mu^T} B \qquad (6.18a)$$

$$B = \frac{d}{s^H} S + \mu^S H \qquad (6.18b)$$

where the definition of the two new constants s^B and μ^S is given as

$$s^B \equiv \frac{\partial S}{\partial T}\Big|_{@B = \text{const.}} = s^H \left(1 - \frac{d^2}{s^H \mu^T}\right) = s^H (1 - k^2) \qquad (6.19a)$$

$$\mu^S \equiv \frac{\partial B}{\partial H}\Big|_{@S = \text{const.}} = \mu^T \left(1 - \frac{d^2}{s^H \mu^T}\right) = \mu^T (1 - k^2) \qquad (6.19b)$$

and

$$\mu^S s^H = \mu^T s^B \qquad (6.19c)$$

where k^2 is the coupling factor defined earlier (see Equation (6.13)) for the x_3 direction.

Another appearance of the linear constitutive equations for magnetostrictive materials makes use of the following variables (named the reluctivity at constant stress):

$$g = \frac{d}{\mu^T}; \quad v^T \equiv \frac{1}{\mu^T} \tag{6.20}$$

Then the constitutive equation will be written as

$$S = s^B T + g \cdot B \tag{6.21a}$$

$$H = -g \cdot T + v^T B \tag{6.21b}$$

A different coupling constant κ^2 is defined as

$$\kappa^2 = -\frac{g^2}{s^B \cdot v^T} = -\frac{k^2}{(1-k^2)} \tag{6.22}$$

and due to its imaginary nature is seldom used.

The constitutive equations (Equations (6.21a) and (6.21b)) can be further modified to yield the following ones:

$$T = \frac{S}{s^B} - \lambda \cdot B = S \cdot c^B - \lambda \cdot B \quad \text{with} \quad c^B = \frac{1}{s^B} \tag{6.23a}$$

$$H = -\lambda \cdot S + v^S \cdot B \tag{6.23b}$$

where c^B is the stiffness matrix at constant magnetic field. Then the classic magnetostrictive constant, λ, can be written as (by equating Equations (6.23a) and (6.23b) to Equations (6.18a) and (6.18b))

$$\lambda = \frac{d}{\mu^T \cdot s^B} = \frac{d}{\mu^S \cdot s^H} \tag{6.24}$$

and

$$v^S = \frac{1}{\mu^S}$$

To obtain the coupling factor k^2, one has to first evaluate the cross-product ratio of the coefficients in the right-hand side of Equations (6.23a) and (6.23b) to yield $\lambda^2 \mu^S s^B$. Then we can write the following expression (see also Equation (6.16)):

$$\lambda^2 \cdot \mu^S \cdot s^B = \left(\frac{d^2}{\mu^T \cdot s^B \cdot \mu^S \cdot s^H} \right) \mu^S \cdot s^B = \frac{d^2}{\mu^T \cdot s^H} = k^2 \tag{6.25}$$

Finally, one can derive the relation between d and λ to be [8]

$$d = \frac{1}{\lambda} \frac{k^2}{(1-k^2)} \Rightarrow k^2 = \frac{\lambda \cdot d}{(1+\lambda d)} \tag{6.26}$$

6.2.1 One-Dimensional Model

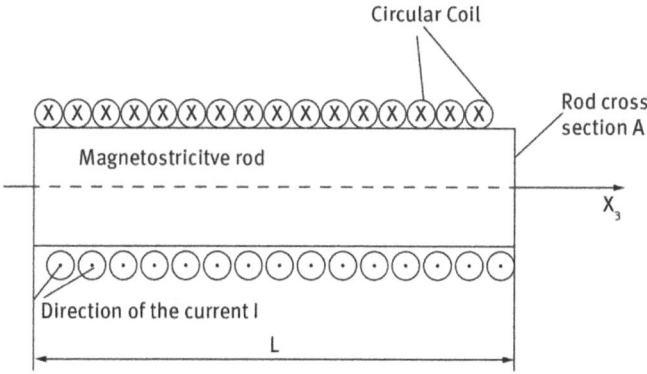

Figure 6.10: One-dimensional schematic model.

Let us assume an 1D model made of magnetostrictive material as depicted in Figure 6.10. The rod has a length L, a cross section A and an Young's modulus E. A tensile force P is applied to the rod, yielding a strain S and a stress $T = P/A$. Let us calculate the various energy contributions of the system presented in Figure 6.10. The strain energy accumulated in the rod due to the strain S and stress T is given by (see the strain–stress relation given in Equation (6.7)):

$$U_{strain} = \frac{1}{2}\int_v S \cdot T \cdot dv = \frac{1}{2}\int_v S \cdot \left(E^H \cdot S - eH\right) \cdot dv =$$

$$= \frac{1}{2}\int_v S \cdot E^H \cdot S \cdot dv - \frac{1}{2}\int_v S \cdot e \cdot H \cdot dv = \frac{A \cdot L}{2}E^H \cdot S^2 - \frac{A \cdot L}{2}S \cdot e \cdot H$$

$$(6.27)$$

where dv is the differential volume of the rod. The magnetic energy in the magnetostrictive rod is calculated as (see the B–H relation given in Equation (6.7))

$$U_{magn.} = \frac{1}{2}\int_v B \cdot H \cdot dv = \frac{1}{2}\int_v \left(e^* \cdot S + \mu^S H\right) \cdot H \cdot dv =$$

$$= \frac{1}{2}\int_v e^* \cdot S \cdot H \cdot dv + \frac{1}{2}\int_v H \cdot \mu^S \cdot H \cdot dv = \frac{A \cdot L}{2}e^* \cdot S \cdot H + \frac{A \cdot L}{2}\mu^S \cdot H^2$$

$$(6.28)$$

The external magnetic work done by a coil with N turns and a current I is given by

$$W_{coil} = I \cdot N \cdot \mu^T \cdot H \cdot A \tag{6.29}$$

Finally, the last contribution would be the mechanical work done by the force P

$$W_{mech} = P \cdot \Delta x_3 = P \cdot S \cdot L \tag{6.30}$$

The total potential energy would be the sum of all the above four contributions, namely

$$\pi = -U_{strain} + U_{magn.} - W_{coil} + W_{mech} \tag{6.31}$$

Substituting the various expressions (Equations (6.27)–(6.30)) into Equation (6.31) yields

$$\pi = -\frac{1}{2} A \cdot L \cdot E^H \cdot S^2 + A \cdot L \cdot e \cdot S \cdot H +$$
$$+ \frac{1}{2} A \cdot L \cdot \mu^S \cdot H^2 - I \cdot N \cdot \mu^T \cdot H \cdot A + P \cdot S \cdot L \tag{6.32}$$

Using a variational principle (like, for instance, the Hamilton principle[7]) with two variables, S and H, would give two linear equations in the following form:

$$-A \cdot L \cdot E^H S + A \cdot L \cdot e \cdot H + P \cdot L = 0 \tag{6.33}$$

$$A \cdot L \cdot e \cdot S + A \cdot L \cdot \mu^S \cdot H - I \cdot N \cdot \mu^T \cdot A = 0 \tag{6.34}$$

Division of Equations (6.32) and (6.33) by AL gives

$$E^H S - e \cdot H = \frac{P}{A} \tag{6.35}$$

$$e \cdot S + \mu^S \cdot H = \frac{I \cdot N \cdot \mu^T}{L} \tag{6.36}$$

Solving for the two unknowns, S and H, while using the relations $\mu^S = \mu^T - d \cdot E^H \cdot d^*$ and $e = E^H \cdot d$ yields

$$S = \frac{I \cdot N \cdot \mu^T \cdot A \cdot e + \mu^S \cdot P \cdot L}{A \cdot L \cdot \mu^T \cdot E^H} \tag{6.37}$$

$$H = \frac{I \cdot N}{L} - \frac{P}{A \cdot e} \left(1 - \frac{\mu^S}{\mu^T} \right) \tag{6.38}$$

Equation (6.37) can be simplified to

$$S = \lambda + S_T \tag{6.39}$$

---

7 $\delta\left(\int^\pi \cdot dt \right) = 0$, where t is time.

where λ is the magnetostriction defined as

$$\lambda = \frac{I \cdot N \cdot \mu^T \cdot A \cdot e}{A \cdot L \cdot \mu^T \cdot E^H} = \frac{I \cdot N \cdot d}{L} \tag{6.40}$$

and the term S_T is the mechanical strain defined as

$$S_T = \frac{\mu^S \cdot P}{A \cdot \mu^T \cdot E^H} = \frac{P}{A} \cdot \frac{1}{\overline{E}^H} \tag{6.41}$$

where

$$\overline{E}^H \equiv E^H \left(\frac{\mu^T}{\mu^S}\right)$$

where $\overline{E}^H$ is the modified Young's modulus. Note that due to the following relation, $\mu^S = \mu^T - d \cdot E^H \cdot d^*$, $\overline{E}^H$ can also be written as

$$\overline{E}^H \equiv E^H + \frac{e^2}{\mu^S} \tag{6.42}$$

Note that the modified Young's modulus $\overline{E}^H$ will be equal to E^H, when the ratio (μ^T/μ^S) will be equal to 1. For all other cases (including the Terfenol-D), this ratio is not equal to 1 and Young's modulus has a different value (larger). Similar to the piezoelectricity, one can define the "blocked force" as the highest force to be achieved. Its expression can be written as

$$F_{\text{blooking}} \equiv \frac{E^H \cdot A}{L} \Delta L = E^H \cdot A \cdot S_{\max} = E^H \cdot A (\lambda + S_T)_{\max} \approx E^H \cdot A \cdot \lambda_{\max} \tag{6.43}$$

In addition to what has been described in this chapter, one should be aware of other models aimed at describing the constitutive equations of a magnetostrictive material (see a concise review in [9]). Such approaches include microscopically motivated models, approximating the switching mechanisms of each single crystal, macroscopically constitutive model based on the phenomenological description of the material (mainly Preisach model [10], which was originally developed to describe magnetization of ferromagnetics; a wide range of hysteresis curves can be dealt by correctly choosing the various parameters of the model) and models based on the principles of thermodynamics.

References

[1] Olabi, A. G. and Grunwald, A., Design and application of magnetostrictive materials, Materials & Design 29(2), 2008, 469–483.

[2] Wan, Y., Fang, D. and Hwang, K.-C., Non-linear constitutive relations for magnetostrictive materials, International Journal of Non-Linear Mechanics 38(7), 2003, 1053–1065.

[3] Dong, S., Li, J. F. and Viehland, D., Longitudinal and transverse magnetoelectric voltage coefficients of magnetostrictive/piezoelectric laminate composite: theory, IEEE Transaction on Ultrasonics, Ferroelectrics, and frequency Control 50(10), October 2003, 1253–1281.

[4] Avakian, A. and Ricoeur, A., Phenomenological and physically motivated constitutive models for ferromagnetic and magnetostrictive materials, 7th ECCOMAS Thematic Conference on Smart Structures and Materials, SMART 2015, Araujo, A. L. and Mota Soares, C. A. et al. (eds.), 13.

[5] Du Trâemolet De Lacheisserie, E., Magnetostriction: Theory and Applications of Magnetoelasticity, Boca Raton, FL: CRC Press, 1993.

[6] Wang, L. and Yuan, F. L., Vibration energy harvesting by magnetostrictive material, Smart Materials and Structures 17(4), 2008, 1–14.

[7] IEEE standard on magnetostrictive materials: Piezomagnetic nomenclature IEEE STD 319–1990, 1991.

[8] Engdahl, G., Modeling of giant magnetostrictive materials, In: Handbook of Giant Magnetostrictive Materials, Engdahl, G. (ed.), Ch. 2, Elsevier, 20 October 1999.

[9] Linnemann, K., Klinkel, S. and Wagner, W., A constitutive model for magnetostrictive and piezoelectric materials, Mitteilung 2(1008), Institut für Baustatik, Universität Karlsruhe, 76128 Karlsruhe, Germany, 40.

[10] Preisach, F., Über die magnetische Nachwirkung, Zeitschrift für Physik A, Hadrons and Nuclei 94(5), 1935, 277–302.

6.3 Behavior of Electrostrictive Materials

Electrostriction is a property found in all dielectric materials, and analogically to magnetostrictive materials, is displaying slight displacements of ions in its crystal lattice when an external electric field is being applied leading to an overall strain in the direction of the applied field.

The resulting strain is proportional to the square of the polarization. Changing the polarity of the electric field (from + to −) does not change the direction of the strain. Mathematically, the relation is written as

$$S_{ij} = Q_{ijkl} \cdot P_k \cdot P_l \tag{6.44}$$

where S_{ij} is a second-order strain tensor, Q_{ijkl} is a fourth-order tensor of the electrostriction coefficients and P_k, P_l are first-order polarization tensors. One should note the quadratic relation to the polarization for electrostrictive materials compared to the linear relation for the piezoelectric materials (see Figure 6.11). Although all dielectrics exhibit some kind of electrostriction, only relaxor ferroelectrics, like lead magnesium niobate (PMN), lead magnesium niobate-lead titanate (PMN-PT) and lead lanthanum zirconate

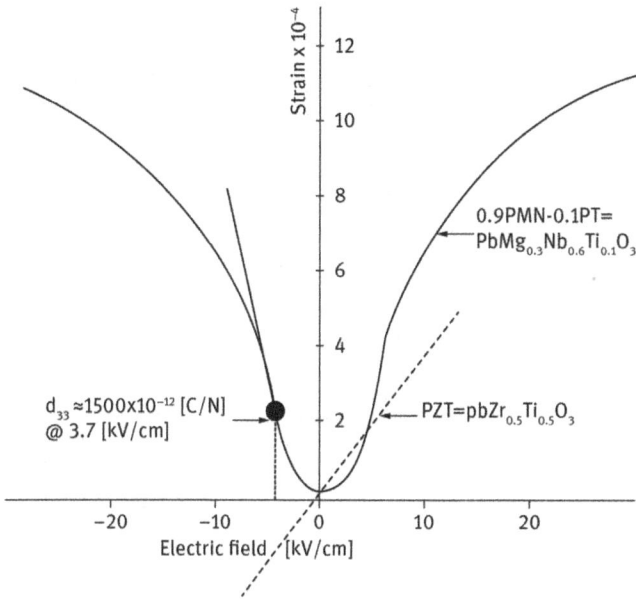

Figure 6.11: Electromechanical coupling in electrostrictive (PMN-PT) and piezoelectric (PZT) ceramics (from [1]).

Table 6.4: Typical electromechanical properties of electrostrictive ceramics in comparison with "soft" PZT (@ 25 °C).

Property	PT EC-97	PMN EC-98	"Soft" PZT EC-65
Density ρ (kg/m^3)	6700	7850	7500
Young's modulus (GPa)	128	61	66
Curie temperature (°C)	240	170	350
Mechanical Q (thin disk)	950	70	100
Dielectric const. @ 1 kHz	270	5500	1725
Dissipation const. @ 1 kHz (%)	0.9	2.0	2.0
k_{31}	0.01	0.35	0.36
k_p	0.01	0.61	0.62
k_{33}	0.53	0.72	0.72
k_{15}	0.35	0.67	0.69
d_{31} ($\times 10^{-12}$ m/V)	−3.0	−312	−173
d_{33} ($\times 10^{-12}$ m/V)	68.0	730	380

Table 6.4 (continued)

Property	PT EC-97	PMN EC-98	"Soft" PZT EC-65
d_{15} ($\times 10^{-12}$ m/V)	67.0	825	584
g_{31} ($\times 10^{-3}$ V m/N)	−1.7	−6.4	−11.5
g_{33} ($\times 10^{-3}$ V m/N)	32.0	15.6	25.0
g_{15} ($\times 10^{-3}$ V m/N)	33.5	17.0	38.2
s_{11}^E ($\times 10^{-12}$ m^2/N)	–	16.3	15.2
s_{12}^E ($\times 10^{-12}$ m^2/N)	–	−5.6	−5.3
s_{33}^E ($\times 10^{-12}$ m^2/N)	7.7	21.1	18.3
s_{11}^D ($\times 10^{-12}$ m^2/N)	–	14.3	13.2
s_{12}^D ($\times 10^{-12}$ m^2/N)	–	−7.6	−7.3
s_{33}^D ($\times 10^{-12}$ m^2/N)	–	10.2	8.8
Poisson's ratio υ	–	0.34	0.31
Aging rate % change per time decade			
Dielectric constant	−0.3	−1.5	−0.8
Coupling constant	−0.4	−0.4	−0.3
Resonant frequency	0.05	0.4	0.2
Electric field dependence			
Maximum positive field (V/mm)	–	900 (DC only)	600
Maximum negative field (V/mm)	–	450 (DC only)	300
Applied field @ 25 °C	79	79	79
Dielectric constant % increase	1.5	22.5	12.0
Dissipation factor	0.8	6.2	7.3

Note: A transverse isotropy is assumed, namely $(\ldots)_{23} = (\ldots)_{13}$.

titanate (PLZT) have very high electrostrictive constants. One should note that due to the relationship presented in Equation (6.44), the application of a mechanical stress would not generate an electric charge, which means that unlike piezoelectric materials, electrostrictive ones cannot be used as sensors. Comparing piezoelectric material to electrostrictive ones reveals that PZTs offer low strains (~0.06%) with significant hysteresis (~15 – 20%), whereas electrostrictive PMN materials would exhibit higher strains (~0.1%) with lower hysteresis (~1 – 4%). However, due to the temperature operating limits PMN would require thermal insulation. For instance, TRS Company[8] developed three PMN-PT compositions for sonar applications, PMN-15, PMN-38 and PMN-85, having operating temperature ranges of 0–30, 10–50 and 75–95 °C, respectively. PMN-15 presents almost 0.14% strain at an electrical field of 30 kV/cm at 2 °C, while at 22 °C the strain drops to 0.115% for the same electrical field.

Actuators   Stacks in preloaded casings

Figure 6.12: Electrostrictive actuators. From Piezomechanik (Piezomechanik Dr. Lutz Pickelmann GmbH, Berg-am-Laim-Str. 64, D-81673 Munich, Germany, www.piezomechanick.com).

Newport AD-30 Ultra-resolution   Newport AD-100 Ultra-resolution
electrostricitve actuator   electrostricitve actuator

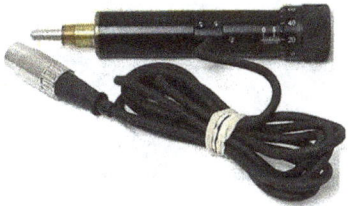

Figure 6.13: Newport electrostrictive actuators.

8 [90] www.trstechnologies.com/Materials/Electrostrictive-Ceramics

Typical electromechanical properties of electrostrictive ceramics can be found in Table 6.4 for lead titanate PT EC-97 and lead magnesium niobate PMN EC-98[9] and for "soft" PZT manufactured by the same company. Compared to the research performed on piezoelectricity, the studies on electrostrictive materials are more limited (see typical references [1–9]). Figure 6.12 presents actuators and stacks manufactured by a German company, Piezomechanik Dr. Lutz Pickelmann GmbH, showing reduced creep and precise reset to zero point by simple voltage control and claiming that these effects are at least 1 order of magnitude smaller than for piezo-elements. The electrostrictive-based stacks are capable to deliver strokes of 6–25 μm with a stiffness ranging from 50 down to 12 N/μm, or higher strokes of 6–40 μm with a stiffness of 120–20 N/μm, while applying voltages up to 120 V. Figure 6.13 presents two types of electrostrictive actuators (from Newport Corporation[10]): one with 30 μm stroke (AD-30) and the second one, AD-100, having 100 μm stroke, a resolution of 0.04 μm and capable of delivering 45 N.

Finally, AOA Xinetics[11] manufactures Lead Magnesium Niobate (PMN: RE) electrostrictive multilayer actuators with PMN layer thickness ranging from 0.1 to 0.15 mm (see Figure 6.14). Typical performance specifications include the following:

– Optimal temperature range: 10–30 °C
– Typical operating voltages: 0–100 V
– Typical hysteresis between 0 and 100 V: <5%
– Optical creep at room temperature: $\sim\lambda/2000$ @ 0.63 μm
– Resolution: μm
– Operating frequency: Driver dependent
– Stroke range: from 3.1 to 30 μm @ 100 V and 25 °C

Figure 6.14: AOA Xinetics electrostrictive multilayer PMN actuators.

9 From EXELIS Inc. (the original data is from EDO Corp., which is currently a subsidiary of Exelis Inc.) Electro-Ceramic Products and Material Specifications Catalogue, www.exelisinc.com
10 www.newport.com
11 http://www.northropgrumman.com/BusinessVentures/AOAXinetics/IntelligentOptics/Products/Pages/Actuators.aspx

References

[1]   Kay, H. F., Electrostriction, Reports on Progress in Physics 18, 1955, 230–250.
[2]   Cross, L. E., Piezoelectric and electrostrictive sensors and actuators for adaptive structures and smart materials, Proc. AME 110th Annual Meeting., San Francisco, USA, December, 1989.
[3]   Zhang, Q., Pan, W., Bhalla, A. and Cross, L. E., Electrostrictive and dielectric response in lead magnesium niobate-lead titanate (0.9PMN, 0.1PT) and lead lanthanum zirconate titanate (PLZT 9.5/65/35) under variation of temperature and electric field, Journal American Ceramic Society 72(4), 1989, 599–604.
[4]   Newnham, R. E., Xu, Q. C., Kumar, S. and Cross, L. E., Smart ceramics, Ferroelectrics 102, 1990, 77–89.
[5]   Cross, L. E., Newnham, R. E., Bhalla, A. S., Dougherty, J. P., Adair, J. H., Varadan, V. K. and Varadan, V. V., Piezoelectric and electrostrictive materials for transducers applications, FINAL REPORT ad-a250 889, Vol. 1, Office of Naval Research, PennState, The Materials Research Laboratory, University Park, PA, USA, June 3, 1992.
[6]   Sherrit, S. and Mukherjee, B. K., Electrostrictive materials: characterization and application for ultrasound, Proc. of the SPIE Medical Imaging Conference, Vol. 3341, San Diego, California, USA, February 1998.
[7]   Waechter, D. F., Liufu, D., Camirand, M., Blacow, R. and Prasad, S. E., Development of high-strain low hysteresis actuators using electrostrictive lead magnesium niobate (PMN), Proc. 3rd CanSmart Workshop on Smart Materials and Structures, 28–29 September 2000, St-Hubert, Quebec, Canada, 31.
[8]   Giurgiutiu, V., Pomirleanu, R. and Rogers, C. A., Energy-based comparison of solid- state actuators, Report # USC-ME-LAMSS-200-102, Laboratory for Adaptive Materials and Smart Structures, University of South Carolina, Columbia, SC 299208, USA, March 1, 2000.
[9]   Ursic, H., Zarnik, M. S. and Kosec, M., $(Pb(Mg_{1/3}Nb_{2/3})O_3$- $PbTiO_3PMN-PT)$ material for actuator applications, Smart Material Research 2011, article ID 452901, 6, doi:10.1155/2011/452901.

6.4 Constitutive Equations of Electrostrictive Materials

As already mentioned in Section 6.3, the electrostrictive effect is manifested as a quadratic dependence of the strain (stress) on the polarization P (or applied electric field) as described schematically in Figure 6.15. Any dielectric material, be it ceramics or polymers possessing a center of inversion or without it, would display the electrostrictive effect.

Mathematically (see [1–10]), the relationship between the strain S_{ij} and the polarization tensor P_k (or P_l) is written as (assuming no mechanical stress is applied)

$$S_{ij} = Q_{ijkl} \cdot P_k \cdot P_l \tag{6.45}$$

where Q_{ijkl} in units of m^4/C^2, is the charge-related electrostrictive coefficients. For isotropic electrostrictive materials (like polymers) Equation (6.45) simplifies into

$$S_1 = Q_{13} \cdot P^2; \quad S_2 = Q_{23} \cdot P^2; \quad S_3 = Q_{33} \cdot P^2 \tag{6.46}$$

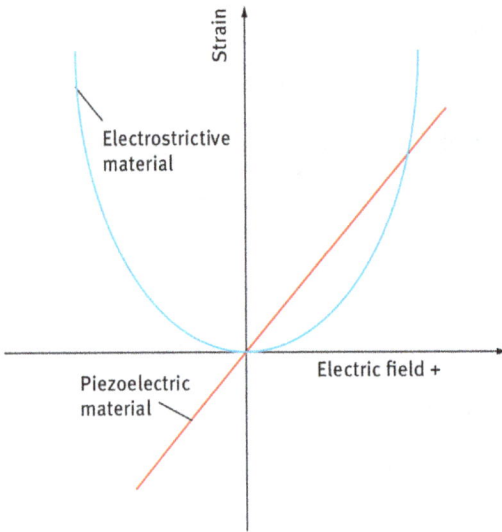

Figure 6.15: Strain versus applied electric field for electrostrictive and piezoelectric materials – a schematic view.

with $S_1, S_2 (S_1 = S_2$ plane isotropy) being the transverse strain (perpendicular to the polarization axis) and s_3 is the longitudinal strain (parallel to the polarization axis). Another formulation (see Figure 6.15) is to connect the strain to the applied electrical field (without mechanical stress) yielding

$$S_{ij} = M_{ijkl} \cdot E_k \cdot E_l \tag{6.47}$$

whereas M_{ijkl} having units of m^2/V^2 is the electrostriction coefficients (in tensor notation), S_{ij} is the strain and E_k (or E_l) is the applied field. For a linear electrostrictive material the coefficients M and Q relate as

$$M_{ijkl} = Q_{mnkl} \cdot \chi_{mi} \cdot \chi_{nj} \tag{6.48}$$

where χ_{mi} (or χ_{nj}) is called the dielectric susceptibility.

One should note that the polarization P_m is related to the electric field E_m by the following relation:

$$P_m = (\varepsilon - \varepsilon_0)E_m = D_m - \varepsilon_0 \cdot E_m \tag{6.49}$$

where ε_0 is the dielectric vacuum permittivity ($= 8.85 \times 10^{-12}$F/m), ε is the dielectric permittivity elsewhere and D_m is the electrical displacement.

Another way of writing the relationship between the polarization and the electrical field is by

$$P_k = \chi_{kl} \cdot E_l \tag{6.50}$$

where χ_{kl} is, as defined before, the dielectric susceptibility tensor [10].

Depending on what independent variables we would like to define and using a phenomenological description we can write, for example, the potential energy as

$$\pi(T, E) = -\frac{1}{2} s_{ijkl} T_{ij} T_{kl} - M_{ijkl} T_{ij} E_k E_l - \frac{1}{2} \varepsilon_{ij} E_i E_j \tag{6.51}$$

leading to the electrostrictive nonlinear coupled electromechanical constitutive equations:

$$S_{ij} = s_{ijkl} \cdot T_{kl} + M_{ijkl} \cdot E_k \cdot E_l$$
$$D_i = \varepsilon_{ij} \cdot E_j + 2 \cdot M_{klij} \cdot T_{kl} \cdot E_j \tag{6.52}$$

with S_{ij} and T_{ij} standing for the strain and the stress, respectively; s_{ijkl}, M_{ijkl} and ε_{ij} represent the elastic compliance, electrostrictive coefficients and the dielectric permittivity matrix. Like for the case of piezoelectric materials, D_i and E_j are the electrical displacement and the electrical field, respectively.

If one is writing the Gibbs free energy expression,[12] with the stress and the polarization (P) being the independent variables, while neglecting higher order terms, we get

$$\Delta G(T, P) = -\frac{1}{2} s_{ijkl} \cdot T_{ij} \cdot T_{kl} - Q_{mnij} \cdot P_m \cdot P_n \cdot T_{ij} + \frac{1}{2} \gamma_{mn} \cdot P_m \cdot P_n$$

where

$$\tag{6.53}$$

$$S_{ij} \equiv -\left(\frac{\partial \Delta G}{\partial T_{ij}}\right)_{@P,\,\text{Temp.}\,=\,\text{const}} \quad ; \quad E_m \equiv \left(\frac{\partial \Delta G}{\partial P_m}\right)_{@T,\,\text{Temp.}\,=\,\text{const}}$$

from which the following constitutive equations can be written as

$$S_{ij} = s_{ijkl} \cdot T_{kl} + Q_{mnij} \cdot P_i \cdot P_j$$
$$P_m = \chi_{mn} \cdot E_n + 2 \cdot M_{mnij} \cdot E_n \cdot T_{ij} \tag{6.54}$$

---

[12] Gibbs free energy or Gibbs energy or Gibbs function or free enthalpy is a thermodynamic potential that measures the maximum or reversible work that may be performed by a thermodynamic system at a constant temperature.

where γ_{mn} is the linear reciprocal[13] dielectric susceptibility. Again using the Gibbs free energy expression with the following definitions:

$$S_{ij} \equiv -\left(\frac{\partial \Delta G}{\partial T_{ij}}\right)_{@E, \text{Temp.} = \text{const}} \quad ; \quad P_m \equiv -\left(\frac{\partial \Delta G}{\partial E_m}\right)_{@T, \text{Temp.} = \text{const}} \tag{6.55}$$

we can get the constitutive equations written as

$$S_{ij} = s_{ijkl} \cdot T_{kl} + M_{mnij} \cdot E_m \cdot E_n$$
$$P_m = \chi_{mn} \cdot E_n + 2 \cdot M_{mnij} \cdot E_n \cdot T_{ij} \tag{6.56}$$

One should note that without external mechanical stress ($T_{kl} = 0$), we obtain from Equations (6.52), (6.54) and (6.56)

$$S_{ij} = M_{ijkl} \cdot E_k \cdot E_l$$

or $\qquad\qquad\qquad\qquad\qquad\qquad\qquad\qquad\qquad\qquad\qquad\qquad\qquad$ (6.57)

$$S_{ij} = Q_{mnij} \cdot P_i \cdot E_j$$

which are exactly Equations (6.47) and (6.45) written at the beginning of this chapter.

Note that the converse effect for electrostrictive materials can be written as

$$Q_{mnij} = -\frac{1}{2}\frac{\partial \gamma_{mn}}{\partial T_{ij}}; \quad M_{mnij} = \frac{1}{2}\frac{\partial \chi_{mn}}{\partial T_{ij}} \tag{6.58}$$

Another converse effect deals with the dependence of the electric field on the piezo-electric coefficient d_{mij}, namely

$$M_{mnij} = \frac{1}{2}\left[\frac{\partial d_{mij}}{\partial E_n}\right]$$

and $\qquad\qquad\qquad\qquad\qquad\qquad\qquad\qquad\qquad\qquad\qquad\qquad\qquad$ (6.59)

$$Q_{mnij} = \frac{1}{2}\left[\frac{\partial d_{mij}}{\partial P_n}\right]$$

leading to the following relations:

$$d_{ijm} \equiv \left[\frac{dS_{ij}}{dE_m}\right] = 2 \cdot M_{ijmn} \cdot E_n$$

$$g_{ijm} \equiv \left[\frac{dS_{ij}}{dP_m}\right] = 2 \cdot Q_{ijmn} \cdot P_n \tag{6.60}$$

$$d_{ijm} = \chi_{mk} \cdot g_{ijm} = 2 \cdot \chi_{mk} \cdot Q_{ijmn} \cdot P_n$$

---

13 [95] Note: [EQN3368].

The explicit expression for Equation (6.45) (for PMN crystals having the point group $m3m$) is given by

$$\begin{Bmatrix} S_1 \\ S_2 \\ S_3 \\ S_4 \\ S_5 \\ S_6 \end{Bmatrix} = \begin{bmatrix} Q_{11} & Q_{12} & Q_{12} & 0 & 0 & 0 \\ Q_{12} & Q_{11} & Q_{12} & 0 & 0 & 0 \\ Q_{12} & Q_{12} & Q_{11} & 0 & 0 & 0 \\ 0 & 0 & 0 & Q_{44} & 0 & 0 \\ 0 & 0 & 0 & 0 & Q_{44} & 0 \\ 0 & 0 & 0 & 0 & 0 & Q_4 \end{bmatrix} \begin{Bmatrix} P_1^2 \\ P_2^2 \\ P_3^2 \\ P_2 P_3 \\ P_3 P_1 \\ P_1 P_2 \end{Bmatrix} \tag{6.61}$$

while

$$Q_{44} = 4Q_{1212} = 2(Q_{11} - Q_{12})$$

Another form for the constitutive equations was proposed in [5–9]. Assuming no saturation of the electric field, which will require the addition of higher order terms, the constitutive equations are written in the following form:

$$\{S\} = [s^D]\{T\} + [g(Q,D)]\{D\}$$
$$\{E\} = -2[g(Q,D)]\{T\} + [\beta^T]\{D\} \tag{6.62}$$

where $[\beta^T] = [\varepsilon^T]^{-1}$ is the dielectric impermeability matrix at constant stress; $\{S\}$, $\{T\}$ are the condensed strain and stress tensors; $\{E\}$ and $\{D\}$ are the electric field and electric displacement vectors, $[s^D]$ is the elastic compliance matrix at constant electrical displacement; and $[g(Q,D)]$ is defined as

$$[g] = \begin{bmatrix} Q_{11}D_1 & Q_{12}D_1 & Q_{12}D_1 & 0 & (Q_{11}-Q_{12})D_3 & (Q_{11}-Q_{12})D_2 \\ Q_{12}D_2 & Q_{11}D_2 & Q_{12}D_2 & (Q_{11}-Q_{12})D_3 & 0 & (Q_{11}-Q_{12})D_1 \\ Q_{12}D_3 & Q_{12}D_3 & Q_{11}D_3 & (Q_{11}-Q_{12})D_2 & (Q_{11}-Q_{12})D_1 & 0 \end{bmatrix} \tag{6.63}$$

It was found that the electrostriction phenomena described schematically in Figure 6.15 is not exactly what happens in experiments, where above a certain electrical field the strain graph flattens out showing saturation of the electrical field. Figure 6.16 shows this behavior, where E_s is the saturated electric field.

For this case, the polarization versus the electrical field is written as [4–11]

$$P = (\varepsilon - \varepsilon_0)E_s \tanh\left[\frac{E}{E_s}\right] = \chi E_s \tanh\left[\frac{E}{E_s}\right] \tag{6.64}$$

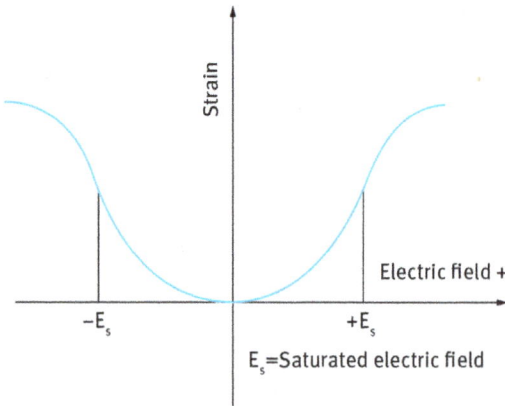

Figure 6.16: Strain versus applied electric field for electrostrictive materials with saturated electric field (E_s) – a schematic view.

Using Equation (6.45) one can write the strain versus the polarization for the saturation case as

$$\{S\} = [Q]\{P\}^2 = [Q](\varepsilon - \varepsilon_0)^2 E_s^2 \tanh^2\left[\frac{E}{E_s}\right] = [Q]\chi^2 E_s^2 \tanh^2\left[\frac{E}{E_s}\right] \tag{6.65}$$

Therefore, the range of the electrical field is divided into two regions as follows:

$$
\begin{aligned}
&\text{I.} \quad \text{For} E \leq E_s \Rightarrow P \approx \chi E \quad \text{and} \quad S = ME^2 \\
&\text{II.} \quad \text{For} E \geq E_s \Rightarrow P \approx \chi E_s \quad \text{and} \quad S = ME_s^2 = \text{const}
\end{aligned}
\tag{6.66}
$$

For the saturation case, the constitutive equations can be written in Voight notation[14] as (see [4–8])

$$T_i = c_{ij}^P \cdot S_j - c_{ij} \cdot Q_{jmn} \cdot P_m \cdot P_n$$

$$E_m = -2c_{ij}^P \cdot Q_{jmn} \cdot S_i \cdot P_n + \chi_m^{-1} \cdot P_m^S \arctan h\left(\frac{P_m}{P_m^S}\right) + 2c_{ij}^P \cdot Q_{jmn} \cdot Q_{ikl} \cdot P_n \cdot P_k \cdot P_l \tag{6.67}$$

where c_{ij}^P is the stiffness matrix at constant polarization P and P_m^S is the saturated polarization. For the simple case of an isotropic material, the stiffness matrix is written as

---

14 Voigt notation: $\{\sigma_{xx}, \sigma_{yy}, \sigma_{zz}, \tau_{yz}, \tau_{xz}, \tau_{xy}\} \equiv \{T_1, T_2, T_3, T_4, T_5, T_6\}$.

$$\left[c_{ij}^P\right] = \frac{E_{\text{Young's}}}{(1+v)(1-2v)} \begin{bmatrix} 1-v & v & v & 0 & 0 & 0 \\ v & 1-v & v & 0 & 0 & 0 \\ v & v & 1-v & 0 & 0 & 0 \\ 0 & 0 & 0 & \frac{1-2v}{2} & 0 & 0 \\ 0 & 0 & 0 & 0 & \frac{1-2v}{2} & 0 \\ 0 & 0 & 0 & 0 & 0 & \frac{1-2v}{2} \end{bmatrix} \quad (6.68)$$

while the electrostrictive coefficients are all zero except

$$Q_{111} = Q_{222} = Q_{333}$$

$$Q_{123} = Q_{133} = Q_{211} = Q_{233} = Q_{311} = Q_{322} \quad (6.69)$$

$$2(Q_{111} - Q_{222}) = Q_{412} = Q_{523} = Q_{613}$$

For a 1D analysis, Equation (6.47) is written as

$$S = ME^2 \quad (6.70)$$

To facilitate the use of electrostrictive materials, the quadratic relation (Equation 6.70) might pose problems in realizing actuators. The aim is to transform the quadratic relation to a linear one. This is done by superimposing on a direct current (dc) field and alternating current (ac), namely

$$E = E_{\text{dc}} + E_{\text{ac}} \quad (6.71)$$

Substituting Equation (6.71) back into Equation (6.70) yields

$$S = M(E_{\text{dc}} + E_{\text{ac}})^2 = M\left(E_{\text{dc}}^2 + 2E_{\text{dc}}E_{\text{ac}} + E_{\text{ac}}^2\right) \quad (6.72)$$

Noting that $ME_{\text{dc}}^2 = \text{const} \equiv S_{\text{dc}}$ one can rewrite Equation (6.72) as

$$S - S_{\text{dc}} \approx M(2E_{\text{dc}}E_{\text{ac}} + E_{\text{ac}}^2) \quad (6.73)$$

Assuming that the superimposed alternating field is much smaller the product of the dc field and the ac field, namely, $E_{\text{ac}}^2 \ll 2E_{\text{dc}}E_{\text{ac}}$ one obtains a linear relationship between the strain and the ac field in the form of

$$S - S_{\text{dc}} \approx M(2E_{\text{dc}}E_{\text{ac}}) \quad (6.74)$$

Defining the term $2ME_{\text{dc}} \equiv d_{\text{eff}}$, where d_{eff} is the effective piezoelectric strain coefficient, one obtains

$$S - S_{\text{dc}} \approx d_{\text{eff}}E_{\text{ac}} \quad (6.75)$$

To compare the properties of the electrostrictive materials with piezoelectric ones, one has to compare d_{eff} to d of a piezoelectric material acting in the same directions.

Finally, a simple 1D solution for the case of an electrostrictive bar presented in Figure 6.17 will be derived using Equations (6.62) (see also [5]). The dimensions of the rod are length L, width b and thickness t, such that $t \ll L$ and $b \ll L$. The only non-zero stress is T_3 ($T_1 = T_2 = T_4 = T_5 = T_6 = 0$). The only electrical displacement is D_3 while $D_1 = D_2 = 0$ (as the electrodes are perpendicular to x_3-axis). For the 1D case, Equation (6.62) simplifies into

$$S_3 = s_{33}^D T_3 + Q_{11} D_3^2$$

$$E_3 = -2Q_{11} T_3 D_3 + \beta_{33}^T D_3$$

(6.76)

From equilibrium of forces along the x_3-axis, one can write that the stress T_3 is constant across all cross sections perpendicular to x_3-axis, namely

$$T_3 = \frac{F}{(b \cdot t)} = \text{const}$$

(6.77)

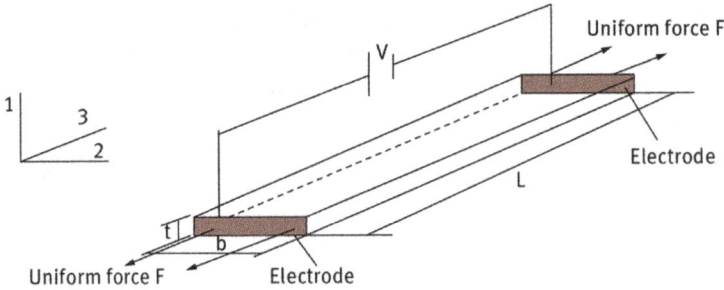

Figure 6.17: A schematic view of the electrostrictive bar with geometric dimensions.

Applying the Gauss theorem $((\partial D_3/\partial x_3) = 0)$ one concludes that $D_3 = \text{const.}$ along the electrostrictive rod. Accordingly, from the second equation in Equations (6.76), the electrical field is also constant and equals to

$$E_3 = -\frac{V}{L}$$

(6.78)

Then we can solve for D_3 to yield

$$D_3 = -\frac{V}{L\left(\beta_{33}^T - \frac{2Q_{11}F}{b \cdot t}\right)}$$

(6.79)

Substituting into the expression of D_3 into the first equation of Equations (6.76) one obtains the strain S_3 as

$$S_3 = s_{33}^D \frac{F}{b \cdot t} + \frac{Q_{11} \cdot V^2}{L^2 \left(\beta_{33}^T - \frac{2Q_{11}F}{b \cdot t} \right)^2} \tag{6.80}$$

References

[1]   Zhang, Q., Pan, W., Bhalla, A. and Cross, L. E., Electrostrictive and dielectric response in lead magnesium niobate-lead titanate(0.9PMN.0.1PT) and lead lanthanum zirconate titanate (PLZT 9.5/65/35) under variation of temperature and electric field, Journal American Ceramic Society 72(4), 1989, 599–604.

[2]   Damjanovic, D. and Newnham, R. E., Electrostrictive and piezoelectric materials for actuator applications, Journal of Intelligent Materials Systems and Structures 3, April 1992, 190–208.

[3]   Sundar, V. and Newnham, R. E., Electrostriction and polarization, Ferroelectrics 135, 1992, 431–446.

[4]   Hom, C. L. and Shankar, N., A fully coupled constitutive model for electrostrictive materials, Journal of Intelligent materials Systems and Structures 5, 1994, 795–801.

[5]   Debus, J.-C., Dubus, B., McCollum, M. and Black, S., Finite element modeling of PMN electrostrictive materials, Paper presented at the 3rd ICIM/ECSSM '96, Lyon, France, 1996.

[6]   Newnham, R. E., Sundar, V., Yimnirun, R., Su, J. and Zhang, Q. M., Electrostriction: nonlinear electromechanical coupling in solid dielectrics. Journal of Physical Chemistry B 101, 1997, 10.141–10.150.

[7]   Coutte, J., Debus, J.-C., Dubus, B., Bossut, R., Granger, C. and Haw, G., Finite element modeling of PMN electrostrictive materials and application to the design of transducers, Proc. of the 1998 IEEE International Frequency Control Symposium, IEEE catalog No. 98CH36165, Ritz-Carlton Hotel, Pasadena, California, USA, 27–29 May, 1998.

[8]   Pablo, F. and Petitjean, B., Characterization of 0.9PMN-0.1PT patches for active vibration control of plate host structures, Journal of Intelligent materials Systems and Structures 11(11), 2000, 857–867.

[9]   Coutte, J., Dubus, B., Debus, J.-C., Granger, C. and Jones, D., Design, production and testing of PMN-PT electrostrictive transducers, Ultrasonics 40, 2002, 883–888.

[10]   Li, J. Y. and Rao, N., Micromechanics of ferroelectric polymer-based electrostrictive composites, Journal of the Mechanics and Physics of Solids 52, 2004, 591–615.

[11]   Lallart, M., Capsal, J.-F., Kanda, M., Galineau, J., Guyomar, D., Yuse, K. and Guiffard, B., Modeling of thickness effect and polarization saturation in electrostrictive polymers, Sensors and Actuators B: Chemical 171–172, 2012, 739–746.

6.5 Appendix IV

6.5.1 Appendix A

To understand the phenomenon of magnetostriction, one needs some background of basic concepts in magnetism and electrical circuit, which is brought in this appendix.

The magnetic field is a vector field and is denoted by two closely related symbols **B** and **H**. **B** is called magnetic field, or alternatively magnetic flux density or magnetic induction and is measured in teslas (T) or newton per ampere per meter (N/A m), in SI units while **H** is named magnetic field, or magnetization field, magnetic field intensity or magnetic field strength and is measured in units of amperes per meter (A/m) in SI units. In CGS units, **B** is measured in gauss (G)[15] and **H** in oersteds[16] (Oe).

B is most commonly defined in terms of the Lorentz force it exerts on moving electric charges, namely

$$\vec{F} = q\left(\vec{E} + \vec{v} \times \vec{B}\right) \tag{6.a}$$

where $\vec{F}, q, \vec{E}, \vec{v}, \vec{B}$ are force, charge of the particle, electric field and magnetic field, respectively, and × denotes the cross product. Accordingly, a charge of 1 C passing through a magnetic field of 1 T, at a speed of 1 m/s, would experience a 1 N force. Therefore, the following relations hold:

$$1\,\text{T} = 1\frac{\text{V}\cdot\text{s}}{\text{m}^2} = 1\frac{\text{N}}{\text{A}\cdot\text{m}} = 1\frac{\text{J}}{\text{A}\cdot\text{m}^2} = 1\frac{\text{H}\cdot\text{A}}{\text{m}^2} = 1\frac{\text{Wb}}{\text{m}^2} = 1\frac{\text{kg}}{\text{C}\cdot\text{s}} = 1\frac{\text{N}\cdot\text{s}}{\text{C}\cdot\text{m}} = 1\frac{\text{kg}}{\text{A}\cdot\text{s}^2} \tag{6.b}$$

where A represents ampere, C represents coulomb, H represents henry (a unit of electrical inductance) and Wb represents weber (a unit of magnetic flux, Φ_B).

Another way to define B is to use the Biot–Savart law, which defines the magnetic field due to a current I on a conducting element having the length dl, namely

$$\vec{B}(r) = \frac{\mu_0}{4\pi} \int_C \frac{I d\vec{l} \times \hat{r}}{r^2} \tag{6.c}$$

where μ_0 is defined as the vacuum permeability and its value is

$$\mu_0 = 4\pi \times 10^{-7} \frac{\text{N}}{\text{A}^2} \approx 1.25666370614\ldots \times 10^{-6} \frac{\text{H}}{\text{m}} \text{ or } \frac{\text{T}\cdot\text{m}}{\text{A}} \text{ or } \frac{\text{Wb}}{\text{A}\cdot\text{m}} \text{ or } \frac{\text{V}\cdot\text{s}}{\text{A}\cdot\text{m}} \tag{6.d}$$

---

15 1 T = 10,000 G.

16 To honor the Danish physicist Hans Christian Ørsted who discovered the relationship between magnetism and electric current when a magnetic field produced in a current-carrying coil deflected an ammeter (an instrument used to measure current) when it was switched on and off.

For an infinite straight conductor (Figure 6.a), with a current I, the magnetic field at a vertical distance d from it will be given as

$$B = \frac{\mu_0 I}{2\pi d}$$ (6.e)

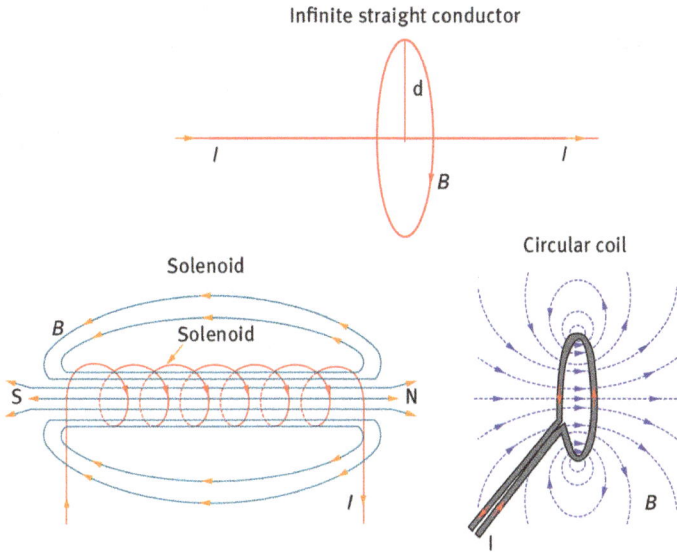

Figure 6.a: Various configurations to define the magnetic field B. Source: ww.thunderbolts.info.

Referring again to Figure 6.a, for the single coil case, having a radius R, the magnetic field at its center would be given as

$$B = \frac{\mu_0 I}{2R}$$ (6.f)

while for a solenoid (Figure 6.a) having N coils the expression is

$$B = \frac{\mu_0 NI}{L} \equiv \mu_0 nI; \quad n = \frac{N}{L}$$ (6.g)

where n is the number of coils per unit length, N is the number of coils and L is its length.

The general relation between H and B can be written as

$$\vec{H} \equiv \frac{\vec{B}}{\mu_0} - \vec{M}$$ (6.h)

where $\vec{M}$ is the magnetization vector field representing how strongly a region of material is magnetized and is defined as the net magnetic dipole moment per unit volume of that region. For highly permeable materials (a material with large induced magnetic flux) the relation reduces to (see Figure 6.b)

$$\vec{B} \equiv \mu\vec{H} \tag{6.i}$$

with μ being the permeability of the material, measured in SI unit by H/m or N/A^2. A relative permeability is defined as

$$\mu_r \equiv \frac{\mu}{\mu_0} \tag{6.j}$$

where $\mu_0 = 4\pi \times 10^{-7} \dfrac{N}{A^2}$. Another term is the magnetic susceptibility defined as

$$\chi_m \equiv \mu_r - 1 \tag{6.k}$$

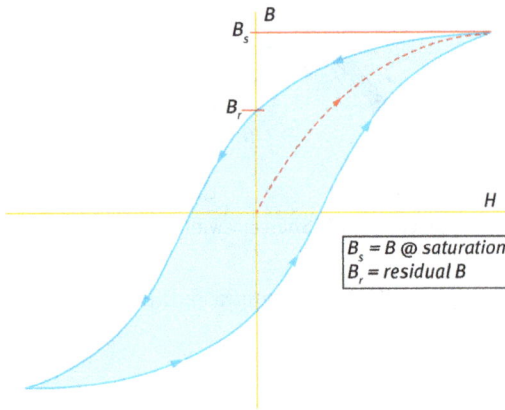

Figure 6.b: A typical B–H hysteresis curve.

6.5.2 Appendix B

Figures 6.c–6.m present data from one of the most known companies, Etrema Products, Inc. for the two giant magnetostrictive material Terfenol-D and Galfenol manufactured by the company. Various influences are presented as a function of various parameters.

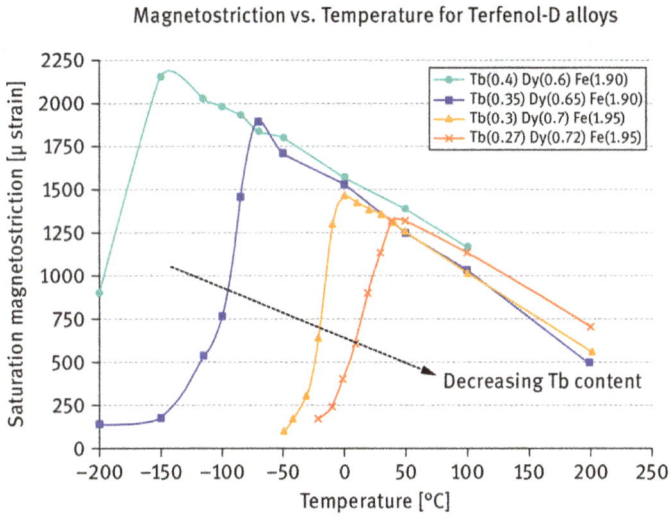

Figure 6.c: Terfenol-D alloys: temperature influence. Source: ETREMA Proprietary.

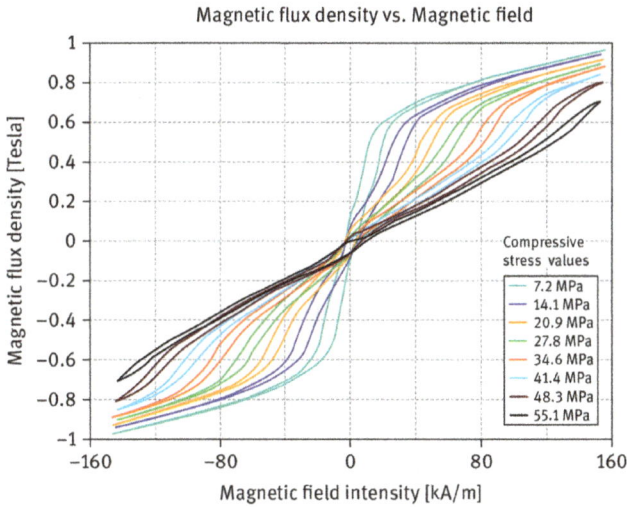

Figure 6.d: Typical Terfenol-D alloy: magnetic flux density versus magnetic intensity for various compressive stresses. Source: ETREMA Proprietary.

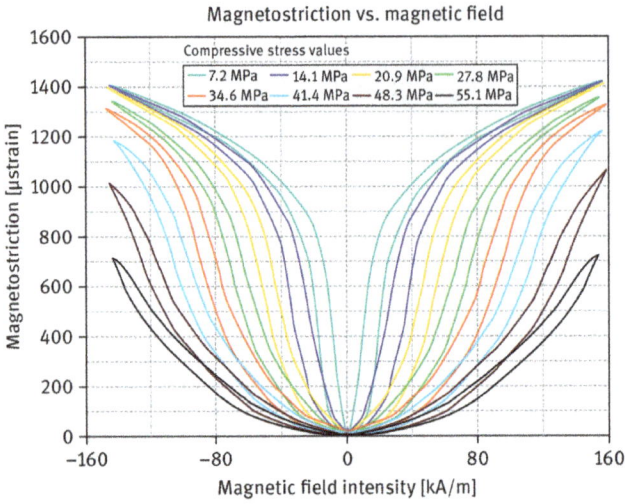

Figure 6.e: Typical Terfenol-D alloy: magnetostriction versus magnetic field for various compressive stresses. Source: ETREMA Proprietary.

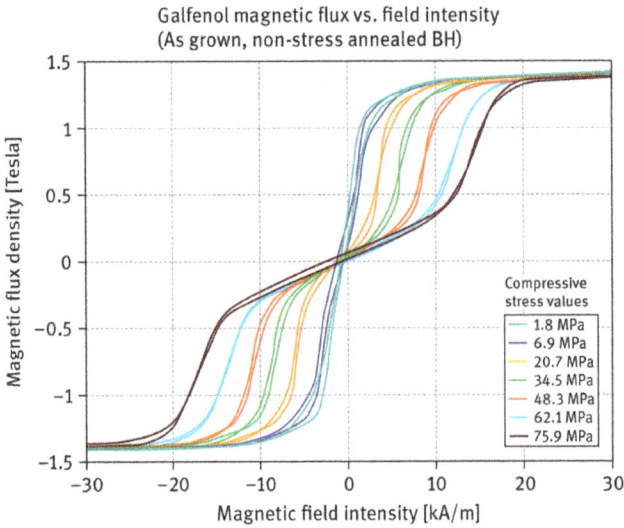

Figure 6.f: Typical Galfenol alloy (as grown, nonstress annealed BH): magnetic flux density versus magnetic field intensity for various compressive stresses. Source: ETREMA Proprietary.

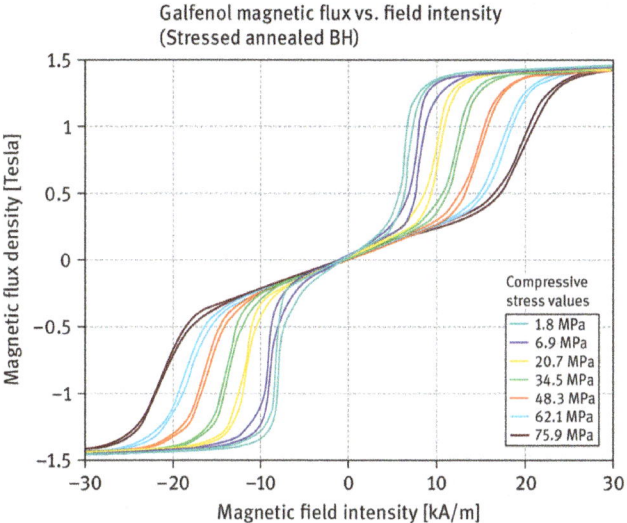

Figure 6.g: Typical Galfenol alloy (stressed annealed BH): magnetic flux density versus magnetic field intensity for various compressive stresses. Source: ETREMA Proprietary.

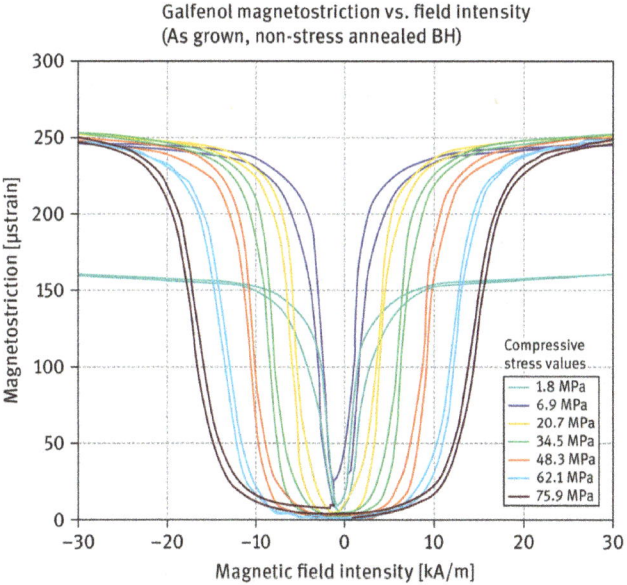

Figure 6.h: Typical Galfenol alloy (as grown, nonstress annealed BH): magnetostriction versus magnetic field intensity for various compressive stresses. Source: ETREMA Proprietary.

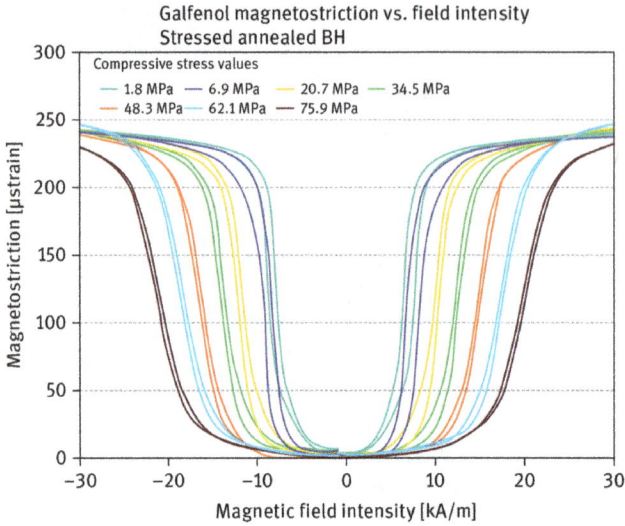

Figure 6.i: Typical Galfenol alloy (stressed annealed BH): magnetostriction versus magnetic field intensity for various compressive stresses. Source: ETREMA Proprietary.

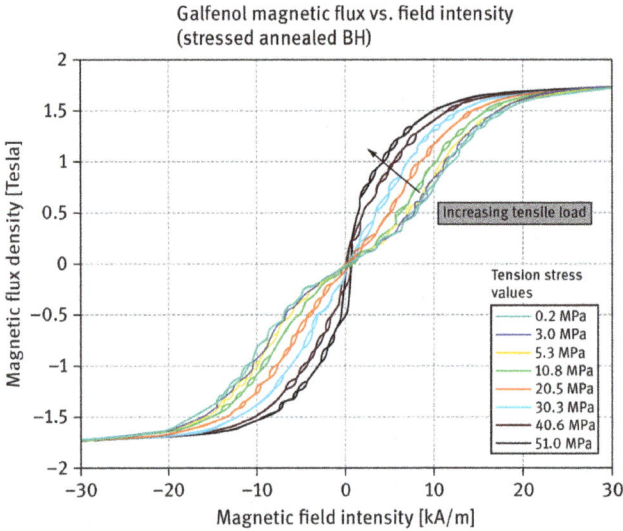

Figure 6.j: Typical Galfenol alloy (as grown, nonstress annealed BH): magnetostriction versus magnetic field intensity for various tensile stresses. Source: ETREMA Proprietary.

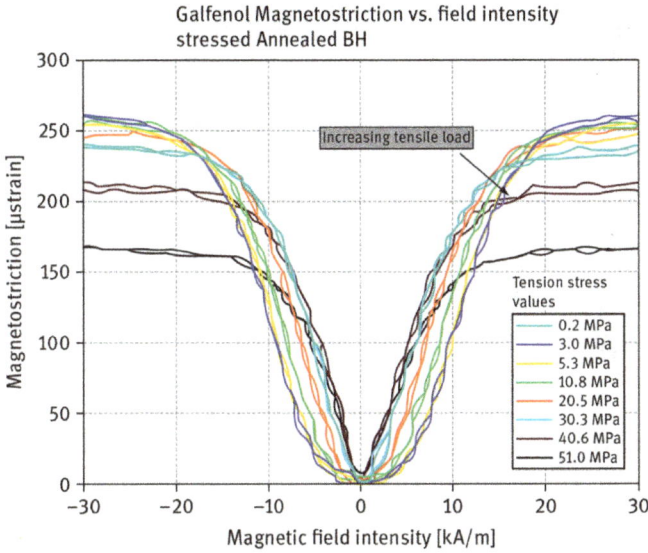

Figure 6.k: Typical Galfenol alloy (stressed annealed BH): magnetostriction versus magnetic field intensity for various tensile stresses. Source: ETREMA Proprietary.

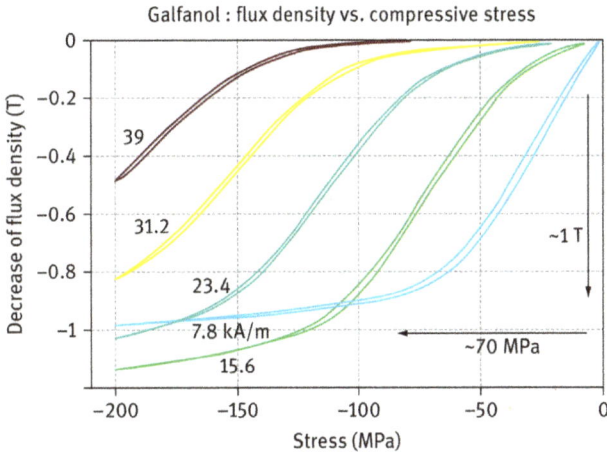

Figure 6.l: Typical Galfenol alloy: decrease in flux density versus stress intensity for various field intensities. Source: ETREMA Proprietary.

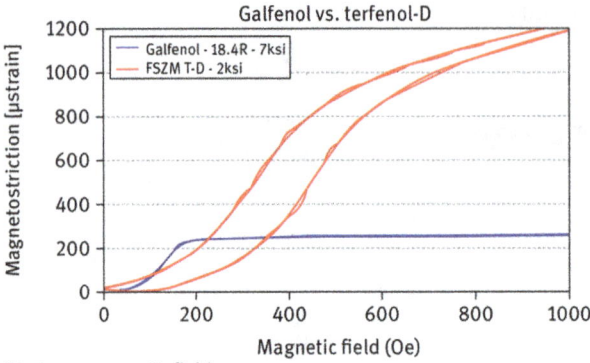

Strain vs. magnetic field

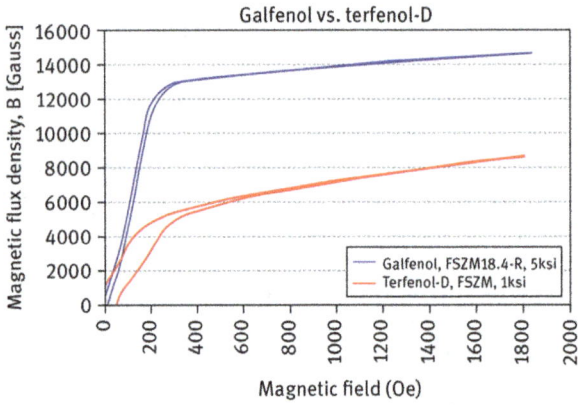

Flux density vs. magnetic field

Figure 6.m: Typical Terfenol-D and Galfenol alloy comparisons. Source: ETREMA Proprietary.

7 Applications of Intelligent Materials in Structures

7.1 Aerospace Sector

One of the large and ambitious projects being conducted in the general area of smart structures during the last years was the European project named Smart Intelligent Aircraft Structures (SARISTU). This project is a level 2, large-scale integrated project aimed at achieving reductions in aircraft weight and operational costs, as well as an improvement in the flight profile specific aerodynamic performance. The project focused on integration activities in three distinct technological areas:
- Airfoil conformal morphing
- Self-sensing
- Multifunctional structures through the use of nano-reinforced resins.

The SARISTU Consortium was coordinated by Airbus and brought together 64 partners from 16 countries. The total budget of the project was €51 million, partially funded by the European Commission under FP7-AAT-2011-RTD-1 (grant agreement number 284562). The project started on September 1, 2011 and completed by August 31, 2015. The final conference was held in May in Moscow, Russia, and the manuscripts (a total of 55 papers) presented there were collected and published (see [1]).

The SARISTU[1] objects are focused on the cost reduction of air travel through a variety of technologies. Various conformal morphing concepts in a laminar wing would improve aircraft performance through a 6% drag reduction, yielding a better fuel consumption and required take-off fuel load. It is believed that the airframe-generated noise will be reduced to 6 dB, reducing the impact of air traffic noise in the vicinity of airports and thus increasing the use of air travel. Another important objective is the integration of the structural health monitoring (SHM) in the manufacturing chain yielding a cost reduction of 1% for the in-service inspection. Finally, it is believed that the incorporation of carbon nanotubes into aeronautical resins would enable weight savings of up to 3% when compared to the present skin/stringer/frame system, while a combination of technologies is expected to decrease electrical structure network installation costs by up to 15%. In parallel, it is claimed that benefits regarding damage tolerance and electrical conductivity improvements would be realized at sub-assembly level.

---

1 www.saristu.eu

https://doi.org/10.1515/9783110726701-007

The SARISTU project is schematically depicted in Figure 7.1, with three main pillars: morphing, integrated sensing and multifunctional concepts [1]. The various work-packages of the project are

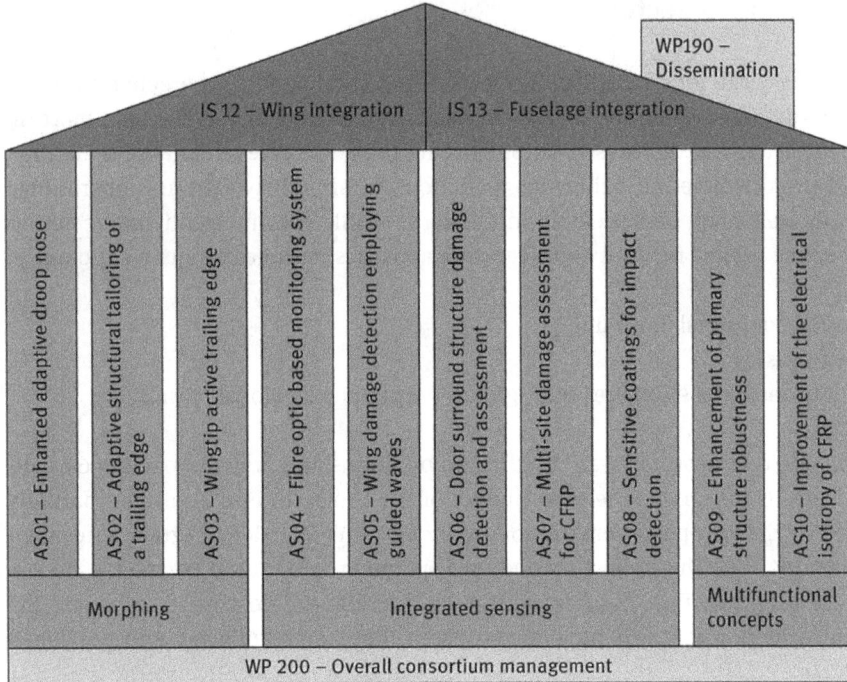

Figure 7.1: The SARISTU project structure [1].

- *ASO1 Enhanced Adaptive Droop Nose (EADN) for a Morphing Wing*, with DLR[2] being its leader. Within this work-package a 3D large-scale adaptive droop nose was designed, calculated, and manufactured and integrated in the morphing wing of IS12 for wing validation and verification based on structural and wind tunnel testing. The result was a morphing and gapless droop nose device, able to provide a laminar flow, thus reducing drag during take-off, cruise, landing and decreasing the overall noise emission. Other technologies (see Figure 7.2) were incorporated in the EADN device to provide bird strike protection, de-icing, surface and lightning protection. Furthermore a low-complexity actuation system, light weight structures, advanced manufacturing solutions, function integration and fatigue resistance were achieved.

---

2 www.dlr.de

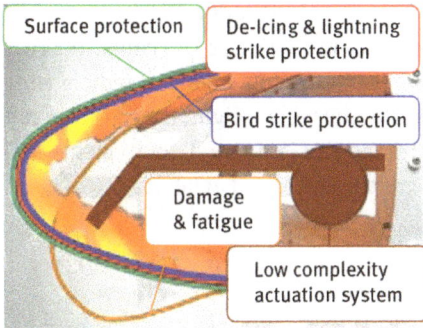

Figure 7.2: The enhanced adaptive droop nose (ETAD) including its requirements – a schematic view.

– **AS02 – Adaptive Structural Tailoring of a Trailing Edge Device**, with CIRA – The Italian Aerospace Research Centre,[3] as its leader. A morphing trailing edge device was calculated, designed, manufactured and tested to yield 10% L/D ratio improvement by chord-wise actuation and an equivalent 10% root bending moment (RBM) reduction by span-wise lift redistribution. The adaptive trailing edge device (ATED) might also lead to both cruise fuel consumption (approx. t 3%) and overall weight reduction, as a consequence of produced drag and RBM decrease. The ATED architecture is based on a multirib SDOF system, each activated by rod-like load-bearing actuators. The ribs are made of rigid parts, linked by stiff rods, any of which could work as independent actuator (one per rib). Conformal or differential actuation produces camber (wing polar) or sweep (lift distribution) variation. The skin is seamless and uses polymer thin layer and foam. The schematic view of the ATED is given in Figure 7.3.

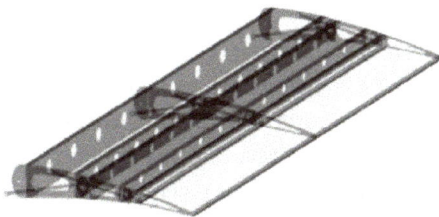

Figure 7.3: The schematic view of the ATED.

---

3 www.cira.it/en

– **AS03 Wingtip Active Trailing Edge**, leader: EADS Innovation Works.[4] An engineering solution was realized for an active winglet (see Figure 7.4) called WATE – Wingtip Active Trailing-Edge. Such active wingtip trailing edge can reduce wing loads at key flight conditions and thus reducing the wing weight. Based on aircraft structural and aerodynamic/aeroelastic requirements structural components and the systems of a WATE were designed, integrated, and tested at lab level.

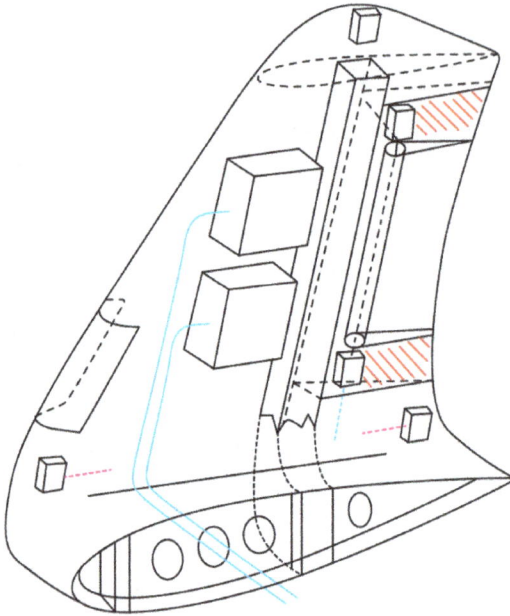

Figure 7.4: A schematic view of the WATE with active tab and morphing (red shaded) transition area.

– **AS04 Fiber-Optic-Based Monitoring System – Concept Assessment, Value, Risk and Exploitation** with INASCO Hellas[5] its leader. A fiber-optic-based monitoring system to be used in composite stiffened aircraft structures was developed within this AS.[6] The system is capable to sense strain on wing and fuselage applications; the strain readings of the system will enable shape sensing of conformal morphing structures, using the trailing edge shape monitoring as an application driver and it will be able to detect debonding of stiffeners. The monitoring system based on fiber optics is implemented on the composite structure during the manufacturing procedure and monitors the assemblies' status throughout its entire

4 www.inmaproject.eu
5 www.inasco.com
6 Application Scenario (similar to work-package in other projects).

life. The multiplexing capability of fiber Bragg grid (FBG) gives the capability to have more than one strain readings leading to smaller and lighter monitoring systems.

- *AS05 Wing Damage Detection Employing Guided Wave Techniques*; Leader: Department of Aerospace Engineering – Università di Napoli "Federico II" (DIAS).[7] This AS is aimed at designing, manufacturing and implementation of a composite wing damage detection system based on guided ultrasonic wave measurements. Propagation characteristics of ultrasonic guided waves are related to the elastic properties of composites through theoretical modeling leading to their nondestructive characterization. Analysis of the interaction of elastic waves with defects is used to improve the effectiveness of ultrasonic NDE for hidden flaw detection in composite structures and help in the development of an effective SHM system. Clusters of high-frequency ultrasonic sensor arrays are located in critical areas of a given structure to analyze the characteristics of the guided waves propagating from a controlled source. The system should be able to detect barely visible damage (BVID) and visible damage (VID) on reinforced skin of a composite wing. The sensors are integrated within the structural subcomponents during the manufacturing phase and thus are reducing drastically the needed inspection time of the selected part yielding a real reduction of lifecycle costs related to maintenance and inspections, through reduced maintenance time and/or inspection intervals.

- *AS06 Impact Damage Assessment Using Integrated Ultrasonic Sensors;* Leader: Martin Bach (EADS Deutschland GmbH).[8] The work in this AS is toward providing self-sensing functionality to the fuselage by permanently embedding piezoelectric transducers into the structure. By actuation and sensing of guided waves in pitch-catch mode, damage will be detected and assessed (Figure 7.5). The self-sensing capabilities regarding damage detection, sensor integration, quality insurance, contacting and data transfer were assessed.

- *AS07 Multi-site Damage Assessment of CFRP Structures* with Airbus-D[9] as its leader. The main objectives of AS07 were the development of a prediction method capable to assess airworthiness of aeronautic structures (typical fuselage panels) in the presence of multisite damages. Virtual testing using FE-modeling was compared with the test results of the physical tests. These results were implemented into a multisite damage prediction method. The prediction method was used

---

7 www.dias.unina.it
8 martin.bach@eads.net
9 www.airbusgroup.com/int/en/group-vision/global-presence/germany.html

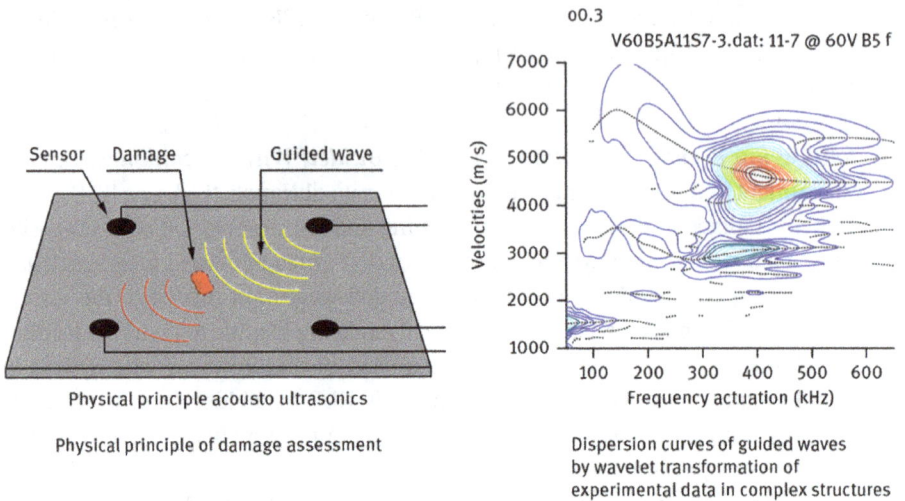

Physical principle acousto ultrasonics

Physical principle of damage assessment

Dispersion curves of guided waves
by wavelet transformation of
experimental data in complex structures

Figure 7.5: Damage assessment using embedded piezo-transducers. From Marti Bach[8].

to develop a damage assessment tool capable to provide a decision support, combining SHM systems information with a simplified prediction method model. Use was made of fiber optics and PZT sensors. Finally the selected SHM technologies were integrated and validated in combination with the assessment tool for the multisite impact damage test articles for the demonstration tasks.

– **AS08 Sensitive Coating for Impact Detection**; Leader: EADS France.[10] It is known that when an impact is implied on a metallic structure, a local shape deformation can be detected visually. However on composite structures, an impact may imply an internal damage without visual indication on the surface. These accidental impacts may occur inside the aircraft fuselage during assembly or maintenance tasks. That is why current composite made airplanes are designed with structure thickness margin and withstand potentially missed damages. Within this AS, the main objective is thus to ease the impact detection on composite structure by improving the BVID (Barely Visible Impact Damage) threshold using a smart coating, which is revealing visually impact occurrence from a defined energy threshold. Reducing the BVID threshold enables the reduction of the thickness margin and thereby save weight (Figure 7.6).

---

10 EADS – European Aeronautic Defence and Space Company NV, the name of Airbus Group SE prior to 2014-www.airbusgroup.com

Figure 7.6: AS08 – Impact revealed by blue microcapsules included in a paint.

– *AS09 Enhancement of Primary Structure Robustness by Improved Damage Tolerance*; Leader: Tecnalia Research & Innovation.[11] This scenario is to evaluate toughening of composite materials through toughening additives such as nano-materials and various types of polymer interleaf materials. This scenario worked alongside AS10 to deliver multifunctional composite materials and potentially improved adhesives. The material properties targeted for improvement within this AS, and identified to be important for damage tolerance, will primarily be mode 1 and 2 fracture toughness, compression after impact, and interlaminar shear strength (Figure 7.7). Improving the damage tolerance of materials can result in reduction in component thickness and weight. These later benefits in turn can lead to reduction in fuel consumption and therefore provide environmental and monetary benefits (see a schematic drawing in Figure 7.8).

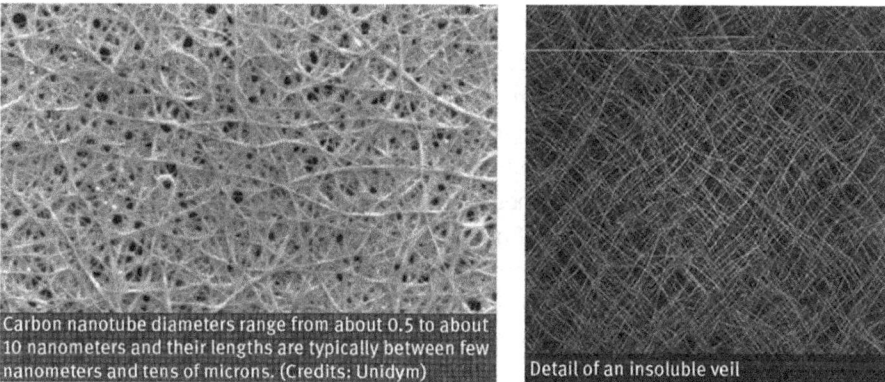

Carbon nanotube diameters range from about 0.5 to about 10 nanometers and their lengths are typically between few nanometers and tens of microns. (Credits: Unidym)

Detail of an insoluble veil

Figure 7.7: Carbon nanotubes and an insoluble veil used in AS09.

11 www.tecnalia.com/en/2

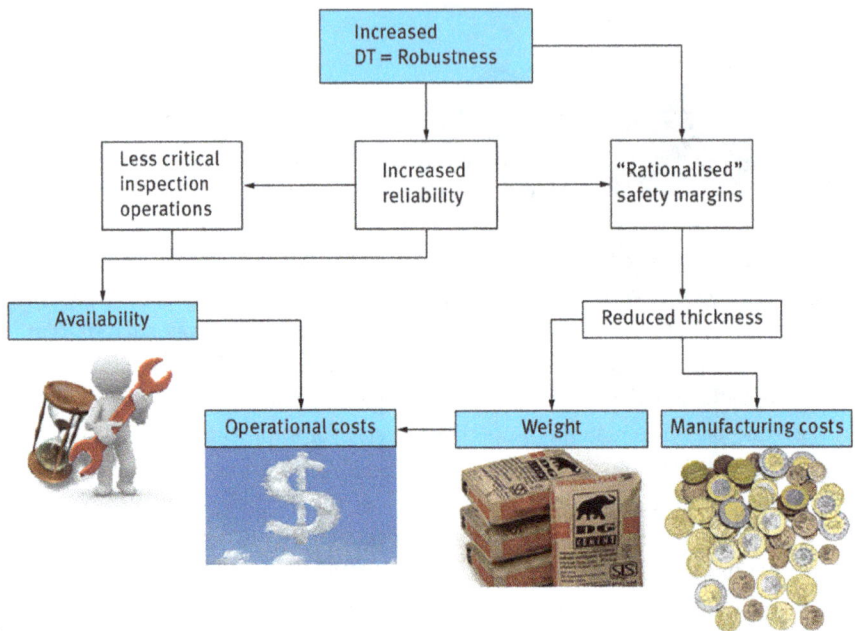

Figure 7.8: Increased reliability and it consequences – a block diagram from AS09.

- **AS10 Improvement of the Electrical Isotropy of Composite Structures** with Tecnalia Research & Innovation as its leader. This application scenario was established to evaluate technologies that can be integrated into structures with the aim of creating an electrical structural network, thus reducing weight and improving survivability of composite structures. These evaluations included various technologies and materials from nanomaterials to metallic strips and coatings (see Figure 7.9). The research done in AS09 on damage tolerance was incorporated in the present evaluation and a validation of the manufacturability of multifunctional materials that show improved damage tolerance and electrical survivability was conducted.

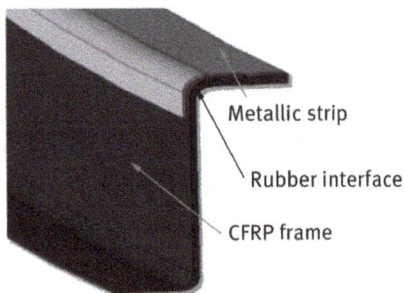

Figure 7.9: A metallic strip for electrical structural network (ESN) used with special permission from EADS Innovation Works.

- *IS12 **Wing Assembly Integration and Testing*** leaded by Alenia.[12] This Integration Scenario (IS) was responsible for integrating the conformal morphing and SHM (based on fiber optics and PZT) into a wing demonstrator while checking and validating various methodologies for robust structural sizing (see Figures 7.10 and 7.11). A demonstrator was manufactured and it figures out a full-scale wing model, 4.5 m span, including winglet, leading edge, and trailing edge with innovative 3D actuators (Figure 7.10). Full-scale conformal morphing devices have structural and actuation functionalities, continuous for trailing edge and winglet, nonautomated mechanical, discrete steps, for conformal morphing leading edge. Low-speed wind tunnel and numerical activities were performed to validate implemented architectures. Figure 7.12 summarizes the various technologies within this IS.

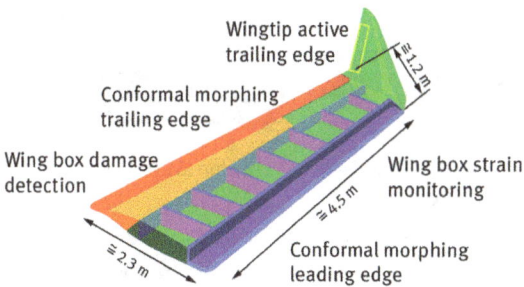

Figure 7.10: Some of the technologies developed and implemented in the SARISTU wing within IS12.

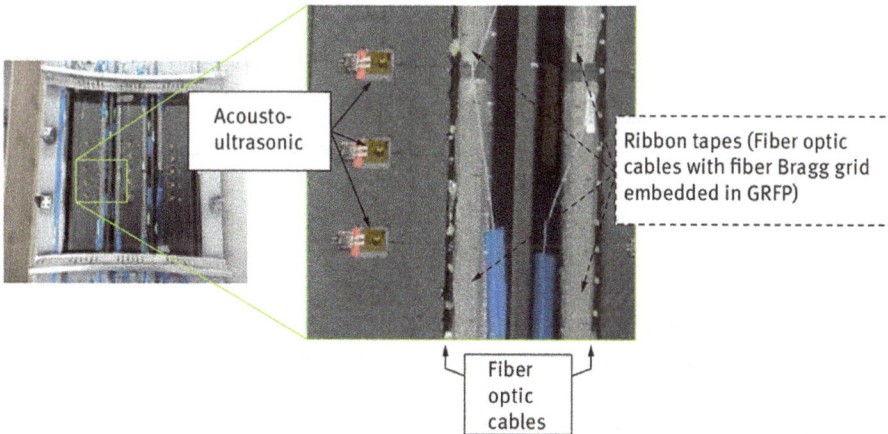

Figure 7.11: SHM methodology applied on the SARISTU wing within IS12.

---

12 Alenia Aermacchi was a subsidiary of Finmeccanica S.p.A., active in the Aeronautics sector. From 1st of January 2016, the activities of Alenia Aermacchi merged into Finmeccanica's Aircraft and Aerostructures Divisions, within the Aeronautics Sector, www.finmeccanica.com.

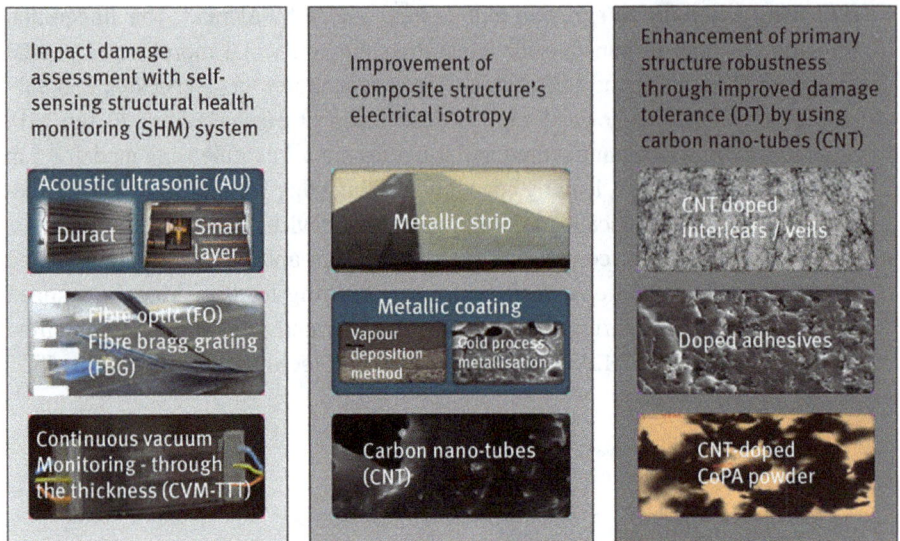

Figure 7.12: Some of the technologies being investigated in SARISTU [1] within IS12.

- *IS13 Fuselage Assembly, Integration, Testing and Validation*; Leader: Airbus Operations GmbH[13] represented by Ben Newman, TECCON Consulting and Engineering GmbH.[14] The main goal of this IS was to provide representative real aircraft structures base on technologies developed and designed in the Application Scenarios leading to the evaluation of the potential of those technologies to provide the following SARISTU project targets:
  - (a) Multi-site damage detection capabilities for operational cost reductions regarding structure inspection by 1%.
  - (b) Improved damage characterization through advanced algorithms for operational cost reductions regarding structure inspection by 1%.
  - (c) The feasibility of combining SHM with nano-reinforced resins for damage tolerance improvement for weight savings exceeding 3%.
  - (d) Electrical structure network installation cost reduced by 15%.

This was done by using four promising technologies (see Figure 7.13): (1) door surround structure and mini-door, focusing on acoustic-ultrasonic SHM; (2) lower panel, focusing on damage tolerance, low level conductivity from nano-reinforced resins, low-cost electrical structure network (ESN) installation and limited SHM; (3) side panel, focusing on

---

13 Formerly known as Airbus Deutschland GmbH, http://www.airbus.com/

14 As of April 2013, TECCON Consulting & Engineering GmbH became ALTRAN Aviation Engineering GmbH, http://www.altran.de/

low-cost ESN for medium conductivity and low cost, high-performance lightning strike protection (LSP); and (4) upper panel, focusing on multiple SHM capabilities. The tested structure was #1.

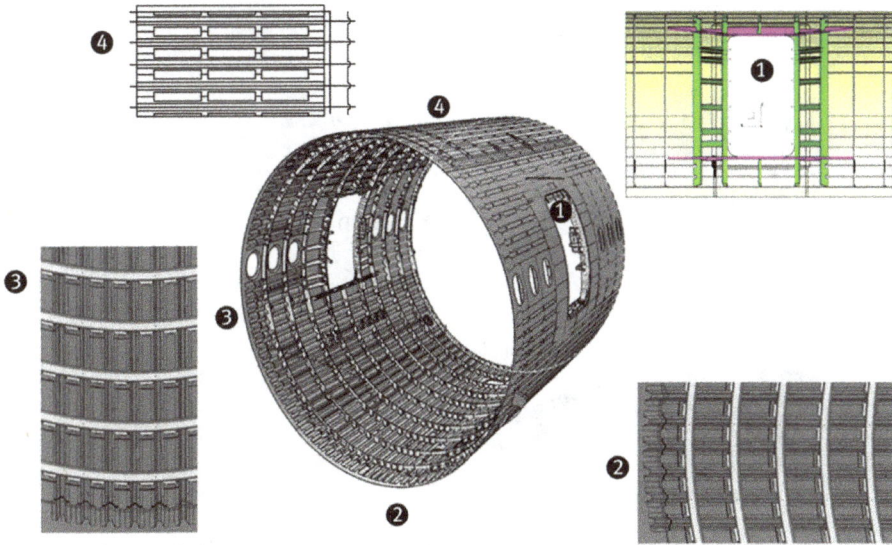

Figure 7.13: Four generic fuselage panels used in IS13 to assess the most promising Application Scenario technologies.

- **WP 190 – Dissemination** leaded by EASN-TIS.[15] This work-package was responsible for various dissemination activities to highlight the SARISTU project achievements and results.

- **WP 200 – Overall Project Management** leaded by AIRBUS Germany with support from TECCON Consulting & Engineering GmbH. Its tasks were to provide the SARISTU project management including monitoring the achievements and meeting the objective of the projects throughout its time period.

After presenting the various activities performed within the project, some highlights connected to its smart structure nature are described next. All those activities are well presented in [1].

Within the SARISTU project an advanced morphing wing trailing edge device was designed, calculated for flutter, actuation strength and aerodynamic performance, and

---

15 The European Aeronautic Science Network- Technology Innovation Services (EASN-TIS), is a Belgium based SME strongly involved in aeronautics and air transport related research projects, http://easn-tis.com/

finally manufactured and tested [3]. A compact light-weight lever driven by electromechanical actuators provided the morphing of the trailing edge (TE), while a seamless
skin [4–14] was used to cover the ribs of the TE (see Figure 7.14). The configuration of
the advanced trailing edge is given schematically in Figure 7.15.

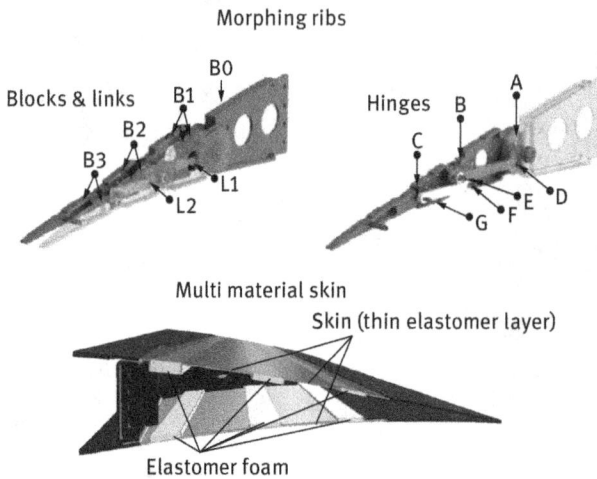

Figure 7.14: A schematic view of the adaptive trailing edge device (ATED) basic components [3].

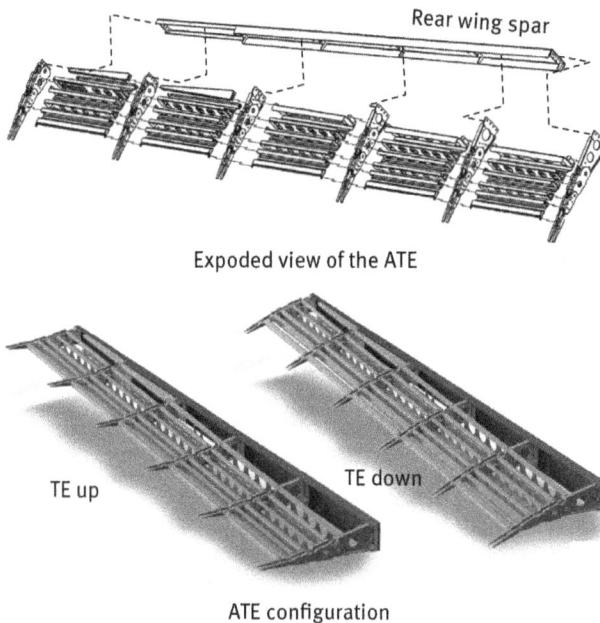

Figure 7.15: The morphing trailing edge configuration [3].

One of the great achievements of the SARISTU project was a seamless skin [4] made of polymer layer and foam, which enabled the movement of the trailing edge in a morphing way, leading to an up, and down-deflection of the trailing edge (see Figures 7.16 and 7.17).

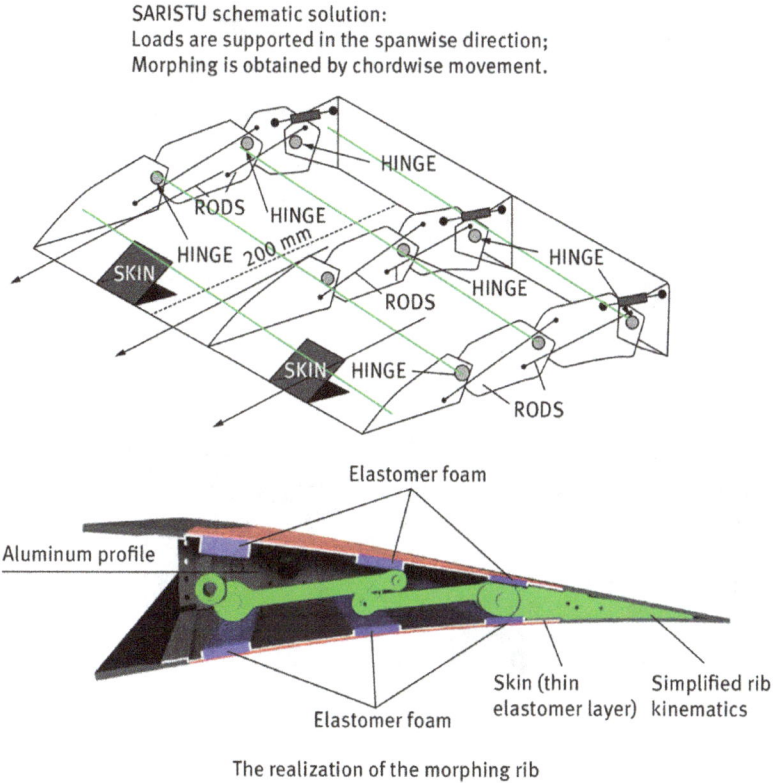

SARISTU schematic solution:
Loads are supported in the spanwise direction;
Morphing is obtained by chordwise movement.

The realization of the morphing rib

Figure 7.16: SARISTU morphing trailing wing – a schematic view [4].

Some of the technologies being dealt with the EADN [5] subtopic include surface protection, de-icing procedures, lightning and bird-strike protection, and simple kinematics to obtain the change of the shape. A 4 m long sample of the wing leading edge was manufactured by Invent[16] to perform wind tunnel and ground tests, bird-strike tests while two small models were equipped with titanium foils. Typical test sections are given in Figure 7.18.

---

16 INVENT GmbH, Braunschweig, Germany, http://www.invent-gmbh.de/

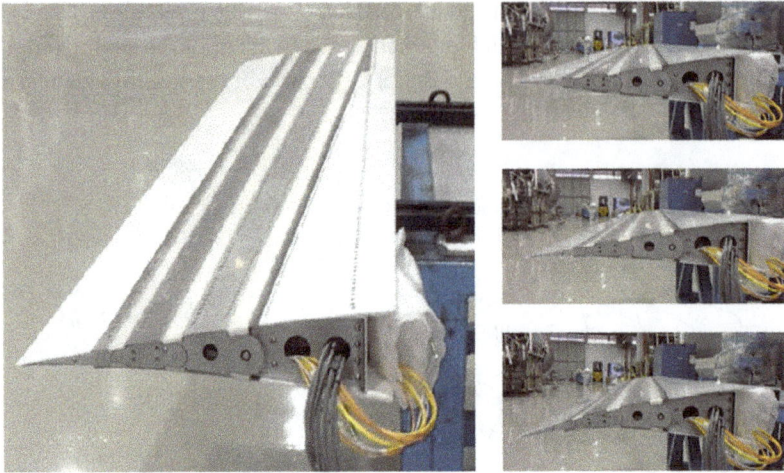

Figure 7.17: SARISTU morphing trailing wing – the manufactured model [4].

The EADN (Enhanced adaptive droop nose)

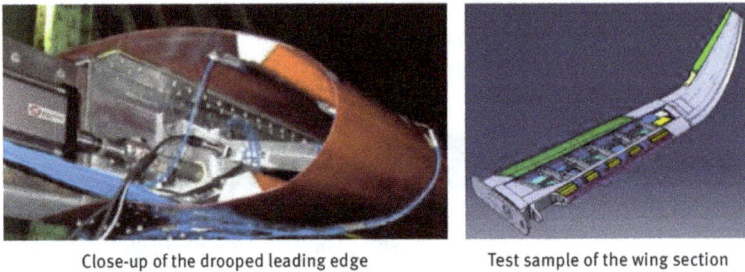

Close-up of the drooped leading edge

Test sample of the wing section

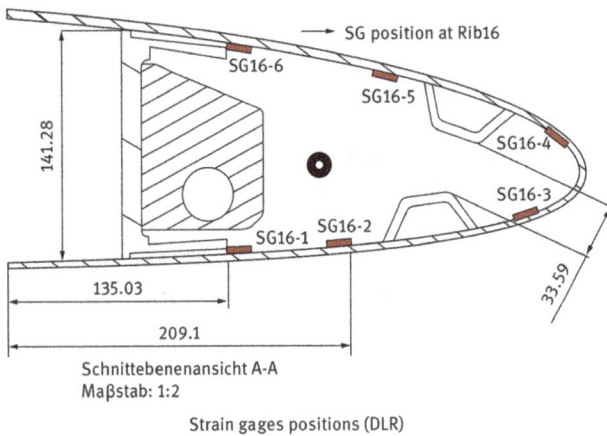

Schnittebenenansicht A-A
Maβstab: 1:2

Strain gages positions (DLR)

Figure 7.18: The EADN typical test sections [5].

The EADN section (Figure 7.19) was further assessed to demonstrate the technology readiness level (RL) of the integration of the various technologies by designing and manufacturing of a test demonstrator investigating the combination of wing bending and the leading edge deployment [6].

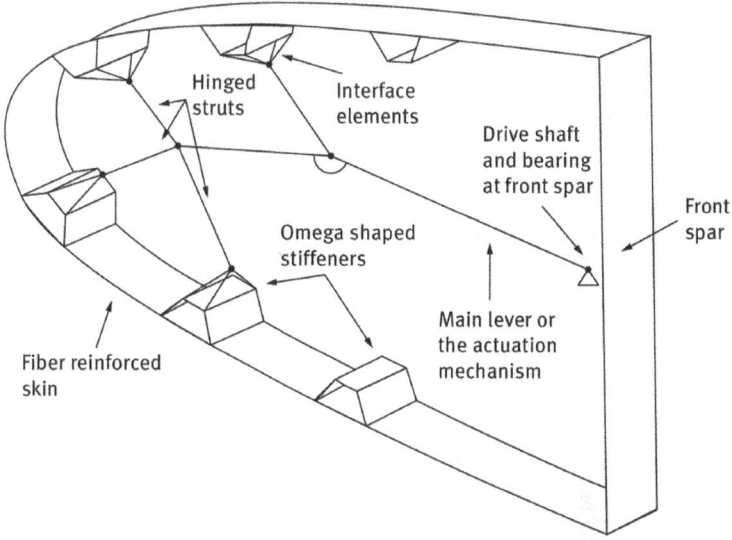

Figure 7.19: A schematic drawing for a monolithic morphing leading edge using conventional skin material.

An advanced way of revealing damage in composite made structures was developed and tested using microcapsules with a manifest color change under visual and UV light [7]. This was done using a leuco dye[17] revealing agent system. Before any impact, the two components are separated, without any coloration. When microcapsules burst upon impact, they free the product that comes into contact with the developer incorporated into the coating. A colored stain appears at the impact location. The concept of using microcapsules, typical results and the multilayer concept developed within the SARISTU project are presented in Figure 7.20.

Work has been done on using optical fibers and strain gages to monitor the loads being applied to the wing spars [8]. The optical fibers were embedded in ribbons and bonded onto the spars. The second monitoring system included conventional strain gages and was designed to measure the loads on the wing during the wind tunnel tests and interacting with the aerodynamic load model (Figure 7.21). Relevant algorithms

---

17 The leuco dye (in greek:leukos = white) is a dye which can switch between two chemical forms, one being colorless and being invisible in acid conditions, is encapsulated, and the reveling agent in the form of a weak acid compound is directly introduced in the coating binder, outside of the microcapsules.

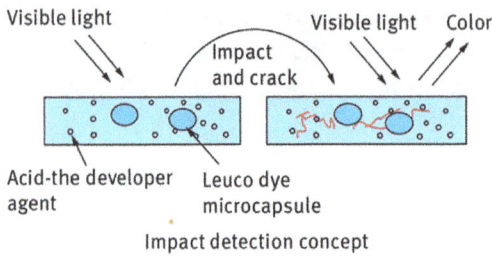

Impact detection concept

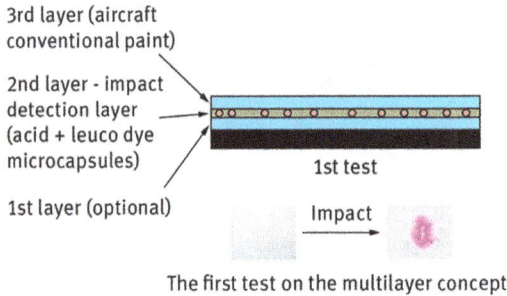

The first test on the multilayer concept

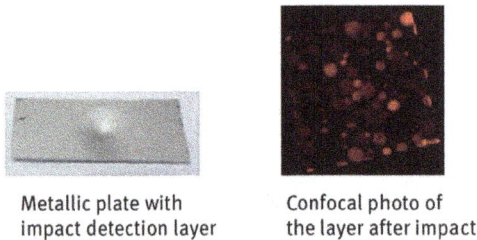

Metallic plate with
impact detection layer

Confocal photo of
the layer after impact

Application of confocal microscopy technology to
study the impacted zone (no destruction of the microcapsules
introduced in the impact detection layer)

Figure 7.20: Revealing impact damage in composite using microcapsules [7].

were developed to identify internal and sectional loads of the wing. The results were implemented in a health and usage monitoring systems to be used in advanced aerospace structures.

An additional approach was investigated in [9] to provide an advanced SHM technology, based on acoustic ultrasonic (AU) realized by permanently embedding a network of piezoelectric transduces in the door surround structure. Ultrasonic guided waves provide information about the integrity of the tested structure (Figure 7.22). The work within [9] included the derivation of the verification approach, which started with the way the damage is introduced in the structure and performing non-destructive inspection (NDI) as a reference. Then the interrogation of the transducers network and the relevant data evaluation were performed, and integrated into a graphical user interface (GUI).

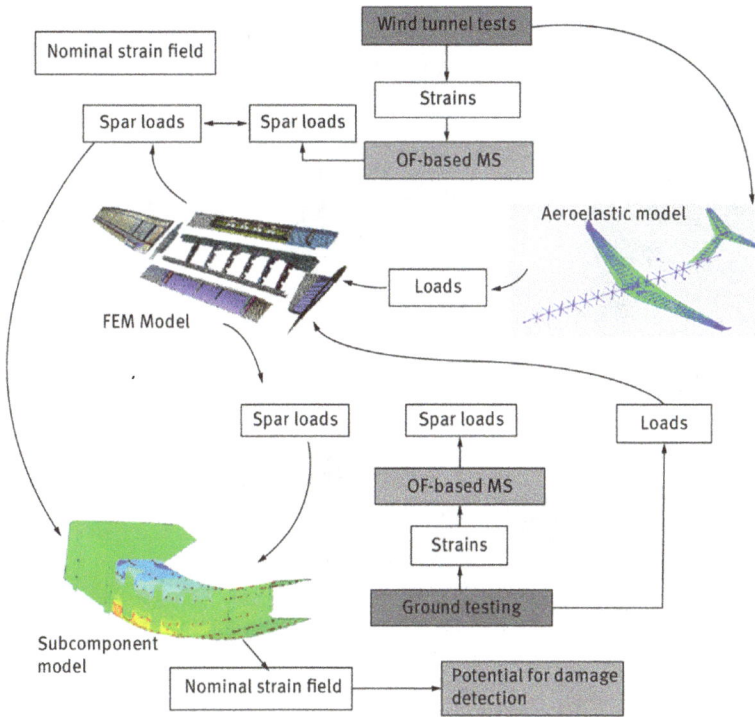

Figure 7.21: The diagram blocks representing the interaction of FE model, aeroelastic model, subcomponent model with the monitoring systems [8].

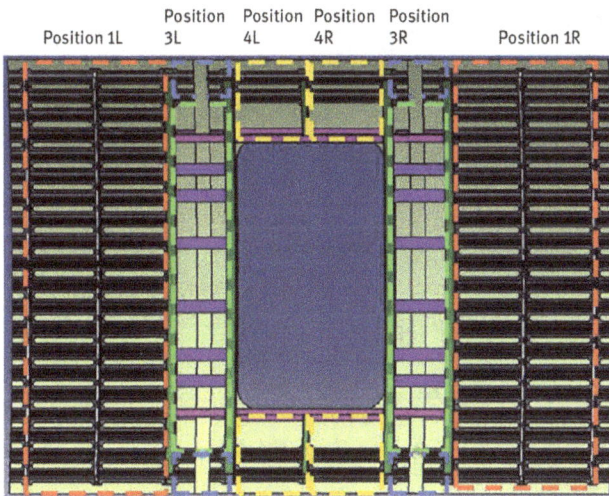

Figure 7.22: Positions of impacts on the completed door surround structure with permanently embedded piezoelectric transducers for acoustic ultrasonic [9].

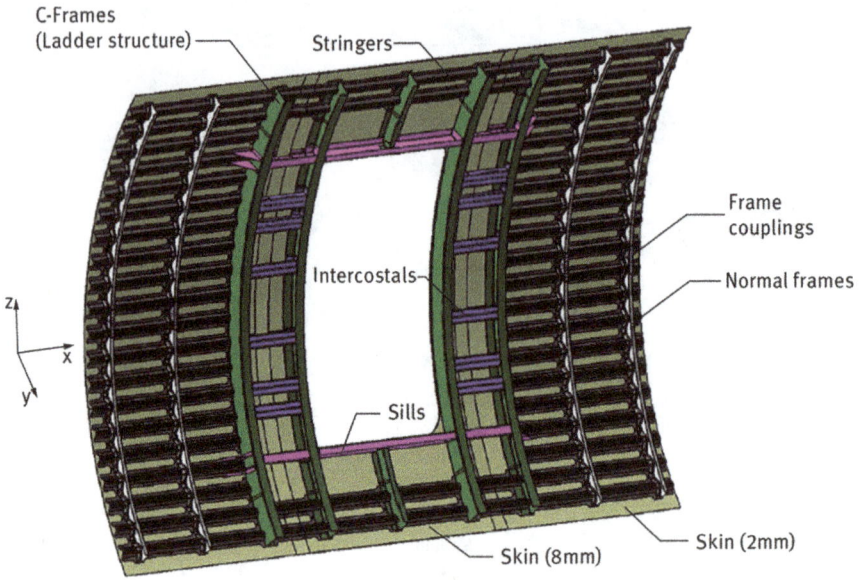

Figure 7.23: Another view of the completed door surround structure without the SHM network [10].

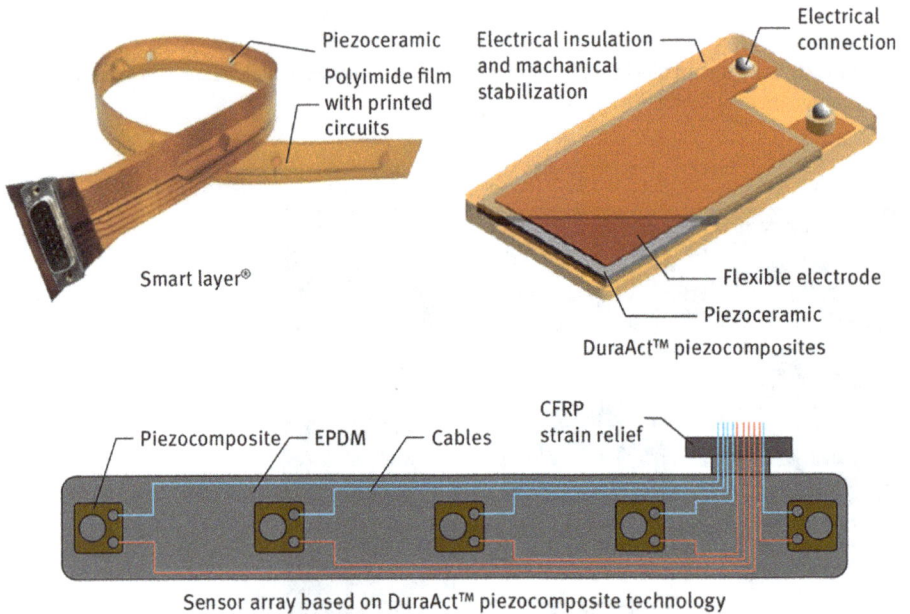

Sensor array based on DuraAct™ piezocomposite technology

Figure 7.24: The piezoelectric transducers used for Lamb-wave-based SHM [10].

Skin lay-up using automated
fiber placement robot

Skin with integrated sensor network

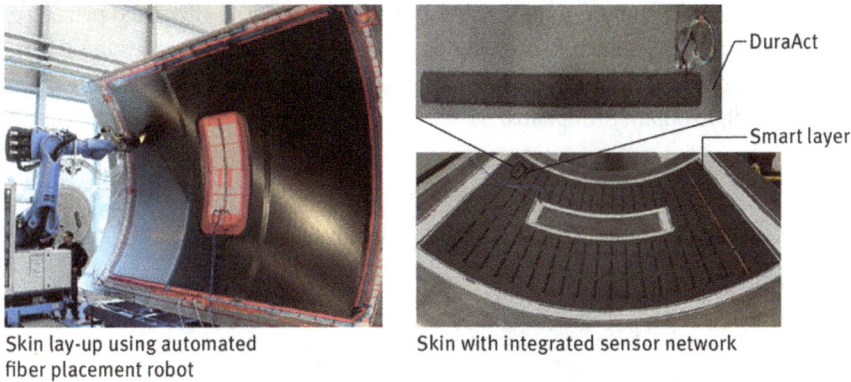

Figure 7.25: The automated manufacture procedure of the door surround structure and the integrated piezoelectric transducers used SHM [10].

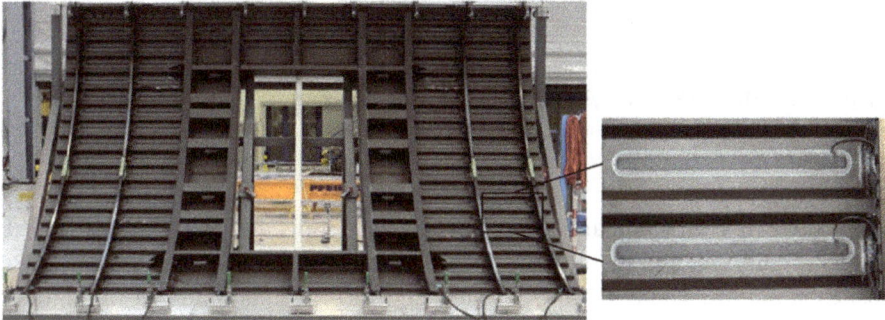

Figure 7.26: Manufacture door surround structure with integrated piezoelectric transducers to form the SHM network [10].

A similar approach was adopted in [10], where the SHM was realized using Lamb waves excited and received by a network of piezoelectric actuators and sensors being embedded in the structure (Figure 7.23). Smart Layer® [11] and DuraAct™ [12–13] type piezoelectric transducers were used for the production and receiving of the Lamb waves (Figure 7.24). The manuscript describes the various development and manufacturing stages of the full-scale composite fuselage door surround structure having an integrated SHM system, which is based on piezoelectric transducers (Figures 7.25 and 7.26). Note the advanced equipment, an automated fiber placement robot, used to manufacture the structure (Figure 7.25) and the advanced sensors (Figure 7.26).

Another interesting methodology for SHM using multiple ultrasonic waves based PAMELA (*P*hased *A*rray *M*onitoring for *E*nhanced *L*ife *A*ssessment) III SHM™ system (Figure 7.27) wirelessly controlled was developed by AERNNOVA [15] within the SARISTU project. The system integrates all the necessary hardware, firmware and software elements to generate excitation signals to be applied to an array of integrated piezoelectric phased array (PhA) transducers embedded in a given structure, acquiring the response signals, sending the data to a central host and/or carrying out the advanced signal processing to obtain SHM maps.

Figure 7.27: PAMELA III SHMTM system with all electronic components inside the composite box placed over the phase array transducer adapter and bonded to the structure to perform SHM testing [15].

The manuscript [15] summarizes the damage test results and the effectivity of the system to interpret and identify the structural damage for both metallic and composites materials. Figures 7.28–7.30 present typical results obtained throughout the project.

Aluminum flat panel with 4 impact damage

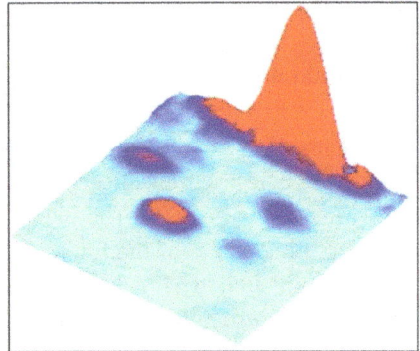

Image of the damage on the left obtained with PAMELA SHM system

Figure 7.28: Damage results image using the PAMELA III SHMTM system on a damaged aluminum flat panel [15].

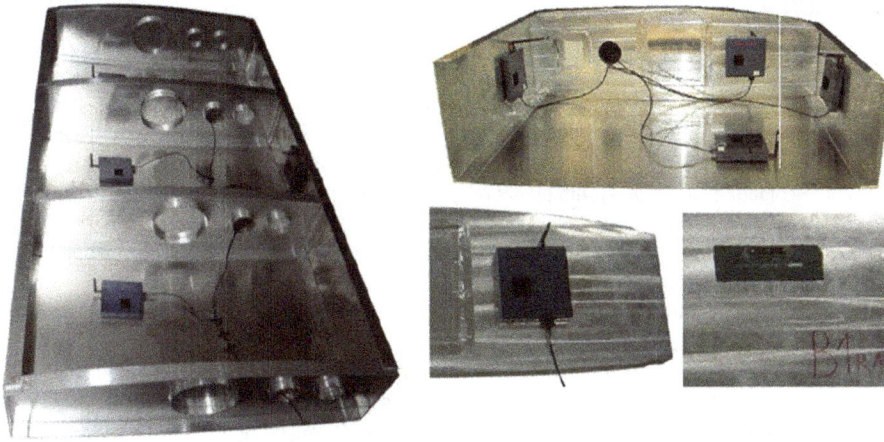

Figure 7.29: The installation of PAMELA III SHMTM system on a 3-bays wingbox demonstrator [15].

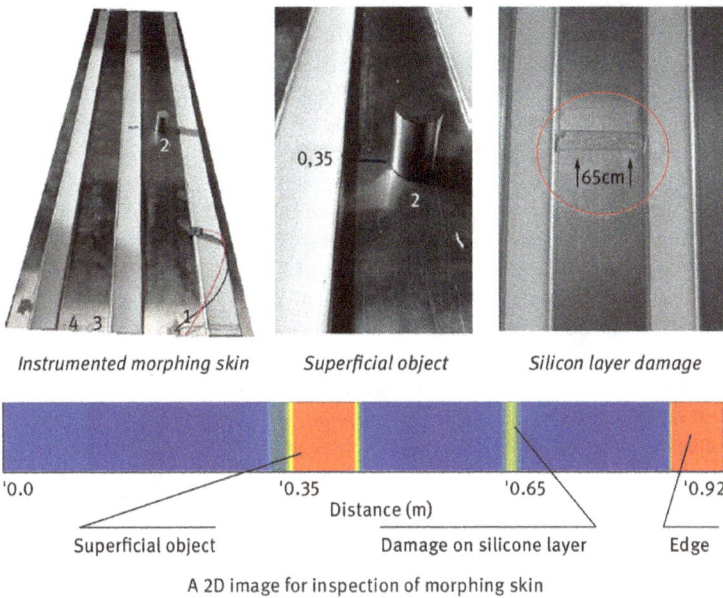

Instrumented morphing skin   Superficial object   Silicon layer damage

'0.0   '0.35   '0.65   '0.92
Distance (m)

Superficial object   Damage on silicone layer   Edge

A 2D image for inspection of morphing skin

Figure 7.30: PAMELA III SHMTM system results for a damage applied on a morphing skin [15].

References

[1] Wolcken, P. C. and Papadopoulos, M., (eds.), Smart Intelligent Aircraft Structures (SARISTU), Proceedings of the Final Project Conference, 19–21 of May 2015, Moscow, Russia, Springer International Publishing AG Switzerland 2016, 1039.

[2] Wolcken, P. C., Kotter, A., Newman, B., Wadleich, R. and Genzel, K., SARISTU: Six years of project management, Smart Intelligent Aircraft Structures (SARISTU), Proceedings of the Final Project Conference, Springer International Publishing AG Switzerland, 2016, 1–40.

[3] Dimino, I., Ciminello, M., Concilio, A., Pecora, R., Amoroso, F., Magnifivo, M., Schueller, M., Gratias, A., Volovick, A. and Zivan, L., Distributed actuation and control of a morphing wing trailing edge, Smart Intelligent Aircraft Structures (SARISTU), Proceedings of the Final Project Conference, Springer International Publishing AG Switzerland, 2016, 171–186.

[4] Schorsch, O., Luhring, A. and Nagel, C., Elastomer-based skin for seamless morphing of adaptive wings, Smart Intelligent Aircraft Structures (SARISTU), Proceedings of the Final Project Conference, Springer International Publishing AG Switzerland, 2016, 187–197.

[5] Snop, V. and Horak, V., Testing overview of the EADN samples, Smart Intelligent Aircraft Structures (SARISTU), Proceedings of the Final Project Conference, Springer International Publishing AG Switzerland, 2016, 85–96.

[6] Kintscher, M., Kirn, J., Storm, S. and Peter, F., Assessment of the SARISTU enhanced adaptive droop nose, Smart Intelligent Aircraft Structures (SARISTU), Proceedings of the Final Project Conference, Springer International Publishing AG Switzerland, 2016, 113–140.

[7] Monier, L., Le Jeune, K., Kondolff, I. and Vilaca, G., Coating for detecting damage with a manifest color change, Smart Intelligent Aircraft Structures (SARISTU), Proceedings of the Final Project Conference, Springer International Publishing AG Switzerland, 2016, 735–743.

[8] Airoldi, A., Sala, G., Evenblij, R., Koimtzoglou, C., Loutas, T., Carossa, G. M., Mastromauro, P. and Kanakis, T., Load monitoring by means of optical fibers and strain gages, Smart Intelligent Aircraft Structures (SARISTU), Proceedings of the Final Project Conference, Springer International Publishing AG Switzerland, 2016, 433–469.

[9] Bach, M., Dobmann, N. and Bonet, M. M., Damage introduction, detection and assessment of CFRP door surrounding panel, Smart Intelligent Aircraft Structures (SARISTU), Proceedings of the Final Project Conference, Springer International Publishing AG Switzerland, 2016, 947–957.

[10] Schmidt, D., Kolbe, A., Kaps, R., Wierach, P., Linke, S., Steeger, S., Dungern, F., Tauchner, J., Breu, C. and Newman, B., Development of a door surround structure with integrated structural health monitoring system, Smart Intelligent Aircraft Structures (SARISTU), Proceedings of the Final Project Conference, Springer International Publishing AG Switzerland, 2016, 935–945.

[11] Acellent Technologies Inc., Smart Layer®, http://www.acellent.com, May 2015.

[12] Wierach, P., Elektromechanisches Funktionmodul, German Patent # DE10051784C1, 2002.

[13] Wierach, P., Development of piezocomposites for adaptive systems, DLR-Forschungsbericht DLR-FB 2010-23, Technical University Braunschweig, Braunschweig, Ph.D. Thesis, 2010.

[14] Schorsch, O., Luhring, A., Nagel, C., Pecora, R. and Dimino, I., Polymer based morphing skin for adaptive wings, Proceedings of the 7th ECCOMAS Thematic Conference on Smart Structures and Materials, SMART 2015, Azores, Portugal, 3–6 June, 2015.

[15] Alcaide, A. and Martin, F., PAMELA SHM system implementation on composite wing panels, Smart Intelligent Aircraft Structures (SARISTU), Proceedings of the Final Project Conference, Springer International Publishing AG Switzerland, 2016, 545–555.

7.2 Medical Sector

The medical sector is one of the sectors with a relative large number of applications using smart materials, mainly Nitinol (SMA) and piezoelectric materials (see typical Refs. [1–8]).

The use of shape memory alloys in the medical sector is widely recognized and accepted, due to its biocompatibility with human body. Normally the Nitinol is a binary chemical system (approx. 50% nickel and 50% titanium), however slightly nickel richer alloys like for example $Ni_{0.53}Ti_{0.47}$, would yield the known phenomenon of SMA, namely shape memory, used for the majority of medical applications [1–3]. The global Nitinol medical devices market for semi-finished goods was evaluated at US$ 1.5 billion in 2012 and forecasted to grow to US$ 2.5 billion in 2019, while the global Nitinol medical devices market for final medical components was worth US$ 8.2 billion in 2012 and is expected US$ 17.3 billion by the end of 2019.[18] Some of the known companies are listed in Table 7.1.

Table 7.1: Worldwide Nitinol medical devices companies.

Company	Headquarter country	Email address
Abbott Laboratories, Inc.	USA	www.abbott.com
Boston Scientific Corporation	USA	www.bostonscientific.com
Cook Medical, Inc.	USA	www.cookmedical.com
Covidien plc	USA[1]	www.medtronic.com/covidien
C. R. Bard, Inc.	USA	www.crbard.com
Custom Wire Technologies, Inc.	USA	www.customwiretech.com
ENDOSMART GmbH	Germany	www.endosmart.com
Medtronic, Inc.	USA	www.medtronic.com
Nitinol Devices & Components, Inc.	USA	www.nitinol.com
Terumo Corporation	Japan	www.terumomedical.com

[1] Covidien Public Limited was an Irish company (headquarters, Dublin, Republic of Ireland). From June 2014, it is a part of Medtronic Inc., USA.

Nitinol succeeded to develop its medical device market place with a broad range of devices and dedicated equipment in the areas of cardiology, neurology, orthopedics and interventional radiology.

---

18 According to a report, entitled: 'Nitinol Medical Devices Market – Global Industry Analysis, Size, Share, Growth, Trends and Forecast, 2013–2019', http://www.transparencymarketresearch.com

Nitinol wires are used in angiography, due to its high recoverable strains, good kink resistance, steerability and torquability. Another wide area of applications is the stents,[19] Nitinol-based devices used to brace the inside circumference of tubular passages (lumens) like the esophagus and biliary ducts, or providing inner support of blood vessels including coronary, carotid, iliac, aorta and femoral arteries. Figures 7.31 and 7.32 present a schematic drawing of a stent in a coronary artery and various other stents.

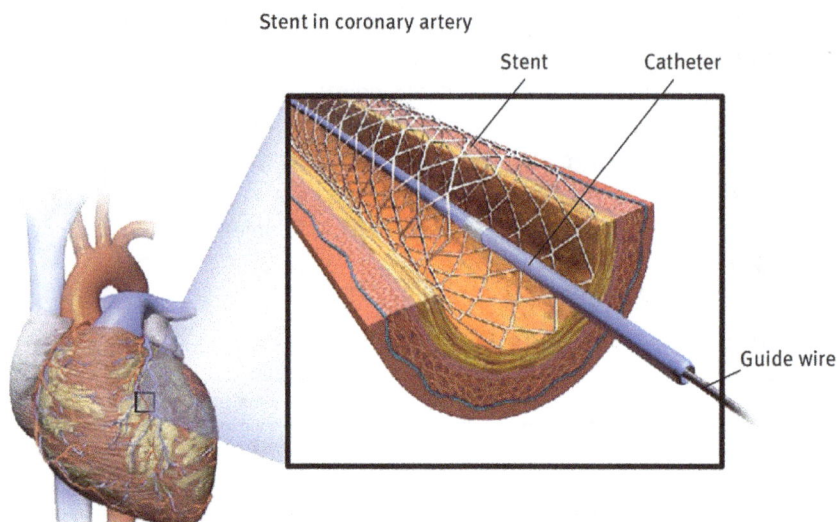

Figure 7.31: Stent in a coronary artery. Source: www.wikipedia.org/wiki/Stent.

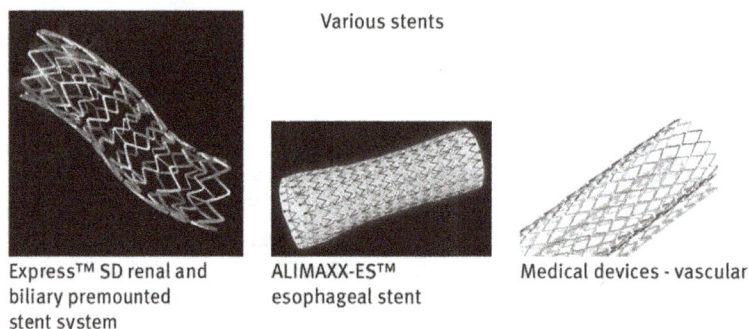

Various stents

Express™ SD renal and biliary premounted stent system

ALIMAXX-ES™ esophageal stent

Medical devices - vascular

Figure 7.32: Various Nitinol made stents. Sources: www.bostonscientific.com; www.endotek.merit.com; Norman Noble Inc. (Norman Noble Inc., Microprecision Medtech Manufacturing; The ALIMAXX-ES Fully Covered esophageal stent is intended for maintaining esophageal luminal patency in esophageal strictures caused by intrinsic and/or extrinsic malignant tumors and for occlusion of esophageal fistulae.).

---

19 The word stent derives from a dentist, Dr. Charles Thomas Stent (1807 to 1885), a dentist in London who developed a dental device to assist in forming an impression of teeth.

Stents is either known as balloon expandable where the angioplasty balloon is employed to both open the blocked vessel and expand the stent, or as self-expanding when the stent is pushed out of the catheter and it immediately opens out to support the already dilated lumen, due to its shape memory property. Most of Nitinol stents are of the self-expanding type. The stent is compressed in its delivery catheter. When the catheter is in the correct position the stent is pushed out and will expand to its original shape against the vessel wall. Other medical applications would include stone extractors from kidneys (Figure 7.33) and implants (Figure 7.34), where the *Conventus DRS*^{TM},[20] made from Nitinol which allows, due to its superelasticity and shape memory, to be collapsed into a small size and inserted in the body through a small incision. Once at the correct location, the surgeon releases it to expand to its original shape.

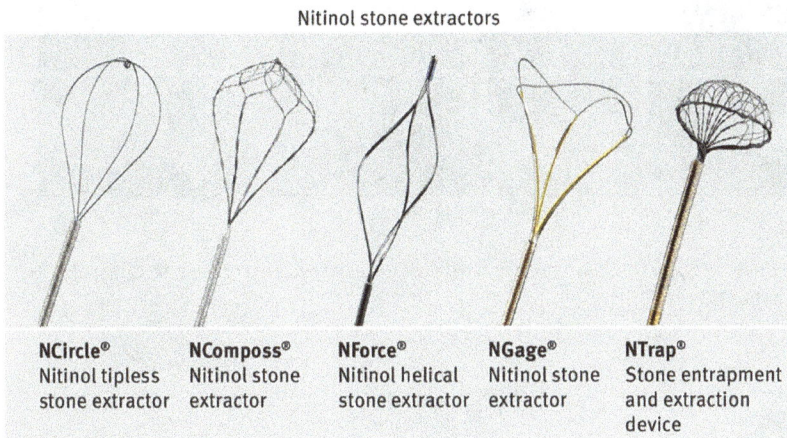

Figure 7.33: Nitinol stone extraction tools. Source: www.cookmedical.com.

Figure 7.35 presents various Nitinol devices for bones fixations, while Figure 7.36 presents typical dental applications for Nitinol and fixation screw for implants of Nitinol.

An interesting critical overview of the compatibility of Nitinol with the human body is given by Shabalovskaya et al. [4]. It is claimed that the present developed surfaces may vary in thickness from a few nanometers up to micrometers, and the release of the nickel (Ni) into the human body can be effectively prevented provided the surface integrity is maintained under strain and if no Ni-enriched sublayers are presented above the Nitinol surface.

A recent application is the piezoelectric bone surgery permitting a selective cut of mineralized tissue while sparing soft tissue. Similar to a dental scaler, a high-frequency vibration, in the range of 25–35 kHz, is transmitted to a metallic tip.

---

20 www.conventusortho.com

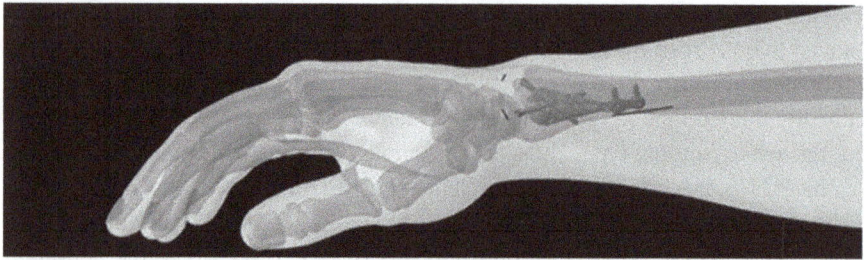

The conventus DRS™ implant

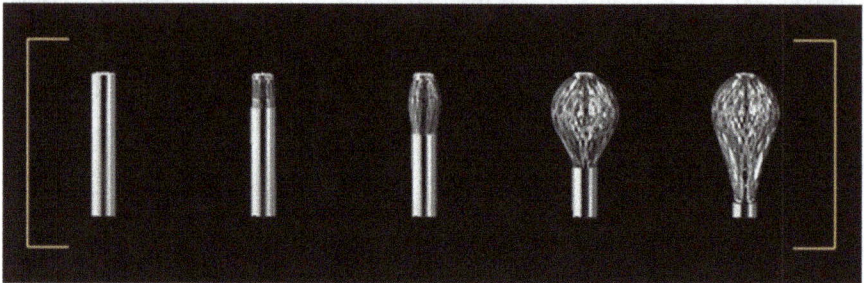

Nitinol implant stages

Figure 7.34: Conventus DRS^{TM} implant for hand wrist. Source: www.conventusortho.com/patients/our-solution-patients/.

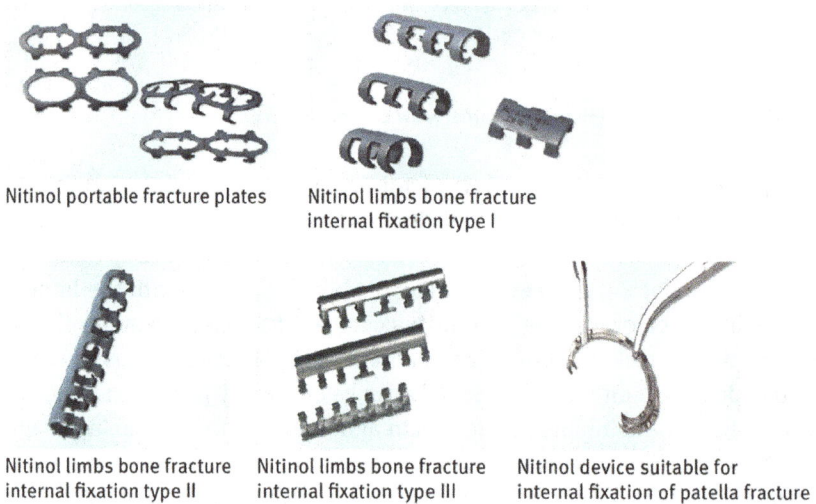

Nitinol portable fracture plates

Nitinol limbs bone fracture internal fixation type I

Nitinol limbs bone fracture internal fixation type II

Nitinol limbs bone fracture internal fixation type III

Nitinol device suitable for internal fixation of patella fracture

Figure 7.35: Nitinol devices for bones fixations from Jiangsu IAWA Biotech Engineering Co., Ltd. China.

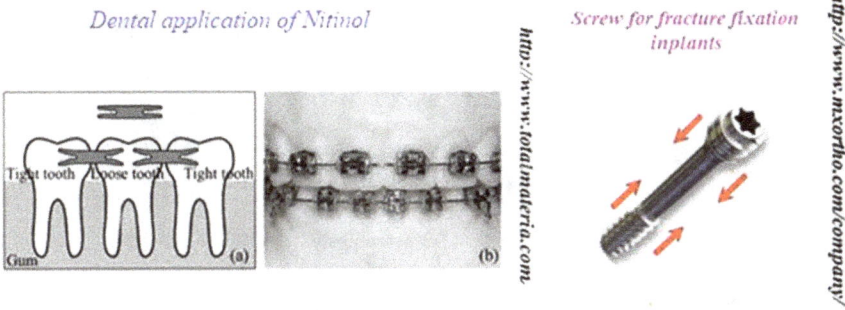

Dental application of Nitinol

Tight tooth Loose tooth Tight tooth

Gum

(a)

(b)

http://www.totalmateria.com

Screw for fracture fixation inplants

https://www.mxortho.com/company/

Figure 7.36: Dental application and fracture fixation screw for implants of Nitinol. Source www.totalmateria.com (From www.totalmateria.com/page.aspx?ID=CheckArticle&site=ktn& NM = 212); www.mxortho.com/company/.

However, the power of the piezosurgical instrument is three to six times higher than that of a dental scaler. The major advantages of this technology include high precision, a design that increases ease of curvilinear osteotomy, less trauma to soft tissue, preservation of neurological and vascular structures, reduced hemorrhage, minimal thermal damage to the bone, as well as overall improvement of healing. Piezoelectric instruments are different from rotary instrumentation or oscillating saws; they require light pressure with constant motion of the tip. Training is required to master the technique (see Figures 7.37–7.39).

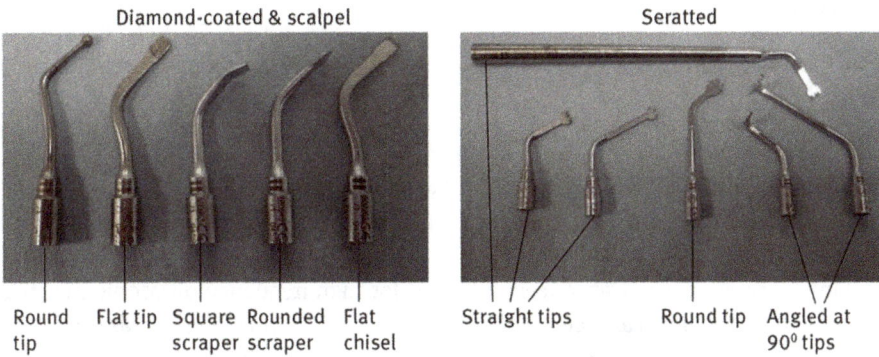

Diamond-coated & scalpel

Seratted

Round tip Flat tip Square scraper Rounded scraper Flat chisel

Straight tips Round tip Angled at 90° tips

Figure 7.37: Various tips for Piezotome 2® system manufactured by Acteon Satalec, France.

When using a surgical bur or saw, the cutting efficiency is linked to the pressure on bone. With a piezoelectric unit, cutting is due to the high-frequency vibration of the tip of the instrument; excessive pressure prevents vibration, decreases efficiency and generates frictional heat. The handle of the instruments is held with a modified pen grasp. A moderate force (1.5–3 N) is used to allow the tip to vibrate comparable to the axial force of handwriting of close to 1 N. A working pressure of 1.5 and 2.0 N with a

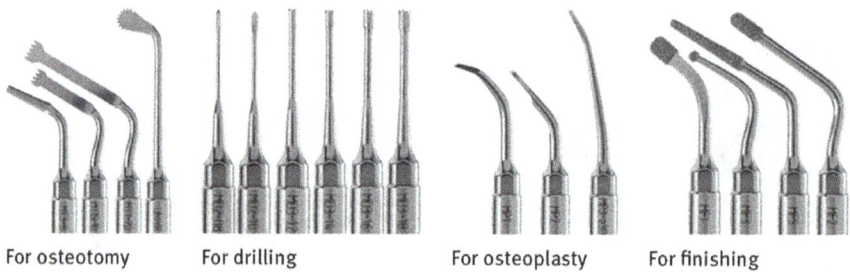

For osteotomy   For drilling   For osteoplasty   For finishing

Figure 7.38: Surgical inserts – Mectron company. Source: www.dental.mectron.com/products/piezosurgery/.

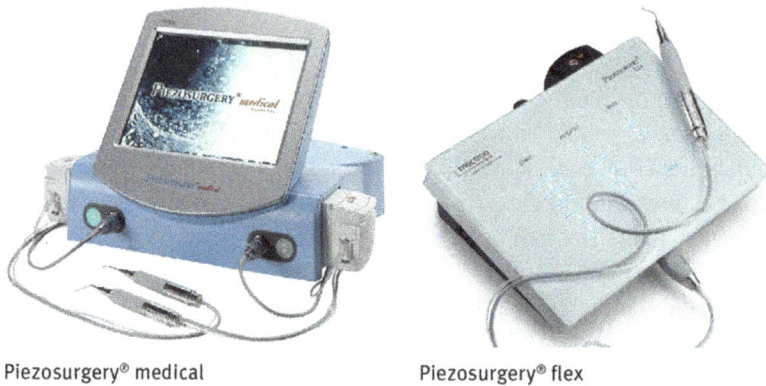

Piezosurgery® medical   Piezosurgery® flex

Figure 7.39: Mectron piezosurgery systems. Source: www.dental.mectron.com/products/piezosurgery/.

minimum of 30 ml/min cooling irrigation has been shown to fulfill the requirements for harmless intraosseous temperature [5–6]. A load of 1.5 N has been therefore recommended for cutting cortical bone.

Finally, piezoelectric transducers are used for high-frequency ultrasonic imaging applications in the medical sector [7–8]. Transducers used in medical ultrasound consist of a thickness mode resonator. The piezoelectric elements are exited and while vibrating they emit ultrasonic waves (see Figures 7.40 and 7.41). The wave propagation velocity in PZT is approximately 4350 m/s, resulting in a nominal thickness ($\lambda/2$) of 435 µm at 5 MHz, for example, though other factors also affect transducer resonance [8]. Trade-offs between desired resonance frequency and element dimensions represent a primary challenge in array design. To increase the choice of available materials, piezocomposites are also manufactured (Figure 7.41)

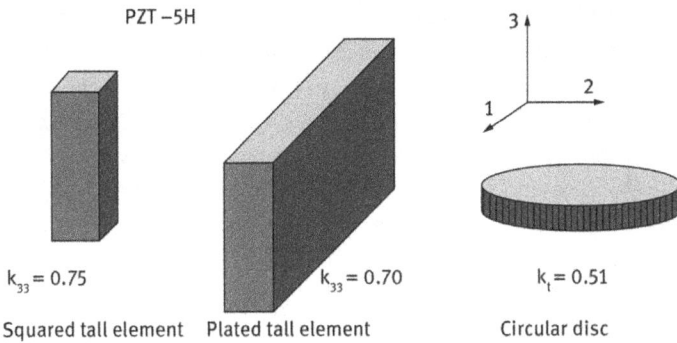

PZT –5H

$k_{33} = 0.75$

$k_{33} = 0.70$

$k_t = 0.51$

Squared tall element   Plated tall element   Circular disc

Figure 7.40: The influence of the geometric dimensions on the electrical mechanical coupling factor [7].

References

[1]   Duerig, T., Pelton, A. and Stockel, D., An overview of Nitinol medical applications, Materials Science and Engineering A 273–275, 1999, 149–160.

[2]   Stockel, D., Nitinol medical devices and implants, Minimal Invasive Therapy & Allied Technology 2(2), 2000, 81–88.

[3]   Morgan, N. B., Medical shape memory alloy applications-the market and its products, Materials Science and Engineering A 378, 2004, 16–23.

[4]   Shabalovskaya, S., Anderegg, J. and Van Humbeeck, J., Critical overview of Nitinol surfaces and their modifications for medical applications, Acta Biomaterialia 4, 2008, 447–467.

[5]   Abella, F., de Ribot, J., Doria, G., Sindreu, D. S. and Roig, M., Applications of piezoelectric surgery in endodontic surgery: a literature review, JOE 40 (3), March 2014, 325–332. doi: http://dx.doi.org/10.1016/j.joen.2013.11.014.

[6]   Hennet, P., Piezoelectric bone surgery: a review of the literature and potential applications in veterinary oromaxillofacial surgery, Frontiers in Veterinary Science 2(8), May 2015, 6. doi: http://dx.doi.org/10.3389/fvets.2015.00008.

[7]   Shung, K. K., Cannata, J. M. and Zhou, Q. F., Piezoelectric materials for high frequency imaging applications: a review, Journal of Electroceramics 19, 2007, 139–145.

[8]   Martin, K. H., Lindsey, B. D., Ma, J., Lee, M., Li, S., Foster, F. S., Jiang, X. and Dayton, P. A., Dual frequency piezoelectric transducers for contrast enhanced ultrasound imaging, Sensors 14, 2014, doi:10.3390/s141120825, 20.825–20.842.

7.3 Piezoelectric Motors

A motor is a device that produces continuous linear or rotary motion. Piezoelectric motors, which make use of the indirect piezoelectric effect to produce motion, have several advantages over conventional electromagnetic motors: do not need strong magnetic fields as conventional electromagnetic motors do, can be miniaturized, can be operated at much lower power and are more reliable. The micromotors satisfy the requirements of precise positioning applications such as mask alignment in IC technology, fiber optic alignment, medical catheter placement, autofocus and optical zoom in mobile phone cameras, pharmaceuticals handling. They can be miniaturized to a size of less than 4 mm and can provide position accuracies up to 0.1 μm or better. They have many other applications in both engineering and medical fields wherever precision movements are required and where a magnetic field in the vicinity is to be avoided. Two types of piezomotors are the linear motor and the rotary one. In these motors, the small displacement of a piezoelectric material due to the converse piezoelectric effect is converted into continuous translatory motion or a rotary motion.

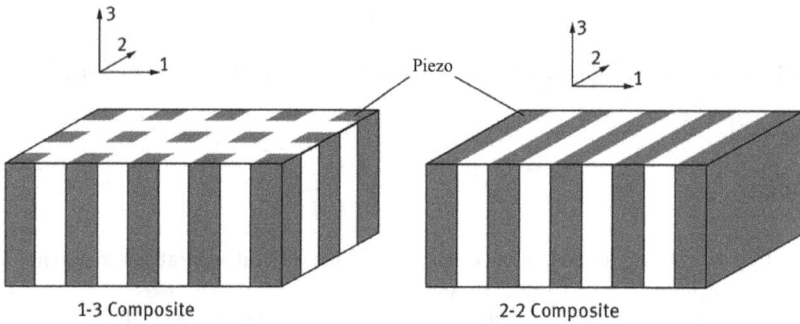

Figure 7.41: Use of piezocomposites (small PZT rods embedded in a low-density polymer) as a source of ultrasonic waves [7].

The converse piezoelectric effect [2] is causing the motor action as electrical energy is converted to mechanical energy. Application of the electric field generates strain in the material. The motor actions due to a voltage V working in parallel expansion and contraction, transverse expansion and contraction, parallel shear and bending (series and parallel connection) modes are schematically illustrated in Figure 7.42. The dimensions of the piezoelectric block, before applying the voltage, are thickness t, width w and a length L. P represents the polarization vector. The change in the dimensions due to the application of the voltage V is given by the following expressions:

$$\Delta t = V \cdot d_{33} \quad \text{for parallel expansion (contraction)} \tag{7.1}$$

$$\frac{\Delta L}{L} = \frac{\Delta w}{w} = \frac{V \cdot d_{31}}{t} \quad \text{for transverse expansion (contraction)} \tag{7.2}$$

$$\Delta x = V \cdot d_{15} \quad \text{for parallel shear} \tag{7.3}$$

$$\Delta x = \frac{3 \cdot L^2 \cdot V \cdot d_{31}}{2 \cdot t^2} \quad \text{for bending in series connection} \tag{7.4}$$

$$\Delta x = \frac{3 \cdot L^2 \cdot V \cdot d_{31}}{t^2} \quad \text{for bending in parallel connection} \tag{7.5}$$

where d_{33}, d_{31} and d_{15} are the piezoelectric strain coefficients (see Chapter 3 of this book).

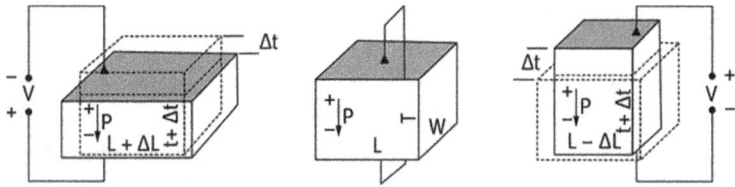

Parallel expansion and contraction

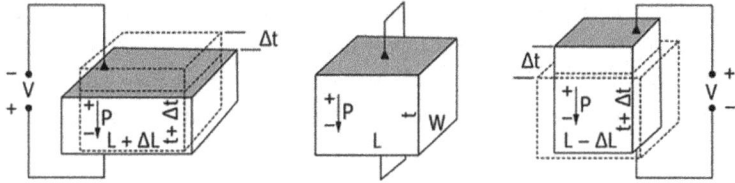

Transverse expansion and contraction

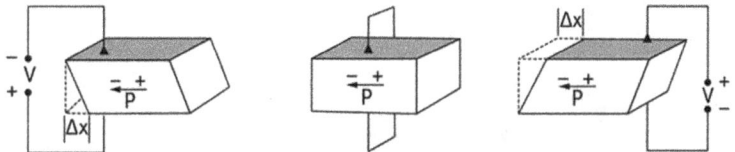

Parallel shear

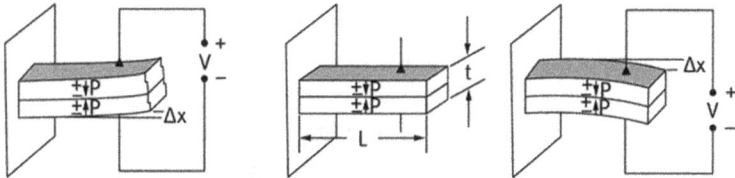

Bending mode-connection in series

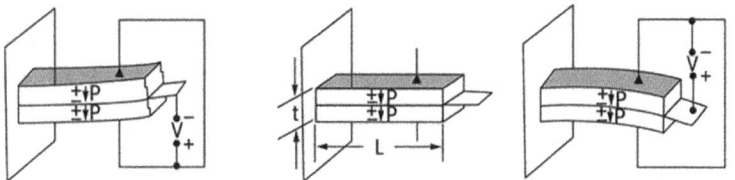

Bending mode-connection in parallel

Figure 7.42: Motor action of a piezoelectric material: parallel expansion and contraction mode; transverse expansion and contraction mode; parallel shear mode; bending mode for connection in series and bending mode for connection in parallel. Source: PIEZO SYSTEMS Inc (PIEZO SYSTEMS, INC – 65 Tower Office Park, Woburn, MA 01801, USA, Phone: 781-933-4850, Fax: 781-933-4173; email: sales@piezo.com).

7.3.1 Linear Piezoelectric Motors

Two types of linear piezoelectric motors are known: one type operates at low frequencies and the other at ultrasonic frequencies.

One type of low-frequency linear piezoelectric motor is the clamping type, which produces linear motion of a slider just like the movement of a worm. This type of piezomotor, called the "inch worm" motor, was first designed and patented by Burleigh Instruments, Inc. The principle of the motor schematically presented in Figure 7.43. The inchworm motor consists of three sets of piezoelectric actuators: two are clamping actuators (1 and 2), and one is the driving actuator (3), as shown in Figure 7.43. The actuators are operated in sequence to achieve motion of the slider, which is to be moved.

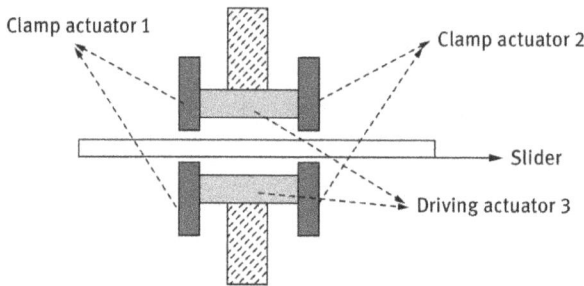

Figure 7.43: A schematic drawing of a piezoelectric inch worm motor.

The two actuators 1 and 2 are the clamp actuators, which hold the slider alternately. Actuator 3 is the drive actuator, which gets extended or contracted in the direction of the slider movement. Initially, all the actuators are disconnected from the slider. When one of the clamp actuators clutches the slider, the other clamp actuator is disconnected. Alternately, one of the clamp actuators clutches the slider. The drive actuator extends or contracts laterally in the direction of the movement of the slider. The steps of operation as illustrated in Figure 7.44 are as follows:

Initially, all the actuators are open and inactive.
Step 1: Clamp actuator 2 is closed.
Step 2: The drive actuator gets extended.
Step 3: Clamp actuator 1 is closed.
Step 4: Clamp actuator 2 is opened.
Step 5: The drive actuator gets contracted.
Step 6: Clamp actuator 2 is closed.
Step 7: Clamp actuator 1 is open.

These steps get repeated to operate the linear motor.

Schematic of a piezoelectric inch worm motor.

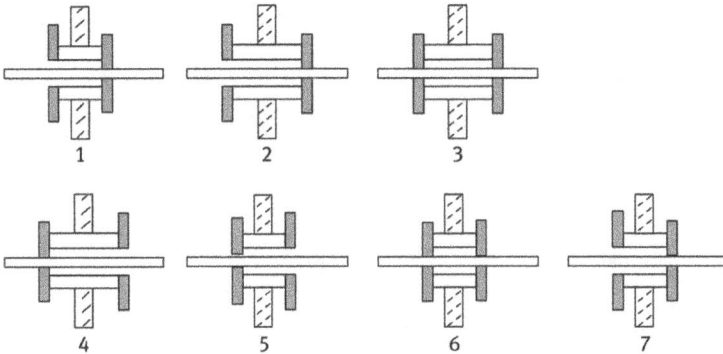

Figure 7.44: The seven steps (one cycle) of operation of an inch worm motor.

7.3.2 Rotary Piezoelectric Motors

There is an interest in developing piezoelectric rotational motors because compared to magnetic micromotors they can easily be scaled down and compared to electrostatic motors they have larger energy density. One of the rotary piezomotors is working on the Piezo LEGS walking principle[21], which is of the nonresonant type, that is, the position of the drive legs is known at any given moment. This assures very good control of the motion over the whole speed range. The performance of a Piezo LEGS motor is different from that of a DC or stepper motor in several aspects. A Piezo LEGS motor is friction based, meaning the motion is transferred through contact friction between the drive leg and the drive disc. For each waveform cycle, the Piezo LEGS motor will take one full step, referred to as one waveform (wfm)-step ($\sim$ 1.5mrad at no load with waveform Rhomb). Figure 7.45 presents schematic illustrations for one complete step. The rotational velocity of the drive axle is the wfm-step angle multiplied with the waveform frequency (1.5 mrad $\times$ 2 kHz = 3 rad/s = 170°/s).

There are other rotational piezoelectric motor architectures as well. One of those architectures is based on progressive waves (Figure 7.46). The stator consists of an elastic ring with an underside glued piezoelectric ceramics. When subjected to an alternating voltage (AC), at high frequency, the piezoelectric ceramics would vibrate. For an exciting frequency equal to the natural frequency of the stator, namely at resonance, mechanical oscillations having the amplitude of 2–3 μm can be reached.

---

21 Micromo, 14881 Evergreen Ave., Clearwater FL 33762, http://www.micromo.com/

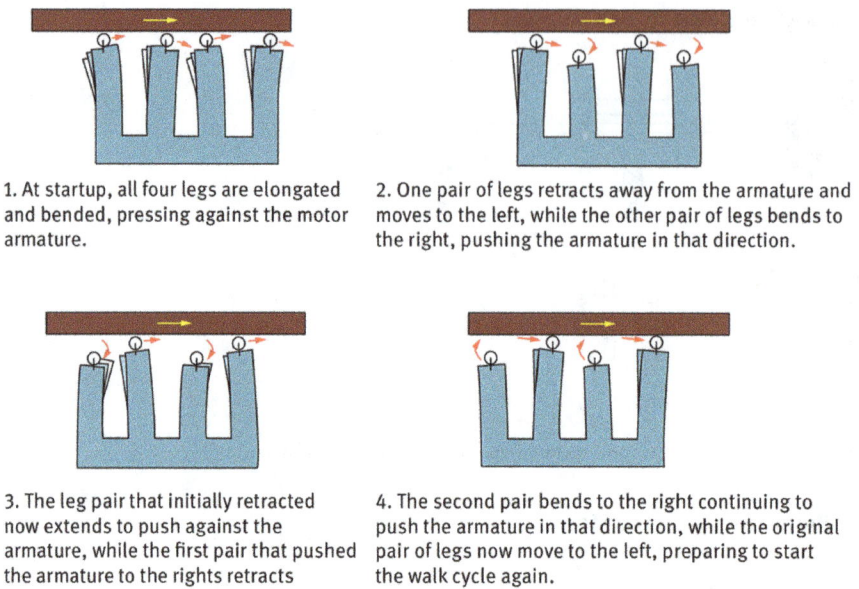

1. At startup, all four legs are elongated and bended, pressing against the motor armature.

2. One pair of legs retracts away from the armature and moves to the left, while the other pair of legs bends to the right, pushing the armature in that direction.

3. The leg pair that initially retracted now extends to push against the armature, while the first pair that pushed the armature to the rights retracts

4. The second pair bends to the right continuing to push the armature in that direction, while the original pair of legs now move to the left, preparing to start the walk cycle again.

Figure 7.45: One full step for the piezo LEGS walking principle – a schematic view.

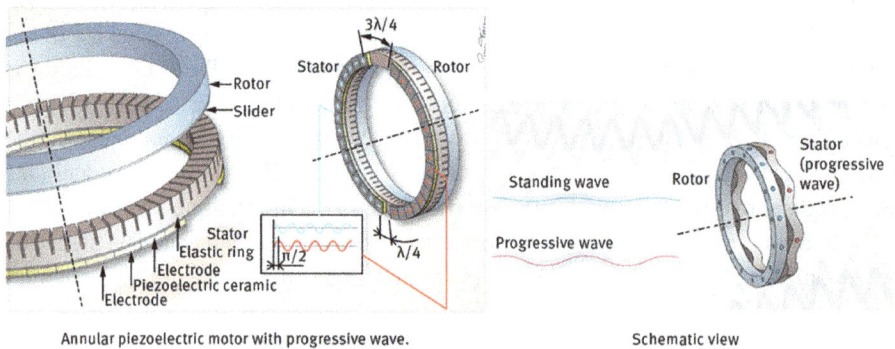

Annular piezoelectric motor with progressive wave.

Schematic view

Figure 7.46: Rotary piezomotors based on progressive waves. Source: www.pierretoscani.com/echo_shortpress.html.

7.3.3 Ultrasonic Piezomotor – Characteristics and Classification

Ultrasonic motors based on piezoelectric materials have many advantages compared with traditional motors based on the electromagnetic effect and it promises a broadband of applications since their invention [1, 3–22].

7.3.3.1 Characteristics of Ultrasonic Motors

Advantages:

- Compact structure, design flexibility, large torque density (torque/weight ratio): Ultrasonic motors have the advantages of compact structure and flexible design because piezoelectric components can excite different types of vibration, including longitudinal, bending, and torsional vibrations. As shown in Table 7.2, their torque density can be three to five times of traditional motor's ones.
- High torque at low speed can directly drive loads without gear; positioning accuracy and response speed are greatly enhanced because this advantage reduces additional volume and weight caused by the gear box, vibrations and noise, energy loss, and position error caused by transmission.
- Motor's moving parts (rotor) featuring small inertia, fast response (microsecond level), self-locking and a high holding torque: Ultrasonic motors can arrive at stable speed in several milliseconds and brake even faster because of friction between the stator and rotor.
- Good controllability of position/velocity and high resolution of displacement: Ultrasonic motors can achieve the control precision of microns and even nanos in the servo system because the operating frequency of the stator is very high and the rotor or the slider is light. Then ultrasonic motor's response is very quick and its displacement resolution is very high.
- No electromagnetic interference: The ultrasonic motors act differently from traditional motors. They do not produce magnetic fields and will not suffer from electromagnetic interference in the process of running.
- Low noise: The operating frequency band of ultrasonic motors is usually more than 20 kHz and beyond the threshold
- of human hearing. In addition, because the motor can directly drive loads, the noise from the gear box for reducing the speed is avoided.
- Operating in extreme environmental conditions: The rational design and appropriate selection of piezoelectric and frictional materials can make ultrasonic motors operate in extreme environmental conditions (vacuum or high/low temperature).

Table 7.2: Comparison between electromagnetic and ultrasonic motors (from [16]).

Motor classification	Manufacturers	Stall torque (N*m)	Rotary speed without load (r/min)	Weight (g)	Torque density (N m/g)	Efficiency (%)
EM, DC, Brush	MICROMO[1]	0.00332	13,500	11	0.0302	71
EM, DC, Brush	MAXON[2]	0.0127	5200	38	0.0334	70
EM, DC, Brush	MABICHI MOTOR[3]	0.0153	14,500	36	0.0425	53

Table 7.2 (continued)

Motor classification	Manufacturers	Stall torque (N*m)	Rotary speed without load (r/min)	Weight (g)	Torque density (N m/g)	Efficiency (%)
EM, DC, Brush	Aeroflex[4]	0.00988	4000	256	0.00386	20
EM, alternating voltage (current), 3 phases	Astro[5]	0.0755	11,500	340	0.0222	20
USM, standing wave, longitudinal-torsional type	Kumada[6]	1.334	120	150	0.889	80
USM, traveling wave type, Φ60	Shinsei[7]	1.0	150	260	0.385	35
USM, traveling wave type, Φ60	PDLab[8]	1.2	180	250	0.522	30

[1] PIEZO SYSTEMS, INC – 65 Tower Office Park, Woburn MA 01801, USA, Phone: 781-933-4850, Fax: 781-933-4173; email: sales@piezo.com
[2] http://www.maxonmotor.com/maxon/view/content/index
[3] http://www.mabuchi-motor.co.jp/en_US/technic/t_0401.html
[4] http://ams.aeroflex.com/motion/motion-motors-brushless.cfm
[5] www.astroflight.com
[6] Kumada, A.; Piezotech Inc., Tokyo, Japan [26]
[7] http://www.shinsei-motor.com
[8] http://jiangsuusm.en.made-in-china.com/company-Jiangsu-TransUSM-Co-Ltd-.html

Disadvantages:
- Small power output, low efficiency: Ultrasonic motors have two energy conversion processes. The first process converts electrical energy into mechanical energy by converse piezoelectric effect. The second process changes vibration of the stator into macro one-direction movement of the rotor by friction between the stator and rotor. Energy loss emerges from these two processes, especially in the latter. As a result, efficiency of ultrasonic motor is low. At present, the efficiency of traveling wave ultrasonic motors is about 30% and output power is less than 50 W.
- A short operational life and unsuitability of continuous operating: Friction and wear problems exist at the interfaces between the stator and rotor in the process of friction drive. In addition, high-frequency vibration can lead to fatigue damage of the rotor and piezoelectric materials, especially when the power output is big and environmental temperature is high. As a result, operational life is shortened, and performance will be reduced after continuous operating for a long time.
- Special requirements for the drive signals: In order to excite the resonance of the stator, the motors have special requirements for amplitude, frequency and phase

of excitation signals. When the motor temperature changes, frequency of excitation signals for piezoelectric elements needs appropriate adjustment to maintain the stability of the output performance. Therefore, the circuit of ultrasonic motor drivers is sometimes complex.

7.3.3.2 Classification of Ultrasonic Motors

Ultrasonic motors with design flexibility and structural diversity have no uniform method of classification. Table 7.3 lists some classifications from different viewing angles.

Table 7.3: Classification of ultrasonic motors.

Viewing angle	Type
Wave propagation method	Traveling wave, standing wave
Movement output way	Linear, rotational
Contact state between stator and rotor	Contact, noncontact
Excitation conditions of stator by piezoelectric components	Resonant, non-resonant
Number of degrees of freedom of the rotor	Single degree of freedom, multi-degree of freedom
Displacement of operating mode in direction	Out-of-plane, in-plane
Geometric shape of stators	Disk, ring, bar, shell
Rotary directions	Unidirectional, bidirectional

As ultrasonic motors are typical products, which utilize vibrations, the classification according to the vibration type can essentially reflect the characteristics of these motors. Herewith, ultrasonic motors can be divided into the following five categories:

7.3.3.3 First Category Based on Longitudinal Vibration

The motor based on the longitudinal vibration mode belongs to the standing wave motor. This kind of motor uses the Langevin vibrator with high converting efficiency from electrical energy to mechanical energy. A Langevin resonator (see Figure 7.47) [25] is a prestressed sandwich transducer: a disk, or paired of disks from piezoelectric ceramic are sandwiched between metal end sections and placed under compression bias by means of bolts drawing up on the metals. The prestress significantly reduces the tension stresses during the piezoceramics operation.

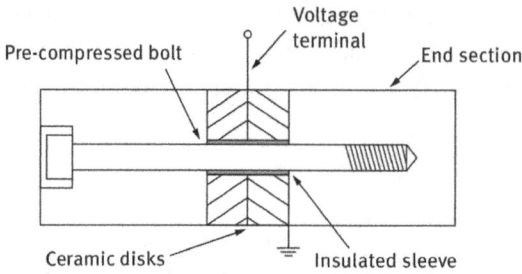

Figure 7.47: A schematic drawing of a Langevin vibrator.

The abrasion of the friction material on the contact surface of the ultrasonic motor is one of the problems the scientists had to deal with. The first standing wave ultrasonic motor proposed by Sashida [27] belongs to this kind of motors. The longitudinal vibration of the stator in one direction can be transformed into the rotating motion of the rotor through the deformation of the flexible sheet. In 1989, Kurosawa [28] proposed a linear ultrasonic motor with higher drive efficiency using the composite mode of two longitudinal vibrations. Experiments proved that the large power density of 76 W/kg can be achieved, maxim urn output force is 51 J and maximum speed is 0.55 m/s.

7.3.3.4 Second Category Based on Composite Mode of Longitudinal-Bending Vibration

In 1989, Tomikawa [29] designed a linear motor based on the longitudinal and bending modes of a rectangular plate, as shown in Figure 7.48. This motor used the first longitudinal vibration mode and the fourth bending mode of rectangular plate to achieve elliptical motion of the driving feet. The efficiency of prototype by experimental measurement was 20.8%. Later, Tomikawa [30] also put forward a flat linear ultrasonic motor using the first longitudinal mode and the eighth bending vibration mode. The experimental result was that no-load maximum speed of prototype was 0.7 m/s, and the maximum thrust was 4 N. The motor was characterized by simple structure: flat shape and fast speed, which were particularly suited for the transmission of light-thin objects such as paper, card and the like. Note that already in 1995, Nikon and NEC Co. Ltd. produced the motor products of this type, respectively.

In 1992, Onishi and Naito [31] designed a IT-shape linear ultrasonic motor with biped structure, as shown in Figure 7.49. Two stacked piezoelectric ceramics in tilt layout stimulated longitudinal and lateral bending modes of the leg parts of the IT-shape elastomer, which synthesizes elliptical motion for driving the guide rails. The excitation frequency was about 90 kHz and the phase difference of voltage imposed was 90°. This motor's no-load speed was 30 cm/s, and the maximum thrust was 10 N. In 1993, SUNSYN Company manufactured and commercialized this IT-shape linear ultrasonic motor for $X–Y$ positioning systems, which became the first application of linear ultrasonic motors.

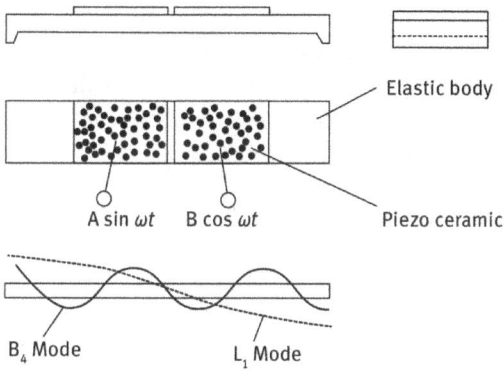

Figure 7.48: Tomikawa's plate-type longitudinal/bending motor.

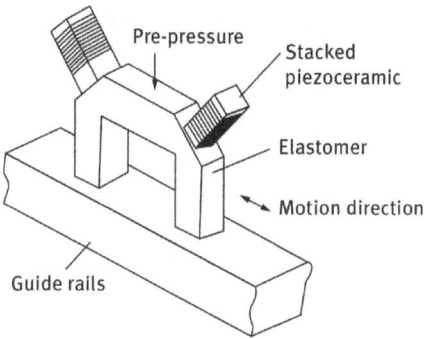

Figure 7.49: PI shape linear motor stator.

7.3.3.5 Third Category Based on Composite Mode of Longitudinal-Torsional Vibration

As shown in Figure 7.50, Kurosawa [32] developed a longitudinal-torsional hybrid motor. The unique characteristic of this motor was that stacked piezoelectric vibrator produced longitudinal vibration. This vibration could possess larger amplitude in conditions of low-voltage and nonresonance. The rotor's diameter was 50 mm and total length was 82 mm. The motor's no-load rotational speed measured was 100 r/min, the maximum torque was 0.7 Nm and the maximum efficiency was 33% when prepressure of 90 N was applied on the rotor, and voltage 34 V (measured as V_{rms}, where rms = root means square), imposed on torsional vibrator. Figure 7.51 shows a longitudinal torsional hybrid ultrasonic motor with a brush developed in a lab. Within the motor longitudinal and torsional piezoelectric ceramics were placed in the stator and rotor, respectively. This design could adjust structural parameters of the stator and rotor individually to keep the modal frequencies of the longitudinal and torsional vibrations as close as possible. The rational design of the structure of the motor could increase the prepressure on the contact surfaces in the operating process, which could improve output performance and operating efficiency. Because the torsional vibration piezoelectric ceramics were placed in the rotor, the brush was used for supplying electric power to the rotor. The motor was called the brush-type longitudinal-torsional hybrid ultrasonic motor. The motor's diameter was 45 mm, length was 210 mm and maximum output torque was 2.5 Nm.

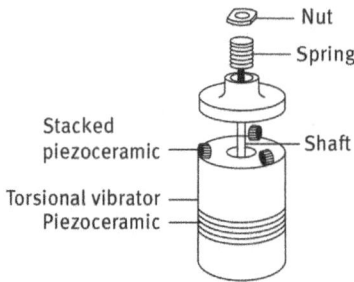

Figure 7.50: Kurosawa's longitudinal-torsional motor.

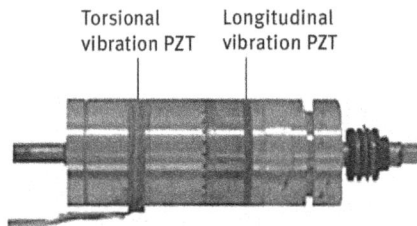

Figure 7.51: Brush-type longitudinal-torsional motor.

7.3.3.6 Fourth Category Based on Bending Vibration

According to the structure of stator, ultrasonic motors based on bending vibration mode can be divided into three categories: bar type, ring type and disk type, which all belong to traveling wave ultrasonic motors. In recent years, USMs with bar-type stator based on modes of the out-of-plane bending vibration have become a hotspot in the research area of micro actuator because of their advantages of simple structure, manufacturing convenience and low cost. Some ultrasonic motors with bar-type stator have been applied to the micro-lens focusing system and medical endoscopy system.

7.3.3.7 Fifth Category Based on In-Plane Vibration

Motors with in-plane vibrations have three types: extension-contraction, bending and torsion. In 1989, Takano used in-plane extension-contraction and bending vibration mode for developing an ultrasonic motor, as shown in Figure 7.52 When alternating voltage signals with phase difference of $\pi/2$ were, respectively, imposed on circular piezoelectric ceramics with two areas, radial and tangential movement of drive point A (A') synthesized the elliptical motion, which drove the rotor. The rotor's diameter was

10 mm, thickness was 2 mm, operating frequency was 43.3 kHz, maximum output torque was 40 mNm and efficiency was 3.5%. In 2005, Zhou et al. [33] developed a linear ultrasonic motor, which used the in-plane extension–contraction vibration modes of a hollow cylinder, as shown in Figure 7.53.

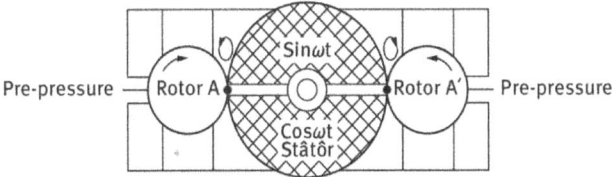

Figure 7.52: USM (ultrasonic motor) based on in-plane modes.

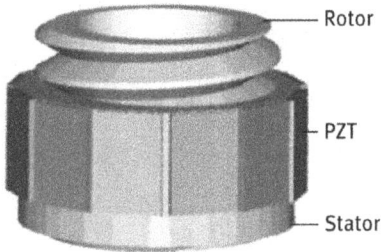

Figure 7.53: Linear USM based on in-plane extension–contraction mode.

Note that in addition to the previous five categories, there are some ultrasonic motors based on other modes such as a composite mode of the torsional and bending vibrations, a composite mode of the longitudinal and shear vibrations.

7.3.4 Ultrasonic Piezomotors: Operating Principles

Piezomotors producing elliptical motion in the contact area between input and output links are the most popular one. For this purpose oblique impact upon the output link or traveling wave is made use of. In piezomotors, making use of oblique impact, friction force transmits motion and energy between input and output links. This may be realized by two oscillatory motions (normal and tangential components) u_y and u_x in the contact area with a phase difference, which is used to change output link motion direction. Both motions can be realized by one or two active links oscillating resonantly. Various oscillations offer possibilities to develop different kinds of piezomotors: longitudinal, transversal, shear and torsional. Piezomotors employing oblique impacts possess a very wide frequency range. Its lower limit is at lower ultrasound frequencies (for elimination of acoustic action), 16–20 kHz, and its upper limit is at several megahertz. Traveling wave motion piezomotors are based on frictional interaction between the traveling wave motion in the elastic body and the output link, that is, its principle of operation is similar to the harmonic traction transmission. Wave propagating along the surface (Rayleigh wave) of the input link forms the elliptical motion in the contact area. Rayleigh wave is a coupled wave of longitudinal and shear waves; thus each surface point in elastic medium moves along an elliptical locus. Flexural, shear, torsional and longitudinal waves are used in piezomotors. Traveling wave in piezoceramic is excited by electrical field. Traveling wave motion piezomotors characteristics (ABB Corporate Research IRCTC/AS) [34] are presented in Table 7.4.

Table 7.4: Properties of some traveling wave piezomotors.

Motor	Unit	USR60	USR45	USR30
Operating frequency	kHz	40	43	42
Operating voltage	V_{rms}	100	100	100
Rated torque	Nm	0.38	0.15	0.04
Rated output	W	4.0	2.3	1.0
Rated rotational speed	rpm	100	150	250
Mechanical time constant	ms	1	1	1
Weight	g	175	69	33
Rotation irregularity	%	2	2	2
Lifetime	h	1000	1000	100
Operating temperature range	°C	−10 + 50	−10 + 50	−10 + 50

References

[1] Sharp, S. L., Design of a linear ultrasonic piezoelectric motor, Master of Science Thesis, Department of Mechanical Engineering, Brigham Young University, Provo, UT, USA, August 2006, 189 pp.

[2] Bansevicius, R. and Tolocka, R. T., Piezoelectric actuator s, Chapter 20.3, The Mechatronics Handbook, Bishop, R. H. – Editor-in-Chief, CRC Press LLC, 2002.

[3] Yoon, M.-S., Khansur, N. H., Lee, K.-S. and Park, Y. M., Compact size ultrasonic linear motor using a dome shaped piezoelectric actuator, Journal of Electroceramics 28(2), 2012, 123–131.

[4] Pirrotta, S., Sinatra, R. and Meschini, A., A novel simulation model for ring type ultrasonic motor, Meccanica 42(2), April 2007, 127–139.

[5] Liu, Y., Chen, W., Liu, J. and Shi, S., A rotary ultrasonic motor using bending vibration transducers, IEEE Transactions on Ultrasonics, Ferroelectrics, and Frequency Control 57(10), October 2010, 2360–2364.

[6] Flynn, A. M., Piezoelectric ultrasonic micromotors, MIT Artificial Intelligence Laboratory, Computer Science and Artificial Intelligence Lab (CSAIL) Artificial Intelligence Lab Publications, AI Technical reports (1964–2004), June 1995, http://hdl.handle.net/1721.1/7086.

[7] Hagood, N. W. and McFarland, A. J., Modeling of a piezoelectric rotary ultrasonic motor, IEEE Transactions on Ultrasonics, Ferroelectrics, and Frequency Control 42(2), March 1995, 210–224.

[8] Kanda, T., Makino, A. and Oomori, Y., A cylindrical micro ultrasonic motor using micromachined piezoelectric vibrator, Okayama University, Japan, Ultrasonics Symposium, 2005 IEEE (Vol. 1), 18–21 September 2005.

[9] Uchino, K., Cagatay, S., Koc, B., Dong, S., Bouchilloux, P. and Strauss, M., Micro piezoelectric ultrasonic motors, ADM001697, ARO-44924.1-EG-CF, International Conference on Intelligent Materials (5th) (Smart Systems & Nanotechnology), State College, PA, 14–17 June 2003.

[10] Watson, B., Friend, J. and Yeo, L., Piezoelectric ultrasonic resonant motor with stator diameter less than 250 μm: the Proteus motor, Journal of Micromechanics and Microengineering 19, 2009, 1–5.

[11] Hirata, H. and Ueha, S., Characteristics estimation of a traveling wave type ultrasonic motor, IEEE IEEE Transactions on Ultrasonics, Ferroelectrics, and Frequency Control 40(4), July 1993, 402–406.

[12] Uchino, K., Piezoelectric ultrasonic motors: overview, Smart Materials and Structures 7(3), 1998, 273–285.

[13] Dong, S., Yan, L., Wang, N. and Viehland, D., A small, linear, piezoelectric ultrasonic cryomotor, Applied Physics Letters 86, 2005, article ID 053501, 3 pp.

[14] Wallaschek, J., Contact mechanics of piezoelectric ultrasonic motors, Smart Materials and Structures 7(3), 1998, 369–381.

[15] Zhang, H., Dong, S.-X., Zhang, S.-Y., Wang, T.-H., Zhang, Z.-N. and Fan, L., Ultrasonic micromotor using miniature piezoelectric tube with diameter of 1.0 mm, Ultrasonics 44, 2006, e603–e606.

[16] Zhao, C. S., Ultrasonic Motors Technologies and Applications, Beijing, Science Press, China, 2010.

[17] Xiaolong, L., Junhui, H., Lin, Y. and Chunsheng, Z., A novel in-plane rotary ultrasonic motor, Chinese Journal of Aeronautics 27(2), 2014, 420–424.

[18] Newton, D., Garcia, E. and Horner, G. C., A linear piezoelectric motor, Smart Materials and Structures 7(3), 1998, 295–305.

[19] Ho, S.-T. and Shin, Y.-J., Analysis of a linear piezoelectric motor driven by a single-phase signal, 2013 IEEE International Ultrasonics Symposium (IUS), 21–25 July 2013, 481–484, Prague, The Czech Republic.

[20] Bauer, M. G., Design of a linear high precision ultrasonic piezoelectric motor, Ph.D. Thesis, Mechanical Engineering Department, Raleigh, North Carolina State University, 2001.

[21] Lopez, J. F., Modeling and optimization of ultrasonic linear motors, Ph.D. Thesis, Faculte des Sciences et Techniques de L'ingenieur, Ecole Polytechnique Federale de Lausanne, Switzerland, Thesis No. 3662, 10th November 2006, 281 pp.

[22] El Ghouti, N., Hybrid Modelling of a Traveling Wave Piezoelectric Motor, Ph.D. Thesis, Department of Control Engineering, Aalborg University, DK-9220 Aalborg Ø, Denmark, Doc. no. D-00-4383, May 2000.

[23] Schaaf, U., Pushy motors (piezoelectric motors), IEEE Review 41(3), May 1995, 105–108.

[24] Shieh, Y. J., Ting, Y., Hou, B.-K. and Yeh, -C.-C., High speed piezoelectric motor, 1–3, applications of ferroelectrics held jointly with 2012 European conference on the applications of polar dielectrics and 2012 International Symposium on Piezoresponse Force Microscopy and Nanoscale Phenomena in Polar Materials (ISAF/ECAPD/PFM), 2012 International Symposium, Aveiro, Portugal, July 2012.

[25] Carotenuto, R., Iula, A. and Pappalardo, M., A displacement amplifier using mechanical demodulation, Applied Physics Letters 73(18), November 1998, 2573–2575.

[26] Kumada, A., Piezoelectric revolving motors applicable for future purpose, IEEE 7th Intenational Symposium on Applications of Ferroelectrics, 213–219, Urbana-Champaign, IL, USA, 6–8 June 1990.

[27] Sashida, T., Motor device using ultrasonic oscillation, US patent #4562374, 16th of May 1984.

[28] Kurosawa, M., Nakamura, K., Okomoto, T. and Ueha, S., An ultrasonic motor using bending vibrations of a short cylinder, IEEE Transactions on Ultrasonics, Ferroelectrics, and Frequency Control 36(5), September 1989, 517–521.

[29] Tomikawa, Y., Takano, T. and Umeda, H., Thin rotary and linear ultrasonic motors using a double-mode piezoelectric vibrator of the first longitudinal and second bending mode, Japan Journal of Applied Physics, 31, 1992, 3073–3076.

[30] Ueha, S. and Tomikawa, Y., Ultrasonic Motors, Theory and Applications, Oxford Science Publication, Clarendon Press, 1993.

[31] Onishi, K. and Naito, K., Ultrasonic linear motor, US patent #5134334 A, July 28th, 1992.

[32] Kurosawa, M. and Ueha, S., Hybrid transducer type ultrasonic motor, IEEE Transactions on Ultrasonics, Ferroelectrics, and Frequency Control 38(3), 1991, 89–92.

[33] Zhou, T., Zhang, K., Chen, Y., Wang, H., Wu, J., Jiang, K. and Xue, P., A cylindrical rod ultrasonic motor with 1 mm diameter and its application in endoscopic OCT, Chinese Science Bulletin 50(8), 2005, 826–830.

[34] Bishop, R. H., The Mechatronics Handbook, CRC Press, 2002, 1290.

8 Energy Harvesting Using Intelligent Materials

8.1 Piezoelectric Energy Harvesting

Energy harvesting or sometimes also called energy scavenging is defined as a way to extract energy from various sources of energy existing in the ambient environment, like ambient light directly from the sun or artificial one, radio frequency (RF) in the form of RF, various thermal sources and mechanical vibration sources and transform it to electrical energy to be provided to a variety of users. The energy extraction can be done using one or more of the following principles [1]: photovoltaic, Seebeck,[1] electromagnetic for both RF and mechanical sources, electrostatic and piezoelectric. The studies dedicated to power harvesting have increased dramatically in recent few decades with the goal of discovering new ways of extraction and storage of the energy available in the ambient environment, with new sources like human body motions or wind-induced vibration being investigated and analyzed. The sources of mechanical energy, therefore, include the transportation area (like cars, trucks, trains and even aircrafts), infrastructures (like bridges, roads, tunnels, farms and houses), industry (like motors, pumps, compressors, vibrations, noise), environment (like wind, ocean current, waves) and human body activities and motion (like breathing, walking, jumping, jogging, exhalation and arm and/or fingers motion). This chapter is dedicated to piezoelectric harvesting from ambient vibrations, where the piezoelectric materials use its d_{31} operating mode and the direct stress application using the d_{33} mode of operation. Roundy [2], Lopes and Gallo [3] gathered a list of various ambient vibration sources as to be eventually harvested by piezoelectric-based harvesters (see Table 8.1).

The diagram showing the way the energy is transformed from its source to electrical energy using a harvester is presented schematically in Figure 8.1. Force and velocity are transformed to voltage and current depending on the type of harvester.

This chapter would discuss scavenging devices based on piezoelectric materials and their capability of transforming ambient mechanical vibrations or direct applied stresses into electrical power (or energy). Figure 8.2 presents a schematic view of the two most popular modes of generation of electrical energy from mechanical energy.

---

1 The Seebeck effect is a phenomenon in which a temperature gradient between two dissimilar electrical conductors or semiconductors produces a voltage difference between the two substances.

https://doi.org/10.1515/9783110726701-008

Table 8.1: Ambient sources of vibration.

Vibration source	Peak acceleration (m/s^2)	Associated frequency (Hz)
Base of three-axis machine tool	10.0	70
Kitchen blender casing	6.4	121
Clothes dryer	3.5	121
Washing machine	0.5	109
Refrigerator	0.1	240
Laptop while a CD is read	0.6	75
Small microwave oven	2.25	121
Door frame as door closes	3.0	125
HVAC[1] vents in office building	0.2–1.5	60
Bread maker	1.03	121
Wooden deck with foot traffic	1.3	385
Car engine compartment	12.0	200
Second story floor of a wood frame office building	0.2	100
External windows (2 ft × 3 ft)[2] next to a busy street	0.7	100
Person nervously tapping their heel	3.0	1
Car instrument panel	3.0	13

[1]HVAC represents heating, ventilation and air conditioning.
[2]1 ft = 0.3048 m.

Figure 8.1: Energy harvesting system – a schematic view.

In Figure 8.2, P represents the polarity direction while σ stands for the mechanical stress applied on a piezoelectric plate having the dimensions $a \times b \times t$ (width × length × thickness). The relationship among the mechanical, electrical and the geometric dimensions, depending on the mode of operations, is given in Equation (8.1) for d_{33} operation mode

$$Q_{@V=0} = d_{33} \cdot \sigma \cdot (a \cdot b)$$
$$V_{@Q=0} = g_{33} \cdot \sigma \cdot t$$

(8.1)

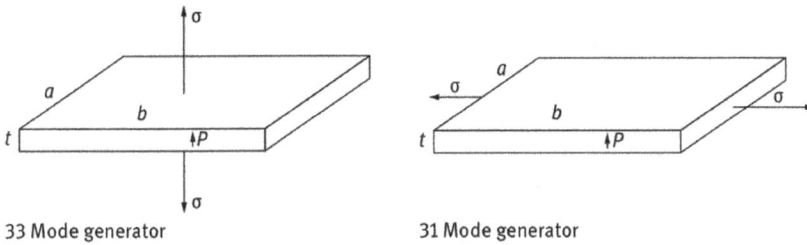

Figure 8.2: Two modes of generation of electric energy – a schematic view.

where Q is the induced charge on the piezoelectric electrodes, V is the induced voltage, and d_{33} and g_{33} are the piezoelectric field constant and the piezoelectric voltage constant, respectively.

For the d_{31} operation mode, we get

$$Q_{@V=0} = d_{31} \cdot \sigma \cdot (a \cdot b)$$
$$V_{@Q=0} = g_{31} \cdot \sigma \cdot t$$

(8.2)

To obtain the electric energy for the two modes of generation we have to multiply the charge by the voltage and divide by 2, yielding for d_{33} operation mode

$$U = \frac{1}{2} Q \cdot V = \frac{1}{2} g_{33} \cdot d_{33} \cdot \sigma_{33}^2 \cdot (a \cdot b \cdot t) = \frac{1}{2} g_{33} \cdot d_{33} \cdot \sigma_{33}^2 \cdot Vol$$

(8.3)

where Vol stands for the volume of the piezoelectric material. Accordingly, for the d_{31} operation mode, the energy will be defined as

$$U = \frac{1}{2} Q \cdot V = \frac{1}{2} g_{31} \cdot d_{31} \cdot \sigma_{11}^2 \cdot (a \cdot b \cdot t) = \frac{1}{2} g_{31} \cdot d_{31} \cdot \sigma_{11}^2 \cdot Vol$$

(8.4)

To perform the harvesting of vibrational energy, a device called harvester, which normally consists of a cantilever-based beam equipped with piezoelectric patches or layers bonded either on one side of the carrying beam or on both sides. To reduce the natural frequency of the piezo-beam, to the ambient frequency, a mass is added to the free end of the cantilever, called the end mass. The base of the cantilever is attached to the vibrating object from which the mechanical energy would then be scavenged into electrical energy using a suitable electrical circuit. The need of a carrying beam is either due to the brittleness of the piezoelectric material (has the characteristics of a ceramic material) in the case of using PZT or due to the flexibility when using PVDF materials. Figure 8.3 presents a typical cantilever piezo-beam, its equivalent 1D of freedom system consisting of a mass, spring, and a damper, and the way the piezoelectric device is connected electrically, by using a diode bridge to the regulator and the storage system be it a battery of a capacitor or even a super-capacitor.

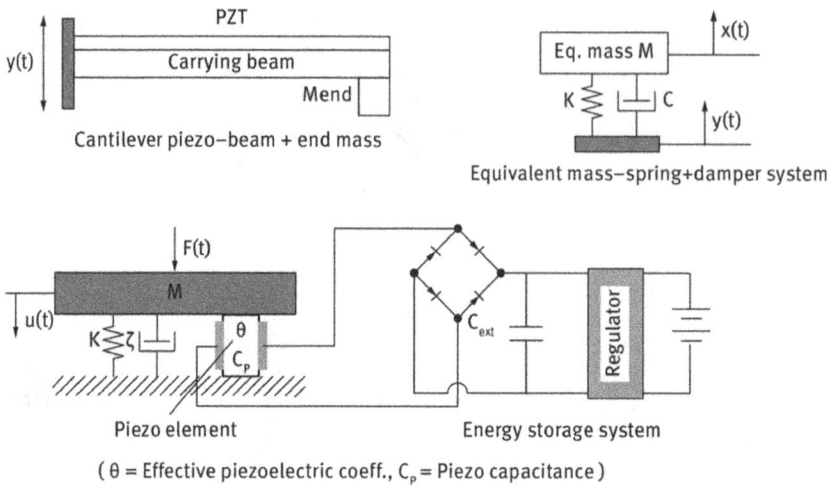

Figure 8.3: A cantilever piezo-beam + end-mass, its equivalent mass, spring, damper system, and the harvester + energy storage system – a schematic view (adapted from [5]).

As described in [8] the model used to describe the motion of the 1D system is attributed to Williams–Yates model [5] (see also a detailed derivation in Section 8.2) and can be written as (see Figure 8.3)

$$M \cdot \ddot{z}(t) + C_d \cdot \dot{z}(t) + K \cdot z(t) = -M \cdot \ddot{y}(t) \tag{8.5}$$

where $z(t) = x(t) - y(t)$ is the relative displacement of the mass M. The total power dissipated on the damper C_d as a function of the vibration frequency ω and the natural frequency of the cantilever piezo beam, ω_n, can be written as (see a detailed derivation in Section 8.2)

$$p_d(\omega) = \frac{2 \cdot \zeta \cdot M \cdot \left(\frac{\omega}{\omega_n}\right)^3 \cdot \omega^3 \cdot Y_0^2}{\left[1 - \left(\frac{\omega}{\omega_n}\right)^2\right]^2 + \left[2 \cdot \zeta \cdot \left(\frac{\omega}{\omega_n}\right)\right]^2} \tag{8.6}$$

where Y_0 is the displacement of the inertial mass M, and ζ is the damping ratio. Taking into account the definitions:

$$\omega_n^2 \equiv \frac{K}{M}; \quad \frac{C_d}{M} \equiv 2 \cdot \zeta \cdot \omega_n \quad \Rightarrow \quad \zeta = \frac{C_d}{2\sqrt{M \cdot K}} \tag{8.7}$$

and tuning the frequency of the cantilever piezo beam to the ambient frequency, namely working at resonance, $\omega = \omega_n$, one can get the maximal available power in the following form

$$p_d(\omega_n)_{\max} = \frac{2 \cdot \zeta \cdot M \cdot \omega_n^3 \cdot Y_0^2}{4 \cdot \zeta^2} = \frac{M \cdot \omega_n^3 \cdot Y_0^2}{2 \cdot \zeta} \tag{8.8}$$

Equation (8.8) can be rewritten using the acceleration of the motion, $\bar{a}$, while $\bar{a} = \omega_n^2 Y_0$ to yield

$$p_d(\omega_n)_{\max} = \frac{M \cdot \bar{a}^2}{2 \cdot \omega_n \cdot \zeta} \tag{8.9}$$

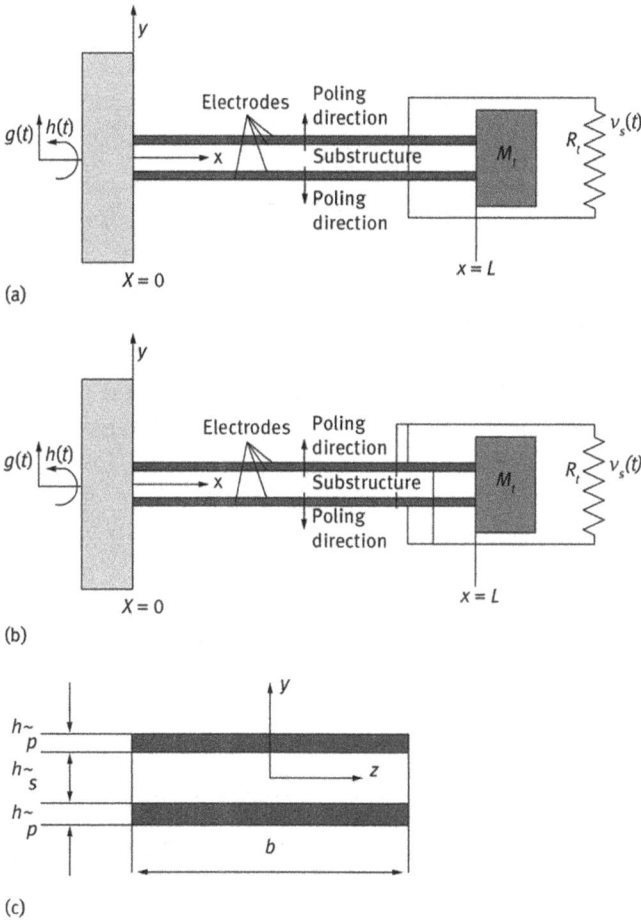

Figure 8.4: Bimorph cantilever configuration: (a) series connection, (b) parallel connection and (c) cross section (from [29]).

The power calculated by Equations (8.8) and (8.9) are usually applied for any linear inertial transducer, although Williams–Yates model was originally derived for an electromagnetic transducer having viscous damping. According to [8], the mechanism of piezoelectric transduction is relatively complicated, and therefore another model (Ertuk–Inman model [27–29]) is to be used when analyzing piezoelectric harvesters. Using Figure 8.4 Ertuk and Inman [29] wrote the following equation of motion for the beam using Euler–Bernoulli beam theory for the series connection:

$$
(\overline{EI})\frac{\partial^4 w_{rel}^s(x,t)}{\partial x^4} + c_s I \frac{\partial^5 w_{rel}^s(x,t)}{\partial x^4 \partial t} + c_a \frac{\partial w_{rel}^s(x,t)}{\partial t} + m \frac{\partial^2 w_{rel}^s(x,t)}{\partial t^2} +
$$
$$
+ \overline{V}_s v_s(t) \left[\frac{d\delta(x)}{dx} - \frac{d\delta(x-L)}{dx}\right] = - \left[m + M_{end}\delta(x-L)\right]\frac{\partial^2 w_{base}(x,t)}{\partial t^2}
$$

(8.10)

where the piezoelectric coupling term $\overline{V}_s$ for the series connection is given by

$$
\overline{V}_s = \frac{\bar{e}_{31}b}{2h_{\tilde{p}}}\left[\frac{h_{\tilde{s}}^2}{4} - \left(h_{\tilde{p}} + \frac{h_{\tilde{s}}}{2}\right)^2\right]
$$

(8.11)

where w_{base} is the effective base displacement having a translational and small rotation component, w_{rel}^s is the relative transverse displacement at series connection, $v_s(t)$ is the voltage response across the resistive load (at series connection), $(\overline{EI})$ is the bending stiffness with I being the moment of inertia, m is the mass per unit length of the beam, c_a is the external viscous damping (due to air or other surrounding fluid), $c_s I$ is the internal strain rate (or Kelvin–Voight damping), M_{end} is the tip mass located at the free end of the cantilever beam, $\bar{e}_{31}$ is the piezoelectric constant and $\delta(x)$ is the Dirac delta function. For the parallel connection case, the equation of motion has the following form:

$$
(\overline{EI})\frac{\partial^4 w_{rel}^p(x,t)}{\partial x^4} + c_s I \frac{\partial^5 w_{rel}^p(x,t)}{\partial x^4 \partial t} + c_a \frac{\partial w_{rel}^p(x,t)}{\partial t} + m \frac{\partial^2 w_{rel}^p(x,t)}{\partial t^2} +
$$
$$
+ \overline{V}_p v_p(t) \left[\frac{d\delta(x)}{dx} - \frac{d\delta(x-L)}{dx}\right] = - \left[m + M_{end}\delta(x-L)\right]\frac{\partial^2 w_{base}(x,t)}{\partial t^2}
$$

(8.12)

and the backward coupling term $\overline{V}_p$ for the parallel connection is given by

$$
\overline{V}_p = 2\overline{V}_s = \frac{\bar{e}_{31}b}{h_{\tilde{p}}}\left[\frac{h_{\tilde{s}}^2}{4} - \left(h_{\tilde{p}} + \frac{h_{\tilde{s}}}{2}\right)^2\right]
$$

(8.13)

while the mass, m and the bending stiffness are written as

$$
m = b(\rho_{\tilde{s}} h_{\tilde{s}} + 2\rho_{\tilde{p}} h_{\tilde{p}})
$$
$$
(\overline{EI}) = \frac{2b}{3}\left\{\overline{E}_{\tilde{s}}\frac{h_{\tilde{s}}^3}{8} + \bar{c}_{11}^E\left[\left(h_{\tilde{p}} + \frac{h_{\tilde{s}}}{2}\right)^3 - \frac{h_{\tilde{s}}^3}{8}\right]\right\}
$$

(8.14)

where $\rho_{\bar{s}}$ and $\rho_{\bar{p}}$ are the mass densities of the carrying structure and the piezoelectric material, respectively, $\bar{E}_{\bar{s}}$ is Young's modulus of the carrying structure, while $\bar{c}_{11}^E$ is the elastic stiffness at constant electrical field. Because the only mechanical strain is that due to the axial strain, the electric displacement can be written as

$$D_S = \bar{e}_{31} S_1^{\bar{p}} + \bar{\varepsilon}_{33}^S E_3 \quad \text{where} \quad \bar{\varepsilon}_{33}^S = \varepsilon_{33}^T - \frac{d_{31}^2}{s_{11}^E} \tag{8.15}$$

and $\bar{\varepsilon}_{33}^S$ is the permittivity component at constant strain (assuming a plane-stress case) and ε_{33}^T is the permittivity at constant stress. Applying the Gauss law on the electrical displacement and assuming the piezoelectric layers operate within a circuit with only resistive load R_{load}, one obtains the circuit equation [8–29]:

$$\frac{\bar{\varepsilon}_{33}^S \cdot b \cdot L}{h_{\bar{p}}} \frac{dv(t)}{dt} + \frac{v(t)}{R_{\text{load}}} = -\bar{e}_{31} h_{\bar{p}_c} b \int_0^L \frac{\partial^3 w_{\text{rel}}(x,t)}{\partial x^2 \partial t} dx \tag{8.16}$$

where b, $h_{\bar{p}}$ and L are the width, the thickness and the length of the piezo layer, respectively, and $h_{\bar{p}_c}$ stands for the distance between the neutral axis and the center of the piezo layer. Substituting in Equation (8.16) a modal representation for the w_{rel} (x,t) in the form

$$w_{\text{rel}}(x,t) = \sum \phi_r(x) \varphi_r(t) \tag{8.17}$$

leads to

$$\frac{\bar{\varepsilon}_{33}^S \cdot b \cdot L}{h_{\bar{p}}} \frac{dv(t)}{dt} + \frac{v(t)}{R_{\text{load}}} = \sum_{i=1}^{\infty} \kappa_r \frac{d\varphi_r(t)}{dt} \tag{8.18}$$

where κ_r is the modal coupling term in the electrical circuit defined as

$$\kappa_r = -\bar{e}_{31} h_{\bar{p}_c} b \int_0^L \frac{d^2 \phi_r(x)}{\partial x^2} dx = -\bar{e}_{31} h_{\bar{p}_c} b \frac{d\phi_r(x)}{\partial x}\bigg|_{x=L} \tag{8.19}$$

According to [8, 29] the coupled voltage response to harmonic base excitation at steady state can be written as

$$v_s(t) = V_s e^{j\omega t} = \frac{\sum_{r=1}^{\infty} \dfrac{j\omega \kappa_r F_r}{\omega_r^2 - \omega^2 + 2\zeta_r \omega_r \omega}}{\dfrac{1}{R_{\text{load}}} + j\omega \dfrac{C_p}{2} + \sum_{r=1}^{\infty} \dfrac{j\omega \kappa_r \chi_r}{\omega_r^2 - \omega^2 + 2\zeta_r \omega_r \omega}} e^{j\omega t} \tag{8.20}$$

where ω is the excitation frequency, $C_p = \bar{\varepsilon}_{33}^S \cdot b \cdot L / h_p$ is the capacitance of the piezoelectric layer, χ_r is the backward modal coupling term, ω_r, ζ_r and F_r are the undamped natural frequency, the mechanical damping ratio and the modal mechanical force at

rth mode, respectively. Accordingly, one can define the $|v(t)|$ as the peak voltage amplitude, the peak power amplitude as $|v(t)|^2/R_{load}$, and the average power amplitude as $|v(t)|^2/(2 \cdot R_{load})$. For the case where $\omega \cong \omega_r$, Equation (8.20) can be simplified to yield a "compact" expression for one mode voltage response and its associated peak power amplitude (see details in [8]).

Typical commercial harvesters based on vibrating benders existing on the market are presented in Figures 8.5–8.7 and 8.9.

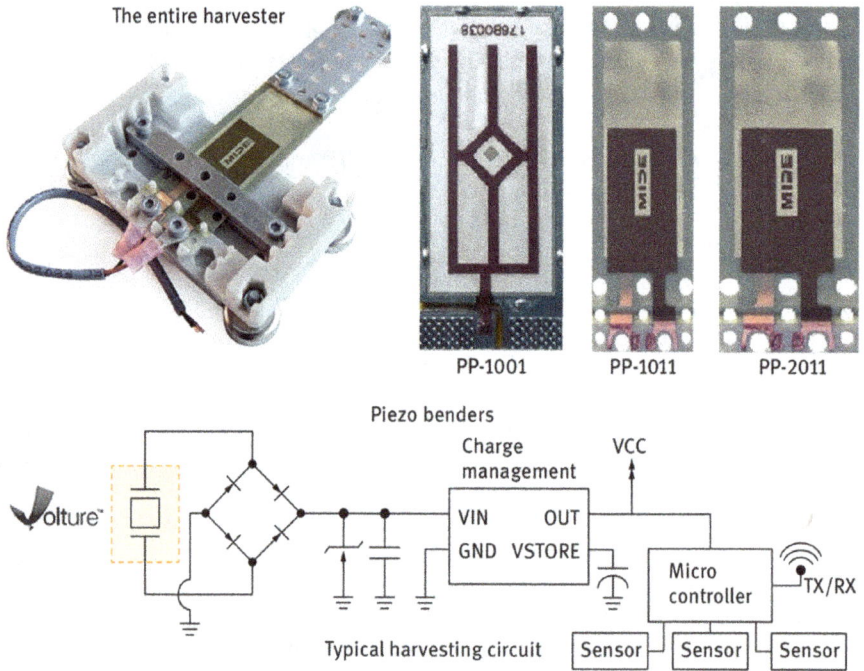

PP-1001   PP-1011   PP-2011

Figure 8.5: Typical Mide's Volture™ (http://www.mide.com/collections/vibration-energy-harvesting-with-protected-piezos) vibration harvesting products.

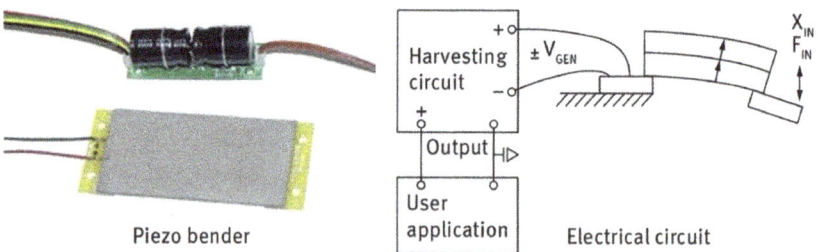

Figure 8.6: Typical Piezo Systems (https://www.piezo.com/prodproto4Ekit.html) harvesting products.

Laminated ceramic layers in a
DuraAct transducer arrangement (array)

Figure 8.7: Piezoelectric bender for harvesting energy from PI (https://www.piceramic.com/applications/piezo-energy-harvesting.html) company.

Voltmeter

Harvester

Shaker

Piezoelectric harvester   Test-set-up

Figure 8.8: Piezoelectric harvester for traffic-induced vibration and a view of its lab test setup (from [17]).

Figure 8.9: The impact battery based on piezoelectric harvester from Ceratec Engineering Corporation (http://www.ceratec-e.com) Japan.

After describing the vibration-based harvesters, the expression for the other type of scavenging energy from direct stress is next presented (see [25, 26]). As the piezoelectric generator based on the direct application of stress in the d_{33} mode of operation, is normally acting far from resonance, the generator acts like a capacitor having the following formula:

$$C_p = \frac{\varepsilon_{33}^T A}{h} \tag{8.21}$$

where ε_{33}^T is the permittivity at constant stress, A and h are the area and the thickness of the capacitor, respectively. Therefore, the energy obtained due to a stress σ_{33} can be written as

$$U = \frac{1}{2}C_p \cdot V_{out}^2 = \frac{1}{2}C_p \cdot g_{33}^2 \cdot \sigma_{33}^2 \cdot h^2$$

with   (8.22)

$$V_{out} = -g_{33} \cdot \sigma_{33} \cdot h = -g_{33} \cdot \frac{F_{33}}{A} \cdot h$$

where A is the area of the piezoelectric generator, F_{33} is the force applied vertical to the area, and g_{33} is the piezoelectric voltage constant.

Using the definition of the capacitance (Equation (8.21)), the relation between the g_{33} and the d_{33}, the piezoelectric field constant, $\varepsilon_{33}^T \cdot g_{33} = d_{33}$ and the volume of the piezoelectric material as $Vol = A \cdot h$ and substituting in Equation (8.22) one obtains

$$U = \frac{1}{2}C_p \cdot V_{out}^2 = \frac{1}{2}(g_{33} \cdot d_{33}) \cdot \sigma_{33}^2 \cdot Vol \qquad (8.23)$$

while the power can be calculated from the following expression:

$$Power = U \cdot f = \frac{1}{2}(g_{33} \cdot d_{33}) \cdot \sigma_{33}^2 \cdot f \cdot Vol = \frac{1}{2}\frac{d_{33}^2}{\varepsilon_{33}^T} \cdot \sigma_{33}^2 \cdot f \cdot Vol \qquad (8.24)$$

To maximize the energy, one would need to apply the highest possible stress, to use a large amount of piezoelectric material and to choose a piezoelectric material with high values for $(g_{33} \cdot d_{33})$. Accordingly, for reaching the maximal power, one would have to use the same values as for maximal energy plus higher frequencies.

A very large number of various studies on harvesting energy were published in the literature. Typical examples of those studies will be next presented. In [1] an extended review is presented on harvesting issues and materials choice. Operating modes and device configurations, from resonant to nonresonant devices including rotational solutions, are reviewed based on power density and bandwidth. Roundy [2] presents a basic study to be latter referenced by many researchers, in which he tries to estimate the effectiveness of vibration-based energy harvesting devices. According to his calculations, the theoretical power density available due to vibrations ranges from 0.5 to 100 mW/cm^3 having an acceleration of 1–10 m/s^2 at 50–350 Hz. His effectiveness rules were derived taking into account coupling coefficients, quality factor of the device, its mass, and the way the electrical external load maximizes the transmission of the electric power. The effectiveness was applied to various ways of scavenging like electromagnetic-, piezoelectric-, magnetostrictive- and electrostatic-based harvesters. Lopes and Gallo [3] present an interesting review on the most important tests and applications being performed up to 2014 for harvesting energy form ambient environment using piezoelectric-based devices. Beeby et al. [8] present an interesting review on energy harvesting vibrational sources for microsystems applications. The review includes besides the piezoelectric-based harvesters working on impact coupled, resonance and human motions, also electromagnetic- and electrostatic-based generators.

Sebald et al. [9] present an experimental studying of a Duffing oscillator with piezoelectric electromechanical coupling, using a nonlinear model, showing that the power frequency bandwidth is multiplied by a factor of 5.45, while the output power is decreased by a factor of 2.4. Vatansever et al. [10] present experimental results of harvesting energy from various wind speeds and water droplets using both PVDF and piezoelectric fiber composite structures. It is claimed that PVDF-based devices are more efficient to harvest energy from these two energy sources. Pozzi and Zhu [11] designed and tested plucked piezoelectric bimorphs to harvest knee-joint energy, while a human being is walking showing a good correlation between the model predictions and the experimental results. Ling et al. [12] present an interesting review of harvesting energy from human activity like shoe-mounted, impulse-excited, and impact-driven harvesters, joint motion having a rotary knee-joint harvester, rotating harvesters, and a flexible wearable harvester. Energy from civil infrastructure and transportation is presented for a case study of bridge, and the way it is harvested. Other interesting sources of energy, like wind flows using flexible piezoelectric films as harvesters and the swaying movement of trees in the wind, harvesting of flow-induced vibrations due to water flow, and energy from rain drops, are also included in the review. Wekin [13] presents a master thesis on how to harvest power from a thermoacoustic engine by characterizing and comparing various piezoelectric materials to be used as harvesters. Another wide review is presented by Bowen et al. [14] summarizing the use of piezoelectric-based devices as potential harvesters in the context of scavenging the mechanical energy from vibrations and presenting various optimization methods for complex piezoelectric devices. The use of ferroelectric and multiferroic materials to convert light into chemical or electrical energy is described and ways of how to implement it into harvesters is discussed. Thermal fluctuations harvesting is also addressed using pyroelectric-based harvesters. Wang et al. [16] in a recent publication investigated the coupling factors to be applied in linear vibration-based harvesters yielding a frequency domain for which the coupling is weak. Dhingra et al. [18] bring in their review various existing projects which used different types of piezoelectric harvesters to scavenge ambient available energies into electrical energy. Sodano et al. [20] add to the experimental database available in the literature by comparing three types of benders: PSI-5H4E (PZT) from piezo Systems Inc. bonded on a 0.0635-mm thick aluminum plate, a $8.255 \times 5.715 \text{cm}^2$ MFC (Macro Fiber Composite)[2] patch bonded on the same aluminum plate, and a Quick Pack actuator[3] (approx. $10.16 \times 2.54 \text{ cm}^2$) acting like a cantilever beam. The efficiencies of the three devices were tested at resonant, chirp 0–500 Hz and random 0–500 Hz

---

2 From Smart Material Corp. www.smart-material.com/MFC-product-main.html
3 From former ACX corporation now owned by Midé corporation.

excitations showing various efficiencies to charge batteries, depending on the type of excitation. Reference [21] presents a review on commercially available piezoelectric ceramics. Different figures of merit are derived and presented, allowing the description of typical load scenarios and commercial piezo-ceramics. Finally, the paper is deriving design rules for improved generator solutions. An interesting application is presented in [23], where piezoelectric harvesters generate power needed for sensing within the helicopter blade itself. Two concepts were tested. The first one was the placing of piezoelectric patches within the slender low-coupled beam-like structure. The second concept was the introduction of a piezoelectric stack in the lag damper of the helicopter rotor, which demanded the modification of the lag damper. According to the study performed, the power is generated in a large quantity at a single location and the analysis could be made using basic analytical equations.

Kim et al. [24] investigated the possibility of harvesting the mechanical vibrations and transform it into electrical energy using a "cymbal" type piezoelectric transducer (see Figure 8.10), which has a circular shape. When the cap is subjected to external axial load and vibrations are induced, radial and axial vibrations are caused to the piezoelectric circular plate yielding high transduction coefficients d_{eff} (effective piezoelectric field constant) and g_{eff} (effective piezoelectric voltage constant) as compared to that of a multilayer stack transducer. This is obtained due to the special structure of the cymbal, which contains cavities allowing the metal end-caps to serve as a mechanical amplificatory of the incident axial stress into radial stresses of opposite signs [24]. The effective piezoelectric field constant can be calculated by the following equation:

$$d_{eff} = d_{33} + Ad_{31} \tag{8.25}$$

where A is the amplification factor which can reach values of 10–100 depending on the design of the caps. Remembering that d_{31} has a negative value and A is a high value, it is clear that the transduction rate of a cymbal is very high. The experimental results performed on a "cymbal" having a diameter of 29 and 1 mm thickness, applying a force of 7.8 N at a frequency of 100 Hz yielded a power of 39 mW measured across a 400 kΩ resistor. The authors report that using a specially designed DC-to-DC convertor allowed the transfer of 30 mW to a 5 kΩ impedance with 2% duty cycle and a switching frequency of 1 kHz. Another possible application is described in [25, 26], where piezoelectric disks are subjected to axial compression stress at a frequency of 5 [Hz]. It was shown that although when increasing the axial compression stress the output power would increase, when subjected to thousands of cycles, the disks cannot withstand the high stresses and therefore the stress has to be reduced, thus yielding lower power. This fact can be solved by adding more disks, thus increasing the volume of the piezoelectric material involved in the harvesting of mechanical energy and its transformation into electrical energy. Table 8.2 summarizes data presented in the literature for various piezoelectric-based harvesters and their power output. The range of the power is usually milliwatts for vibration-based harvesters, and only the results presented in [25, 26] show power of watts (due to relatively high volume of piezoelectric material involved in the transduction).

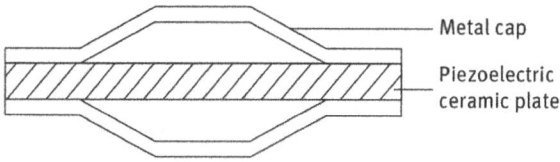

Figure 8.10: A "Cymbal" piezoelectric transducer cross section – a schematic view.

Table 8.2: Typical harvesters and their power output.

Source energy	Type of device	Size	Power harvested (average)	Ref.
Walking of human being	Shoe-mounted harvester: PVDF stave mounted under the insole	Two 8-layer stacks of 28 µm PVDF sandwiching a 2 mm flexible substrate glued with epoxy	1.3 mW @ 250 kΩ load and frequency of 0.9 Hz	[12]
Walking of human being- harvesting the heel strikes	Shoe-mounted harvester: PZT dimorph inserted under the heel	Two PZT transducers of $5 \times 5 \times 0.0381$ cm^3 bonded to a pre-stressed 5×8.5 cm^2 curved sheet of spring steel	8.4 mW @ 500 kΩ load and frequency of 0.9 Hz	[12]
Human body motion	Impulse-excited harvester: a cylindrical proof mass actuates an array of piezoelectric bi-morph beams through magnetic attraction	Eight $72 \times 5 \times 0.5$ mm^3 beams with top and bottom 0.2 mm PZT 507[1]	2.1 mW @ 2.7m/s^2 and frequency of 2 Hz	[12]
Human body motion	Impact-driven harvester: a moving mass impacts piezoelectric benders	Volume 25 cm^3, weight of 60 g	0.047 mW @ frequency of 0.5 Hz 0.6 mW @ frequency of 10 Hz and amplitude of 10 cm	[12]
Human body motion	Rotary knee-joint harvester	Bimorphs made from a 130 µm thick metal shim sandwiched between 2 layers of 125 µm thick, acting in series.	57.6 mW @ 7.1 m/s^2 and frequency of 320 Hz and 10 kΩ load or 3.646 µW @ 7.1 m/s^2 and frequency of 300 Hz and 2 Ω load	[11, 12]

Table 8.2 (continued)

Source energy	Type of device	Size	Power harvested (average)	Ref.
Slow human motion	Flexible wearable PVDF in shell-type structure harvester	The shell structure is made of 127 µm × 30 × 5 mm² polyester film+PVDF film 110 µm × 20 × 2 mm²	0.87 mW @ frequency of 3.3 Hz and 90 Ω load	[12]
Wind	PVDF piezo-film	Piezo-film area 7.44cm² and thickness 64 µm	Energy harvested: 51.66 µJ @ wind speed of 12.3 m/sec on a 150 nF capacitor	[12]
Water flow	Flow induced vibrations using PVDF or PZT-5H thin film	Piezoelectric film including 24 µm PVDF+two 28 µm electrodes layers+125 µm polyester layer	For pressure differences of 1. 790–2.392 kPa yielded peak-to-peak output voltages of 1. 77–2.30 V	[12]
Rain drops	PVDF sheet	25 µm thick mono-stretched PVDF sheet having $d31 = 2$ pJ/N	Up to 1 µW	[12]
Car vibrations	Piezo film element laminated in Mylar (polyester film)+end mass	12.19 × 30 × 1.57 mm³ sheet	Up to 770 mV @ 8.25 m/s²	[19]
Vibrations	PZT (PSI-5H4E²)	40 × 62 × 0.25725 mm³ PZT sheet bonded on a 40 × 80 × 1.016 mm³ aluminum cantilever plate	Maximal instantaneous power of 1.5–2 mW on 1 kΩ and average power of 0.14–0.2 mW on 1 kΩ@ resonance of the plate	[15]
Vibrations	Bimorph mounted as cantilever beam.	Overall size (including the end mass) =1 cm³	0.2 mW on 250 kΩ @ 2.25 m/s² and 85 Hz; 0.38 mW on 250 kΩ @ 2.25 m/s² and 60 Hz	[2]
Traffic-induced bridge vibrations	Cantilever piezo-beam with an end mass (12 g) having a resonant frequency close to 14.5 Hz	Two bimorph patches QP20W (from Midé Corp.) were sandwiching a 40 × 220 × 0.8 mm³ steel plate	0.03 mW at below 15 Hz excitation	[17] (see also Figure 8.8)

Table 8.2 (continued)

Source energy	Type of device	Size	Power harvested (average)	Ref.
Time dependent force	An arch equipped with PVDF and triboelectric layers[3]	Dimensions of the arch: $7 \times 3\text{cm}^2$	The PVDF alone produces 80 V and 7.62 $\mu A/cm^2$; The triboelectricity produces alone 380 V and 4.3 $\mu A/cm^2$	[22]
Time depended force yielding vibrations	Piezoelectric "Cymbal"-type transducer	Dimensions: diameter 29 mm, thickness 1 mm	39 mW due to 7.8 N @ 100 Hz on 400 kΩ	[24] (see Figure 8.10)
Time depended force	Layered disks under compression	12 disks of diameter 10 mm and thickness of 4 mm; 8 disks of diameter 15 mm and thickness of 4 mm	0.14–0.15 W at 75 MPa @ 5 Hz on 2.8–3 MΩ; 0.35 W at 100 MPa @ 5 Hz on 2.8 MΩ.	[25, 26]

[1]From Morgan Electro Ceramics, www.morgantechnicalceramics.com/products
[2]From Piezo Systems Inc.
[3]Triboelectric generator is fabricated by stacking two polymer sheets made of materials with different triboelectric characteristics. Once subjected to mechanical displacement the friction between the two layers due to the nanoscale surface roughness generates equal charges but with opposite signs at the two sides.

References

[1] Caliò, R., Rongala, U. B., Camboni, D., Milazzo, M., Stefanini, C., de Petris, G. and Oddo, C. M., Piezoelectric energy harvesting solutions, Sensors 14, 2014, 4755–4790. doi:10.3390/s140304755.
[2] Roundy, S., On effectiveness of vibration-based energy harvesting, Journal of Intelligent Materials Systems and Structures 16, 2005, 809–823.
[3] Lopes, C. M. A. and Gallo, C. A., A review of piezoelectric energy harvesting and applications, 2014 IEEE 23rd International Symposium on Industrial Electronics (ISIE), 1–4 June, 2014, 1285–1288.
[4] Williams, C. B., Shearwood, C., Harradine, M. A., Birch, T. S. and Yates, R. B., Development of an electromagnetic micro-generator, IEE Proceedings on Circuits Devices Systems 148(6), December 2001, 337–342.
[5] Priya, S. and Inman, D. J. (eds.), Energy Harvesting Technologies, ©Springer Science+Business media, LLC, 2009, 517.
[6] Beeby, S. P., Tudor, M. J. and White, N. M., Energy harvesting vibration sources for microsystems applications, Measurement Science and Technology 17, 2006, 175–195.

[7] Yang, B., Lee, C., Xiang, W., Xie, J., He, J. H., Kotlanka, R. K., Low, S. P. and Feng, H., Electromagnetic energy harvesting from vibrations of multiple frequencies, Journal of Micromechanics and Microengineering 18, 2009, article ID 035001, 8.

[8] Priya, S. and Inman, D. J. (eds.), *Energy harvesting technologies*, Appendix A: First draft of standard on vibration energy harvesting, 507–513,©Springer Science+Business media, LLC 2009, 517 pp.

[9] Sebald, G., Kuwano, H., Guyomar, D. and Ducharne, B., Experimental Duffing oscillator for broadband piezoelectric energy harvesting, Smart Materials and Structures 20, 2011, article ID 102001, 10.

[10] Vatansever, D., Hadimani, R. L., Shah, T. and Siores, E., An investigation of energy harvesting from renewable sources with PVDF and PZT, Smart Materials and Structures 20, 2011, article ID 055019, 6.

[11] Pozzi, M. and Zhu, M., Plucked piezoelectric bimorph for knee-joint energy harvesting: modelling and experimental validation, Smart Materials and Structures 20, 2011, article ID 055007, 10.

[12] Ling, B. K., Li, T., Hng, H. H., Boey, F., Zhang, T. and Li, S., Waste Energy Harvesting Mechanical and Thermal Energies, Chapter 2 – Waste Mechanical Energy Harvesting (I): Piezoelectric Effect, Springer, 2014, 25–133. http://www.springer.com/978-3-642-54633-4.

[13] Wekin, A. B. E., Characterization and comparison of piezoelectric materials for transducing power from a thermoacoustic engine, A master thesis submitted to the School of mechanical and Materials Engineering, Washington State University, August 2008, 130 pp.

[14] Bowen, C. R., Kim, H. A., Weaver, P. M. and Dunn, S., Piezoelectric and ferroelectric materials and structures for energy harvesting applications, Energy Environmental Science 7, 2014, 25–44.

[15] Sodano, H. A., Magliula, E. A., Park, G. and Inman, D. J., Electric power generation using piezoelectric devices, *13th International Conference on Adaptive Structures and Technologies (ICAST13)*, Breitbach, E. J., Campanile, L. F. and Monner, H. P. (eds.), CRC press, October 7–9, 2002, Potsdam, Germany, 153–161.

[16] Wang, X., Liang, X., Shu, G. and Watkins, S., Coupling analysis of linear vibration energy harvesting systems, Mechanical Systems an Signal Processing, 2015. doi:http://dx.doi.org/10.1016/j.ymssp.2015.09.006i.

[17] Peigney, M. and Siegert, D., Piezoelectric energy harvesting from traffic-induced bridge vibrations, Smart Materials and Structures 22, 2013, article ID 095019.

[18] Dhingra, P., Biswas, J., Prasad, A. and Meher, S. S., Energy harvesting using piezoelectric materials, Special Issue of International Journal of Computer Applications (0975–8887), International conference on Electronic Design and Signal Processing (ICEDSP), 2012, 38–42.

[19] Mohamad, S. H., Thalas, M. F., Noordin, A., Yahya, M. S., Hassan, M. H. C. and Ibrahim, Z., A potential study of piezoelectric energy harvesting in car vibration, Journal of Engineering and Applied Sciences 10(19), October 2015, 8642–8647.

[20] Sodano, H. A. and Inman, D. J., Comparison of piezoelectric energy harvesting devices for charging batteries, Journal of Intelligent Material Systems and Structures 16(10), 2005, 799–807.

[21] Rödig, T. and Schönecker, A., A survey on piezoelectric ceramics for generator applications, Journal of American Ceramic Society 93(4), 2010, 901–912.

[22] Jung, W.-S., Kang, M.-G., Moon, H. G., Baek, S.-H., Yoon, S.-J., Wang, Z.-L., Kim, S.-W. and Kang, C.-Y., High output piezo/triboelectric hybrid generator, www.nature.com/scientificreports, Scientific Reports 5, March 2015, doi:10.1038/srep09309, 9309.

[23] De Jong, P. H., Power harvesting using piezoelectric materials-applications in helicopter rotors, Ph.D. Thesis, 160 pp., University of Twente, Enschede, The Netherlands, February 2013.

[24] Kim, H. W., Batra, A., Priya, S., Uchino, K. and Markley, D., Energy harvesting using a piezoelectric "cymbal" transducer in dynamic environment, Japanese Journal of Applied Physics 43(9), 2004, 6178–6183.

[25] Abramovich, H., Tsikhotsky, E. and Klein, G., An experimental determination of the maximal allowable stresses for high power piezoelectric generators, Journal of Ceramic Science and Technology 4(3), 2013, 131–136. doi:10.4416/JCST2013-00006.

[26] Abramovich, H., Tsikhotsky, E. and Klein, G., An experimental investigation on PZT behavior under mechanical and cycling loading, Journal of the Mechanical Behavior of Materials, 22 (3–4), 2013, 129–136.

[27] Ertuk, A. and Inman, D. J., Issues in mathematical modeling of piezoelectric energy harvesters, Smart Materials and Structures 17(6), 2008, article ID 065016.

[28] Ertuk, A. and Inman, D. J., A distributed parameter electromechanical model for cantilevered piezoelectric energy harvesters, Journal of Vibration and Acoustics, Transactions of the ASME 130(4), 2008, article ID 041002.

[29] Ertuk, A. and Inman, D. J., An experimentally validated bimorph cantilever model for piezoelectric energy harvesting from base excitations, Smart Materials and Structures 18(1), 2009, article ID 025009.

8.2 Electromagnetic Energy Harvesting

In principle, the electromagnetic harvesting is performed on a structure vibrating at given frequencies. A generator is placed on those vibrating structures, while a magnet is vibrating inside a coil. The relative motion of the magnet is converted to electrical energy by an electromechanical transducer. The most known model for a single degree of freedom is the model developed by Williams et al. [1], which will be next presented. A mass m, a spring k and a damper C_d are connected in serial as depicted in Figure 8.11. Note that the transducer mentioned above is represented by the damper, C_d, shown in Figure 8.11. The system is located in a case having a displacement, $y(t)$, while the displacement of the mass (relative to an external reference system) is $x(t)$. The mass relative displacement is $z(t)$. Its expression is given by

$$z(t) = x(t) - y(t) \quad \Rightarrow \quad x(t) = z(t) + y(t) \tag{8.26}$$

Now, as described in [1], the single equation of motion for the mass, spring and damper can be written as

$$m \cdot \ddot{x}(t) + C_d[\dot{x}(t) - \dot{y}(t)] + k[x(t) - y(t)] = 0 \tag{8.27}$$

or using Equation (8.26), we can get

$$m \cdot \ddot{z}(t) + C_d \cdot \dot{z}(t) + k \cdot z(t) = -m \cdot \ddot{y}(t) \tag{8.28}$$

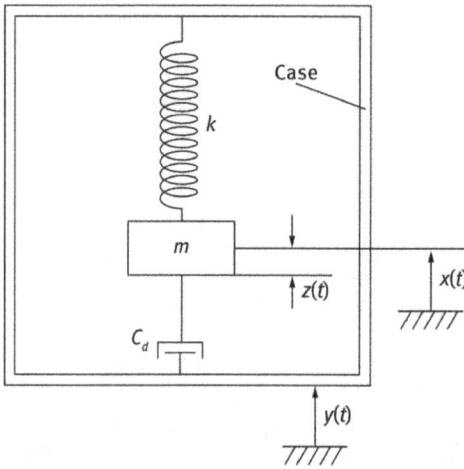

Figure 8.11: A schematic drawing for the linear inertial generator.

Applying the Laplace transform,[4] we obtain

$$-m \cdot s^2 \cdot Y(s) = s \cdot Z(s) \left[m \cdot s + C_d + \frac{k}{s} \right] \tag{8.29}$$

Ref. [1] provides the equivalent electrical circuit for the mechanical system described above as

$$-I(s) = V(s) \left[C_d \cdot s + \frac{1}{R} + \frac{1}{s \cdot L} \right] \tag{8.30}$$

where $I(s)$ and $V(s)$ are the input current and the induced voltage, C is a capacitor, R is a resistor and L is an inductance. Comparing Equations (8.29) and (8.30), while introducing the constant k_e (the transducer electromagnet constant), we obtain the following relations among electrical and mechanical terms, namely

$$I(s) = \frac{m}{k_e^2} s^2 \cdot Y(s), \quad V(s) = k_e \cdot s \cdot Z(s)$$

$$\tag{8.31}$$

$$C = \frac{m}{k_e^2}, \quad R = \frac{k_e^2}{C_d}, \quad L = \frac{k_e^2}{k}$$

The equivalent electrical circuit based on the electrical parameter is shown in Figure 8.12. Note that the damper C_d is "represented" by the resistor R. Therefore, the power being dissipated in the damper has the following expression:

---

4 The Laplace transform is an integral transform that takes a function of a positive real variable t (time) to a function of a complex variable s (frequency).

$$P_d(s) = \frac{V^2(s)}{R} = \frac{\frac{k_e^2 \cdot m^2 \cdot s^6 \cdot Y^2(s)}{[m \cdot s^2 + C_d \cdot s + k]^2}}{\frac{k_e^2}{C_d}} = \frac{C_d \cdot m^2 \cdot s^6 \cdot Y^2(s)}{[m \cdot s^2 + C_d \cdot s + k]^2} \tag{8.32}$$

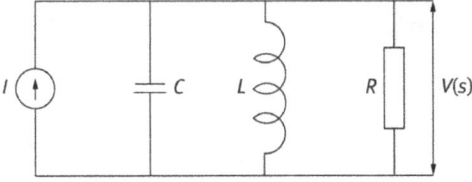

Figure 8.12: A schematic drawing for the equivalent electrical circuit.

Assuming a harmonic vibration having the following expression, $Y(t) = Y_0 \cdot \cos(\omega \cdot t)$, $\omega = 2 \cdot \pi \cdot f$, where f is frequency, and substituting in Equation (8.32), we obtain the dissipated power in the frequency domain, namely

$$p_d(\omega) = \frac{C_d \cdot m^2 \cdot \omega^6 \cdot Y_0^2}{\left[(k - m \cdot \omega^2)^2 + (\omega \cdot C_d)^2\right]} \tag{8.33}$$

Let us refer to the natural frequency of mass, spring and damper system. The resonance frequency, ω_n, and the associated damping ratio ζ are defined as

$$\omega_n^2 \equiv \frac{k}{m} = \frac{1}{L \cdot C}, \quad \frac{C_d}{m} \equiv 2 \cdot \zeta \cdot \omega_n \quad \Rightarrow \quad \zeta = \frac{C_d}{2\sqrt{m \cdot k}} = \frac{1}{2 \cdot R}\sqrt{\frac{L}{C}} \tag{8.34}$$

Using the natural frequency and the damping ratio, Equation (8.33) can be modified to obtain

$$p_d(\omega) = \frac{C_d \cdot \left(\frac{\omega}{\omega_n}\right)^3 \cdot \omega^3 \cdot Y_0^2}{\omega_n \left\{\left[1 - \left(\frac{\omega}{\omega_n}\right)^2\right]^2 + \left[2 \cdot \zeta \cdot \left(\frac{\omega}{\omega_n}\right)\right]^2\right\}} = \frac{2 \cdot \zeta \cdot m \cdot \left(\frac{\omega}{\omega_n}\right)^3 \cdot \omega^3 \cdot Y_0^2}{\left[1 - \left(\frac{\omega}{\omega_n}\right)^2\right]^2 + \left[2 \cdot \zeta \cdot \left(\frac{\omega}{\omega_n}\right)\right]^2} \tag{8.35}$$

One should note that Equation (8.35) is applicable to any linear inertial generator, regardless of the type of the electromechanical transducer. If we design the generator such that the natural frequency would be as the excitation one, namely, $\omega = \omega_n$, and substituting in Equation (8.35) we get the dissipated power at the resonance frequency

$$p_d(\omega_n) = \frac{2 \cdot \zeta \cdot m \cdot \omega_n^3 \cdot Y_0^2}{4 \cdot \zeta^2} = \frac{m \cdot \omega_n^3 \cdot Y_0^2}{2 \cdot \zeta} \tag{8.36}$$

To maximize the dissipated power, one would have to increase the mass, have a high natural frequency and a high displacement, while trying to minimize the damping ratio.

To realize the electromagnetic harvester, a permanent magnet is attached to the vibrating mass, m, while a coil is wounded around the circular case. Due to the relative displacement of the mass, z, inside the coil, a voltage $V_c(t)$ will be induced in it, having the following expression:

$$V_c(t) = k_e \cdot \dot{z}(t) \tag{8.37}$$

Therefore, the force acting on the vibrating magnet, due to a current $I_c(t)$, will be written as

$$F_c(t) = k_e \cdot I_c(t) \tag{8.38}$$

Assuming that the coil has a resistance R_c, a self-inductance L_c, and an electrical load R_L is connected to the circuit, a "new" equivalent electrical circuit having four parts, the force input, the mechanical part, the electrical part and the external electrical load, is obtained as depicted in Figure 8.13.

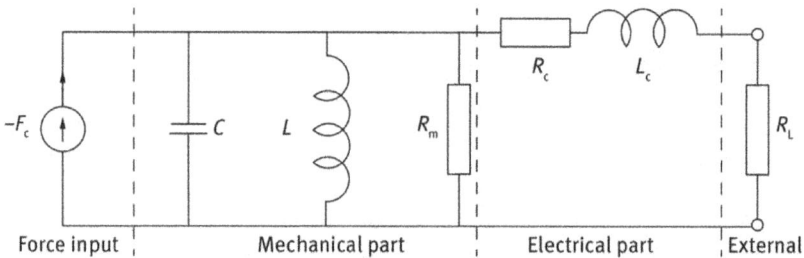

Figure 8.13: A schematic drawing for the equivalent electrical circuit + the electrical part of the coil and the external electrical loading.

One should note that the representing resistor in the mechanical part, R_m, is defined as

$$R_m = \frac{k_e^2}{C_m} \tag{8.39}$$

where C_m stands for both the mechanical damping and the electromagnetic iron and eddy current losses. One should note that at resonance ($w = w_n$), the parallel impedances C and L become infinite [1]. Thus, Thévenin's[5] equivalent circuit for the

---

5 The Thévenin's theorem states that it is possible to simplify any linear circuit, no matter how complex, to an equivalent circuit having a single voltage source and series resistance connected to an electrical load.

system at resonance simplifies to a drawing shown in Figure 8.14. According to [1] the voltage V (in Figure 8.14) is given by

$$V = -\frac{k_e \cdot m \cdot \ddot{y}(t)}{C_m} = \frac{k_e \cdot m \cdot w_n^2 \cdot Y_0}{C_m} \tag{8.40}$$

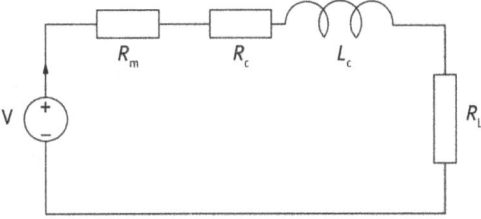

Figure 8.14: Thévenin's equivalent circuit for the system at resonance.

To evaluate the power dissipated on the external resistor, the current on R_L is calculated as

$$I \cdot Z_N = V \quad \Rightarrow \quad I = \frac{k_e \cdot m \cdot w_n^2 \cdot Y_0}{C_m \cdot Z_N} = \frac{k_e \cdot m \cdot w_n^2 \cdot Y_0}{C_m \cdot \sqrt{(R_c + R_L + R_m)^2 + (w_n L_c)^2}} \tag{8.41}$$

Then the power at resonance can be written as

$$p_L(w_n) = I^2 \cdot R_L = \frac{k_e^2 \cdot m^2 \cdot w_n^4 \cdot Y_0^2}{C_m^2 \cdot \left[(R_c + R_L + R_m)^2 + (w_n L_c)^2 \right]} R_L \tag{8.42}$$

To maximize the power $p_L(w_n)$, one has to use an external load such that $R_L \gg R_c$ and $R_L \gg w_n \cdot L_c$. Also by matching the electrical impedance to the mechanical damping impedance, namely, $k_e = \sqrt{R_L \cdot C_m}$, Equation (8.42) is rewritten as

$$p_L(w_n)_{max} = \frac{m^2 \cdot w_n^4 \cdot Y_0^2}{4 \cdot C_m} \tag{8.43}$$

Remembering that, $2 \cdot \zeta_m \cdot w_n = \frac{C_m}{m}$, Equation (8.43) transforms into

$$p_L(w_n)_{max} = \frac{m \cdot w_n^3 \cdot Y_0^2}{8 \cdot \zeta_m} \tag{8.44}$$

where ζ_m is the mechanical damping ratio. Based on Equation (8.44), some design rules can be written to maximize the power harvested in an electromagnetic linear inertial generator [1]:

- The inertial mass, m, should have a high value depending of the system to be designed.
- Once the value of the mass was chosen, the spring coefficient, k, should be adjusted to yield $\omega_n = \omega_{base}$ (ω_{base} is the excitation frequency).
- As the generator is built to work at resonance, one has to be sure that the inner dimension of the case allows a maximal displacement of the vibrating mass.
- Minimize the coil impedance in such a way that it will be less than 1 order of magnitude that the electrical load (R_L).
- To reduce the damping ratio, the mechanical damping and the stray[6] losses should be a small as possible.
- Finally one has to choose the electromagnet factor k_e such that it fulfills the relation presented in Equation (8.39), namely

$$R_m = \frac{k_e^2}{C_m}.$$

One should note that in practice not all the above design rules can be implemented, and a compromise should be reached to obtain an optimal design. For example, if k_e is chosen as a large value, while R_c is kept as a small value, this requires an increase in the number of turns of the coil, which in turn will increase the value of R_c if the space for the coil is limited.

Based on the model described above, Williams et al. [1] designed, built and tested a demonstrator (a cylinder with a total diameter of 5 mm, and a total height of 700 μm) yielding a power of 0.3 μW at a frequency of 4 MHz.

Other researchers present similar results. Roundy [2] estimated the harvested power using electromagnetic, piezoelectric, magnetostrictive and electrostatic transducer technologies. It is claimed that depending on material types and operation parameters, relatively high coupling coefficients (0.6–0.8) are possible for each of these technologies. The most suitable technology, therefore, needs to be decided upon based on the operating environment and the constraints of the design problem. Based on a range of common vibrations measured by the author, the maximum theoretical power density ranges from 0.5 to 100mW/cm^3 for vibrations in the range at 50–350 Hz having an acceleration of 1–10m/s^2. Beeby et al. [3] reviewed the state of the art in vibration energy harvesting for wireless, self-powered microsystems. The vibration-powered generators consisting of inertial mass, spring and a damper were considered, while considering three types of transduction mechanisms: piezoelectric, electromagnetic and electrostatic. A comprehensive review of existing piezoelectric generators is presented, including impact coupled, resonant and human-based devices.

---

[6] Stray losses are all the losses that cannot be accurately determinate. For our electromagnet generator this includes losses due to undetermined fields, harmonic flux pulsations in the iron circuit and eddy currents in the windings.

Electromagnetic generators are also reviewed including large-scale discrete devices and wafer-scale integrated versions. Electrostatic-based generators[7] are also presented under the classifications of in-plane overlap varying, in-plane gap closing and out-of-plane gap closing. Arnold [4] presents an interesting review on compact magnetic power generation systems (less than a few cm^3), while presenting ways of miniaturizing magnet generators, evaluating the design and the performance of the devices, while integrating high-performance hard magnetic materials, microscale core laminations, low-friction bearings, high-speed rotor dynamics, and compact, high-efficiency power converters. Another manuscript of Beeby et al. [5]. presents a compact electromagnetic generator (component volume $0.1cm^3$, practical volume $0.15cm^3$) harvesting 46 µW on a resistive load of 4 kΩ at an acceleration of $0.59m/s^2$ at a low-resonant frequency of 52 Hz. The micro-generator having 2300 turns in its coil, showed a voltage of 428 mV (rms), which turned to be sufficient for subsequent rectification and voltage step-up circuitry. Another team of researchers, Yang et al. [6] presented the experimental results of an electromagnet harvester, capable of harvesting at the three first natural frequencies of the system ($f_1 = 369$ Hz, $f_2 = 938$ Hz and $f_3 = 1184$ Hz). The maximum output voltage and power reported are 1.38 mV and 0.6 µW for f_1 and 3.2 mV and 3.2 µW for the second mode, with a 14-µm exciting vibration amplitude and a 0.4-mm gap between the magnet and coils. A similar approach was adopted by the team of Liu et al. [7] in which a 3D excitation was used to harvest energy using a novel electromagnetic energy harvester. The interesting results showed the first vibration mode of 1285 Hz is an out-of-plane motion, while the second and third modes of 1470 and 1550 Hz, respectively, are in-plane at angles of 60°(240°) and 150°(330°) to the horizontal (x-) axis. Applying an excitation acceleration of 1 g m/s^2, the maximum power density achieved was reported to be 0.444, 0.242 and 0.125 µW/cm^3 at vibration modes at first, second and third frequency, respectively. Another advanced research is presented by Illy et al. [8], in which energy is harvesting from human motion by exploiting swing and shock excitations. The harvesting was done by two types of inductive energy harvesters: a multicoil topology harvester using the swing motion of the foot and the second device is a shock type harvester being excited into resonance upon heel strike. Both devices were modeled and designed with the key constraint of device height in mind, in order to facilitate the integration into the shoe sole. The two harvesters were characterized under different motion speeds and with two test subjects on a treadmill yielding an average power output of up to 0.84 mW using the swing harvester (total device volume + the housing of 21 cm^3), while for the shock harvester (total device volume of 48cm^3) the output power was up to 4.13 mW. The power density for the

---

7 Electrostatic generators utilize the relative movement between electrically isolated charged capacitor plates to generate energy. The work done against the electrostatic force between the plates provides the harvested energy.

first device was reported to be 40 µW/cm^3 and 86 µW/cm^3 for the second device, much better than the results presented in [7].

A typical picture of an electromechanical harvester manufactured by Perpetuum Ltd.,[8] England, is shown in Figure 8.15.

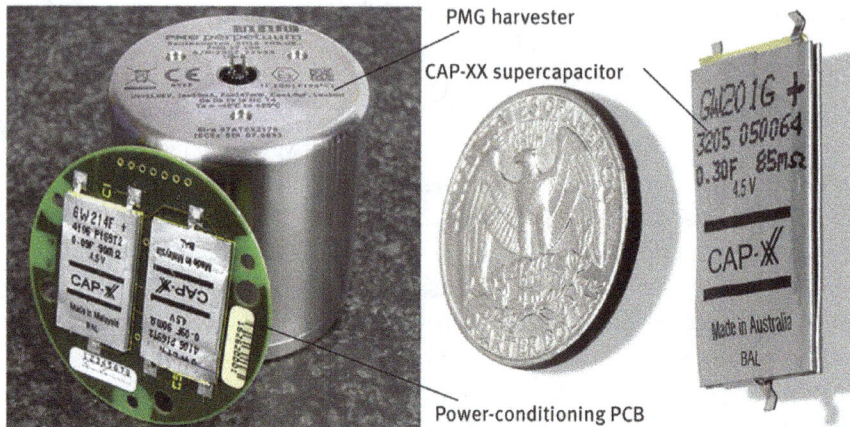

Figure 8.15: The PMG17 harvester, its PCB, and the used supercapacitor (from Perpetuum Ltd., England). The quarter US dollar is brought for sizing the objects.

References

[1]   Williams, C. B., Shearwood, C., Harradine, M. A., Birch, T. S. and Yates, R. B., Development of an electromagnetic micro-generator, IEE Proceedings on Circuits Devices Systems 148(6), December 2001, 337–342.

[2]   Roundy, S., On effectiveness of vibration-based energy harvesting, Journal of Intelligent Materials Systems and Structures 16, 2005, 809–823.

[3]   Beeby, S. P., Tudor, M. J. and White, N. M., Energy harvesting vibration sources for microsystems applications, Measurement Science and Technology 17, 2006, 175–195.

[4]   Arnold, D. P., Review of microscale magnetic power generation, IEEE Transactions on Magnetics 43(11), 2007, 3940–3951.

[5]   Beeby, S. P., Torah, R. N., Tudor, M. J., Glynne-Jones, P., O'Donnell, T., Saha, C. R. and Roy, S., A micro electromagnetic generator for vibration energy harvesting, Journal of Micromechanics and Microengineering 17, 2007, 1257–1265.

[6]   Yang, B., Lee, C., Xiang, W., Xie, J., He, J. H., Kotlanka, R. K., Low, S. P. and Feng, H., Electromagnetic energy harvesting from vibrations of multiple frequencies, Journal of Micromechanics and Microengineering 19, 2009, article ID 035001, 8.

[7]   Liu, H., Soon, B. W., Wang, N., Tay, C. J., Quan, C. and Lee, C., Feasibility study of a 3D vibration-driven electromagnetic MEMS energy harvester with multiple vibration modes, Journal of Micromechanics and Microengineering 22, 2012, article ID 125020, 8.

---

8 www.perpetuum.com

[8]   Ylli, K., Hoffmann, D., Willman, A., Becker, P., Folkmer, B. and Manoli, Y., Energy harvesting from human motion: exploiting swing and shock excitations, Smart Materials and Structures 24(2), 2015, article ID 025029, 12.

8.3 Bimorph Electrical Power Under Vibration Excitation

The piezoelectric bimorph, shown schematically in Figure 8.16 consists of a pair of identical piezoelectric twin layers, poled along the thickness direction and separated by a metallic carrying beam in the middle. The bimorph is a cantilever beam, with the clamped side being excited harmonically in the vertical direction (z) with a known amplitude, A, and a given frequency, ω. At the free side of the cantilever beam, a mass M_0 is connected. As shown in Figure 8.16 the piezoelectric electrodes are connected in parallel to an external impedance Z_L.

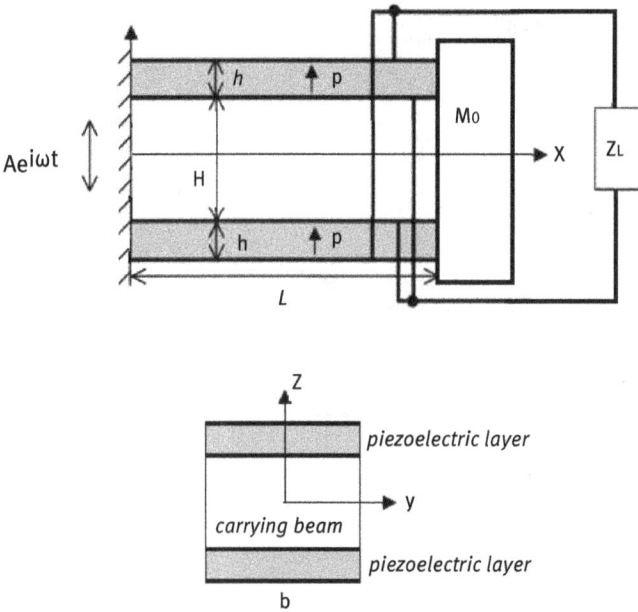

Figure 8.16: The piezoelectric bimorph schematic model.

Assuming that the bimorph is a slender beam, we can write the following piezoelectric equations for the present case (where the x, y, z system is the usual 1, 2 and 3):

$$S_x = s_{11}T_x + d_{31}E_z$$

$$D_z = d_{31}T_x + \varepsilon_{33}E_z \tag{8.45}$$

where S_x and T_x are the strain and the stress in the x direction (along the beam – Figure 8.16), respectively, D_3 is the electrical displacement and E_z is the electrical field induced due to the vibrations. The constants, s_{11}, d_{31} and ε_{33} are the axial elastic compliance measured at constant electric field, the transverse piezoelectric constant and the transverse dielectric constant measured at fixed stress, respectively. For the present case, we have the following expressions for the strain and the electric field:

$$S_x = -zw_{,xx}$$
$$E_z = -V/h \tag{8.46}$$

where w is the transverse displacement, V is the voltage accumulated on the piezoelectric electrodes due to the induced vibrations and h is the thickness of piezoelectric layer. Solving for T_x from the first equation in Equation (8.45), and then substituting into the second one yielding D_z, ones gets

$$T_x = s_{11}^{-1}S_x - s_{11}^{-1}d_{31}E_z$$
$$D_z = s_{11}^{-1}d_{31}S_x + \bar{\varepsilon}_{33}E_z \tag{8.47}$$

where

$$\bar{\varepsilon}_{33} = \varepsilon_{33}\left(1 - k_{31}^2\right)$$
$$k_{31}^2 = \frac{d_{31}^2}{\varepsilon_{33} \cdot s_{11}} \tag{8.48}$$

One should note that the stress in the middle layer (the metal layer) is given by

$$T_x = s_{11}^{-1}S_x = E_{substrate}S_x = -E_{substrate}zw_{,xx} \tag{8.49}$$

where $E_{substrate}$ is the Young's modulus of the metallic layer. The bending moment is given by

$$M = \int zT_x dydz = -Dw_{,xx} + s_{11}^{-1} \cdot d_{31} \cdot \frac{V}{h} \cdot (H+h) \cdot h \cdot b =$$
$$= -Dw_{,xx} + E_{piezo} \cdot d_{31} \cdot V \cdot (H+h) \cdot b \tag{8.50}$$

where

$$D = \left\{\frac{2}{3} \cdot E_{substrate} \cdot \left(\frac{H}{2}\right)^3 + \frac{2}{3} \cdot E_{piezo}\left[\left(\frac{H}{2}+h\right)^3 - \left(\frac{H}{2}\right)^3\right]\right\} \cdot b \tag{8.51}$$

The shear force in the beam, is given by the differentiation of the moment, namely

$$\breve{V} = -M,_x = +D \cdot w,_{xxx} \tag{8.52}$$

The flexural vibration of the slender bimorph beam is then written as

$$M,_{xx} = m\ddot{w}(x,t) \quad\Rightarrow\quad -D \cdot w(x,t),_{xxxx} = m\ddot{w}(x,t) \tag{8.53}$$

with the mass per unit length being given as

$$m = H \cdot b \cdot \rho_{Al} + 2 \cdot h \cdot b \cdot \rho_{piezo} \tag{8.54}$$

Let us calculate the electric charge accumulated on the top electrode at $z = (h + H/2)$; integrating the electrical displacement on the relevant surface gives

$$Q_{top} = -\int_0^L\int_0^b D_z{}_{@\left(z=\frac{H}{2}+h\right)} dx \cdot dy = -\left\{ b \cdot E_{piezo} \cdot d_{31}\left(\frac{H}{2}+h\right)[w,_x(L,t) - w,_x(L,t)] + b \cdot \bar{\varepsilon}_{33}\frac{V}{h}L \right\} \tag{8.55}$$

The relation between the charge and the current is known to be as

$$I = \frac{dQ}{dt} \tag{8.56}$$

which leads for our case to the relation of the voltage as a function of the two piezo-electric layers:

$$2I = \frac{\hat{V}}{z_L} \quad\Rightarrow\quad \hat{V} = 2 \cdot I \cdot z_L \tag{8.57}$$

The boundary conditions of the cantilever beam, having a piezoelectric bimorph can be written as

$$@x=0 \quad w(0,t) = Ae^{i\omega t}; \quad w,_x(0,t) = 0 \tag{8.58}$$

$$@x=L \quad M(L,t) = 0; \quad \breve{V}(L,t) = M_0\frac{\partial^2 w(L,t)}{\partial t^2} \tag{8.59}$$

Using Equation (8.53), together with the boundary conditions given by Equations (8.58) and (8.59), we can suggest a solution for the posed problem, having the following form:

$$\begin{Bmatrix} w(x,t) \\ \hat{V}(t) \\ Q(t) \\ I(t) \end{Bmatrix} = \begin{Bmatrix} W(x) \\ \tilde{V} \\ \tilde{Q} \\ \tilde{I} \end{Bmatrix} e^{i\omega t} \tag{8.60}$$

Substituting the proposed solution into Equation (8.53) and the associated boundary conditions, yields

$$D\frac{d^4 W}{dx^4} - m \cdot \omega^2 W = 0 \tag{8.61}$$

The solution of Equation (8.61) has the following form:

$$W = A_1 \cosh(\beta x) + A_2 \sinh(\beta x) + A_3 \cos(\beta x) + A_4 \sin(\beta x) \tag{8.62}$$

with A_1–A_4 coefficients to be determined from the boundary conditions and

$$\beta^4 \equiv \frac{m\omega^2}{D} \tag{8.63}$$

The associated boundary conditions presented by Equations (8.58) and (8.59) transform into the following expressions:

$$@x = 0 \quad W(0) = A; \quad W_{,x}(0) = 0 \tag{8.64}$$

$$@x = L \quad -D \cdot W_{,xx}(L) + E_{piezo} d_{31} \frac{\tilde{V}}{h} \cdot (H + h) \cdot h \cdot b = 0$$

$$D \cdot W_{,xxx}(L) = -M_0 \cdot \omega^2 \cdot W(L) \tag{8.65}$$

The expression for the current will have the following form:

$$\tilde{I} = i \cdot \omega \cdot b \cdot$$

$$\left\{ E_{piezo} \cdot d_{31} \left(\frac{H}{2} + h\right) \cdot \beta \cdot [-A_1 \sinh(\beta L) - A_2 \cosh(\beta L) + A_3 \sin(\beta L) - A_4 \cos(\beta L)] - \bar{\varepsilon}_{33} \frac{\tilde{V}}{h} L \right\} \tag{8.66}$$

Applying the boundary conditions (Equations (8.64) and (8.65)) we get the four equations for the four constants A_1–A_4:

$$A_1 + A_3 = A \tag{8.67}$$

$$A_2 \beta + A_4 \beta = 0 \tag{8.68}$$

$$-D \cdot [A_1 \cdot \beta^2 \cdot \cosh(\beta L) + A_2 \cdot \beta^2 \cdot \sinh(\beta L) - A_3 \cdot \beta^2 \cdot \cos(\beta L) - A_4 \cdot \beta^2 \cdot \sin(\beta L)] +$$
$$+ E_{piezo} d_{31} \cdot \tilde{V} \cdot (H + h) \cdot b = 0 \tag{8.69}$$

$$D \cdot [A_1 \cdot \beta^3 \cdot \sinh(\beta L) + A_2 \cdot \beta^3 \cdot \cosh(\beta L) + A_3 \cdot \beta^3 \cdot \sin(\beta L) - A_4 \cdot \beta^3 \cdot \cos(\beta L)] =$$
$$= -M_0 \cdot \omega^2 \cdot [A_1 \cdot \cosh(\beta L) + A_2 \cdot \sinh(\beta L) + A_3 \cdot \cos(\beta L) + A_4 \cdot \sin(\beta L)] \tag{8.70}$$

plus an expression for the voltage generated

$$\frac{\tilde{V}}{2 \cdot Z_L \cdot i \cdot \omega \cdot b} = E_{piezo} \cdot d_{31} \left(\frac{H}{2} + h\right) \cdot \beta \cdot [-A_1 \sinh(\beta L) - A_2 \cosh(\beta L) + A_3 \sin(\beta L) -$$

$$- A_4 \cos(\beta L)] - \bar{\varepsilon}_{33} \frac{\tilde{V}}{h} L \tag{8.71}$$

or

$$\tilde{V} \cdot \left(\frac{1}{2Z_L} + \frac{1}{Z_0}\right) = E_{piezo} \cdot d_{31} \left(\frac{H}{2} + h\right) \cdot \beta \cdot [-A_1 \sinh(\beta L) - A_2 \cosh(\beta L) + A_3 \sin(\beta L) -$$

$$- A_4 \cos(\beta L)] \tag{8.72}$$

where

$$C_0 = \frac{\bar{\varepsilon}_{33} \cdot b \cdot L}{h}, \quad Z_0 = \frac{1}{i \cdot \omega \cdot C_0} \tag{8.73}$$

Note that the impedances of Z_L and Z_0 depend on the input excitation frequency, ω, whose specific form depends on the specific structure of the output circuit. The explicit expressions for the four constants and the voltage are next presented:

$$A_1 = \frac{A}{\Delta}[-2 \cdot M_0 \cdot \omega^2 \cdot D \cdot \beta \cdot \cos(\beta \cdot L) \cdot \sinh(\beta \cdot L) - D \cdot \beta^4 \cdot \cos^2(\beta \cdot L)$$

$$M_0 \cdot \omega^2 \cdot \delta \cdot [1 - \cos(\beta \cdot L) \cdot \cosh(\beta \cdot L) - \sin(\beta \cdot L) \cdot \sinh(\beta \cdot L)] +$$

$$- D^2 \cdot \beta^4 \cdot [\sin^2(\beta \cdot L) + \cos(\beta \cdot L) \cdot \cosh(\beta \cdot L) + \sin(\beta \cdot L) \cdot \sinh(\beta \cdot L)] +$$

$$- D \cdot \beta^3 \cdot \delta \cdot \sin(\beta \cdot L) \cdot \cosh(\beta \cdot L) - \beta^3 \cdot \delta \cdot \sin(\beta \cdot L) \cdot \cos(\beta \cdot L)] \tag{8.74}$$

$$A_2 = \frac{A}{\Delta}[2 \cdot M_0 \cdot \omega^2 \cdot D \cdot \beta \cdot \cos(\beta \cdot L) \cdot \sinh(\beta \cdot L) - \beta^4 \cdot \sin^2(\beta \cdot L) +$$

$$M_0 \cdot \omega^2 \cdot \delta \cdot [\sin(\beta \cdot L) \cdot \cosh(\beta \cdot L) + \cos(\beta \cdot L) \cdot \sinh(\beta \cdot L)] +$$

$$+ D^2 \cdot \beta^4 \cdot [\cos(\beta \cdot L) \cdot \sinh(\beta \cdot L) + \cosh(\beta \cdot L) \cdot \sin(\beta \cdot L) + \cos(\beta \cdot L) \cdot \sin(\beta \cdot L)] +$$

$$+ D \cdot \beta^3 \cdot \delta \cdot [2 \cdot \sin(\beta \cdot L) \cdot \sinh(\beta \cdot L) + \sin^2(\beta \cdot L)] - D \cdot \beta^4 \cdot \delta \cdot \sin(\beta \cdot L) \cdot \cos(\beta \cdot L)] \tag{8.75}$$

$$A_3 = A - A_1 = \frac{A}{\Delta}[-2 \cdot M_0 \cdot \omega^2 \cdot D \cdot \beta \cdot \cos(\beta \cdot L) \cdot \sinh(\beta \cdot L) - D \cdot \beta^4 \cdot \cos^2(\beta \cdot L)$$

$$+ M_0 \cdot \omega^2 \cdot \delta \cdot [1 - \cos(\beta \cdot L) \cdot \cosh(\beta \cdot L) - \sin(\beta \cdot L) \cdot \sinh(\beta \cdot L)] +$$

$$- D^2 \cdot \beta^4 \cdot [\sin^2(\beta \cdot L) + \cos(\beta \cdot L) \cdot \cosh(\beta \cdot L) + \sin(\beta \cdot L) \cdot \sinh(\beta \cdot L)] +$$

$$- D \cdot \beta^3 \cdot \delta \cdot \sin(\beta \cdot L) \cdot \cosh(\beta \cdot L) - \beta^3 \cdot \delta \cdot \sin(\beta \cdot L) \cdot \cos(\beta \cdot L)] \tag{8.76}$$

$$A_4 = -A_2 = -\frac{A}{\Delta}[2 \cdot M_0 \cdot \omega^2 \cdot D \cdot \beta \cdot \cos(\beta \cdot L) \cdot \sinh(\beta \cdot L) - \beta^4 \cdot \sin^2(\beta \cdot L) +$$
$$M_0 \cdot \omega^2 \cdot \delta \cdot [\sin(\beta \cdot L) \cdot \cosh(\beta \cdot L) + \cos(\beta \cdot L) \cdot \sinh(\beta \cdot L)] +$$
$$+ D^2 \cdot \beta^4 \cdot [\cos(\beta \cdot L) \cdot \sinh(\beta \cdot L) + \cosh(\beta \cdot L) \cdot \sin(\beta \cdot L) + \cos(\beta \cdot L) \cdot \sin(\beta \cdot L)] +$$
$$+ D \cdot \beta^3 \cdot \delta \cdot [2 \cdot \sin(\beta \cdot L) \cdot \sinh(\beta \cdot L) + \sin^2(\beta \cdot L)] - D \cdot \beta^4 \cdot \delta \cdot \sin(\beta \cdot L) \cdot \cos(\beta \cdot L)]$$

$$(8.77)$$

and

$$\tilde{V} = \frac{\gamma}{\alpha}\left(\frac{H}{2} + h\right) \cdot \beta \cdot [-A_1 \sinh(\beta L) - A_2 \cosh(\beta L) + (A - A_1) \sin(\beta L) + A_2 \cos(\beta L)]$$

$$(8.78)$$

where the constants A_1–A_4 are defined in Equations (8.74)–(8.77) and the Δ is defined as

$$\Delta = -2 \cdot (D^2 \cdot \beta^4 - M_0 \cdot \omega^2 \cdot \delta) \cosh(\beta \cdot L) \cdot \cos(\beta \cdot L) - 2 \cdot D^2 \cdot \beta^4 + M_0 \cdot \omega^2 \cdot \delta$$
$$- (D^2 \cdot \beta^3 \cdot \delta - \beta^3 \cdot \delta - 2 \cdot M_0 \cdot \omega^2 \cdot \beta \cdot D) \cdot \sinh(\beta \cdot L) \cdot \sin(\beta \cdot L) \qquad (8.79)$$
$$- (D^2 \cdot \beta^3 \cdot \delta - \beta^3 \cdot \delta) \cdot \sin(\beta \cdot L) \cdot \cosh(\beta \cdot L) + 2 \cdot M_0 \cdot \omega^2 \cdot \delta \cdot \beta \cdot D$$

with

$$\alpha = \left(\frac{1}{2 \cdot Z_L} + \frac{1}{Z_0}\right); \quad \gamma = E_{piezo} \cdot d_{31}; \quad \delta = \frac{\gamma^2}{\alpha} \cdot (H + h) \cdot \left(\frac{H}{2} + h\right) \cdot b$$

$$(8.80)$$

Having the explicit expressions, one can calculate the behavior of the vibrating bimorph for various input amplitudes, A, and excitation frequency, ω, with the external impedance Z_L being a parameter.

It is instructive to understand the different types of electrical connections for a bimorph. Figure 8.17 presents the two types of connection: parallel and series. One should be aware that those electrical connections assume that between the piezoelectric material and the metal substrate a nonconductive layer exists.

(a) Parallel connection

(b) Series connection

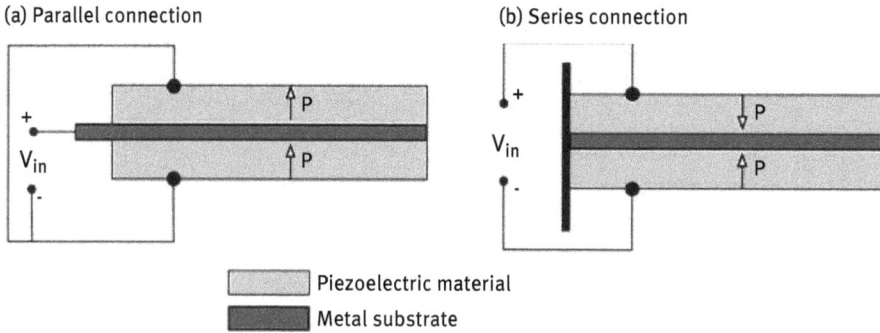

Piezoelectric material

Metal substrate

Figure 8.17: Piezoelectric bimorph connected in parallel (a) or in series (b).

8.4 Piezoelectric Harvester with Enhanced Frequency Bandwidth

8.4.1 Introduction

The present section is based on the master study of my student, Mr. Idan Har-nes [1, 2].

Piezoelectric vibration energy harvesting is a technique to accumulate electrical energy from mechanical vibrations. Converting mechanical vibrations from the ambient environment into electrical power enables us to operate remote small electrical consumers such as wireless sensors or low duty cycle radio transmitters.

The main problem with ambient vibration energy harvesting is the random characteristics of both the frequencies and their associated amplitudes. A basic harvester is a single bimorph with an end mass. The use of a single cantilever harvester as a harmonic oscillator to harvest vibrational energy is not effective due to its inherent narrow frequency bandwidth stemmed from the need to adjust the natural frequencies of the harvester to the platform excitation frequencies.

Browsing the open literature presents many interesting applications for piezoelectric harvesters with wideband vibration and the reader is invited to browse Refs. [1–41].

Using the basic concept proposed in [26], the present study focuses on an advanced system based on three identical bimorphs bonded on three different cantilever beams with different end masses that are connected by various springs. Figure 8.18 presents a schematic model of the proposed harvester.

From an electrical point of view, it is known that a series connection between two piezo strips is used for sensors, while for energy harvesting, the parallel connection is applied. A comparison between series and parallel connections can be found in [18]. The advantages of the proposed concept are:

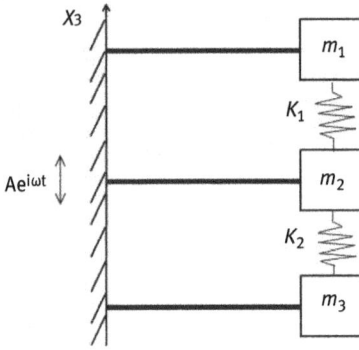

Figure 8.18: The present proposed concept.

The frequency bandwidth is wider than for a single bimorph design.

Contrary to single bimorph solutions, the present designed system is less sensitive to changes in the input frequency.

The connection of the bimorphs by springs adds degrees of freedom to the designer by allowing the choice of bimorphs and substrate beams dimensions, end masses sizes, and spring constants.

8.4.2 Derivation of the Equations of Motion and Their Solution

To derive the equations of motions for the system presented in Figure 8.18, it is assumed that the beam can be modelled using the Euler-Bernoulli theory, a perfect bonding between the piezoelectric layer and the substrate beam and linear springs between the end masses.

Based on Figure 8.19, which presents three bimorphs having a length of L, each with an end mass and two linear springs, K_1 and K_2, connecting the three masses, M_1, M_2 and M_3, the equations of motion, using the Newton's second law, for the three attached masses on the right end of the beams can be written in the following form:

$$\breve{V}_1(L,t) - K_1[w_1(L,t) - w_2(L,t)] = M_1 \cdot \ddot{w}_1(L,t) \tag{8.81a}$$

$$\breve{V}_2(L,t) + K_1[w_1(L,t) - w_2(L,t)] - K_2[w_2(L,t) - w_3(L,t)] = M_2 \cdot \ddot{w}_2(L,t) \tag{8.81b}$$

$$\breve{V}_3(L,t) + K_2[w_2(L,t) - w_3(L,t)] = M_3 \cdot \ddot{w}_3(L,t) \tag{8.81c}$$

where $\breve{V}_1$, $\breve{V}_2$ and $\breve{V}_3$ represent the shear force at each free end of the three bimorphs, respectively. Assuming that the variables in Equations (8.81a)–(8.81c) can be written as

$$\left\{ \begin{array}{c} w_i(L,t) \\ \breve{V}_i(L,t) \end{array} \right\} = \left\{ \begin{array}{c} W_i(L) \\ \widehat{V}_i(L) \end{array} \right\} e^{i\omega t}, \quad i = 1,2,3 \tag{8.82}$$

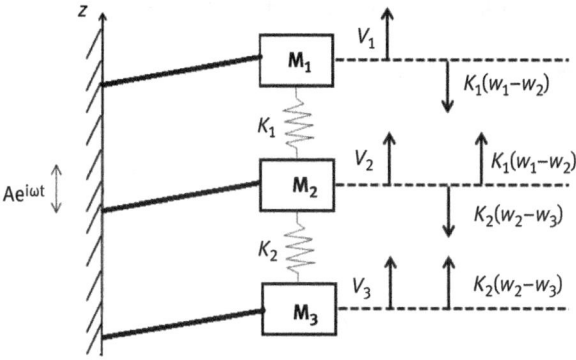

Figure 8.19: The schematic model for the analytical solution.

with ω^2 being the angular frequency squared of the system. Substituting Equation (8.82) into Equations (8.81a)–(8.81c) yields the shear forces at $x = L$

$$\breve{V}_1(L) = K_1[W_1(L) - W_2(L)] - \omega^2 \cdot M_1 \cdot W_1(L) \qquad (8.83a)$$

$$\breve{V}_2(L) = -K_1[W_1(L) - W_2(L)] + K_2[W_2(L) - W_3(L)] - \omega^2 \cdot M_2 \cdot W_2(L) \qquad (8.83b)$$

$$\breve{V}_3(L) = -K_2[W_2(L,t) - W_3(L,t)] - \omega^2 \cdot M_3 \cdot W_3(L) \qquad (8.83c)$$

Based on the derivation presented in Section 8.3, we can use Equation (8.62) for each beam of the system to yield

$$W_i(x) = A_{1i} \cosh(\beta_i x) + A_{2i} \sinh(\beta_i x) + A_{3i} \cos(\beta_i x) + A_{4i} \sin(\beta_i x), \quad i = 1,2,3 \quad (8.84a)$$

where

$$\beta_i^4 \equiv \left(\frac{m\omega^2}{D} \right)_i, \quad i = 1,2,3 \qquad (8.84b)$$

with D_i and m_i being defined like in Equations (8.51) and (8.54) (Section 8.3), respectively, namely

$$D_i = \left\{ \frac{2}{3} \cdot E_{\text{substrate}} \cdot \left(\frac{H}{2} \right)^3 + \frac{2}{3} \cdot E_{\text{piezo}} \left[\left(\frac{H}{2} + h \right)^3 - \left(\frac{H}{2} \right)^3 \right] \right\}_i \cdot b_i, \quad i = 1,2,3 \quad (8.84c)$$

where H and h represent the substrate thickness and the piezo layer thickness, respectively:

$$m_i = \left(H \cdot b \cdot \rho_{\text{substrate}} + 2 \cdot h \cdot b \cdot \rho_{\text{piezo}} \right)_i, \quad i = 1,2,3 \qquad (8.84d)$$

while $\rho_{substrate}$ and ρ_{piezo} are the substrate layer density and piezo layer density, respectively, and b is the bimorphs width.

The associated boundary conditions for the three equations presented by Equation (8.84a) can be written as

$$@x = 0$$

$$W_1(0) = A \quad \text{and} \quad W_{1,x}(0) = 0 \tag{8.85a}$$

$$W_2(0) = A \quad \text{and} \quad W_{2,x}(0) = 0 \tag{8.85b}$$

$$W_3(0) = A \quad \text{and} \quad W_{3,x}(0) = 0 \tag{8.85c}$$

$$@x = L$$

$$-D_1 \cdot W_{1,xx}(L) + \left[E_{piezo} d_{31} \frac{\tilde{V}}{h} \cdot (H+h) \cdot h \cdot b \right]_1 = 0 \quad \text{and} \tag{8.85d}$$

$$D_1 \cdot W_{1,xxx}(L) = K_1[W_1(L) - W_2(L)] - \omega^2 \cdot M_1 \cdot W_1(L)$$

$$-D_2 \cdot W_{2,xx}(L) + \left[E_{piezo} d_{31} \frac{\tilde{V}}{h} \cdot (H+h) \cdot h \cdot b \right]_2 = 0 \quad \text{and}$$
$$\tag{8.85e}$$
$$D_2 \cdot W_{2,xxx}(L) = -K_1[W_1(L) - W_2(L)] + K_2[W_2(L) - W_3(L)] - \omega^2 \cdot M_2 \cdot W_2(L)$$

$$-D_3 \cdot W_{3,xx}(L) + \left[E_{piezo} d_{31} \frac{\tilde{V}}{h} \cdot (H+h) \cdot h \cdot b \right]_3 = 0 \quad \text{and}$$
$$\tag{8.85f}$$
$$D_3 \cdot W_{3,xxx}(L) = -K_2[W_2(L,t) - W_3(L,t)] - \omega^2 \cdot M_3 \cdot W_3(L)$$

where, as in Section 8.3, $\tilde{V}$ represents the bimorph's voltage.

The voltage induced on each bimorph, due to the wall excitation is written as (see Section 8.3, Equation (8.72))

$$\tilde{V}_i = \Gamma_i \cdot \beta_i \cdot [-A_{i1} \sinh(\beta_i L) - A_{i2} \cosh(\beta_i L) + A_{i3} \sin(\beta_i L) - A_{i4} \cos(\beta_i L)], \quad i = 1,2,3 \tag{8.86a}$$

where

$$\Gamma_i = \frac{\left[E_{piezo} \cdot d_{31} \left(\frac{H}{2} + h \right) \right]_i}{\left(\frac{1}{2z_L} + \frac{1}{z_0} \right)_i}, \quad i = 1,2,3 \tag{8.86b}$$

and z_L, z_0 being the external impedance and internal impedance of the bimorphs system respectively.

Applying the boundary conditions presented above, we obtain a set of regular equations with 12 unknowns, written as

$$
\begin{bmatrix}
1 & 0 & 1 & 0 & 0 & 0 & 0 & 0 & 0 & 0 & 0 & 0 \\
0 & \beta_1 & 0 & \beta_1 & 0 & 0 & 0 & 0 & 0 & 0 & 0 & 0 \\
a_{31} & a_{32} & a_{33} & a_{34} & 0 & 0 & 0 & 0 & a & 0 & 0 & 0 \\
a_{41} & a_{42} & a_{43} & a_{44} & a_{45} & a_{46} & a_{47} & a_{48} & 0 & 0 & 0 & 0 \\
0 & 0 & 0 & 0 & 1 & 0 & 1 & 0 & 0 & 0 & 0 & 0 \\
0 & 0 & 0 & 0 & 0 & \beta_2 & 0 & \beta_2 & 0 & 0 & 0 & 0 \\
0 & 0 & 0 & 0 & a_{75} & a_{76} & a_{77} & a_{78} & 0 & 0 & 0 & 0 \\
a_{81} & a_{82} & a_{83} & a_{84} & a_{85} & a_{86} & a_{87} & a_{88} & a_{89} & a_{810} & a_{811} & a_{812} \\
0 & 0 & 0 & 0 & 0 & 0 & 0 & 0 & 1 & 0 & 1 & 0 \\
0 & 0 & 0 & 0 & 0 & 0 & 0 & 0 & 0 & \beta_3 & 0 & \beta_3 \\
0 & 0 & 0 & 0 & 0 & 0 & 0 & 0 & a_{119} & a_{1110} & a_{1111} & a_{1112} \\
0 & 0 & 0 & 0 & a_{125} & a_{126} & a_{127} & a_{128} & a_{129} & a_{1210} & a_{1211} & a_{1212}
\end{bmatrix}
\begin{pmatrix}
A_{11} \\ A_{12} \\ A_{13} \\ A_{14} \\ A_{21} \\ A_{22} \\ A_{23} \\ A_{24} \\ A_{31} \\ A_{32} \\ A_{33} \\ A_{34}
\end{pmatrix}
=
\begin{pmatrix}
A \\ 0 \\ 0 \\ 0 \\ A \\ 0 \\ 0 \\ 0 \\ A \\ 0 \\ 0 \\ 0
\end{pmatrix}
$$

$$(8.87)$$

The various terms appearing in the matrix presented by Equation (8.87) are

$$a_{31} = -D_1 \cdot \beta_1^2 \cdot \cosh(\beta_1 L) - \chi_1 \cdot \sinh(\beta_1 L), \quad a_{32} = -D_1 \cdot \beta_1^2 \cdot \sinh(\beta_1 L) - \chi_1 \cdot \cosh(\beta_1 L)$$

$$a_{33} = +D_1 \cdot \beta_1^2 \cdot \cos(\beta_1 L) + \chi_1 \cdot \sin(\beta_1 L), \qquad a_{34} = +D_1 \cdot \beta_1^2 \cdot \sin(\beta_1 L) - \chi_1 \cdot \cos(\beta_1 L)$$

with $\quad \chi_1 = \left[E_{piezo} \times d_{31} \cdot (H + h) \cdot b \right]_1 \cdot \Gamma_1$

$$(8.88a)$$

$$a_{41} = +\alpha_1, \quad a_{42} = +\alpha_2, \quad a_{43} = +\alpha_3, \quad a_{44} = +\alpha_4, \quad a_{45} = +K_1 \cdot \cosh(\beta_2 L)$$

$$a_{46} = +K_1 \cdot \sinh(\beta_2 L), \quad a_{47} = +K_1 \cdot \cos(\beta_2 L), \quad a_{48} = +K_1 \cdot \sin(\beta_2 L)$$

with

$$\alpha_1 = D_1 \cdot \beta_1^3 \cdot \sinh(\beta_1 L) - \left(K_1 - \omega^2 \cdot M_1 \right) \cdot \cosh(\beta_1 L)$$

$$\alpha_2 = D_1 \cdot \beta_1^3 \cdot \cosh(\beta_1 L) - \left(K_1 - \omega^2 \cdot M_1 \right) \cdot \sinh(\beta_1 L)$$

$$\alpha_3 = D_1 \cdot \beta_1^3 \cdot \sin(\beta_1 L) - \left(K_1 - \omega^2 \cdot M_1 \right) \cdot \cos(\beta_1 L)$$

$$\alpha_4 = -D_1 \cdot \beta_1^3 \cdot \cos(\beta_1 L) - \left(K_1 - \omega^2 \cdot M_1 \right) \cdot \sin(\beta_1 L)$$

$$(8.88b)$$

$$a_{75} = -D_2 \cdot \beta_2^2 \cdot \cosh(\beta_2 L) - \chi_2 \cdot \sinh(\beta_2 L), a_{76} = -D_2 \cdot \beta_2^2 \cdot \sinh(\beta_2 L) - \chi_2 \cdot \cosh(\beta_2 L)$$

$$a_{77} = +D_2 \cdot \beta_2^2 \cdot \cos(\beta_2 L) + \chi_2 \cdot \sin(\beta_2 L), \quad a_{78} = +D_2 \cdot \beta_2^2 \cdot \sin(\beta_2 L) - \chi_2 \cdot \cos(\beta_2 L)$$

with $\quad \chi_2 = \left[E_{piezo} \cdot d_{31} \cdot (H + h) \cdot b \right]_2 \cdot \Gamma_2$

$$(8.88c)$$

$$a_{81} = +K_1 \cdot \cosh(\beta_1 L), \quad a_{82} = +K_1 \cdot \sinh(\beta_1 L), \quad a_{83} = +K_1 \cdot \cos(\beta_1 L)$$

$$a_{84} = +K_1 \cdot \sin(\beta_1 L), \quad a_{85} = \alpha_5, \quad a_{86} = \alpha_6, \quad a_{87} = \alpha_7, \quad a_{88} = \alpha_8$$

$$a_{89} = -K_2 \cdot \cosh(\beta_3 L), \quad a_{810} = -K_2 \cdot \sinh(\beta_3 L), \quad a_{811} = -K_2 \cdot \cos(\beta_3 L)$$

$$a_{812} = -K_2 \cdot \sin(\beta_3 L)$$

with

$$\alpha_5 = D_2 \cdot \beta_2^3 \cdot \sinh(\beta_2 L) - \left[(K_1 + K_2) - \omega^2 \cdot M_2 \right] \cdot \cosh(\beta_2 L)$$

$$\alpha_6 = D_2 \cdot \beta_2^3 \cdot \cosh(\beta_2 L) - \left[(K_1 + K_2) - \omega^2 \cdot M_2 \right] \cdot \sinh(\beta_2 L)$$

$$\alpha_7 = D_2 \cdot \beta_2^3 \cdot \sin(\beta_2 L) - \left[(K_1 + K_2) - \omega^2 \cdot M_2 \right] \cdot \cos(\beta_2 L)$$

$$\alpha_8 = -D_2 \cdot \beta_2^3 \cdot \cos(\beta_2 L) - \left[(K_1 + K_2) - \omega^2 \cdot M_2 \right] \cdot \sin(\beta_2 L) \tag{8.88d}$$

$$a_{119} = -D_3 \cdot \beta_3^2 \cdot \cosh(\beta_3 L) - \chi_3 \cdot \sinh(\beta_3 L), \quad a_{1110} = -D_3 \cdot \beta_3^2 \cdot \sinh(\beta_3 L) - \chi_3 \cdot \cosh(\beta_3 L)$$

$$a_{1111} = +D_3 \cdot \beta_3^2 \cdot \cos(\beta_3 L) + \chi_3 \cdot \sin(\beta_3 L), \quad a_{1112} = +D_3 \cdot \beta_3^2 \cdot \sin(\beta_3 L) - \chi_3 \cdot \cos(\beta_3 L)$$

with $\quad \chi_3 = \left[E_{piezo} \cdot d_{31} \cdot (H+h) \cdot b \right]_3 \cdot \Gamma_3 \tag{8.88e}$

$$a_{125} = +K_2 \cdot \cosh(\beta_2 L), \quad a_{126} = +K_2 \cdot \sinh(\beta_2 L)$$

$$a_{127} = +K_2 \cdot \cos(\beta_2 L), \quad a_{128} = +K_2 \cdot \sin(\beta_2 L)$$

$$a_{129} = +\alpha_9, \quad a_{1210} = +\alpha_{10}, \quad a_{1211} = +\alpha_{11}, \quad a_{1212} = +\alpha_{12}$$

with

$$\alpha_9 = D_3 \cdot \beta_3^3 \cdot \sinh(\beta_3 L) - \left(K_2 - \omega^2 \cdot M_3 \right) \cdot \cosh(\beta_3 L)$$

$$\alpha_{10} = D_3 \cdot \beta_3^3 \cdot \cosh(\beta_3 L) - \left(K_2 - \omega^2 \cdot M_3 \right) \cdot \sinh(\beta_3 L)$$

$$\alpha_{11} = D_3 \cdot \beta_3^3 \cdot \sin(\beta_3 L) - \left(K_2 + \omega^2 \cdot M_3 \right) \cdot \cos(\beta_3 L)$$

$$\alpha_{12} = -D_3 \cdot \beta_3^3 \cdot \cos(\beta_3 L) - \left(K_2 + \omega^2 \cdot M_3 \right) \cdot \sin(\beta_3 L) \tag{8.88f}$$

Once the coefficients of the three equations appearing in Equation (8.84a), A_{1i}, A_{2i}, A_{3i} and A_{4i} ($i = 1, 2, 3$–12 terms) are found, the voltages generated on each bimorph (Equation (8.86a)) can be evaluated and the harvested power can be calculated for a given excitation frequency, ω.

The power is calculated using the following expression

$$P_i = I_i \cdot V_i \tag{8.89}$$

The natural frequencies of the system can be found by demanding the vanishing of the determinant of the coefficients matrix. Accordingly, a code was written within MATLAB[9] software program, which calculates the natural frequencies of the system, the various coefficients and the harvested power under a given excitation frequency.

Damping is included in the analytical model by allowing the elastic compliance to have complex values. Therefore, s_{11} (see Equation (8.15) will be replaced by $s_{11}(1 - iQ^{-1})$ (see a discussion in [26] and [31]), where Q is the quality factor of the bimorph (assumed to be $Q = 100$, as often used in preliminary calculations).

Due to the complexity of the quality factor, the power calculation will be updated to

$$P_i = \frac{1}{2}\left(\bar{I}_i \cdot V_i + I_i \cdot \bar{V}_i\right) \tag{8.90}$$

where $\bar{I}$ and $\bar{V}$ are the conjugate numbers of I_i and V_i, respectively.

To validate the present concept, a degenerated two bimorphs system was used as is presented in Figure 8.20.

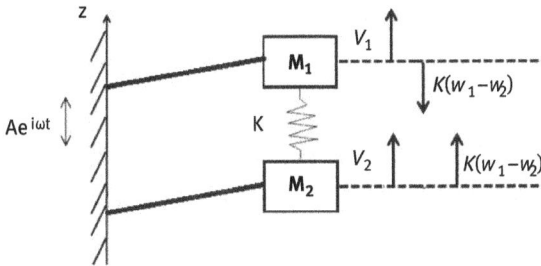

Figure 8.20: The schematic two bimorphs spring-connected analytic model.

For this system, the boundary conditions and the generated voltage will have the following form:

$$@x = 0$$

$$W_1(0) = A, \quad W_{1,x}(0) = 0 \tag{8.91a}$$

$$W_2(0) = A, \quad W_{2,x}(0) = 0 \tag{8.91b}$$

---

9 www.mathworks.com/products/matlab/

$@x = L$

$$- D_1 \cdot W_{1,xx}(L) + \left[E_{piezo} d_{31} \frac{\tilde{V}}{h} \cdot (H+h) \cdot h \cdot b \right]_1 = 0$$

$$D_1 \cdot W_{1,xxx}(L) = K_1 [W_1(L) - W_2(L)] - w^2 \cdot M_1 \cdot W_1(L) \qquad (8.91c)$$

$$- D_2 \cdot W_{2,xx}(L) + \left[E_{piezo} d_{31} \frac{\tilde{V}}{h} \cdot (H+h) \cdot h \cdot b \right]_2 = 0$$

$$D_2 \cdot W_{2,xxx}(L) = - K_1 [W_1(L) - W_2(L)] - w^2 \cdot M_2 \cdot W_1(L) \qquad (8.91d)$$

$$\tilde{V}_i = \Gamma_i \cdot \beta_i \cdot [-A_{i1} \sinh(\beta_i L) - A_{i2} \cosh(\beta_i L) + A_{i3} \sin(\beta_i L) - A_{i4} \cos(\beta_i L)], \quad i = 1,2$$
$$(8.91e)$$

8.4.3 Numerical Validation

To compare the present results to those of [26], the following data was used:

$$\text{Spring const. } K_0 = \tfrac{3EI}{L^2}, \quad \text{impedances } Z_0 = \tfrac{1}{i w C_0} \quad Z_L = i \cdot Z_0$$

$$\text{with } I = \tfrac{b(\pi h + 2c)^3}{12}, \quad C_0 = \tfrac{\varepsilon_{33} \cdot b \cdot L}{h}$$

The bimorphs were made from PZT 5H: $S_{11} = 16.5 \times 10^{-12}$ [m^2/N],

$$d_{31} = -274 \times 10^{-12} [C/N], \quad \varepsilon_{33} = 3400 \varepsilon_0, \quad \rho = 7500 \text{ [kg/m}^3\text{]}.$$

The quality factor was taken as $Q = 10^2$, while L = 25[mm], b=8[mm], h = 2[mm], c = 2[mm], the acceleration amplitude: $w^2 A = 1$ [m/s^2] and for the elastic substrate

$$E = 70 \text{ [GPa] and } \rho = 2700 \text{ [kg/m}^3\text{]}.$$

As can be seen from Figure 8.21a to d, a very good correlation between the two models exists, although the one in Ref. [26] has a mistake at the X axis values: the values defined as w[Hz], are actually w[rad/s] (see Figures 8.21b and d). Another important issue is the electrical connections between the two bimorphs: The two bimorphs in [26] are electrically connected together to yield a single output voltage for the harvester. As described further in the present section, this actually reduces the output power. The correct solution, as used further in this section, should be to connect each bimorph to the storage device and control the power using a smart design of the electric card.

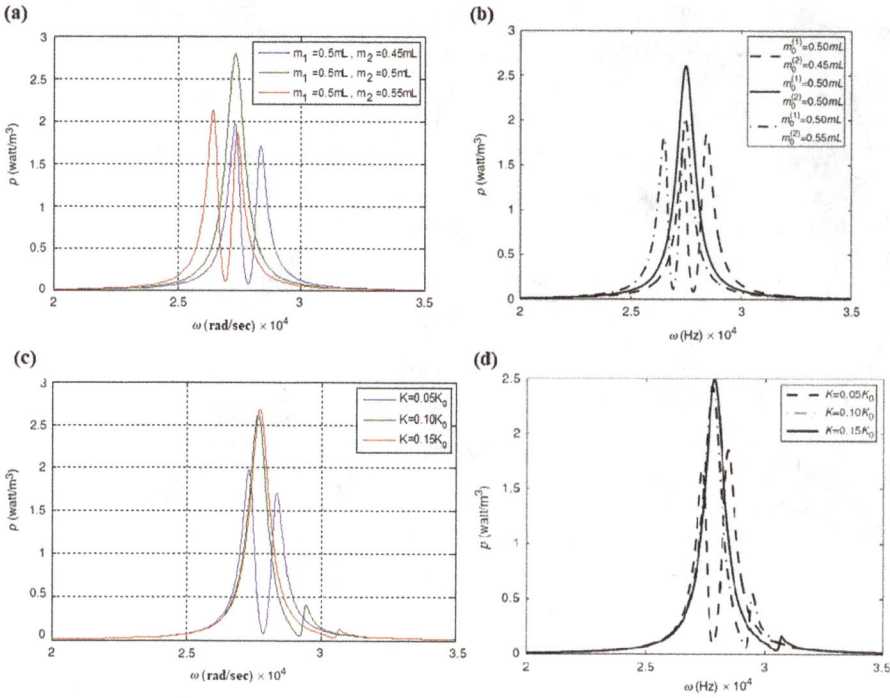

Figure 8.21: Two bimorphs interconnected by a spring-analytic power density model results for various end masses: (a) present model; (b) from [26]; for various spring constants; (c) present model; (d) from [26].

8.4.4 Experimental Validation

This section is aimed at presenting the experimental validation of the analytical model. Three test cases were carried out: (a) testing of the three bimorphs system, (b) testing of the two bimorphs arrangement and (c) the testing of the three bimorphs system with no spring connections. An experimental setup was designed and built and various configurations of the bimorphs were tested and their results were recorded for further processing.

The experimental setup is presented in Figures 8.22 and 8.23a and b. Its main components are:

- The piezo harvesting bimorph system
- 1 kΩ resistance load (for each bimorph)
- A laser sensor (LG5A65PU, BANNER)
- An oscilloscope
- A laser sensor power supply
- A shaker table

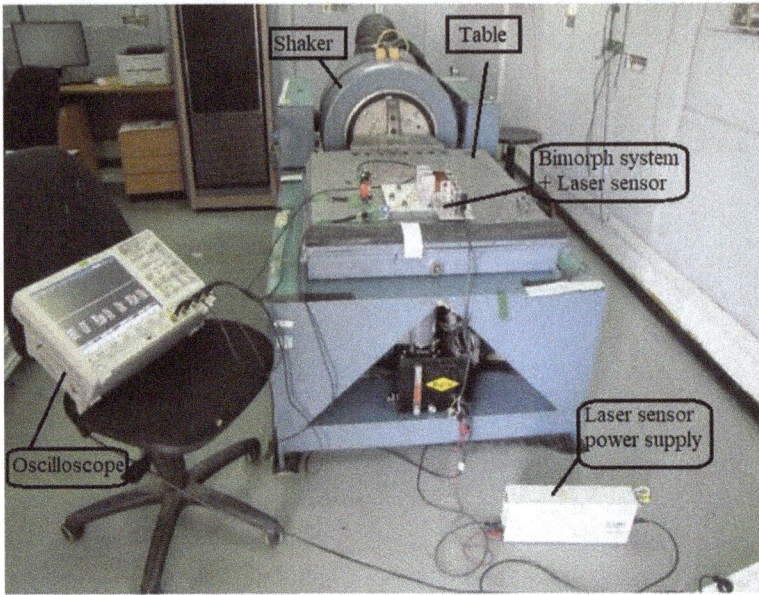

Figure 8.22: The experimental test setup.

Each bimorph was connected separately to the oscilloscope measuring and recording its output voltage. Another channel of the oscilloscope was used to measure the laser sensor output. The laser sensor could measure only the responses of two external masses (it could measure the response of the middle mass due to no direct line to it). Typical configurations of the three bimorphs spring connected system are presented in Figure 8.23a and b.

Each bimorph was constructed of two piezo layers and a 301 stainless steel[10] strip. The piezo layer was PI Ceramic[11] P-876.A11, which has a 0.1 mm piezo thickness (PIC255) and an electronic insulation (Kapton[12] tape) that protects the piezo material and preloads the layer. The dimensions of the piezo patch are $61 \times 35[mm^2]$ while the piezo layer is only $50 \times 30[mm^2]$ – see Figure 8.24.

The end masses were manufactured from 303 stainless steel[13] and were fixed to the bimorph with a screw and a nut to a counter plate, as can be seen in Figure 8.23b.

---

10 Type 301 is an austenitic chromium-nickel stainless steel that provides high strength and good ductility when cold worked. It is a modification of Type 304 in which the chromium and nickel contents are lowered to increase the cold work-hardening range.

11 /www.piceramic.com/en/

12 Kapton is a polyimide film developed by DuPont in the late 1960s that remains stable across a wide range of temperatures, from −269 to +400 °C.

13 Alloy 303 was specially designed to exhibit improved machinability while maintaining good mechanical and corrosion resistant properties Due to the presence of sulphur in the steel composition,

(a)   (b)

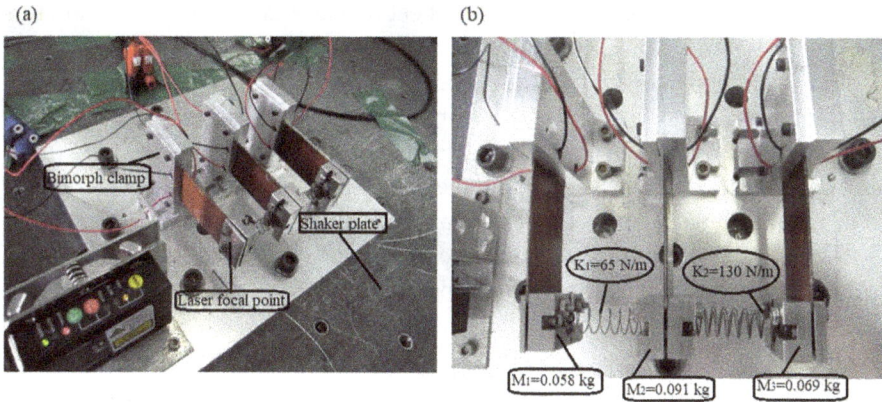

Figure 8.23: (a) A close-up on the experimental three bimorph systems and (b) upper view of the system.

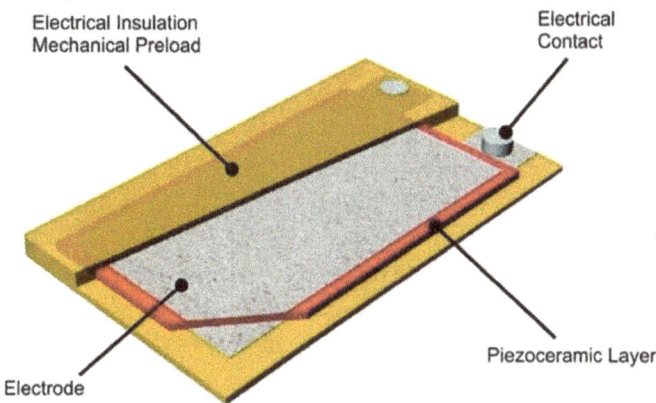

Figure 8.24: Piezo patch, PI Ceramic P-876.A11, from [2].

The end mass + counter plate + screw and nut have been weighted before assembly on the bimorphs yielding a total mass of 58 [g] and 69 [g] for the two external bimorphs and 91 [g] for the middle bimorph.

The springs were manufactured from 302 stainless steel.[14] Fixing the spring to the end mass was done by adding a small rectangle sheet that could enter between

---

Alloy 303 is the most readily machineable austenitic stainless steel; however, the sulphur addition does lower Alloy 303's properties.

14 Alloy 302 is a variation of the 18% chromium / 8% nickel austenitic alloy, which is the most familiar and the most frequently used in the stainless steel family. Alloy 302 is a slightly higher carbon version of 304, often found in strip and wire forms.

the first and final coils of the spring. The sheet had been attached to the end mass by two screws (see Figure 8.23b).

Inherent differences between the analytical model and the tested one might lead to different performance results. Every bimorph (even designed to be similar) has its own stiffness due to gluing variance of the piezo layer to the substrate (position and amount of glue), a variation in the dimensions of the substrate due to inaccurate cutting, and disparity between the piezo layers.

The assembly of the whole system also causes some variance between the models due to the way the bimorphs are clamped and due to different assemblies of the end masses and the springs.

To be able to use the analytical model in a reliable way, the model was tuned and adjusted, using the experimental results, by multiplying the assumed stiffness of each bimorph by a stiffness factor (experimentally found). Additionally, the Q factor (quality factor) was also adjusted to give a range of 10–20. One should note that the Q factor mainly influences the height and width of the frequency response graph.

Figure 8.25 presents the generated voltage versus input frequency for each bimorph separately ($M_1 = 0.058$ kg, $M_2 = 0.091$ kg and $M_3 = 0.069$ kg), with no spring connection, $K_1 = K_2 = 0$.

Note that the analytic curve presented in Figure 8.25 is after calibration of the Q factors and stiffness factors. The stiffness and Q factors as obtained from the calibration process are presented in Table 8.3 and the comparison between the analytical tuned model and the experimental results in Table 8.4.

As presented in Table 8.4, a good match between the analytical model (after its tuning process) and the experimental results was obtained. The difference between the analytic model and the experimental result is in the range of 4% (maximal), which is small enough for engineering purposes.

Figure 8.26 presents the generated voltage versus the input excitation frequency for each bimorph separately. The system has three end masses and two springs. The end masses weights are 0.058 kg and 0.069 kg, for the two external bimorphs, and 0.091 kg for the middle bimorph. The spring constant connecting the external mass (0.058 kg) to the middle mass (0.091 kg) is 65 N/m. The spring constant connecting the external mass (0.069 kg) to the middle mass is 130 N/m.

Note that the electrical connection for each bimorph (connecting the two piezoelectric layers sandwiching the substrate beam) was in parallel, as was discussed and presented at the end of Section 8.3. The results are summarized also in Table 8.5.

As shown in Table 8.5, the best fit for each bimorph is when the peak natural frequency is the natural frequency the bimorph is "responsible" for. Looking at the 0.058 kg bimorph for example, shows that the best correlation is for the second peak at the 18.8 Hz which is close to its own natural frequency (17.8 Hz). At the first peak (16 Hz) the correlation is not so good, however it appears at minimal voltage

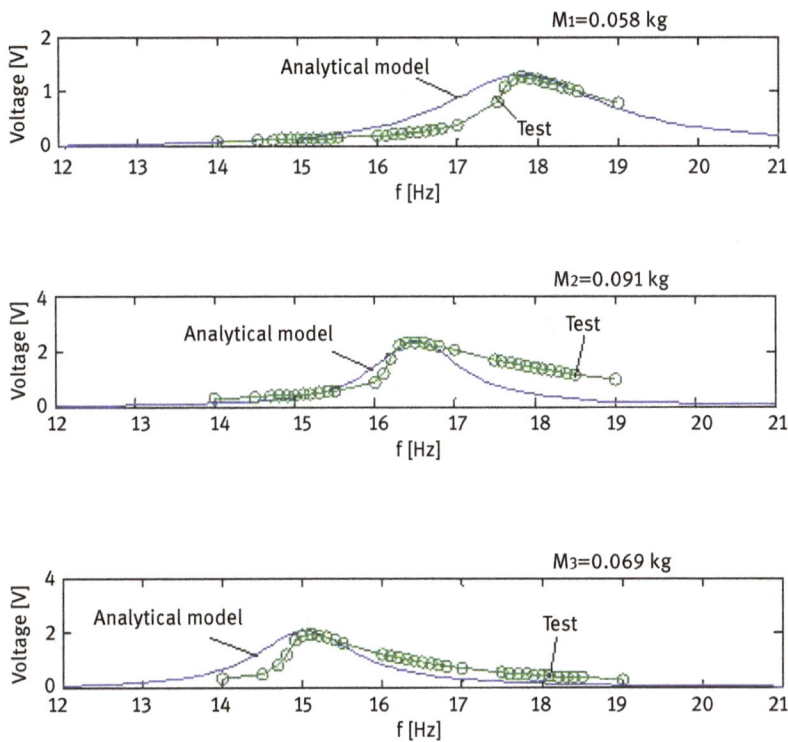

Figure 8.25: Three bimorphs without spring connection ($K_1 = K_2 = 0$): analytical tuned model versus experimental results.

Table 8.3: Three bimorphs system without spring connections ($K_1 = K_2 = 0$) – stiffness and Q factors.

Bimorph #	Stiffness factor	Quality factor
1	1.13	5.85
2	1.50	10.00
3	0.95	7.90

amplitude, therefore the influence of the miss-correlation has a small influence on the total voltage correlation.

Figure 8.27 displays the generated tuned voltage versus excitation input frequency for each bimorph. The same system presented above for the parallel electrical was again tested, this time with series connections.

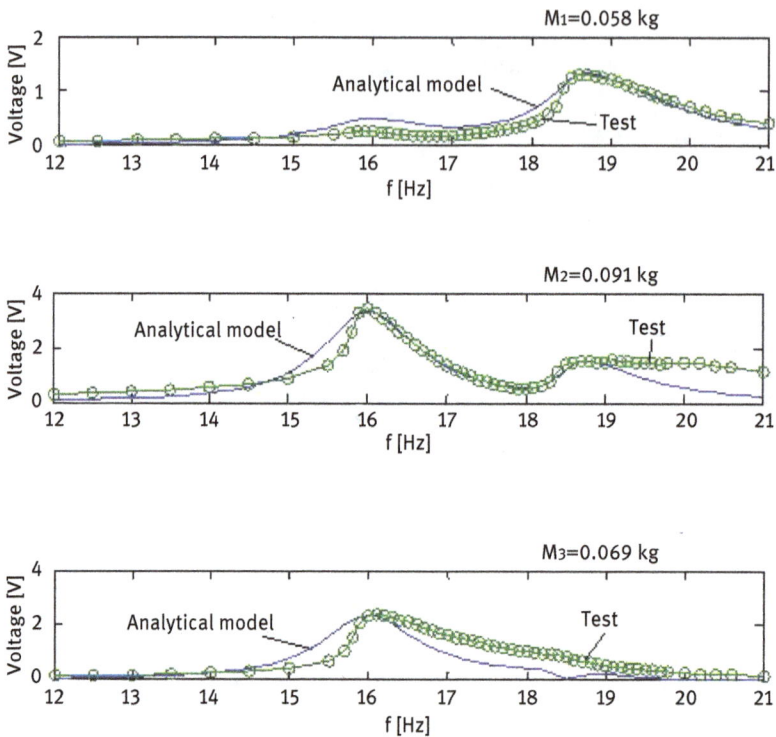

Figure 8.26: Three bimorphs (parallel electrical connections) with spring connection ($K_1 = 65$ N/m and $K_2 = 130$ N/m): analytical tuned model versus experimental results.

Table 8.4: Three bimorphs system without spring connections ($K_1 = K_2 = 0$): analytical predictions versus experimental results.

Mass, kg	0.058	0.091	0.069
Natural frequency, Hz	17.8	16.5	15.1
Predicted voltage, V – analytical model	1.304	2.368	2.042
Experimental voltage, V	1.260	2.320	1.965
Analytical model prediction to experimental result ratio	1.035	1.020	1.039

Table 8.6 displays a relatively good comparison between the predictions of the analytical tuned model and the experimental results. The generated voltage for each bimorph, and thus the total voltage of the system, when each bimorph has a series connection, is smaller than the parallel connection case: 1 V for the series case, compared to 2 V for the parallel connection. The natural frequencies remain the same.

Table 8.5: Experimental results versus analytical tuned predictions: the three bimorphs system with parallel electrical connections.

Mass, kg	0.058		0.091		0.069	
Natural frequency, Hz	First peak 16	Second peak 18.8	First peak 16	Second peak 18.8	First peak 16	Second peak 18.8
Predicted voltage, V – analytical model	0.4942	1.303	1.661	0.7619	2.326	0.1895
Experimental voltage, V	0.2500	1.280	1.710	0.7650	2.335	0.6050
Analytical model prediction to experimental result ratio	1.9768	1.017	0.971	0.996	0.996	0.110

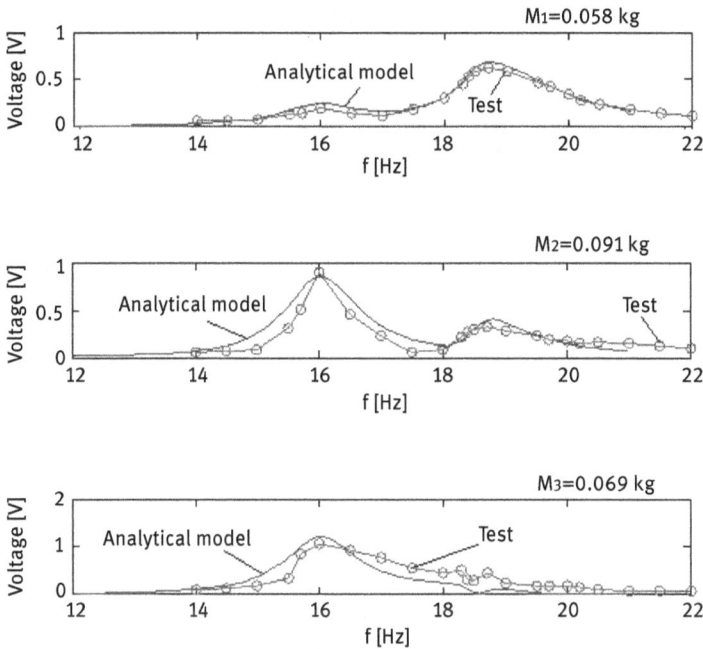

Figure 8.27: Three bimorphs (series electrical connections) with spring connection ($K_1 = 65$ N/m and $K_2 = 130$ N/m): analytical tuned model versus experimental results.

The better output-generated voltage when using the parallel connection was the main reason for choosing the parallel connection throughout all the calculations presented in the present research.

Figure 8.28 compares the total generated voltage for a three versus two bimorphs systems as obtained during the test series. One can observe that the three bimorphs system has lower natural frequencies as compared to the two bimorphs system, due to

Table 8.6: Experimental results vs. analytical tuned predictions – the three bimorphs system with series electrical connections.

Mass, kg	58		91		69	
Natural frequency, Hz	First peak 16	Second peak 18.8	First peak 16	Second peak 18.8	First peak 16	Second peak 18.8
Predicted voltage, V – analytical model	0.2459	0.6812	0.856	0.3931	1.202	0.06639
Experimental voltage, V	0.195	0.625	0.900	0.325	1.055	0.450
Analytical model prediction to experimental result ratio	1.260	1.089	0.950	1.200	1.130	0.147

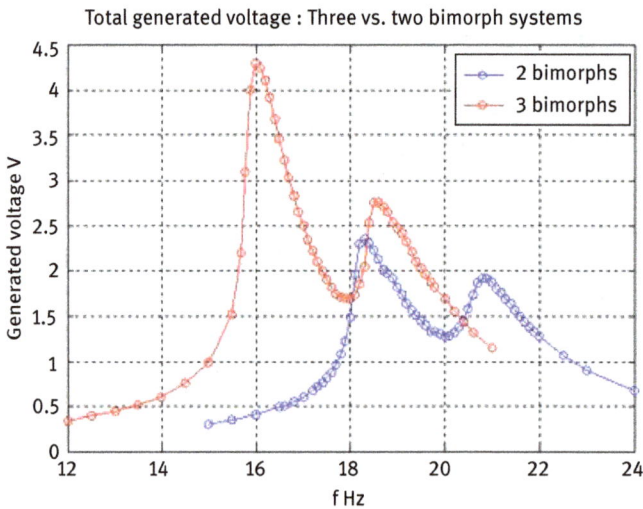

Figure 8.28: Three bimorphs vs. two bimorphs system – experimental results (voltage).

its high mass middle mass and the stiffness factor that is larger than in the two bimorphs system. As expected, the voltage and the bandwidth for the three bimorphs system are larger than for the two bimorphs system.

Before reaching to conclusions and recommendation, it is worth to mention the parametric investigation being performed for the two and three bimorphs system and presented in [1,2].

8.4.5 Three Versus Two Bimorph Systems

As stated at the beginning of Section 8.4, the main objective of the present research was to expend the power output bandwidth. A system having more bimorphs it is expected to have a wider bandwidth. Figure 8.29 shows a typical performance comparison between three and two bimorphs systems. The numerical comparison was made for $M_1 = 0.03$ kg, $M_2 = 0.06$ kg, $M_3 = 50$ kg, $K_1 = 40$ N/m and $K_2 = 80$ N/m for the three bimorphs system, while the data for the two bimorphs system was $M_1 = 0.03$ kg, $M_2 = 0.05$ kg and $K = 50$ N/m.

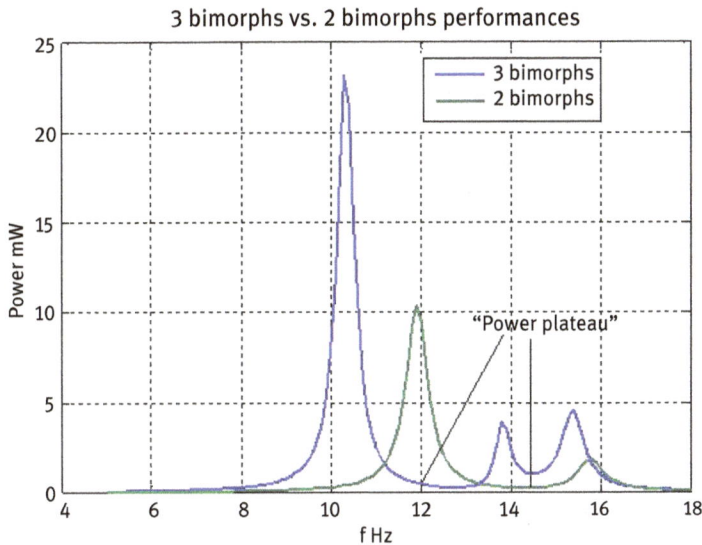

Figure 8.29: Three bimorphs versus two bimorphs system performances: numerical results (voltage).

It can be seen that the three bimorphs system has a wider bandwidth compared to that of two bimorphs system. Moreover, the three bimorphs system has a larger output power and by adding an additional bimorph (the middle one) the width of the "power pit" is reduced. Similar results were published in [28].

Another important issue is the electrical connections between the bimorphs. The bimorphs can be electrically connected together and then to the electrical circuit, or each bimorph can be connected individually to the electrical circuit and the output from all the bimorphs directed to the storing system.

The connected systems assume that all the plus signs are connected to each other and all the minus signs connected to each other. The individual systems assume that every bimorph is electrically on its own and the electric circuit would be responsible for the sum of the voltages, currents and power output.

As shown in Figure 8.30 the power output obtained by "all together" electrical connection of the three bimorphs is less than the separate type connection (except for the first and third natural frequencies). Although the output power for these natural frequencies is higher, the bandwidth is very narrow, the "power plateau" between them is wider and the power for the second natural frequency is almost negligible. Therefore, to reach an extended frequency bandwidth and a high total harvested power, the individual electric connection is recommended.

One should note that the data used for the results presented in Figure 8.30 were bimorph length, width and height, $L = 0.05$ m, $b = 0.03$ m and $h = 0.0001$ m, respectively. The masses were $M_1 = 0.03$ kg, $M_2 = 0.06$ kg, $M_3 = 0.05$ kg, and the spring coefficients were $K_1 = 40$ N/m and $K_2 = 60$ N/m.

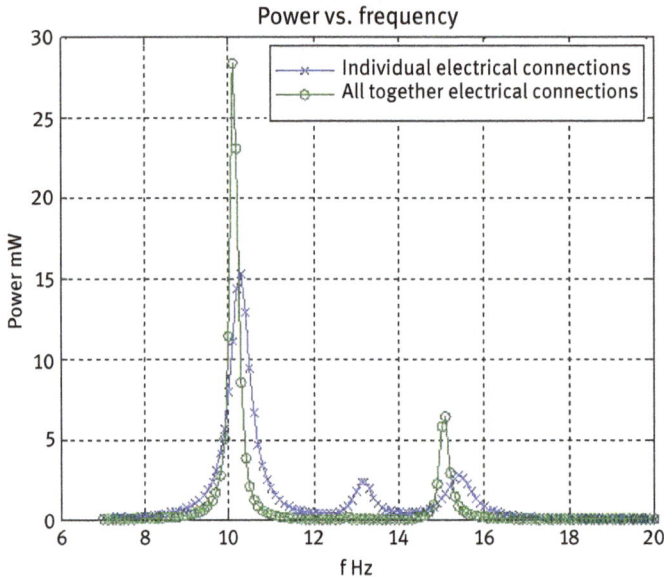

Figure 8.30: Three bimorphs system vs. frequency-numerical results (power in mW) for individual and all together electrical connections.

8.4.6 Conclusions and Recommendations

Based on the analytical and experimental study exhibited above some interesting conclusions and recommendations can be drawn (see also [1–2]).
- It turned out that the connection spring constant is crucial. If the spring constant is too small compared to the bimorph's stiffness than, its influence is negligible, and the system acts like there is no spring ($K = 0$). On the other hand, spring constant too stiff compared to the bimorph's stiffness causes the spring acting like a rigid bar. Thus, for the two bimorphs system, for example, the second natural frequency

would vanish. A preferred spring constant would be about 15% from the stiffness of the bimorph ($0.15 \left(3EI/L^2 \right)$).

- As the mass ratio between the two bimorphs increases, the "power plateau"[15] (see Section 8.5.5) increases, which reduces the system effectiveness. Decreasing the mass ratio decreases the "power plateau" but narrows the frequency bandwidth. Thus, the selection of the mass ratio is critical. A preferred end mass ratio should be 10–20, depending on system geometry and spring rate. The same conclusion can be found in [28, 39].
- Note that a complete disappearance of the "power plateau" can be achieved only for the case of two very close natural frequencies, yielding a narrow bandwidth, and therefore not applicable. The designer would have to choose the harvesters parameters in such a way to minimize the width of the "power plateau" while maximizing the bandwidth.
- As expected and predicted, the three bimorphs system generates higher power and wider bandwidth compared to the two bimorphs system. Adding more bimorphs and springs to the system is expected to increase even more the generated output power as well as the bandwidth of the system.
- The three bimorphs system presents larger degrees of freedom for the designer as compared with the two bimorphs system. Thus, the designer will be able to deal better with the "power plateau" issue and to correctly adjust the natural frequencies to the expected excitation input.
- The "all together" versus individual electrical connection for the various bimorphs forming the harvester system which was parametrically investigated in [1, 2] is showing that bimorphs' individual electrical connections and then summing up their generated power would yield a better harvester. This conclusion is in line to a similar statement presented in [37].

References

[1]   Har-nes, I., Bandwidth expansion for piezoelectric energy harvesting, Master thesis, Faculty of Mechanical Engineering, Technion, I.I.T., 32000, Haifa, Israel, August 2016, 170.

[2]   Abramovich, H. and Har-nes, I., Analysis and experimental validation of a piezoelectric harvester with enhanced frequency bandwidth, Materials 2018, 11(1243), 41. doi:10.3390/ma11071243 WWW.MDPI.COM/JOURNAL/MATERIALS.

[3]   Arms, S. W., Townsend, C. P., Churchill, D. L., Galbreath, J. H. and Mundell, S. W., Power management for energy harvesting wireless sensors, Proceedings of the SPIE International Symposium on Smart Structures and Smart Materials, San Diego, CA, USA, 7–10 March, 2005.

[4]   Murimi, E. and Neubauer, M., Piezoelectric energy harvesting: an overview, Proceedings of Sustainable Research and Innovation Conference, 2014, 4, 117–121.

---

15 A "power plateau" is defined as the null generated power between two adjacent natural frequencies.

[5]   Tang, L., Yang, Y. C. and So, K., Broadband vibration energy harvesting techniques. In: Advances in Energy Harvesting Methods, New York, NY, USA, Springer Science Business Media, 2013, 17–61.

[6]   Caliò, R., Rongala, U. B., Camboni, D., Milazzo, M., Stefanini, C., De Petris, G. and Oddo, C. M., Piezoelectric energy harvesting solutions, Sensors 2014, 14, 4755–4790.

[7]   Kong, L. B., Li, T., Hong, H. H., Boey, F., Zhang, T. and Li, S., Waste Mechanical Energy Harvesting (I): Piezoelectric Effect, Waste Energy Harvesting-Mechanical and Thermal Engines, Lecture Notes in Energy, Berlin/Heidelberg, Germany, Springer, 2014, 19–133.

[8]   Challa, V. R., Prasad, M. G., Shi, Y. A. and Fisher, F. T., A vibration energy harvesting device with bidirectional resonance frequency tenability, Smart Materials and Structures 2008, 17 (015035), 10. doi:10.1088/0964-1726/01/015035.

[9]   Challa, V. R., Prasad, M. G. and Fisher, F. T., Towards an autonomous self-tuning vibration energy harvesting device for wireless sensor network applications, Smart Materials and Structures 2011, 20(025004), 11. doi:10.1088/0964-1726/20/2/025004.

[10] Challa, V. R., Prasad, M. G. and Fisher, F. T., High efficiency energy harvesting device with magnetic coupling for resonance frequency tuning, Proceedings of SPIE Sensors and Smart Structures Technologies for Civil, Mechanical, and Aerospace Systems 6932, 2008, 69323Q. doi:10.1117/12.776385.

[11] Ferrar, M., Ferrari, M., Guizetti, M., Ando, B., Baglio, S. A. and Trigona, C., Improved energy harvesting from wideband vibrations by nonlinear piezoelectric converters, Sensors and Actuators A, Physical 162(2), 2010, 425–431.

[12] Cottone, F., Gammaitoni, L., Vocca, H. A. and Ferrari, V., Piezoelectric buckled beams for random vibration energy harvesting, Smart Materials and Structures 21(3), 2012, 035021, 11.

[13] Eichhorn, C., Goldschmidtboeing, F. A. and Woias, P. A frequency tunable piezoelectric energy converter based on a cantilever beam, Proceedings of the PowerMEMS2008+microEMS2008, Sendai, Japan, 9–12 November 2008, 309–312.

[14] Eichhorn, C., Goldschmidtboeing, F., Porro, Y. and Woias, P., A piezoelectric harvester with an integrated frequency tuning mechanism. In Proceedings of the PowerMEMS, Washington, DC, USA, 1–4 December 2009, 45–48.

[15] Eichhorn, C., Tchagsim, R., Wilhelm, N., Biancuzzi, G. and Woias, P., An energy-autonomous self-tunable piezoelectric vibration energy harvesting system. In Proceedings of the 2011 IEEE 24th International Conference Micro Electro Mechanical Systems (MEMS), Cancun, Mexico, 23–27 January 2011, 1293–1296.

[16] Wu, X., Lin, J., Kato, S., Zhang, K., Ren, T. and Liu, L., Frequency adjustable vibration energy harvester, Proceedings of the PowerMEMS 2008+microEMS2008, Sendai, Japan, 9–12 November 2008, 245–248.

[17] Ko, S. C., Je, C. H. A. and Jun, C. H., Mini piezoelectric power generator with multi-frequency response, Procedia Engineering 5, 2010, 770–773.

[18] Aridogan, U., Basdogan, I. and Erturk, A., Multiple patch–based broadband piezoelectric energy harvesting on plate-based structures, Journal of Intelligent Material Systems and Structures 25(4), 2014, 1664–1680. doi:10.1177/1045389X14544152.

[19] Berdy, D. F., Srisungsitthisunti, P., Jung, B., Xu, X., Rhoads, J. F. and Peroulis, D., Low-frequency meandering piezoelectric vibration energy harvester, IEEE Transactions on Ultrasonics, Ferroelectrics, and Frequency Control 59(5), 2012, 846–858.

[20] Miah, H., Heum, D. A. and Park, J. Y., Low frequency vibration energy harvester using stopper-engaged dynamic magnifier for increased power and wide bandwidth, Journal of Electrical Engineering and Technology 11(3), 2016, 707–714.

[21] Erturk, A., Renno, J. M. and Inman, D. J., Modeling of piezoelectric energy harvesting from an L-shaped beam-mass structure with an application to UAVs, Journal of Intelligent Material Systems and Structures 20, 2009, 529–544.

[22] Wang, H. Y., Tang, L. H., Guo, Y., Shan, X. B. and Xie, T., A 2DOF hybrid energy harvester based on combined piezoelectric and electromagnetic conversion mechanisms, Journal of Zhejiang University-Science A (Applied Physics & Engineering) 15(9), 2014, 711–722.

[23] Alghisi, D., Dalola, S., Ferrari, M. and Ferrari, V., Ball-impact piezoelectric converter for multi-degree-of-freedom energy harvesting from broadband low-frequency vibrations in autonomous sensors, Procedia Engineering 87, 2014, 1529–1532.

[24] Ferrari, M., Bau, M., Cerini, F. and Ferrari, V., Impact-enhanced multi-beam piezoelectric converter for energy harvesting in autonomous sensors, Procedia Engineering 47, 2012, 418–421.

[25] Aryanpur, R. M. and White, R. D., Multi-link piezoelectric structure for vibration energy harvesting. Proc. SPIE 8341, Active and Passive Smart Structures and Integrated Systems, 2012, 83411Y, April 26, 2012, doi:10.1117/12.915438.

[26] Yang, Z. and Yang, J., Connected vibrating piezoelectric bimorph beams as a wide-band piezoelectric power harvester, Journal of Intelligent Material Systems and Structures 20, March 2009, 569–574. doi:10.1177/104389X0800042.

[27] Lee, P., A wide band frequency-adjustable piezoelectric energy harvester, Master's Thesis, University of North Texas, Denton, TX, USA, 2012, 27.

[28] Zhang, H. and Afzalul, K., Design and analysis of a connected broadband multi-piezoelectric bimorph beam energy harvester, IEEE Transactions on Ultrasonics, Ferroelectrics, and Frequency Control 61(6), 2014, 1016–1023.

[29] Meruane, V. and Pichara, K., A broadband vibration-based energy harvester using an array of piezoelectric beams connected by springs, Shock and Vibration, 2016, 9614842, 13. doi:10.1155/2016/9614842.

[30] Shahruz, S. M., Limits of performance of mechanical band-pass filters used in energy scavenging, Journal of Sound and Vibration 293(1–2), 2006, 449–461.

[31] Shahruz, S. M., Design of mechanical band-pass filters for energy scavenging, Journal of Sound and Vibration 292(3–5), 2006, 987–998.

[32] Xue, H., Hu, Y. and Wang, Q.-M., Broadband piezoelectric energy harvesting devices using multiple bimorphs with different operating frequencies, IEEE Transactions on Ultrasonics, Ferroelectrics, and Frequency Control 55, 2008, 2104–2108.

[33] Qi, S., Investigation of a novel multi-resonant beam energy harvester and a complex conjugate matching circuit, Ph.D. Thesis, University of Manchester, Manchester, UK, 2011.

[34] Lien, I. C. and Shu, Y. C., Array of piezoelectric energy harvesting by the equivalent impedance approach, Smart Materials and Structures 21, 2012, 082001. doi:10.1088/0964-1726/21/8/082001.

[35] Lin, H. C., Wu, P. H., Lien, I. C. and Shu, Y. C., Analysis of an array of piezoelectric energy harvesters connected in series, Smart Materials and Structures 22, 2013, 094026. doi:10.1088/0964-1726/22/9/094026.

[36] Lien, I. C. and Shu, Y. C. Piezoelectric array of oscillators with respective electrical rectification, Proc. Vol. 8688, Active and Passive Smart Structures and Integrated Systems, Proceedings of the 2013 SPIE Smart Structures and Materials+Nondestructive Evaluation and Health Monitoring, San Diego, CA, USA, 10–14 March 2013.

[37] Al-Ashtari, W., Hunsting, M., Hemsel, T. and Sextro, W., Enhanced energy harvesting using multiple piezoelectric elements: theory and experiments, Sensors and Actuators A: Physics 200, 2013, 138–146.

[38] Wu, P. H. and Shu, Y. C., Finite element modeling of electrically rectified piezoelectric energy harvesters, Smart Materials and Structures 24, 2015, 094008. doi:10.1088/0964-1726/24/9/094008.

[39] Dechant, E., Fedulov, F., Fetisov, L. Y. and Shamonin, M., Bandwidth widening of piezoelectric cantilever beam arrays by mass-tip tuning for low-frequency vibration energy harvesting, Applied Science 7, 2017, 1324. doi:10.3390/app7121324.

[40] Yang, Y., Wu, H. and Soh, C. K., Experiment and modeling of a two-dimensional piezoelectric energy harvester, Smart Materials and Structures 24, 2015, 125011. doi:10.1088/0964-1726/24/12/125011.

[41] Miller, L. M., Elliot, A. D. T., Mitcheson, P. D., Halvorsen, E., Paprotny, I. and Wright, P. K., Maximum performance of piezoelectric energy harvesters when coupled to interface circuits, IEEE Sensors Journal 16, 2016, 4803–4815.

9 Introduction to Fiber Optic

9.1 Geometrical Optics: Basic Concepts

Geometrical optics is a model of optics that describes light propagation in terms of rays. The ray in geometric optics is useful for approximating the paths along which light propagates.

One should remember that the speed of light in vacuum and its associated frequency are

$$c = 2.99792458 \times 10^8 \text{ m/s}$$

$$f = c/\lambda \cong 10^{15} \text{ Hz}$$

When the light is traveling in a material, like water for instance, it will slow down, leading to a term called index of refraction, $n = c/v$, where v is the speed of the light through that material (see Table 9.1 for various values of n).

Table 9.1: Index of refraction for common materials (from *CRC Handbook of Chemistry and Physics*).

Material	Air	Water	Fused quartz	Whale oil	Crown glass	Salt	Asphalt	Diamond	Lead
n	1.0003	1.33	1.4585	1.46	1.52	1.54	1.635	2.42	2.6

Remembering that the wave speed (v) is related to the frequency (f) and wavelength (λ), $v = f/\lambda$ and that the light frequency does not change with the medium it travels through, we obtain the following relation between the index of refraction and the wavelength of the light

$$\lambda_0 = c/f \quad @\text{vacuum}$$

$$\lambda = v/f \quad @\text{other medium} \tag{9.1}$$

$$\Rightarrow \frac{\lambda_0}{\lambda} = \frac{c}{v} = n$$

9.1.1 Reflection of Light

When the light hits glossy surfaces, like for example mirrors, it will be reflected back, as shown schematically in Figure 9.1. The law of reflection states that the incident ray, the normal and the reflected ray lay in the same plane and $\theta_i = \theta_r$.

https://doi.org/10.1515/9783110726701-009

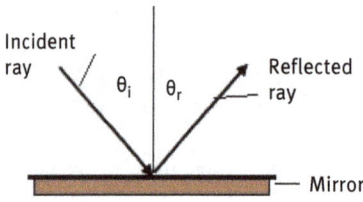

Figure 9.1: Schematic view of a reflected ray.

9.1.2 Snell's Law – Refractions

When light travels from one medium with index of refraction n1 to a second medium having an index of refraction n_2 it changes its direction, as shown schematically in Figure 9.2. Snell's law[1] can be written as

$$n_1 \sin \theta_1 = n_2 \sin \theta_2$$
$$\Rightarrow v_1 \sin \theta_2 = v_2 \sin \theta_1 \tag{9.2}$$

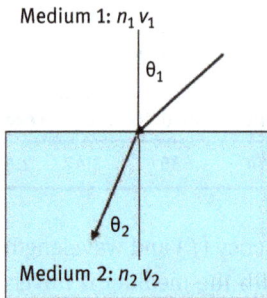

Figure 9.2: Schematic view of a refracted ray – Snell's law.

9.1.3 Critical Angle and Total Internal Reflection

When light travels from a medium with high value of n to another medium having lower n, the application of the Snell's law might require that the sine of the angle would be larger that unity. This of course cannot happen and therefore for such cases the light will be totally reflected, which is known in literature as *total internal reflection*. The *critical angle* is defined as the largest possible angle of incidence, which still allows refraction. This can be written as

$$n_1 \sin \theta_1 = n_2 \sin \theta_2$$

---

1 Snell is the English family name version of Willebrord Snellius (June 13, 1580–October 30, 1626) a Dutch astronomer and mathematician.

$$\theta_{\mathrm{crit}} = \arcsin\left(\frac{n_2}{n_1}\sin\theta_2\right)$$

but $\quad \sin\theta_2 = 1$

$$\Rightarrow \theta_{\mathrm{crit}} = \arcsin\left(\frac{n_2}{n_1}\right) \tag{9.3}$$

9.1.4 Numerical Aperture and Acceptance Angle of a Fiber

Numerical aperture (NA) is a nondimensional quantity related to the acceptance angle of a given angle. The term is written as

$$\mathrm{NA} = n_o \sin\theta_a \tag{9.4}$$

where θ_a is the maximum half of the acceptance angle of a fiber, and n_o is the index of refraction of the material outside the fiber. Typically the outside material is air, and therefore $n_o=1$, yielding $\mathrm{NA} = \sin\theta_a$. Another way of defining the NA is using the schematic drawing presented in Figure 9.3. A ray at the maximum half acceptance angle propagates into a fiber. It will hit the interface cladding-core of the finer and it will be reflected. According to Snell's law, one can write the following

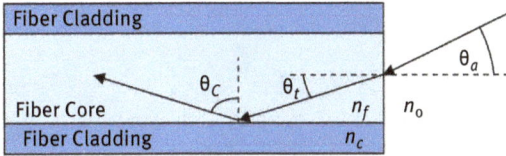

Figure 9.3: A light ray impinging the fiber at half its acceptance angle.

$$n_o \sin\theta_a = n_f \sin\theta_t \quad \text{(air core)}$$

$$n_f \sin\theta_c = n_c \sin\theta_r \quad \text{(core cladding)}$$

but $\quad \sin\theta_{\mathrm{crit},\,c} = \left(\frac{n_c}{n_f}\sin\theta_r\right) \quad$ and $\quad \sin\theta_r = 1$

$$\Rightarrow \sin\theta_{\mathrm{crit},\,c} = \left(\frac{n_c}{n_f}\right)$$

however $\quad \theta_{\mathrm{crit},\,c} + \theta_t = 90^0 \Rightarrow \sin\theta_{\mathrm{crit},\,c} = \cos\theta_t$

$$\Rightarrow \sin\theta_{\mathrm{crit},\,c} = \left(\frac{n_c}{n_f}\right) = \cos\theta_t = \sqrt{1 - \sin^2\theta_t}$$

or $\quad \sin\theta_t = \sqrt{1 - \frac{n_c^2}{n_f^2}} \tag{9.5}$

Using the first row in Equation (9.5) and substituting the equation in the last row of Equation (9.5) yields

$$n_o \sin \theta_a = n_f \sin \theta_t \quad \text{(air \quad core)}$$

$$\text{but} \quad \sin \theta_t = \sqrt{1 - \frac{n_c^2}{n_f^2}}$$

$$\Rightarrow n_o \sin \theta_a = n_f \sqrt{1 - \frac{n_c^2}{n_f^2}} = \sqrt{n_f^2 - n_c^2} = \text{NA}$$

$$\text{or} \quad \sin \theta_{a, \max} = \sqrt{\frac{n_f^2 - n_c^2}{n_o^2}} \Rightarrow \theta_{a, \max} = \arcsin\left(\sqrt{\frac{n_f^2 - n_c^2}{n_o^2}}\right)$$

(9.6)

The total angle of acceptance for the light conus would be twice the last expression of Equation (9.6).

The refractive index of the core changes from its center toward the cladding. The profile of the refractive index is generally given as

$$n^2(r) = n_f^2 \left[1 - 2\Delta\left(\frac{r}{0.5d}\right)^\alpha\right] \quad r < 0.5d \quad (@\text{core})$$

$$n^2(r) = n_c^2 = \text{constant} \quad r > 0.5d \quad (@\text{cladding})$$

$$\text{where} \quad \Delta = \frac{n_f^2 - n_c^2}{2n_f^2}$$

and α for special cases is

$\alpha = 1$ for triangular profile

$\alpha = 2$ for parabolic profile

$\alpha \rightarrow \infty$ for Step – Index profile

(9.7)

Another parameter is the normalized frequency, V is given as

$$V = 2\pi \frac{0.5d}{\lambda} \text{NA} = \pi \frac{d}{\lambda} \text{NA}$$

where

$$\text{NA} = n_f \sqrt{2\Delta}$$

(9.8)

where d is the core diameter and λ is the light wavelength. The V parameter determines the fiber-operating regime: mono-mode propagation or multi-mode propagation.

The number of modes guided through the core is a function of the above profile, namely

$$N \approx \frac{V^2}{2} \frac{\alpha}{\alpha+2} \qquad (9.9)$$

For a step-index profile if $V < 2.405$, then only a single mode will propagate. This is also called the fundamental mode LP_{01} with a cut-off frequency of $V_C = 0$.

For a graded-index profile ($\alpha = 2$) we get

$$V_c \approx 2.405 \cdot \sqrt{2} \approx 3.4 \qquad (9.10)$$

9.1.5 Interference

When two rays have the same wavelength and are in phase (see Figure 9.4), they will add, yielding a wave having the sum of the peak-to-peak amplitudes of the input rays thus leading to what it is called a *Constructive Interference*.

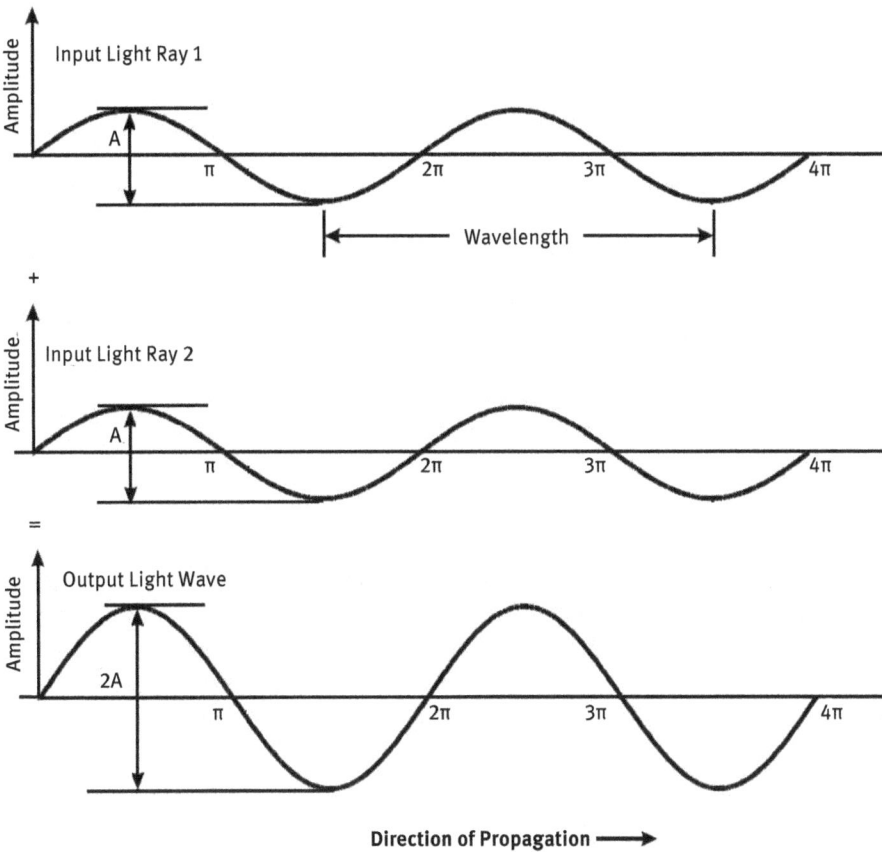

Figure 9.4: Constructive interference – a schematic view.

The opposite case leading to *Destructive Interference* (see Figure 9.5) is due to the "out of phase" state of one of the rays compared to the other. If the phase is π, then the output light peak-to-peak amplitude would vanish.

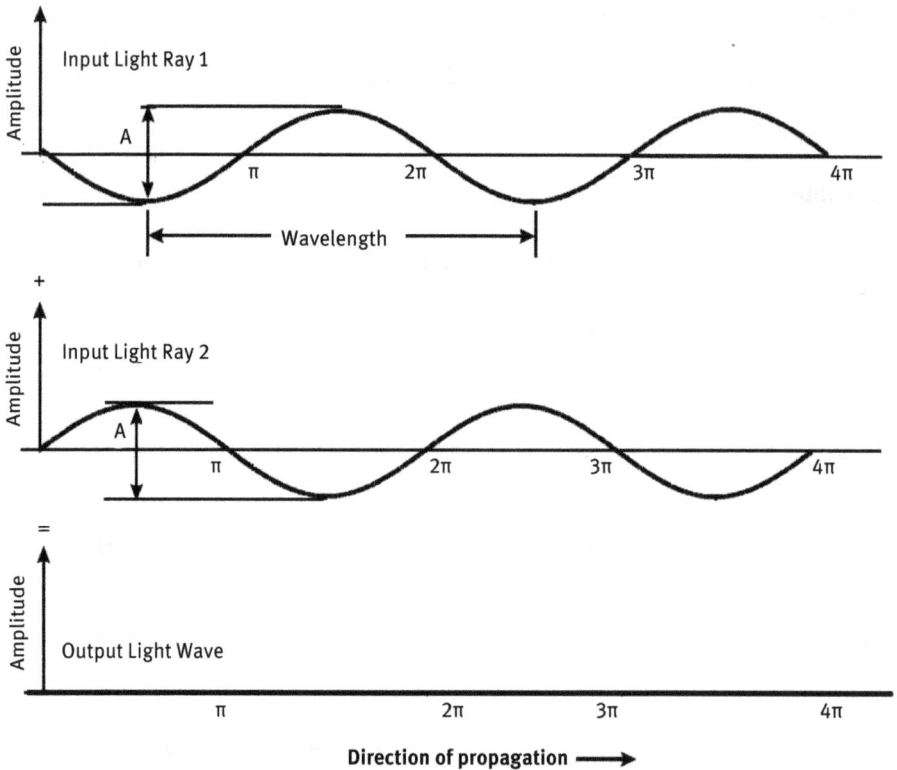

Figure 9.5: Destructive interference (out-of-phase = π) – a schematic view.

9.1.6 Diffraction Grating for Light Waves

Diffraction grating or simply a grating is a thin film made of glass or even plastic material that has a large number of lines per mm etched on it. The density of the grating can be 250 lines/mm up to 3000 lines/mm and even higher. When light emanating from a bright and small source would pass through the grating it will generate a large number of independent sources (see Figure 9.6). Those sources are in-phase ones or *coherent sources*. Each source would send rays in all directions. Placing a screen at a D distance from the grating would reveal bright fringes on it (Figure 9.6). The principal maxima would appear on both sides of the central maxima. Note that as all the sources are in-phase at fringe P_0, the brightest one, and as not all the waves are in phase at $P3$, it is much dimmer than P_0.

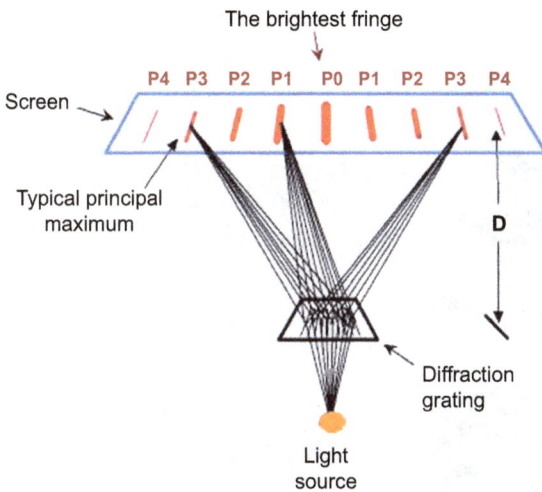

Figure 9.6: Schematic view of a diffraction grating and the associated screen (not to scale).

Assuming that the distance between two adjacent lines to be d and N being the number of lines per mm of grating one can write

$$d = \frac{1}{N} \quad \text{or} \quad N = \frac{1}{d} \tag{9.11}$$

Then the expression to find the principal maxima for a diffraction grating would be

$$d \sin \theta_n = n\lambda, \quad n = 1, 2, 3, \ldots$$

where

$$\lambda = \text{wavelength} \tag{9.12}$$

9.2 Basic Characteristics of a Fiber Optic

Fiber optic is defined as the science of light transmission through very fine glass or plastic fibers (see Figure 9.7). Currently it is used in more and more applications due to its advantages over copper conductors. These advantages include: large bandwidth (>10 GHz per km for single mode fibers or between 200 and 600 MHz per km for multimode fibers compared to 10–25 MHz per km for electrical conductors), immunity to electromagnetic/radio frequency interference, easiness to install, requires less space and the weigh is 10–15 times lower than copper leading to reduced cost for installation. Moreover, the fibers are electrically insulated, do not emit sparks or cause short circuits and do not pose any shock hazards.

Figure 9.7: Various fiber optic cables.

A fiber optic link (see Figure 9.8) is composed of four main parts:

– Transmitter – This device converts electrical signals to optical ones. It contains a light source like a LED (Light Emitting Diode) or a LASER (Light Amplification by Stimulated Emission of Radiation) or a VCSEL (Vertical Cavity Emitting Laser). These light sources use an infrared band (approx. 850 nm[2]).

– Fiber – The optical fiber transmits the light using a special designed tube consisting of five different layers (see schematic view in Figure 9.9)

  a. Core – This is the central tube, in which the light is propagating and usually is made from either fused silica[3] (amorphous SO_2) or doped silica (like Germania GeO_2 or phosphorous pentoxide P_2O_5 as dopers).

  b. Cladding is the second layer engulfing the core. It is also made from doped silica (doped with Boria B_2O_3 or fluorine) to yield a refractive index (n-see Appendix A) lower that the refractive index of the core. This allows the light to be confined in the core by internal reflections at the core-cladding interface.

  c. Buffer coating protects the fiber during its installation and has a thickness of 900 μm.

---

2 1 nm = 10^{-9} m.

3 Silica = silicon dioxide, SiO_2, found in the nature as quartz. Sand contains a large amount of silica.

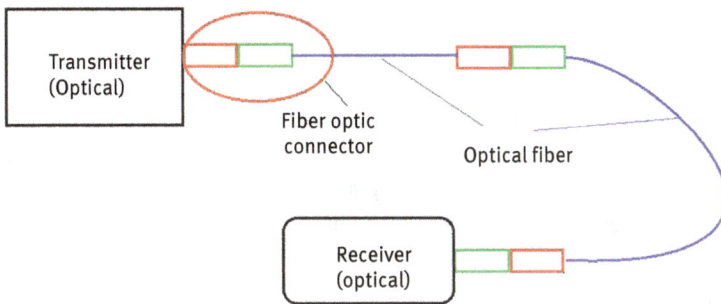

Figure 9.8: Schematic fiber optic link.

d. The fourth layer, usually made of Kevlar,[4] provides the required strength to the fiber cable.
e. The external layer is a jacket made from PVC or plenum coated by Teflon[5] to prevent fire protection.

– Connectors – To operate smoothly connectors are used to link the fiber optic cable to the transmitter and the receiver. Two method exist:

a. Direct connectivity – This method directly welds two fibers with an electric arc or what is known in the literature as *fusion splicer*. The good thing of this simple procedure is that there is almost no loss of light when passing from one fiber to the other; however, the disadvantages like permanent connection, fragility and cost of the machine (fusion splicer) make the method less attractive to users.

b. Physical connectors – This method involves connecting two fibers by physically connecting two fibers using two plugs. The connection can be mated and unmated yielding a robust connection with thousands of times of connections and disconnections. The disadvantages of this method are: higher loss of light as compared to the fusion splicer requires special tools and the process is longer.

One can find two types of connectors (see Table 9.2):

1. Physical contact connectors – Use is made of a cap bonded to a polished fiber using epoxy and ensures alignment between the two fibers. The result is a robust connection with low light loss (approx. 0.3 dB) easy to be cleaned and cost effective.

---

4 Kevlar® = created by Stephanie Kwolek at Du Pont™ is a para-amid synthetic fiber (hear-resistant and very strong), which is known for its high tensile strength-to-weight ratio.
5 Teflon = a synthetic fluoropolymer of tetrafluoroethylene, or polythtrafluoroethylene having high melting point (327 °C), thermal conductivity of 0.25 W/(moK) and a chemical formula of $(C_2F_4)_n$.

2. Expanded beam connectors – Lens are placed at the exit of each fiber to widen the ray and collimate the light passing through the fiber. The light loss is higher (0.8–2.5 dB) and the whole process is more complicated than the previous type of connectors, and therefore it is less found in the market.

– Receiver – The last part of the chain aiming at converting the optical signals using a photodiode in electrical ones. The photodiodes can either be of type PIN (Positive Intrinsic Negative) or APD (Avalanche Photo Diode).

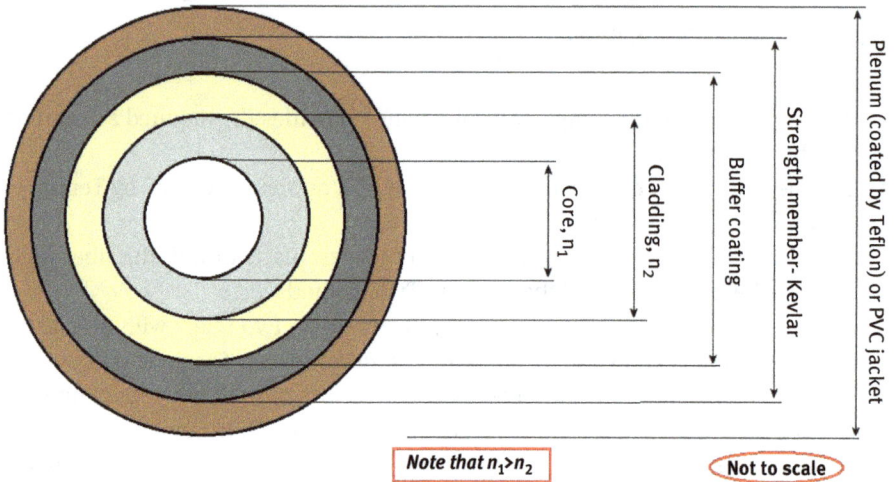

Figure 9.9: Typical cross section of a fiber optic cable.

Another important property is the type of the glass used to manufacture the optic fiber. In today's high-speed networks, Graded Index Multimode fiber (the refractive index of the core is gradually decreased toward the cladding using often a parabolic law yielding a "bend" of the light as it travels along the fiber) or Step Index Multimode or Single Mode fiber (the light is reflected at the cladding-core interfaces due to the abrupt n in the material refractive index, n) is used to improve light transmission over long distances. Multimode fiber has a larger core and is typically used for short distances within buildings. Single mode fiber has a smaller core and is used for long distances typically outside between buildings. As can be seen in Figure 9.10, the signal is the same for the Step-index single-mode (a), while for the Step-index multimode, there is a considerable attenuation of the output signal compared with the input one and large dispersion (b). Using Graded-index multimode reduces the dispersion while improving the output signal (reduced attenuation) (c).

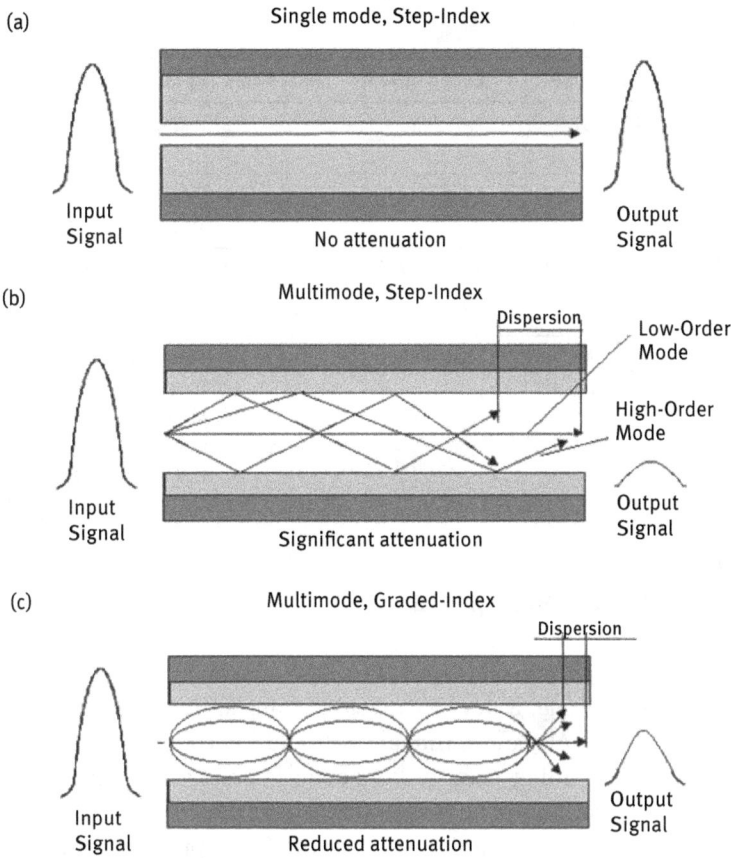

Figure 9.10: The influence of the glass type on the fiber optic performance – a schematic drawing (adapted from L-com® Global Connectivity).

Table 9.2: Fiber optic connectors (adapted from L-com® Global Connectivity).

Connector	Type	Coupling Type	# of fibers	Applications
	LC	Snap on RJ45 style	1	Gigabit Ethernet,[6] Video multimedia
	SC	Snap on	1	CATV,[7] Test equipment

---

6 Ethernet is a family of computer networking technologies commonly used in local area networks.
7 CATV represents Community Antenna Television, a cable television.

Table 9.2 (continued)

Connector	Type	Coupling Type	# of fibers	Applications
	ST	Twist on	1	LANs,[8] Military
	FC	Screw on	1	Datacom, Telecommunications
	MT-RJ	Snap on RJ45 style	2	Gigabit Ethernet ATM[9]
	MPO(MTP)	Push/Pull	6 or 12	Active device transceiver,[10] O/E[11] modules interconnections, QSFP[12] transceivers

9.3 Fiber Optic – Historical Aspects

It is instructive to understand the evolution of fiber optic from its historical aspects. It is agreed, that the *Photophone*, a device capable of transmitting of sound on a light ray, created in 1880 by Alexander Graham Bell and his co-inventor Charles Summer Tainter [1], as the starting point of fiber-optic communication. Although the test performed by Bell and Tainter on June 3, 1880, failed due to the atmosphere medium, it can be regarded as the starting point of the fiber-optic application to communication.

Later on, Harold Hopkins and Narinder Singh Kapany[13] (the latter also known as *Father of Fiber Optics*) at Imperial College, London, Great Britain, showed that even rolled fiberglass allows the transmission of light in contrast to the previous assumption that light can propagate only in a straight medium. Following the development

---

8 LANs represents local area networks, a series of computers linked together to form a network in a prescribed location.

9 ATM represents asynchronous transmission mode.

10 Transceiver is a device that is able to both transmit and receive.

11 O/E represents opto-electronic.

12 QSFP represents Quad Small Form-factor Pluggable is used at the interface of hardware to the fiber optic cable.

13 Hopkins, H. H. and Kapany, N. S., A flexible fibrescope, using static scanning, Nature, Vol. 173, 1954, pp. 39–41.

of optical cladding by the Dutch scientist van Heel,[14] Kapany [2] was the first person to conceive the term "fiber optics" in an article named Fiber Optics in *Scientific American* in November 1960.

A few years later, in 1966, Charles K. Kao and George Hockham proposed optical fibers while working at STC Laboratories (STL) at Harlow, England. While Kao concluded that the fundamental limit on glass transparency is below 20 decibels per kilometer, which would be practical for communications. Hockham calculated that clad fibers should not radiate much light. They prepared together a paper proposing for the first time fiber-optic communications [3]. Kao is known in the literature as the Godfather of Broadband, the father of Fiber Optics, and the Father of Fiber Optic Communications, was awarded the 2009 Nobel Prize in Physics for *groundbreaking achievements concerning the transmission of light in fibers for optical communication* and was knighted by Queen Elizabeth II *for services to fiber optic communications.*

One should note that 10 years before, in 1958, Schawlow and Townes [4] published a paper, which is considered to be the first evidence for the laser operation but it wasn't until 1960 that the first light-emitting maser – better known as the laser – was constructed [5], enabling latter the extensive use of optical fibers for communications.

Interesting to mention that C. H. Townes shared with N. G. Basov & A.M. Prokhorov the 1964 Nobel Prize in physics *for fundamental work in the field of quantum electronics, which has led to the construction of oscillators and amplifiers based on the maser-laser principle.*

Corning Glass Works successfully developed optical fibers in 1970, with a low attenuation of about 20 dB/km for communications purposes. At the same time GaAs compact semiconductor lasers were developed by researchers (Zhores Alferov's group) at the Ioffe Institute in Leningrad (today St. Petersburg), Russia and in parallel and independently by Panish & Hayashi at Bell Labs, NJ, USA, suitable for transmitting light through fiber optic cables for long distances.

Since 1970 and up today, the transmission capability had enormously improved as can be seen in Table 9.3 and Figure 9.11.

Note that the straight line in Figure 10.5 denotes that the BL term will double each year.

It is worth to remember the capability of wireless networks and their dramatic advances through the years, like the first generation (1G in 1980s), second generation (2G in 1990s), third generation (3G in 2000s), fourth generation (4G in 2010s) and the fifth generation (5G in 2020s). The 5G is expected to display enhanced mobile broadband (EMBB), massive machine type communications (MMTC) and ultra-reliable and low-latency communication (URLLC) [1].

---

14 Simons, C. A. J., A.C.S. van Heel: teacher and inspirator of technical optics, Proc. SPIE 3190, Fifth International Topical Meeting on Education and Training in Optics, (8 December 1997); https://doi.org/10.1117/12.294379

Table 9.3: Transmission capability versus years.

Generation	Year realized	Bit rate	Repeater[15] spacing	Operating wavelength
First (graded-index fibers)	1980	45 Mb/s	10 km	0.8 µm
Second (single-mode fibers)	1985	100 Mb/s up to 1.7 Gb/s	50 km	1.3 µm
Third (single-mode lasers)	1990	10 Gb/s	100 km	1.55 µm
Fourth (optical amplifiers)	1996	10 Tb/s	>10,000 km	1.45–1.62 µm
Fifth (Raman amplification)	2002	40–160 Gb/s	24,000–35,000 km	1.53–1.57 µm

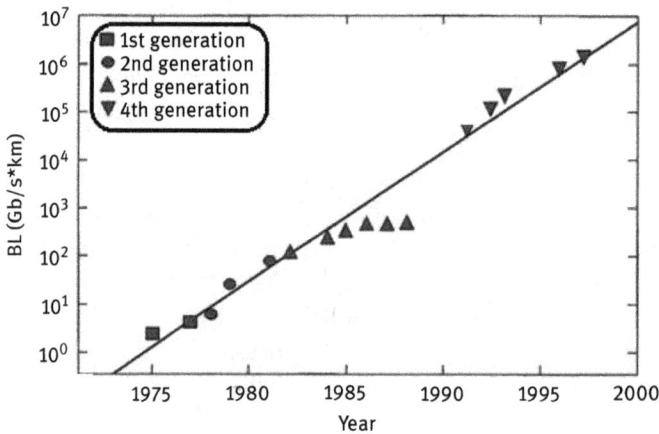

Figure 9.11: BL (Bit[16]*length) versus year for the first four generations (adapted from [1]).

Surely, it can be stated that the topic of fiber optic is today well established and documented in many books (for instance [6–16]) and thousands of manuscripts. In what follows, we shall restrict ourselves only to Bragg gratings and their applications.

---

15 Repeater –a wireless network device that repeats wireless signals to extend range. It is not connected with cable to a router/modem or users.

16 Bit – a basic unit of information having the value of either 0 or 1, used in computing and digital communications.

9.4 Fiber Bragg Grating (FBG)

Fiber Bragg grating (FBG) is an optical fiber sensor capable of measuring accurately mechanical strains and temperature induced strains. The name Bragg was given to the device following the Bragg law.[17] The FBG discovery is attributed to Kenneth Hill and his co-investigators, who discovered it already in 1978 while working at Communication Research Center, Canada [17, 18]. Besides sensing strain and temperature, it can detect also pressure, stress and refractive index of glass. FBG can be found in many industries, like aerospace, civil and maritime as well as in biochemical, health and biomedical devices.

The schematic structure of a typical FBG is presented in Figure 9.12.

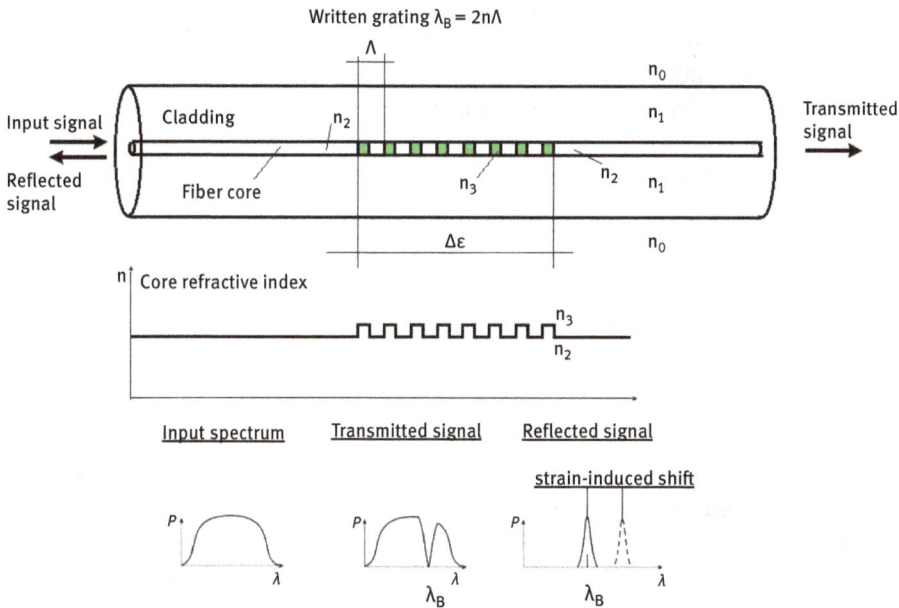

Figure 9.12: A schematic view of the transmitted and reflection spectra for a FBG.

The manufacturing process to induce grating into the core of the fiber is named *inscription* or *writing*. There are a few ways to introduce the grating [19]:

---

17 Lawrence Sir William Bragg proposed in 1912 what is known as the Bragg Law of x-ray diffraction. The law is used for the determination and investigation of crystals and thin films. Sir Bragg and his father William Henry Bragg won the Nobel Prize in Physics in 1915, *for their services in the analysis of crystal structures by means of x-rays.*

a. An interferometric procedure-in which the photo-sensible region of the core is ex-
    posed to an interferometric pattern to yield the whole grating. This is done by
    illuminating a proper mask having the FBG period. Note that this approach pro-
    cedure can be done also without a mask.
b. The direct point – by point method – a laser is used to write the grating, while
    controlling its parameter and the nonlinear absorption of the laser pulse. A transla-
    tional platform having the fiber on it is moving and the laser is writing one point
    after the other to form the grating.
c. The continuous core – scanning procedure – in this method the only parameters
    to be controlled are connected to the moving of the translational platform on
    which a fiber is attached.

As reported in [20], besides the uniform grating described above, there are also
chirped type grating, long period grating, tilted grating, phase-shifted grating and
supper-structure FBG grating.

For a simple FBG with periodic, equally spaced perturbations (see Figure 9.12)
the strength of the modulation is only positive [19]. The refractive index along the
grating will vary as a square wave having a duty cycle of half-period Λ of the grating
and can be written as

$$n_{core} = n_{core\,0} \qquad 0 < x < \Lambda/2$$

$$n_{core} = n_{core\,0} + \Delta n \qquad \Lambda/2 < x < \Lambda \tag{9.13}$$

where Δn is the modulation strength and n_{coreo} is the refractive index at the un-
touched core.

The operating system is as follows: part of the light traveling forward through the
core of the fiber is reflected back due to the grating and the rest will pass over. This
cases the light beam to have two components that passes in two opposite directions.
At certain wavelengths decided by the grating parameters, the Bragg condition is sat-
isfied yielding a constructive interference of the reflected light. The spectrum of the
reflected light would present a peak at the Bragg wavelength, λ_B, while the transmitted light will show a notch filter (like a transfer function-see Figure 9.12). Those light
components far from λ_B will travel unperturbed.

Remembering that the Bragg law is given by

$$2d \sin \theta = n\lambda, \qquad n = 1, 2, 3, 4, \ldots \tag{9.14}$$

where d is the distance between two adjacent crystalline planes, θ is the incidence
angle and λ is the wavelength, one can obtain the expression for λ_B. The angle θ is
90^0 and d is the distance between adjacent peaks (see Section 9.1), and for $n = 1$ we
get $\lambda = 2d$, and one has to take into account the refraction index of the fiber (as the
original Bragg law was derived for vacuum) leading to

$$\lambda_B = 2n_{\text{eff}}\Lambda \tag{9.15}$$

The Bragg is therefore a function of the effective refractive index of the fiber, n_{eff}, and the grating pitch, Λ.

As described in [19], application of the coupled-mode theory (see [21]), enables the calculation of the reflectivity and transmissivity spectrum. Defining the following parameters

$$\alpha \equiv \frac{2\pi}{\lambda}\overline{\Delta n}_{\text{eff}} \quad \kappa \equiv \frac{\pi}{\lambda}\nu\overline{\Delta n}_{\text{eff}} \quad L \equiv N\Lambda \tag{9.16}$$

where α is the DC self-coupling coefficient, $\overline{\Delta n}_{\text{eff}}$ is the DC index change spatially averaged over a grating period, κ is the AC coupling coefficient, ν is the FBG order (unit for an uniform FBG) and N is the number of periods that leads to the reflected amplitude

$$R_{\text{am.}} = -\frac{\kappa \sinh\left(\sqrt{\kappa^2 - \alpha^2}L\right)}{\alpha^2 \sinh\left(\sqrt{\kappa^2 - \alpha^2}L\right) + i\left(\sqrt{\kappa^2 - \alpha^2}\right)\cosh\left(\sqrt{\kappa^2 - \alpha^2}L\right)} \tag{9.17}$$

and power

$$\text{Power} = |R_{\text{am.}}^2| = -\frac{\sinh^2\left(\sqrt{\kappa^2 - \alpha^2}L\right)}{\cosh^2\left(\sqrt{\kappa^2 - \alpha^2}L\right) - \frac{\alpha^2}{\kappa^2}} \tag{9.18}$$

The full-width-half-maximum (FWHM) or the bandwidth (see Figure 9.13) is written as

$$\text{FWHM}_\lambda = \lambda_B\beta\sqrt{\left(\frac{\Delta n}{2n_{c0}}\right)^2 + \left(\frac{1}{N}\right)^2}, \tag{9.19}$$

where $\beta \approx 1$ for high reflectivity grating

$\beta \approx 0.5$ for weak reflectivity grating

As can be seen in Figure 9.13, the spectrum is composed from a central peak having a variable bandwidth according to the given parameters of the FBG, embraced from both sides by side lobes.

For different uses [20, 21] other grating profiles can be considered like chirped or phased-shifted grating which might reduce the side lobes. Another interesting profile named an apodized grating profile, which consists in a non-uniform strength of the modulation along the grating (like for raised-cosine or Gaussian profile), is shown to drastically reduce the amplitudes of the side lobes.

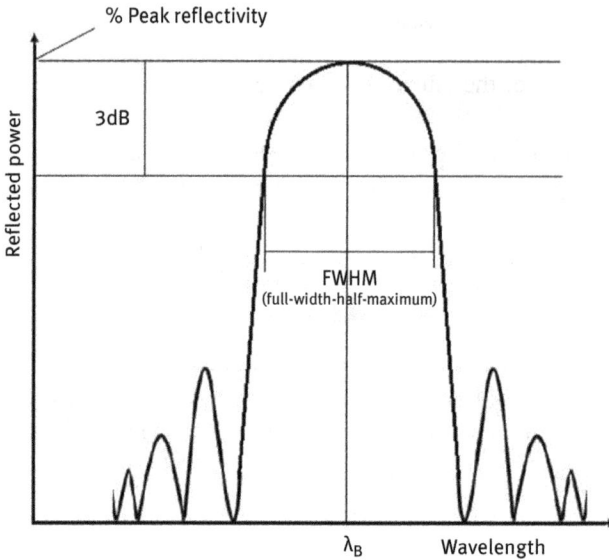

Figure 9.13: Schematic reflectivity of a FBG.

9.5 Fiber Bragg Grating Used as Strain Sensor

As described in Figure 9.12, strain or temperature will induce a shift in the Bragg wavelength. This shift can be calculated as follows: partial differentiating Equation (9.15) yields

$$\Delta\lambda_B = 2\left(\Lambda\frac{\partial n_{\text{eff}}}{\partial\varepsilon} + n_{\text{eff}}\frac{\partial\Lambda}{\partial\varepsilon}\right)\Delta\varepsilon$$

$$+ 2\left(\Lambda\frac{\partial n_{\text{eff}}}{\partial T} + n_{\text{eff}}\frac{\partial\Lambda}{\partial T}\right)\Delta T =$$

$$= (\Delta\lambda_B)_{\text{strain}} + (\Delta\lambda_B)_{\text{temp.}} \tag{9.20}$$

The first part of the left-hand side of Equation (9.20) presents the shift in Bragg wavelength due to strain, while the second part is due to temperature.

Equation (9.19) can be re-written in another useful form to yield [19]

$$\frac{\Delta\lambda_B}{\lambda_B} = (1-\rho_E)\varepsilon + (\beta+\chi)\Delta T \tag{9.21}$$

where ρ_E is the photo-elastic coefficient stemming from the Pockel's stress-optic tensor [23], ε is the strain in the direction of the fiber, β is the thermal expansion of the material of the fiber (like silica) and χ is the thermo-optic coefficient of the material [22] (the temperature dependence of the refractive index, dn_{eff}/dT). The authors in

[22] present the following data (fiber silica with germanium doped core) for the constants in Equation (10.9): $\rho_E = 0.22$, $\beta = 0.55 \times 10^{-6}/°C$ and $\chi = 8.6 \times 10^{-6}/°C$ leading to $\Delta\lambda_B/\Delta\varepsilon = 1.2$ pm/μs and $\Delta\lambda_B/\Delta T = 14.18$ pm/°C. Note that in [24] a different value is presented for the relationship strain-Bragg shift: $\Delta\lambda_B/\lambda_B = 0.79\varepsilon$.

One should note that using FBG as strain sensor, the strain output would normally include the added temperature strain. A compensation is needed to subtract the influence of the temperature so that the output will provide only mechanical strains. An interesting method is presented in [22], in which two parallel mounted FBGs, one bonded while the other is un-bonded and provides the temperature induced strain, which is subtracted from the reading of the first FBG to yield the net strain encountered at the particular part of the structure (on which the FBGs are attached). Typical studies on the use of FBG to sense strain are presented in [25–32].

9.6 Interrogators for Fiber Bragg Grating

The information gathered by the use of FBG should be decoded to enable its use. This process is named demodulation technique or interrogation. There are several techniques for FBG based sensors ([19, 20]): Bulk optics, Filtering (having either passive edge or active bandpass filtering), interferometric, laser sensing and others. According to [19] the most used interrogation techniques are the filtering and interferometric methods. The filtering method converts the wavelength shift of the FBG output into a change in an electrical signal using a photo-diode which collects the optical signal of the FBG reflected spectrum (instead of using a conventional optical spectrum analyzer – OSA, which was found to be slow and expensive to use [19]). The interferometric methods allocate an optical phase difference to the Bragg wavelength shift. An unbalanced asymmetric Mach Zehnder interferometer is normally used, to receive the reflected spectrum of the sensing grating and passed to a phase modulator [24] located on one of the arms of the interferometer. The change in the measured intensity can be transformed into a phase difference once the interferometer phase and its associated intensity output is known. Then the required wavelength shift is obtained using an adequate signal processing applied on the photodiode voltage, leading to resolutions in the range of $pm/\sqrt{Hz}$ [19].

More on interrogators can be found in [19, 20] or in books dealing with optical fibers, like [6–16].

References

[1] Liu, X., Evolution of fiber-optic transmission and networking toward the 5G era, iScience 22, December 20, 2019, 489–506.
[2] Kapany, N. S., Fiber Optics Principles and Applications, Academic Press, 1967, 429.

[3] Kao, K. C. and Hockham, G. A., Dielectric fibre surface waveguides for optical frequencies, Proceedings of the IEEE 113, 1966, 1151–1158.

[4] Schawlow, A. L. and Townes, C. H., Infrared and optical masers, Physical Review 112(6), 1958, 1940–1949.

[5] Maiman, T., Stimulated optical radiation in ruby, Nature 187, 1960, 493–494.

[6] Saleh, B. E. A. and Teich, M. C., Fundamentals of Photonics, Chapter 8: Fiber Optics, John Wiley & Sons, Inc., 1991, 947.

[7] Ansari, F., (ed.), Applications of Fiber Optic Sensors in Engineering Mechanics, American Society of Civil Engineers, 1993, 230.

[8] Reese, R. T. and Kawahara, W. A., editors, Handbook on Structural Testing, Society for Experimental Mechanics (US), Fairmont Press, 1993, 402.

[9] Udd, E., editor, Fiber Optic Smart Structures, John Wiley & Sons Inc., 1995, 688.

[10] Culshaw, B., Smart Structures and Materials, Artech House, 1996, 207.

[11] Van Steenkiste, R. J. and Springer, G. S., Strain and Temperature Measurement with Fiber Optic Sensors, Technomic Publishing Co, 1997, 294.

[12] Ansari, F., editor, Fiber Optic Sensors for Construction Materials and Bridges, Technomic Publishing Co., 1998, 267.

[13] Ghatak, A. and Thyagarajan, K., Introduction to Fiber Optics, Cambridge University Press, 1998, 581.

[14] Measures, R. M., Structural Monitoring with Fiber Optic Technology, Academic Press, 2001, 716.

[15] Lopez-Higuera, J. M. editor, Handbook of Optical Fibre Sensing Technology, John Wiley & Sons Inc., 2002, 828.

[16] Al-Amri, M., El-Gomati, M. and Zubairy, M. S., Optics in our Time, Chapter 4: Applications, Springer Open, 2016, 504.

[17] Hill, K. O., Fujii, Y., Johnson, D. C. A. and Kawasaki, B. S., Photosensitivity in optical fiber waveguides: application to reflection fiber fabrication, Applied Physics Letters 32(10), 1978, 647–649.

[18] Hill, K. O. and Meltz, G., Bragg grating technology fundamentals and overview, Journal of Lightwave Technology 15(8), 1997, 1263–1276.

[19] Campanella, C. E., Cuccovillo, A., Campanella, C., Yurt, A. and Passaro, V. M. N., Fibre Bragg grating based strain sensors: review of technology and applications, Sensors 18, 2018, Id paper 3115, 27. doi:10.3390/s18093115.

[20] Sahota, J. K., Gupta, N. and Dhawan, D., Fiber Bragg grating sensors for monitoring of physical parameters: a comprehensive review, Optical Engineering 59(6), June 2020, Id paper 060901.

[21] Pierce, J. R., Coupling modes of propagation, Journal of Applied Physics 25, 1954, 179–183.

[22] Werneck, M. M., Allil, R. C. S. B., Ribeiro, B. A. and De Nazaré, F. V. B., A guide to fiber Bragg grating sensors, INTECH Open Science book, Chapter 1, 2013, 25, http://dx.doi.org/10.5772/54682.

[23] Nye, J. F., Physical properties of Crystals: Their Representation by Tensors and Matrices, Oxford University Press, 1957, 324.

[24] Gagliardi, G., Salza, M., Avino, S., Ferraro, P. and De Natale, P., Probing the ultimate limit of fiber-optic strain sensing, Science 330, 19th November. 2010, 1081–1084.

[25] Fidanboylu, K. and Efendioğlu, H. S., Fiber optic sensors and their applications, 5th International Advanced Technologies Symposium (IATS'09), May 13–15, 2009, Karabuk, Turkey.

[26] Pevec, S. and Donlagić, D., Multiparameter fiber-optic sensors: a review, Optical Engineering 58(7), Id paper 072009, 2019, 26. doi:10.1117/1.OE.58.7.072009.

[27] Guemes, A., Fernandez-Lopez, A. and Soller, B. J., Optical fiber distributed sensing-physical principles and applications, Structural Health Monitoring 9(3), 2010, 233–245. doi:10.1177/1475921710365263.

[28] Sánchez, D. M., Gresil, M. and Soutis, C., Distributed internal strain measurement during composite manufacturing using optical fibre sensors, Composite Science and Technology 120, 2015, 49–57. doi:10.1016/j.compscitech.2015.09.023.

[29] Chandarana, N., Sánchez, D. M., Soutis, C. and Gresil, M., Early damage detection in composites by distributed strain and acoustic event monitoring, Procedia Engineering 188, 2017, 88–95, doi:10.1016/j.proeng.2017.04.515.

[30] Caucheteur, C., Guo, T. and Albert, J., Polarization-assisted fiber Bragg grating sensors, Tutorial and review, Journal of Lightwave Technology 35(16), 2027, 3311–3322.

[31] Sano, Y. and Yoshino, T., Fast optical wavelength interrogator employing arrayed waveguide grating for distributed fiber Bragg grating sensors, Journal of Lightwave Technology 21(1), 2003, 132–139.

[32] Freydin, M., Ratner, M. K. and Raveh, E. D., Fiber-optics-based aeroelastic shape sensing, AIAA Journal 57(12), 2019, 5094–5103.

10 Miscellaneous Topics

10.1 Enhanced Flexural Behavior of Plates Equipped with SMA Wires

10.1.1 The Use of SMA Properties for Actuation: The Restrained Recovery Phenomena

As presented in Chapter 4, the shape memory effect is triggered by the loading of the SMA alloy, yielding a plastic residual strain S_{res}, as shown in Figure 10.1. The amount of residual strain is a function of the martensitic fraction (ξ) of the material after unloading. For $\xi = 1$, the residual strain would be equal to the maximum recoverable strain S_L (Figure 10.1a). Note that, heating beyond the austenite start temperature (A_s) will cause strain recovery due to transformation of stress-induced martensite to austenite. Referring to Figure 10.1a, if the stress is high enough to induce full martensitic phase deformation, the residual strain (S_{res}) will be equal to the maximum recoverable strain (S_L), however, in case the stress is smaller, the martensite fraction will be

$$\xi = \frac{S_{res}}{S_L} \tag{10.1}$$

Three basic mechanical boundary conditions can be examined when the material is heated to recover the plastic induced strain. The first one will be a free strain recovery which will occur when one of the mechanical boundaries is not constrained (see Figure 10.1b). For this case, the SMA material will contract as a function of the temperature during the heating process. Restrained recovery will occur if both boundary conditions are constrained to move axially, yielding a zero strain, while the material is heated (see Figure 10.1c). Preventing the contraction of the material generates a recovery stress, σ^r in the SMA material as a function of the temperature during the heating process. However, as presented in Figure 10.1d, when instead of an unmovable end side of the SMA wire, a spring is connected, and as the wire is heated above the $M \rightarrow A$ (martensite to austenite) transformation temperature, the wire recovers strain and tends to contract. This causes the extension of the linear spring and a corresponding increase in the stress in the wire.

To quantify the generated stress, the Liang and Rogers model, presented in Chapter 4, will be used for the three basic boundary conditions.

As presented in Chapter 4, the constitutive relation is presented by Equation (10.2), assuming no differentiation between temperature- and stress-induced martensite is done:

$$\sigma - \sigma_0 = Y(\xi)(S - S_L\xi) - Y(\xi)\left(S_0 - S_L\xi_0\right) \tag{10.2}$$

https://doi.org/10.1515/9783110726701-010

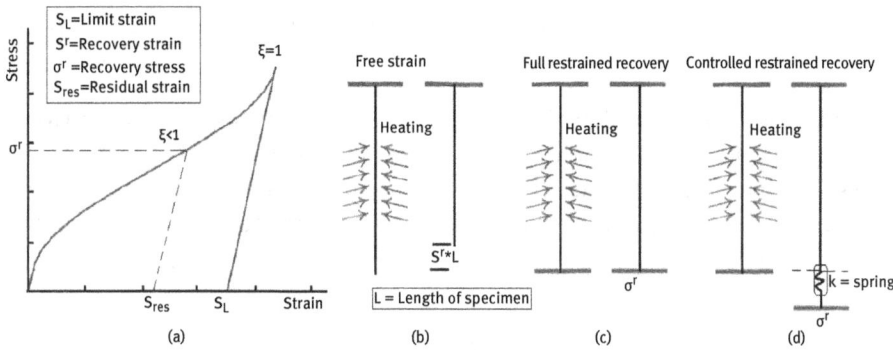

Figure 10.1: (a) Stress versus strain for a typical SMA material (shape memory effect), (b) heating the specimen in a free strain mode, (c) heating the specimen in a full restrained recovery mode and (d) heating the specimen in a controlled restrained recovery mode.

10.1.1.1 Free Strain Recovery

For the case of free recovery, the assumption is that the SMA material has been loaded and unloaded resulting in a residual strain, S_{res}, as presented in Figure 10.1a.

Assuming the stress is zero (the initial stress and strain are also zero), we get from Equation (10.2)

$$S = S_L \xi \tag{10.3}$$

If the temperature is above the martensite start temperature, M_s, with an initial martensite fraction, $\xi_0 = S_{res}/S_L$, and substituting Equation (4.9) into Equation (10.3) yields for the Liang and Rogers model, the following expression and the rest of variable are defined in Chapter 4:

$$S^r = \frac{S_{res}}{2}\{\cos[\alpha_A(T - A_s)] + 1\} \tag{10.4}$$

where

$$\alpha_A = \frac{\pi}{A_f - A_s} \tag{10.4}$$

10.1.1.2 Restrained Recovery

Assuming that $\sigma_0 = 0$, $S_0 = 0$ and $\xi_0 = 1$, and neglecting thermal expansion, Equation (10.2) has the following form:

$$\sigma = Y(\xi)(S - S_L\xi) + Y(\xi)(S_L) = Y(\xi)[S - S_L(\xi - 1)] \tag{10.5}$$

It can be shown that the restrained recovery stress, σ^r, can be written as

$$\sigma^r = \frac{Y(\xi)S^r}{2}\left\{1 - \cos\left[\alpha_A(T - A_s) - \frac{\alpha_A}{C_A}\sigma^r\right]\right\}$$

(10.6)

where $Y(\xi)$ is a linear function of the martensitic fraction defined in Equation (10.3). One should note that Equation (10.6) is an iterative equation since the recovery stress, σ^r, appears on both sides of the equation. The simplest way to solve Equation (10.6) is to choose a temperature and iterating to determine the stress that corresponds to the assumed temperature.

10.1.1.3 Controlled Restrained Recovery

As shown in Figure 10.1d, when instead of an unmovable end side of the SMA wire, a spring is connected, and as the wire is heated above the $M \rightarrow A$ transformation temperature, the wire recovers strain and tends to contract. This causes the extension of the linear spring and a corresponding increase in stress in the wire. The stress and the strain in the wire are related by the displacement compatibility of the SMA wire and the linear spring. Therefore, the expression for the strain is given by

$$S = -\frac{\Delta L}{L} = -\frac{F^r}{kL} = -\frac{\sigma^r A}{kL}$$

(10.7)

where k is the spring coefficient, L is the length of the wire and A is its cross-sectional area.

The controlled recoverable stress can be written as

$$\sigma^r = S_L Y(\xi)\frac{1 - \xi}{\left[1 + \frac{AY(\xi)}{kL}\right]} = \frac{Y(\xi)S^r}{2\left[1 + \frac{AY(\xi)}{kL}\right]}\left\{1 - \cos\left[\alpha_A(T - A_s) - \frac{\alpha_A}{C_A}\sigma^r\right]\right\}$$

(10.8)

One should note that for an infinite stiffness ($k \rightarrow \infty$), Equation (10.8) will converge to Equation (10.6).

Note that for a pre-design quick calculation of the stress as a function of the plastic strain, a linear behavior can be assumed as depicted in Figure 10.2.

10.1.2 Experimental Results

An extensive experimental campaign was initiated and performed at the Structures Laboratory, Faculty of Aerospace Engineering, Technion, I.I.T., 32000 Haifa, Israel, under the leadership of Prof. Haim Abramovich, the author of this book.

The tests were intended to prove experimentally the capability of Nitinol wire to enhance the flexural behavior of a loaded plate.

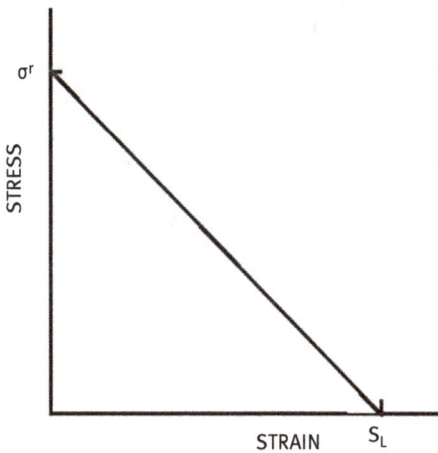

Figure 10.2: Schematic linear graph – recoverable stress versus strain.

Figure 10.3 presents the various types of Nitinol structures used for the tests, namely rods, springs and wires. The characterization of the SMA is done in two stages, as presented in.

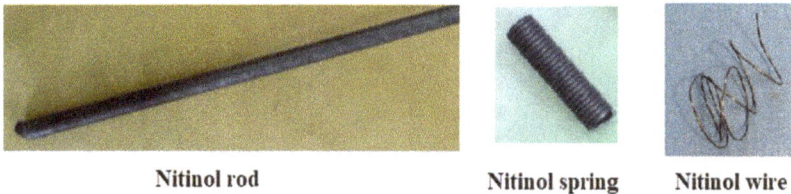

Nitinol rod   Nitinol spring   Nitinol wire

Figure 10.3: The three types of Nitinol structures used for the tests: a rod, a spring and a wire.

Figure 10.4. The first stage, consists of a SMA wire (or rod, or spring) being axially tensioned in its plastic region, yielding very high values of plastic strains (see Figure 10.4a and Table 10.1). Typical graphs of the monitored behavior are presented in Figure 10.5a–d (note that the graphs do not present the return to zero load). This stage ends at reducing the axial load to zero and a plastic strain remaining in the tested specimen.

The second step of the characterization is displayed in Figure 10.4b, where the Nitinol specimen, with a plastic deformation is again clamped at its ends. Attaching two electrical wires (Figure 10.4b) causes its heating above the transformation temperature. As the two ends prevent the expansion of the wire, considerable forces are developed within the wire (Figure 10.4c), to recover the plastic strain, as were summarized in Table 10.1. Typical results are shown in Figures 10.6a–d. One should note that after the

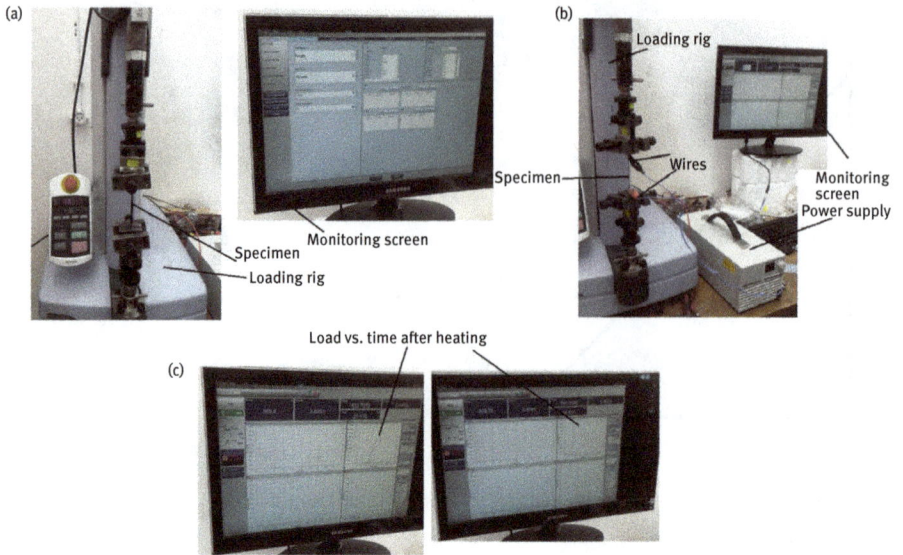

Figure 10.4: The test setup: (a) first stage – tension, (b) second stage – heating and (c) a typically generated graph of load versus time after heating.

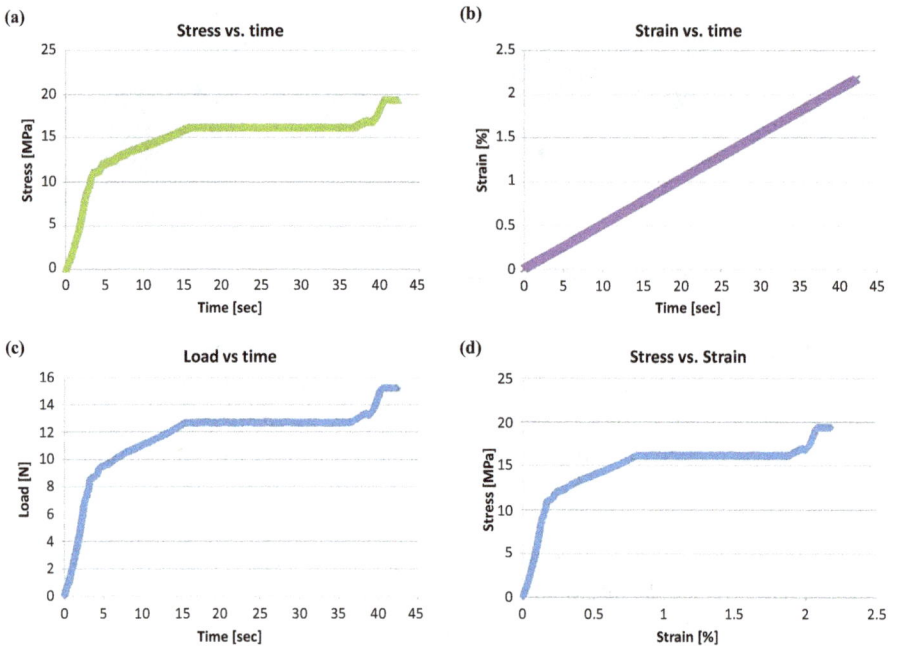

Figure 10.5: Specimen A1 @ first stage – loading: (a) stress versus time, (b) strain versus time, (c) load versus time and (d) stress versus strain.

load reached its peak the heating was stopped, causing the reduction of the load. Heating again, as presented in Figure 10.6a, brings back the load's peak.

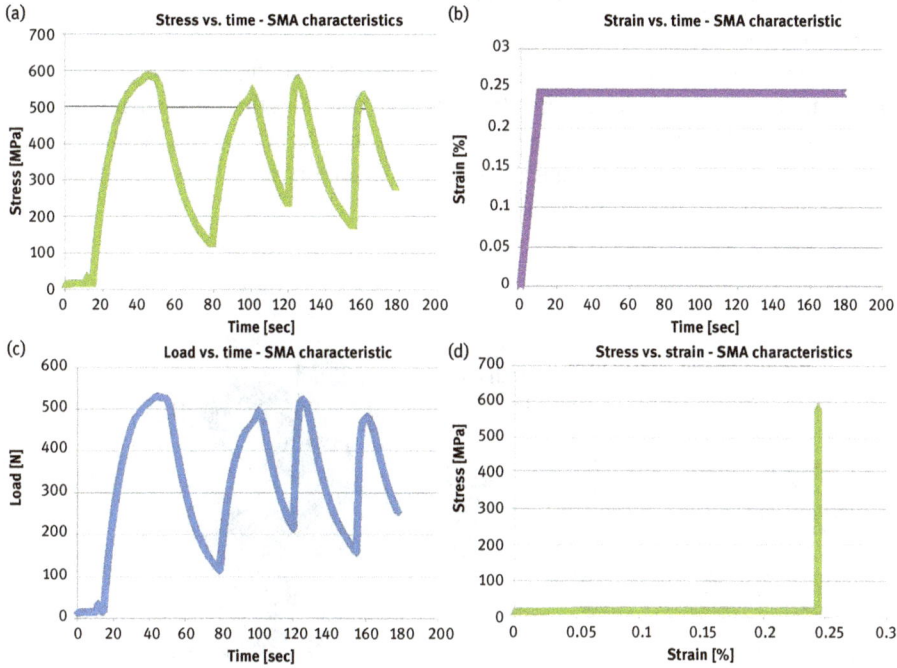

Figure 10.6: Specimen A1 @ second stage – loading: (a) stress versus time, (b) strain versus time, (c) load versus time and (d) stress versus strain.

Table 10.1: Experimental data for the various tested Nitinol structures.

Specimen #	Cross section area (mm²)	Transformation temperature (°C)	Max. strain (%)	Force generated (N)
A1 (wire)	0.900	44	2.1583	525.0000
B1 (wire)	0.785	75	0.7130	616.9300
B2 (wire)	0.785	75	0.5500	611.7856
C1 (wire)	0.900	37	7.7430	591.0000
C2 (wire)	0.900	37	8.4680	706.2384
C3 (wire)	0.900	37	7.6130	651.5474
D1 (rod)	0.196	37	13.000	12.3000
E1 (spring)	–	37	3.8500	15.8000

To show the capability of Nitinol wires to enhance the bending characteristics of a plate, a demonstrator was designed and is presented in Figures 10.7 and 10.8. It consists of nine Nitinol wires, with a 0.5 mm diameter capable of being tensioned before the tests to a predefined loading, using a wrench. The wires are connecting two plates. The lower plate is loaded by a dead weight. The heating is provided in two ways: either using electric current or heating the surrounding air by a blower, thus providing the required balance force to cancel the dead weight, thus leaving the bottom plate flat. The out-of-plane displacements of the bottom plate are monitored during the tests using two strain gages, bonded on the bottom plate. Figure 10.9 presents additional details of the realized demonstrator.

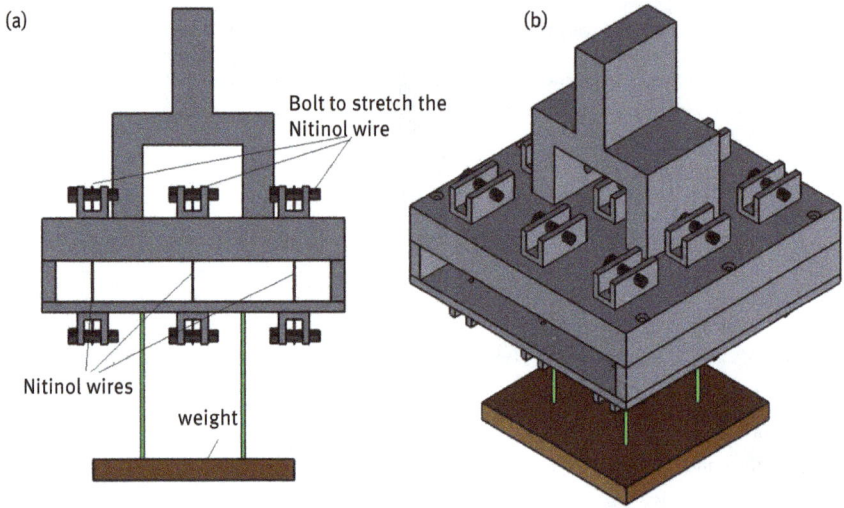

Figure 10.7: The demonstrator drawing: (a) front view and (b) a 3D view.

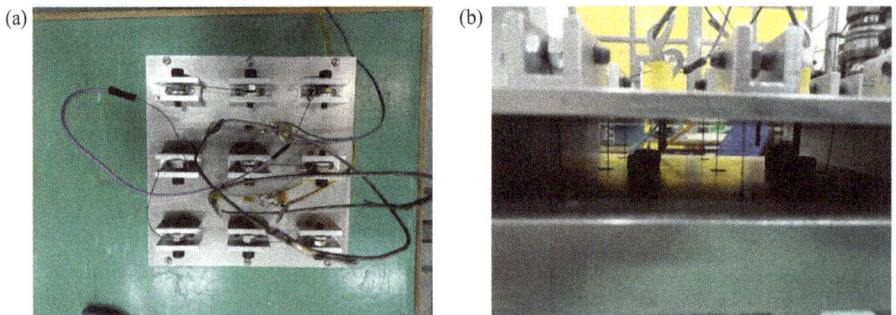

Figure 10.8: The realized demonstrator: (a) top view and (b) front view.

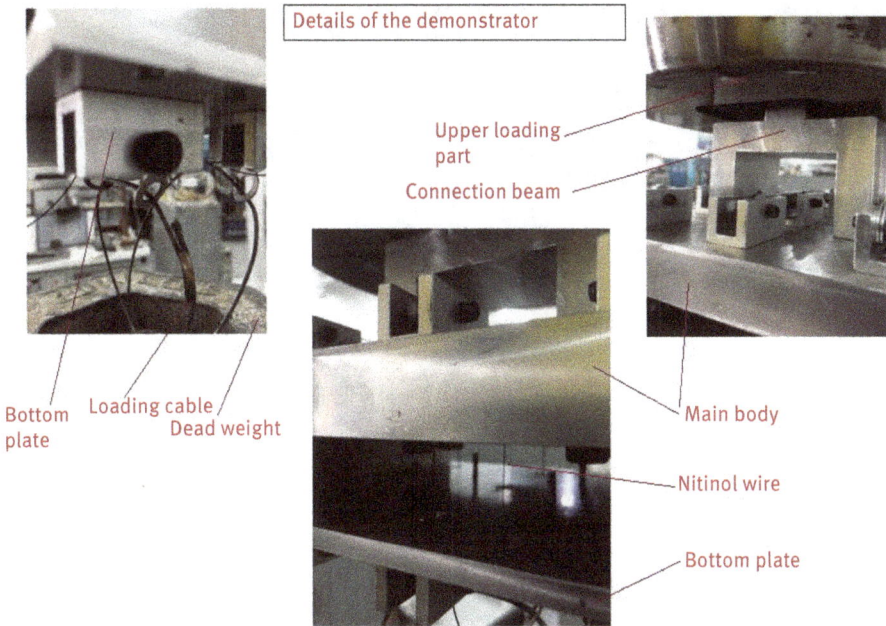

Figure 10.9: Additional details of the realized demonstrator.

Prior to performing the acceptance tests, FE simulations were performed using the ANSYS code. Figure 10.10a displays the deflection of the bottom plate due to the dead load, while Figure 10.10b displays the net forces due to the heating of the Nitinol wires.

The maximal deflection in Figure 10.10a is 0.023188 mm compared with the maximal inflection of 0.022573 mm in Figure 10.10b. The two values are very close, namely the activation of the Nitinol wires is capable to balance the deflection induced by the application of the dead weight.

The tests on the demonstrator were mainly performed by using a blower, thus obtaining an average temperature of 45 °C, while the transformation temperature of the Nitinol wires being at 37 °C. By heating the wires to 45 °C, the phase of Austenite was surely achieved. Typical tests results can be seen in Figure 10.11, where a maximal strain of +20 µstrain (read by strain gage #1) is canceled leading to a strain of −30 µstrain (average) obtained due to the Nitinol activation. Stopping the heating brings again the mechanical peaks of +40 till 80 µstrain, which again are canceled to −30 µstrain due to resuming of the activation of the Nitinol wires. Same tendency can be seen also for the second strain gage (#2).

Based on the above presented results, one can summarize the following conclusions:
1. Considerable loads can be obtained using the shape memory effect in constrained state using various Nitinol wires and rods specimens in tension.

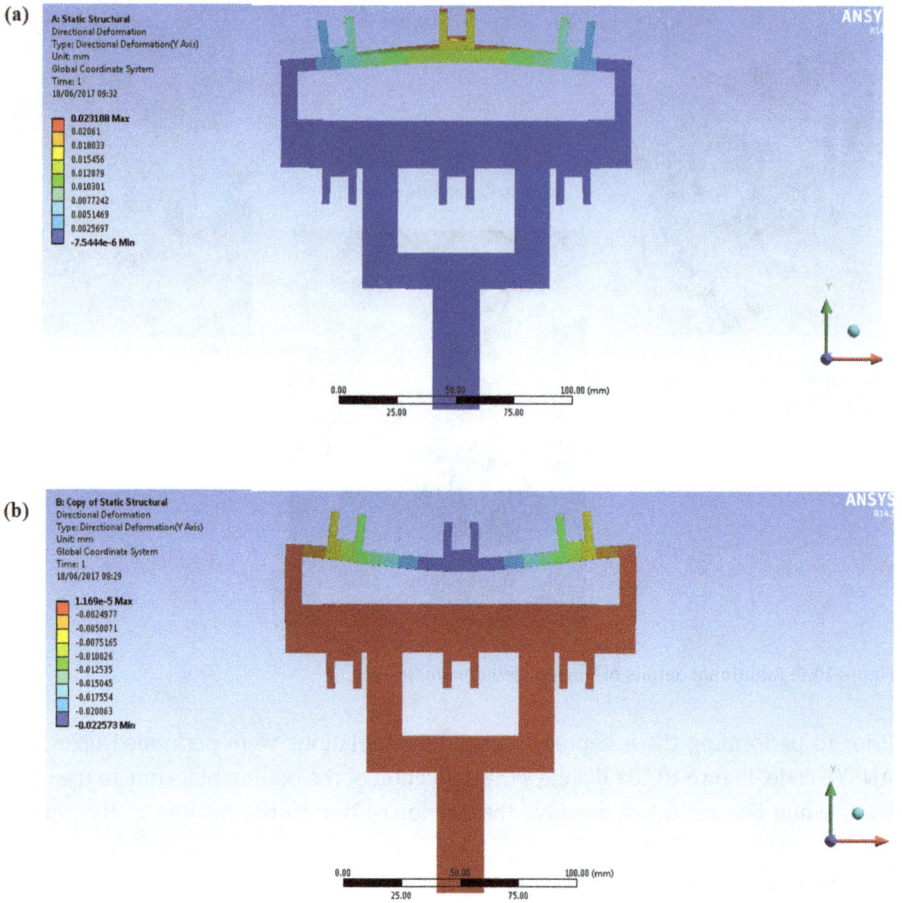

Figure 10.10: ANSYS Finite element simulation: (a) under static dead load and (b) activation of the Nitinol wires.

2. The higher is the plastic deformation; the higher is the induced load.
3. The tests performed on the demonstrator showed promising results. The activation of the Nitinol wires succeeded in balancing the deflection due to the mechanical dead weight, leaving a very small net deflection of the bottom plate.

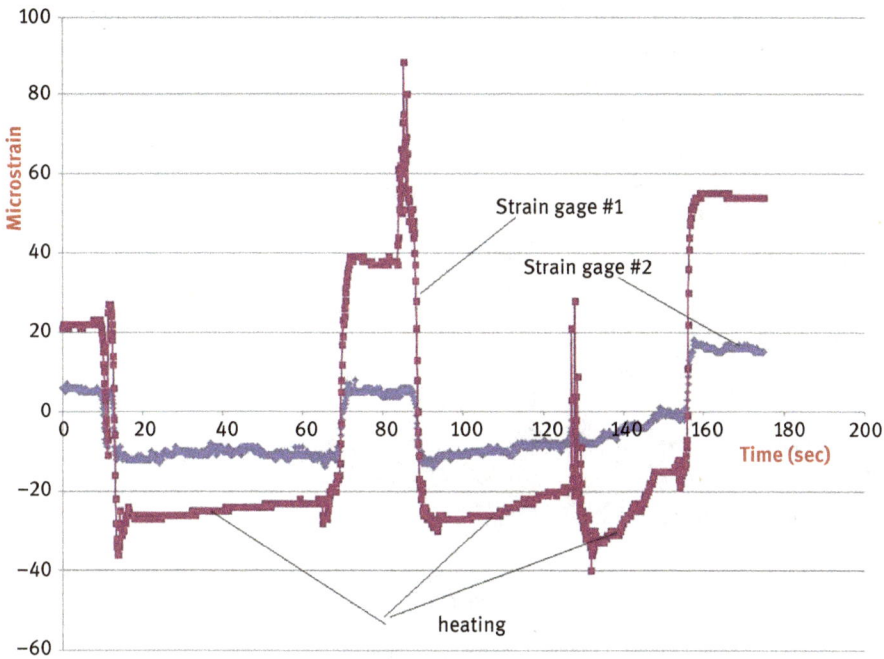

Figure 10.11: Strain gages readings versus time: activation of the Nitinol wires.

10.2 Piezoelectric Fiber Composite

Ceramic monolithic piezoelectric material is brittle and susceptible to accidental breakage during handling and bonding procedures or during their service. Moreover, ceramic monolithic piezoelectric transducers have poor ability to conform to curved surfaces (see Section 1.1). Those limitations have encouraged researchers to develop alternative manufacturing methods yielding piezoelectric fibers. One should remember that besides the ceramic fibers, PVDF copolymers fibers and microfibers are also available on the market. References [1–19] are typical research studies concerning the capabilities of various types of fiber to act as sensors and/or actuators.

10.2.1 PZT Fibers

The insertion of PZT fibers (with a diameter of below 250 μm) in a polymer matrix, equipped with interdigitated electrodes are known in the literature as active fiber composites (AFC) or macrofiber composites (MFC). A schematic view is presented in Figure 10.12. Another configuration of PZT fiber in a polymer matrix is shown in Figure 10.13 and is known as 1–3 piezocomposite. Note that around the year 2000 the MFC has been developed by NASA [7], where the PZT fibers are rectangular been diced from a regular piezoceramic plate.

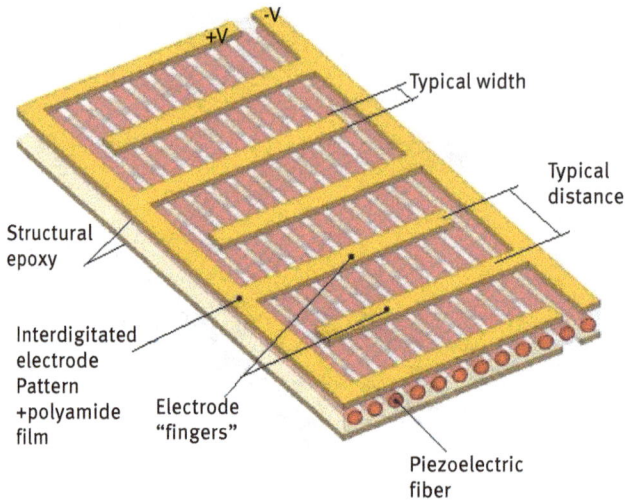

Figure 10.12: A schematic drawing of the active fiber composite.

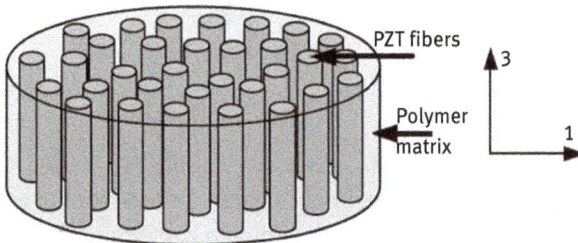

Figure 10.13: A schematic drawing of the 1–3 piezocomposite.

Currently, Smart Material Company, Dresden, Germany [7], manufactures the MFC. Schönecker [6] describes other methods of fabrication of PZT fibers.

The poling of the MFC and 1–3 piezocomposite configurations is challenging. Due to the geometrical dimensions of a piezoelectric fiber (one large dimension in the longitudinal direction and the other two dimensions been much smaller the fiber cross section), the poling is done using the IDE (interdigitated electrodes) fingers, as presented in Figure 10.14 leading to a longitudinal electrical field, that although is not constant along the fiber, provides the d_{33} constant for the transducer. Note that directly between the electrodes, a "dead zone" is generated, and by a careful design of both the width of the electrodes and the distance between two adjacent electrodes, it can be minimized (see [3]).

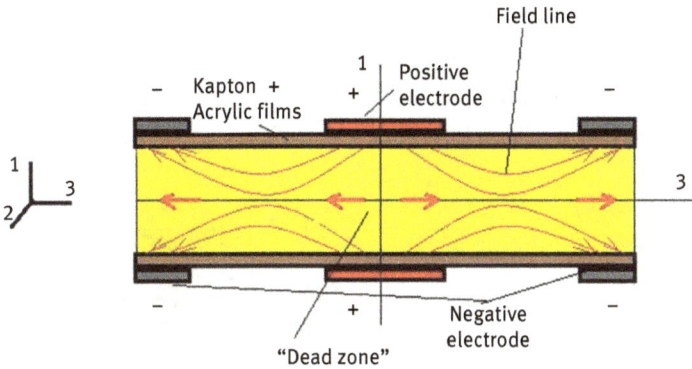

Figure 10.14: A schematic drawing for an MFC patch.

As the fibers are embedded in a polymer matrix, analytical mixing rules for MFCs should provide the homogenized properties of the active layer, based on the material properties from the constituents (matrix and PZT) and the volume fraction of fibers V_f. The equivalent properties for the d_{31} and d_{33} MFC unit volumes are presented in Table 10.2.

The equivalent mechanical, piezoelectric and dielectric properties for the 1–3 piezo-composites (see Figure 10.13) can be calculated by the following equations (see also [6] and [8]), where $()^m$ represents a matrix property and $()^p$ stands for a piezo-fiber property.

The equivalent density is given by

$$\rho_{eq.} = V_f \rho^f + (1 - V_f)\rho^m \tag{10.9}$$

where V_f is the piezoelectric volume fraction. The equivalent radial permittivity has the following expression

$$(\bar{\varepsilon}_r)_{eq.} = V_f(\bar{\varepsilon}_r)^f + (1 - V_f)(\bar{\varepsilon}_r)^m \tag{10.10}$$

The expression for the equivalent permittivity at constant stress after polarization has the following form

$$(\bar{\varepsilon}_{33}^\sigma)_{eq.} = V_f \left[(\bar{\varepsilon}_{33}^\sigma)^f - \frac{\left(d_{33}^f\right)^2}{(1 - V_f)(s_{33}^E)^f + V_f(s_{33})^m} \right] + (1 - V_f)(\bar{\varepsilon}_{11})^m \tag{10.11}$$

where $(s_{33}^E)^f$ is the elastic compliance of the piezo-fiber at constant electrical field (E) and $(s_{33})^m$ is the elastic compliance of the matrix (the polymer surrounding the piezoelectric fibers).

The equivalent 33 piezoelectric charge constant can be written as

Table 10.2: Mixing rules for d_{31} and d_{33} MFC unit volumes.

d_{31} MFC unit volume	d_{33} MFC unit volume

electrodes	polymer
polymer	electrodes
	PZT

Equivalent mechanical properties

$$E_1 = V_f E_1^p + (1 - V_f) E_1^m \quad *$$

$$E_2 = \frac{E_2^m}{V_f \frac{E_2^m}{E_2^p} + (1 - V_f)}$$

$$v_{12} = V_f v_{12}^p + (1 - V_f) v_{12}^m$$

$$v_{21} = v_{12} \frac{E_2}{E_1}$$

$$G_{12} \equiv G_{21} = \frac{G_{12}^m}{V_f \frac{G_{12}^m}{G_{12}^p} + (1 - V_f)} =$$

$$= \frac{G_{21}^m}{V_f \frac{G_{21}^m}{G_{21}^p} + (1 - V_f)}$$

$$G_{13} \equiv G_{31} = V_f G_{13}^p + (1 - V_f) G_{13}^m =$$
$$= V_f G_{31}^p + (1 - V_f) G_{31}^m$$

$$G_{23} \equiv G_{32} = \frac{G_{23}^m}{V_f \frac{G_{23}^m}{G_{23}^p} + (1 - V_f)} =$$

$$= \frac{G_{32}^m}{V_f \frac{G_{32}^m}{G_{32}^p} + (1 - V_f)}$$

$$E_3 = V_f E_3^p + (1 - V_f) E_3^m \quad *$$

$$E_2 = \frac{E_2^m}{V_f \frac{E_2^m}{E_2^p} + (1 - V_f)}$$

$$v_{32} = V_f v_{32}^p + (1 - V_f) v_{32}^m =$$

$$v_{23} = v_{32} \frac{E_2}{E_3}$$

$$G_{32} \equiv G_{23} = \frac{G_{32}^m}{V_f \frac{G_{32}^m}{G_{32}^p} + (1 - V_f)} =$$

$$= \frac{G_{23}^m}{V_f \frac{G_{23}^m}{G_{23}^p} + (1 - V_f)}$$

$$G_{31} \equiv G_{13} = V_f G_{31}^p + (1 - V_f) G_{31}^m =$$
$$= V_f G_{13}^p + (1 - V_f) G_{13}^m$$

$$G_{21} \equiv G_{12} = \frac{G_{21}^m}{V_f \frac{G_{21}^m}{G_{21}^p} + (1 - V_f)} =$$

$$= \frac{G_{12}^m}{V_f \frac{G_{12}^m}{G_{12}^p} + (1 - V_f)}$$

Equivalent piezoelectric properties

$$d_{31} = \frac{V_f d_{31}^p E_1^p}{E_1}$$

$$d_{32} = -d_{31} v_{12} + V_f d_{31}^p (1 + v_{12}^p)$$

$$d_{33} = \frac{V_f d_{33}^p E_3^p}{E_3}$$

$$d_{32} = -d_{33} v_{32} + V_f \left(d_{32}^p + d_{33}^p v_{32}^p \right)$$

Equivalent dielectric properties

$\bar{\varepsilon}_{33}^2 = V_f \left(\bar{\varepsilon}_{33}^2 \right)^p + (1 - V_f) \left(\bar{\varepsilon}_{33}^2 \right)^m$	$\bar{\varepsilon}_{33}^2 = V_f \left(\bar{\varepsilon}_{33}^2 \right)^p + (1 - V_f) \left(\bar{\varepsilon}_{33}^2 \right)^m$

* $()^m$ represents matrix property; $()^p$ represents piezo-fiber property.

$$(d_{33})_{eq.} = \frac{(d_{33})^f}{1 + \frac{(1 - V_f)(s_{33}^E)^f}{V_f(s_{33})^m}}$$ (10.12)

The equivalent electromechanical coupling factor in the thickness mode is given by

$$(k_t)_{eq.} = \frac{(e_{33})^f}{\sqrt{(c_{33}^D)_{eq.}(\bar{\varepsilon}_{33}^S)_{eq.}}}$$ (10.13)

while the equivalent acoustic impedance has the following expression

$$(Z_{acoustic})_{eq.} = \sqrt{(c_{33}^D)_{eq.} \rho_{eq.}}$$ (10.14)

where

$$(e_{33})_{eq.} = V_f \left\{ (e_{33})^f - \frac{2(1 - V_f)(e_{31})^f \left[(c_{13}^E)^f - (c_{12})^m \right]}{V_f[(c_{11})^m + (c_{12})^m] + (1 - V_f)\left[(c_{11}^E)^f + (c_{12}^E)^f \right]} \right\}$$

$$(c_{33}^E)_{eq.} = V_f \left\{ (c_{33}^E)^f - \frac{2(1 - V_f)\left[(c_{13}^E)^f - (c_{12})^m \right]^2}{V_f[(c_{11})^m + (c_{12})^m] + (1 - V_f)\left[(c_{11}^E)^f + (c_{12}^E)^f \right]} \right\} +$$
$$+ (1 - V_f)(c_{11})^m$$

$$(\bar{\varepsilon}_{33}^S)_{eq.} = V_f \left\{ (\bar{\varepsilon}_{33}^S)^f - \frac{2(1 - V_f)\left[(e_{31})^f \right]^2}{V_f[(c_{11})^m + (c_{12})^m] + (1 - V_f)\left[(c_{11}^E)^f + (c_{12}^E)^f \right]} \right\} +$$
$$+ (1 - V_f)(\bar{\varepsilon}_{11})^m$$

$$(c_{33}^D)_{eq.} = (c_{33}^E)_{eq.} + \frac{\left[(e_{33})_{eq.} \right]^2}{(\bar{\varepsilon}_{33}^S)_{eq.}}$$ (10.15)

where e_{ij}, s_{ij} and c_{ij} are the piezoelectric strain constant, the elastic compliance and the elastic stiffness, respectively. Note that the expression $(\bar{\varepsilon}_{33}^S)$ is the permittivity after polarization at constant strain.

10.2.2 PVDF Copolymer Fibers

In general, there are two types of manufacturing methods for piezoelectric fibers: melt-spinning and electrospinning. The spinning process is a manufacturing method for creating polymer fibers like PVDF. It is using a specialized form of extrusion using a spinneret to yield multiple continuous filaments. Note that to be able to spin the polymer it must be converted into a fluid state either by heating or using chemical reactions to dissolve it in a solvent.

The melt spinning process uses a melted and stretched PVDF, which is heated to yield a suitable viscosity to produce the fibers. The melted polymer is pushed through a spinneret having small holes. Note that each hole would produce an individual fiber while the spinneret number of holes defines the number of fibers in a yarn. To pole the PVDF yarn, a high electrical energy is required (up to 20 kV). Typical conditions of poling are 80–90 °C and the drawing ratio between the fast and slow rollers is 5:1, thus inducing stretching of the yarn [14].

A schematic drawing of the PVDF melt spinning process is depicted in Figure 10.15. Electrospinning process is a fiber production method which uses electric field to draw charged threads of PVDF solutions or polymer melts to up to hundred nanometers of fiber diameter.

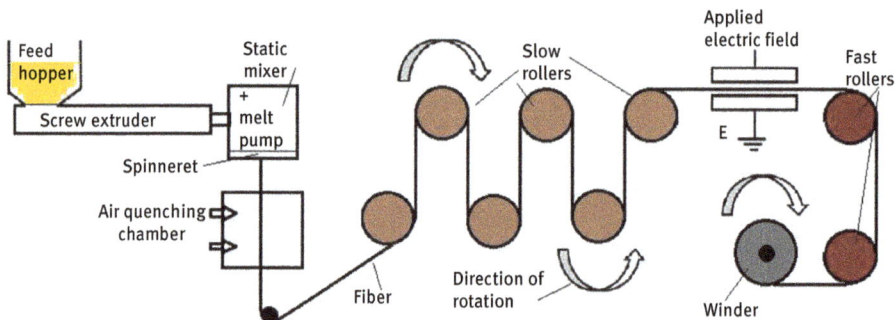

Figure 10.15: A schematic drawing for the continuous process of PVDF melt-spinning fibers (adapted from [14]).

It is known that when a high voltage is applied to a liquid droplet, its body becomes charged leading to electrostatic repulsion which acts against the droplet surface tension yielding a stretched body. Increasing the voltage up to a critical point, a stream of liquid would erupt from the surface (the point of eruption is known in the literature as the Taylor cone – see Figure 10.16).

For a liquid having sufficient high molecular cohesion, the stream is not broken and a jet of charged liquid is formed. The liquid jet will dry in flight and the type of the current flow would change from ohmic one to convective, as the charges migrate to the surface of the fiber. A whipping process (see Figure 10.16) caused by electrostatic repulsion at small bends of the fiber leads to the elongation of the jet till it is deposited on the grounded collector.

The PVDF fibers would then be used for various applications, like wearable textile, sensors or harvesters. The piezoelectric properties of the PVDF fibers should be experimentally measured to be able to calculate its properties as sensors or actuators, as found in Refs. [13–19].

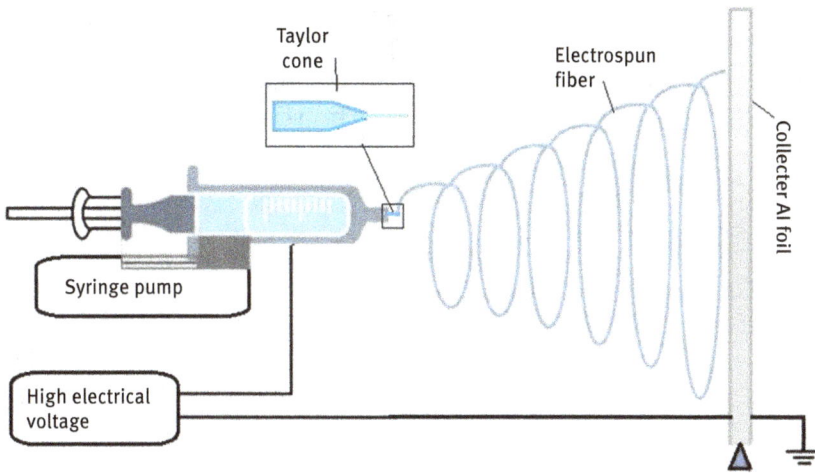

Figure 10.16: A schematic drawing for the production of PVDF electrospinning fibers.

10.3 Acoustic Energy Harvesting

Acoustic energy is the energy generated by noise. It is a parasitic environmental energy source and it can be harvested into useful electrical energy. Common noise sources would include vehicles (trucks and cars, motorcycles), airplanes, power plants, machines, trains and loudspeakers. It is interesting to note that the acoustic energy will eventually propagate and dissipate in the form of thermal energy. One should remember that audible frequency range to humans is 20 Hz up to 20 kHz, and the noise in this range should be transformed to electrical energy to reduce the environmental pollution due to this nuisance. The topic had been addressed in the literature, with typical studies being presented in [20–29].

10.3.1 Acoustic Basics

Sound is propagating in the air in a longitudinal waveform, which can be written for a plane wave case as

$$\frac{\partial^2 p}{\partial x^2} = \frac{1}{c^2}\frac{\partial^2 p}{\partial t^2} \tag{10.16}$$

where p is the sound pressure, c is the propagation speed in the air and x and t are the one dimensional special coordinate and time, respectively. Remember that $c = 343$ m/s at °C and can also be expressed as $c = f \cdot \lambda$, where f is frequency and λ is the wavelength. Equation (10.16) assumes that the sound propagation is adiabatic and inviscid. However, for certain cases the neglecting of viscosity would lead to wrong conclusions (see discussion in [29]).

Due to its great variability, the audible sound pressure is expressed by a logarithmic index SPL (sound-pressure-level) measured in decibel (*dB*)

$$\text{SPL} = 20\log_{10}\left(\frac{p}{p_r}\right) \quad [\text{dB}] \tag{10.17}$$

where p is the root-mean-square sound pressure and $p_r = 20\ \mu\text{Pa}$, the reference sound pressure in air. Another important expression is the sound power given by multiplication of the sound intensity (I_s) by the incident surface area (A_s) yielding

$$\text{Power} = I_s \cdot A_s = (p \cdot v) \cdot A = \left(\frac{p^2}{Z}\right) \cdot A_s \tag{10.18}$$

where v is the particle velocity and Z is the specific acoustic impedance.

10.3.2 Acoustic Power Augmentation

To be able to harvest the acoustic energy, one has to amplify the acoustic sound pressure, which is low. For $SPL = 114\ dB$, the corresponding sound pressure would be 10 Pa (see [29]). One of the most effective acoustic resonators is the Helmholtz resonator consisting of a cavity and a neck. The sound propagates inside the resonator, the pressure increases gradually and generates at the bottom of the cavity enough pressure to excite it. Placing a flexible wall at the bottom of the cavity and a piezoelectric disk (see Figure 10.17) would allow the bending of the transducer transforming the sound energy into electrical energy.

More examples can be found in [20–22, 27, 29]. Monthéard et al. [23] present a very promising report on powering a commercial data-logger using harvested aero acoustic noise, while Noh [26] studied the way to harvest acoustic energy in a railway environment. It was found that a high-speed train generates 50–200 Hz in the passenger car and between two adjacent cars. A Helmholtz resonator was designed for a 174 Hz target noise. The experimental results show a generated voltage of 0.7 V for a sound pressure level of 100 dB. Finally, two master theses, Monroe [24] and Mir [28] address the acoustic energy harvesting (AEH). Monroe [24] studied size and bandwidth of AEH devices and investigated a large-scale acoustic energy harvester based on piezoelectric PVDF film, having an area of 100 cm². Mir [28] proposed a novel metamaterial providing enhanced sound isolation, while transforming the isolated noise into usable electrical energy. Numerical calculations predict approximately 2 mW electrical power on 10 kΩ from 100 Hz sound frequency.

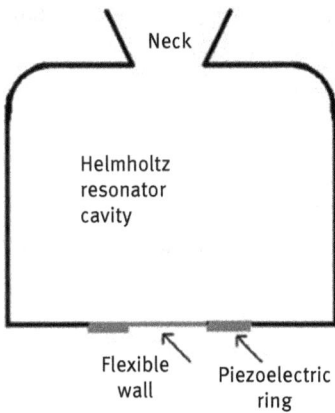

Neck

Helmholtz
resonator
cavity

Flexible
wall

Piezoelectric
ring

Figure 10.17: A schematic drawing for the acoustic harvester.

10.4 Harvesting Using SMA

Shape memory alloys (SMA) had been discussed in Chapter 4, presenting their properties and their applications. It was shown that their dynamic performance couldn't be compared with piezoelectric materials, as it demands heating and cooling cycles which inherently takes time. Still the literature presents studies on SMA with piezoelectric harvesters or magnetic shape memory alloys (MSMA)[1] transducers to scavenge parasitic energy.

10.4.1 Magnetic Shape Memory Alloys

Magnetic shape memory alloys (first discovered by Dr. Kari Ullakko and his colleagues at MIT in 1996) are ferromagnetic materials that can produce displacements and forces under moderate magnetic fields. Typically, MSMAs are alloys of nickel, manganese and gallium (Ni–Mn–Ga). The magnetic shape memory effect occurs in the low-temperature martensite phase of the alloy, where the elementary cells composing the alloy have tetragonal geometry. If the temperature increases beyond the martensite–austenite transformation temperature, the alloy would go to the austenite phase where the elementary cells have cubic geometry. The switching between these two phases is commonly triggered by a magnetic field but it can also be driven by a change in temperature or mechanical deformation as in conventional SMAs.

The large magnetically induced strain as well as the short response times make the MSMA technology very attractive for the design of innovative actuators to be applied

---

1 Ullakko, K., (1996). Magnetically controlled shape memory alloys: A new class of actuator materials, Journal of Materials Engineering and Performance, Vol. 5, No.3, 1996, pp. 405–409. doi:10.1007/BF02649344.

in pneumatics, robotics, medical devices and mechatronics. MSM alloys change their magnetic properties depending on the deformation. This companion effect is useful for the energy harvesters or design of displacement, speed or force.

Table 10.3: Properties of MSMA NiMnGa single crystal.[2]

Property	Value
Elongation in magnetic field	Up to 6%, typically 3–5%
Response	Up to 1–2 kHz
Force density	~2 MPa
Work output (force × stroke)	Max. 100 kJ/m^3
Fatigue life	Several hundred million cycles
Magnetic field	<0.8 T
Upper temperature limit	Transformation of martensite to austenite at 70 °C
Curie temperature	95–105 °C

10.4.2 SMA and MSMA Harvesting

SMA or MSMA are used to harvest parasitic fluctuations of the temperature into electrical energy. References [30–43] present typical research studies for this type of harvester.

Avirovik et al. [30] present a harvester based on a piezoelectric bimorph cantilever beam configuration. The free end of the cantilever is connected to a 100 μm diameter preloaded SMA wire which is heated by a laser up to 110 °C, leading to the transformation of the SMA yielding a force tending to bend the cantilever. The harvested power as a function of the preloaded SMA and its position along the cantilever was measured to be in the range of 0.003–0.006 μW (RMS).

Avirovik et al. [31] present another interesting device in the form of a small SMA heat engine to power wireless sensors nodes. The device includes a hot water reservoir, a pulley system and a SMA wire. The electrical power is generated by a micro-generator due to the SMA wire transformation from martensite to austenite when being heated which causes a torque on the micro-generator and thus it is rotating, generating electrical power. A value of 1.8 mW reported has been measured on an 80 Ω resistance.

The use of SMA and piezoelectric patches are presented in [32–36, 38, 39, 42] for thermal energy harvesting, while Davidson and Mo [37] present a review on energy

---

2 Goodfellow-Supplier of materials for research and development – https://www.goodfellow.com

harvesting technologies for structural health monitoring, which includes an MSMA-based thermal harvester.

Use of MSMA as the material to construct vibration based harvester can directly transform the vibrational energy to electrical energy based on the magnetic flux gradient. Thus, if one uses copper coils been wrapped around a MSMA element in a constant magnetic field, a strain (or a stress) being applied on the MSMA device, would induce electrical current in the coils (see [39, 41, 43]).

Finally, Zacharov et al. [40] advocates the use of combined pyroelectric, piezoelectric and shape memory effects for thermal energy harvesting by using a hybrid laminated composite material for harvesting quasi-static temperature variations. The harvester is consisted of a MFC piezoelectric patch and a TiNiCu shape memory alloy, an while the pyroelectric effect of the MFC piezoelectric patch is taken into account.

10.5 Road Traffic Harvesting Using Piezoelectric Transducers

As described above, the harvesting of parasitic mechanical energy and its transformation into usable electrical energy using piezoelectric material operating in the d_{33} mode, led to a series of manuscripts [44–45] describing piezoelectric actuators embedded on driving roads, and scavenging the energy from passing motor traffic. The idea is to build a piezoelectric generator which will sense the passing vehicle, deform due to its weight and thus produce electrical energy due to the direct piezoelectric effect. The two early papers by Abramovich et al. [44, 45] report laboratory experiments on those generators (see Figure 10.18) and discuss the high-power multi-element piezoelectric generators made from PZT ceramic under long-term cyclic external mechanical loading and showing that a safe-side stress level of 30 MPa can be used with a small reduction in the generated power as can be seen in Figure 10.19. To compensate for the relatively low electric power, due to relatively reduced mechanical stresses applied on the PZT disks, one can increase the volume of the material used by placing layers of piezoelectric material one on top of the other, each subjected to the same mechanical stress. This will yield the required electric power from a safe given mechanical stress without reduction in its output. The follow-up manuscripts (see [46–52]) mainly review the state of the art for piezoelectric road generators, discussing the prospects of realizing it, while presenting new results on this topic. Kim et al. [46] review what had been done on piezoelectric road generators and intend to perform tests on Georgia, USA highways, Duarte and Ferreira [47] also outline the development of energy harvesting technologies for road pavements, trying to divide them into various classes, while discussing and presenting a technical analysis and comparison for those technologies, using the results achieved with available prototypes. Kour and Charif [48] aim to assess the functionality of piezoelectricity in roads to use the energy due to the moving vehicles. The energy is converted into electrical energy using piezoelectric technology to replace fossil fuel in streetlight applications, advocating this new technology claiming that piezoelectric road is a new energy evolution to provide a sustainable solution in terms of environment, economy and social needs.

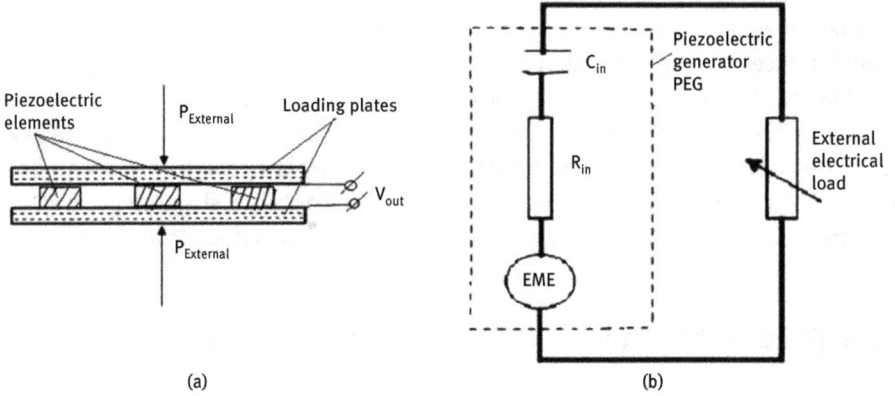

Figure 10.18: The piezoelectric generator (PEG): (a) a schematic drawing of the mechanical layout and (b) the equivalent electrical circuit and its associated external electrical load (adapted from [44]).

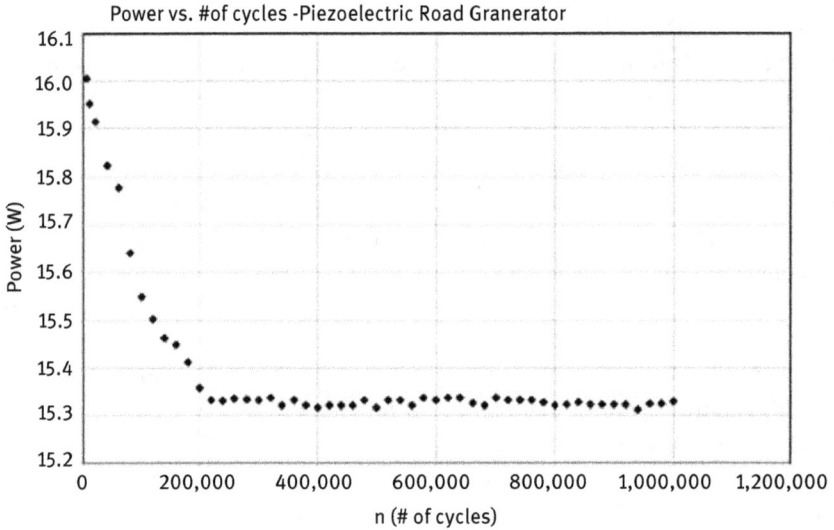

Figure 10.19: The power output of a piezoelectric road generator loaded at 30 MPa at a frequency of 5 Hz (adapted from [45]).

Papagiannakis et al. [49] present the development of several piezoelectric prototypes capable of harvesting energy from the action of traffic on highways. The amount of energy available for harvesting is explored through finite element simulations of the strain energy in pavements under moving tire loads. The prototypes developed involve various configurations of cylindrical and prismatic piezoelectric elements. Exploratory analysis shows that it is desirable to alternate the polarity of stacked piezoelectric elements and connect them in parallel to avoid generating unmanageably high voltages. The prototypes are tested in uniaxial compression under sinusoidal loading at a frequency of 10 Hz. Curves are fitted to the electrical power versus stress laboratory data. For a single pass of a 44.48-kN truck tire load, the electrical power generated is estimated to be between 1.0 and 1.8 W, far from the experimental results presented in [44–45]. Yang et al. [50] in a recent manuscript (2017) present the results for a road generator using a stacked mode. The capability of their model to withstand repeated loading up to 150 kN, yielded only 100,000 cycling for the piezoelectric transducer, without reduction of the output power. The authors quote an open circuit voltage of 280 V, without mentioning the achieved power. It is worth mentioning that the amount of piezoelectric material used for the piezoelectric road generator was small, expecting a small output power. Another new paper (from 2018) on piezoelectric road generators is the one presented by Qabur and Alshammari [51] in which they review again the state of the art for harvesting energy from roadways, claiming that PZT can be considered as the most efficient material due to its unique features according to many references and real or practical applications that various factors such as geometry, thickness and structure affect the output of the piezoelectric process and that energy harvesters have the ability to be enhanced or improved by minimizing the mechanical and electrical losses of the harvesting stages in order to achieve the highest efficiency possible. Their sum up by saying that the implementation of piezoelectric materials on roadways needs further research and a more comprehensive analysis of different data that are obtained from real life or practical applications. Walubita et al. [52] review in their article the state of the art in road energy harvesting technology, with a focus on piezoelectric systems, including an analysis of the impact of the technology from social and environmental standpoints. Overall, their literature findings indicate that the expansion of the roadway energy harvesting technology to a large practical scale is feasible, but such an undertaking should be wisely weighed from broader perspectives. Ultimately, the article provides a positive outlook of the potential contributions of road energy harvesting technologies to the ongoing energy and environmental challenges of human society. The article highlights an experimental study carried out by the Texas A&M Transportation Institute (TTI) in which a prototype of a piezoelectric module, named Highway Sensing and Energy Conversion (HiSEC) was experimented in a laboratory at TTI. The HiSEC module is made up of three pairs of piezo discs stacked and connected with two diodes. The montage is thereafter enclosed in a metal case covered on the top with an impact cap. The maximal power obtained by this module was approximately 13.5 mW at a mechanical stress of 0.827 MPa (120 psi) and a frequency of 1 Hz. Note the small amount of the piezoelectric material, the low mechanical stress and the low frequency employed in their experiment, leading to low performances.

References

[1]   Williams, R. B., Park, G., Inman, D. J. and Wilkie, W. K., An overview of composite actuators with piezoceramic fibers, International modal analysis conference, Proceedings of IMAC-XX, the Westin Los Angeles Airport, Los Angeles, CA, USA, 4–7, February 2002, 421–427.

[2]   Mallik, N. and Ray, M. C., Effective coefficients of piezoelectric fiber-reinforced composites, AIAA Journal 41(4), 2003, 704–710.

[3]   Nelson, L. J., Bowen, C. R., Stevens, R., Cain, M. and Stewart, M., Modelling and measurement of piezoelectric fibres and interdigitated electrodes for the optimization of piezofibre composites, Proceeding Vol. 5053, Smart Structures and Materials, 2003: Active Materials: Behavior and Mechanics, 2003, San Diego, CA, US., 12, doi:10.1117/12.484738.

[4]   Berger, H., Kari, S., Gabbert, U., Ramos, R. R., Guinovart, R., Otero, J. A. and Catillero, J. B., An analytical and numerical approach for calculating effective material coefficients of piezoelectric fiber composites, International Journal of Solids and Structures 42, 2005, 5692–5714.

[5]   Ralf, S., Modelling and characterization of piezoelectric 1–3 fibre composites., In: Chapter A8.4 from Piezoelectric and acoustic materials for transducer applications, Safari, A. and Akdoğan, E. K. (eds.), Springer Science+Business Media, LLC, 2008, 483.

[6]   Schönecker, A., Piezoelectric fiber composite fabrication, Chapter 13., From Piezoelectric and acoustic materials for transducer applications, Safari, A. and Akdoğan, E. K. (eds.), Springer Science+Business Media, LLC, 2008, 483.

[7]   Deraemaeker, A., Nasser, H., Benjeddou, A. and Preumont, A., Mixing rules for the piezoelectric properties of Macro Fiber Composites (MFC), Journal of Intelligent Material Systems and Structures 20(2), 2009, 1475–1482. doi:10.1177/1045389X09335615.

[8]   Smith, W. A., Modeling 1–3 composite piezoelectrics: hydrostatic response, IEEE Transactions on Ultrasonics, Ferroelectrics, and Frequency Control 40(1), 1993, 41–49.

[9]   Kerur, S. B. and Ghosh, A., Active vibration control of composite plate using AFC actuator and PVDF sensor, International Journal of Structural Stability and Dynamics 11(2), 2011, 237–255. doi:10.1142/S0219455411004075.

[10]   Lin, X.-J., Zhou, K.-C., Zhang, X.-Y. and Zhang, D., Development, modeling and application of piezoelectric fiber composites, Transactions of Nonferrous Metals Society of China 23, 2013, 98–107.

[11]   Nilson, E., Mateu, L., Spies, P. and Hagström, B., Energy harvesting from piezoelectric textile fibers, Procedia Engineering 87, 2014, 1569–1572.

[12]   Jemai, A., Najar, F., Chafra, M. and Ounaies, Z., Mathematical modeling of an active-fiber composite energy harvester with interdigitated electrodes, Shock and Vibration 2014, 2014, 9, Paper Id 971597.

[13]   Ji, S. H., Cho, J. H., Jeong, Y. H., Paik, J.-H., Yun, J. D. and Yun, J. S., Flexible lead-free piezoelectric nanofiber composites based on BNT-ST and PVDF for frequency sensor applications, Sensors and Actuators. A, Physicals 247, 2016, 316–322.

[14]   Matsouka, D. and Vassiliadis, S., Piezoelectric melt-spun textile fibers: technological overview, Chapter 4 in Piezoelectricity-organic and inorganic materials and application, Intech-Open 2016, 66–82. doi:http://dx.doi.org/10.5772/intechopen.78389.

[15] Kumar, R. S., Sarathi, T., Venkataraman, K. K. and Bhattacharyya, A., Enhanced piezoelectric properties of polyvinylidene fluoride nanofibers using carbon nanofiber and electrical poling, Material Letters 255, 2019, 4, Paper Id 126515.

[16] Park, S., Kwon, Y., Sung, M., Lee, B.-S., Bae, J. and Yu, W.-R., Poling-free spinning process of manufacturing piezoelectric yarns for textile applications, Materials and Design 179, 2019, 10, Paper Id. 107889.

[17] Ghafari, E. and Lu, N., Self-polarized electrospun polyvinylidene-fluoride (PVDF) nanofiber for sensing applications, Composites Part B 160, 2019, 1–9.

[18] Lam, T.-N., Wang, -C.-C., Ko, W.-C., Wu, J.-M., Lai, S.-N., Chuang, W.-T., Su, C.-J., Ma, C.-Y., Luo, M.-Y., Wang, Y.-J. and Huang, E.-W., Tuning mechanical properties of electrospun piezoelectric nanofibers by heat treatment, Materialia 8, 2019, 8, Paper Id 100461.

[19] Cui, N., Jia, X., Lin, A., Liu, J., Bai, S., Zhang, L., Qin, Y., Yang, R., Zhou, F. and Li, Y., Piezoelectric nanofiber/polymer composite membrane for noise harvesting and active acoustic detection, Nanoscale Advances 1, 2019, 4909–4914. doi:10.1039/c9na00484j.

[20] Sherrit, S., The physical acoustics of energy harvesting, 2008 IEEE International Ultrasonics Symposium Proceedings, Beijing International Convention Center (BICC) Beijing, China, November 2–5, 2008, 1046–1055. doi:10.1109/ULTSYM.2008.0253.

[21] Lin, J.-T., Lee, B. and Alphenaar, W., Non-linear energy harvesting with random noise and multiple harmonics, Chapter 12, Small-scale energy harvesting, Intech-Open 2012, 283–302.

[22] Rahman, A. and Hoque, M. E., Harvesting energy from sound and vibration, Paper ID: Am-22, International Conference on Mechanical, Industrial and Materials Engineering 2013 (ICMIME2013), 1–3 Nov. 2013, RUET, Rajshahi, Bangladesh, 6.

[23] Monthéard, R., Airiau, C., Bafleur, M., Boities, V., Dilhac, -J.-J., Dollat, X., Nolhier, N. and Piot, E., Powering a commercial datalogger by energy harvesting from generated aeroacoustic noise, Journal of Physics: Conference Series 557, 2014, 5, Paper Id: 012025, doi:10.1088/1742-6596/557/1/012025.

[24] Monroe, N. M., Broadband Acoustic Energy Harvesting via Synthesized Electrical Loading, Master of Engineering Thesis, Department of Electrical Engineering and Computer Science, Massachusetts, USA, MIT, Cambridge, 2017, 149.

[25] Zhao, N., Zhang, S., Yu, F. R., Chen, Y., Nallanathan, A. and Leung, V. C. M., Exploiting interference for energy harvesting: a survey, research issues, and challenges, IEEE Access 5, 2017, 10403–10421. doi:10.1109/ACCESS.2017.2705638.

[26] Noh, H.-M., Acoustic energy harvesting using piezoelectric generator for railway environmental noise, Advances in Mechanical Engineering 10(7), 2018, 1–9. doi:10.1177/1687814018785058.

[27] Choi, J., Jung, I. and Kang, C.-Y., A brief review of sound energy harvesting, Nano Energy 56, 2019, 169–183.

[28] Mir, F., Acoustoelastic metamaterial with simultaneous noise filtering and energy harvesting capability from ambient vibrations, Master of engineering thesis, Mechanical Engineering, College of Engineering and Computing, University of South Carolina, Columbia, South Carolina, USA, 2019, 62.

[29] Yuan, M., Cao, Z., Luo, J. and Chou, X., Recent developments of acoustic energy harvesting: a review, Micromachines 10(1), 2019, 48, 21. doi:10.3390/mi10010048.

[30] Avirovik, D., Kumar, A., Bodnar, R. J. and Priya, S., Remote light energy harvesting and actuation using shape memory alloy-piezoelectric hybrid transducer, Smart Materials and Structures 22, 2013, 6, Paper Id: 052001. doi:10.1088/0964-1726/22/5/052001.

[31] Avirovik, D., Kishore, R. A., Vuckovic, D. and Priya, S., Miniature shape memory alloy heat engine for powering wireless sensor nodes, Energy Harvesting and Systems 1(1–2), 2014, 13–18.

[32] Gosliga, J. S. and Ganilova, O. A., Energy harvesting based on the hybridization of two smart materials, Paper # 170, EACS 2016-6th European Conference on Structural Control, Sheffield, England, 11–13 July 2016, 12.

[33] Zakharov, D., Lebedev, G., Cugat, O., Delamare, J., Viala, B., Lafont, T., Gimeno, L. and Shelyakov, A., Thermal energy conversion by coupled shape memory and piezoelectric effects, Journal of Micromechanics and Microengineering 22, 2012, 7, Paper Id: 094005. doi:10.1088/0960-1317/22/9/094005.

[34] Zakharov, D., Gusarov, B., Gusarova, E., Viala, B., Cugat, O., Delamare, J. and Gimeno, L., Combined pyroelectric, piezoelectric and shape memory effects for thermal energy harvesting, Journal of Physics: Conference Series 476, 2013, 5, Paper Id:012012. doi:10.1088/1742-6596/476/1/012021.

[35] Namli, O. C. and Taya, M., Design of piezo-SMA composite for thermal energy harvester under fluctuating temperature, Journal of Applied Mechanics 70, 2011, 8, paper Id: 03101.

[36] Namli, O. C., Jae-Kon, L. and Taya, M., Modeling of piezo-SMA composites for thermal energy harvester, Proc. SPIE 6526, Behavior and Mechanics of Multifunctional and Composite Materials 2007, 65261L, 12 April 2007, San Diego, California, USA., 12. doi:10.1117/12.715786.

[37] Davidson, J. and Mo, C., Recent advances in energy harvesting technologies for structural health monitoring applications, Smart Materials Research 2014, 2014, 14, Paper Id: 410316. doi:10.1155/2014/410316.

[38] Todorov, T., Nikolov, N., Todorov, G. and Ralev, Y., Modelling and investigation of a hybrid thermal energy harvester, MATEC Web of Conferences, Vol. 148, 2018, International Conference of Engineering Vibration (ICoEV), Paper Id. 12002, 6. doi:10.1051/matecconf/201814812002.

[39] Viet, N. V., Zaki, W. and Umer, R., Analytical investigation of an energy harvesting shape memory alloy-piezoelectric beam, Archive of Applied Mechanics 90(12), 2020, 2715–2738. doi:10.1007/s00419-020-01745-9.

[40] Zakharov, D., Gusarov, B., Gusarova, E., Viala, B., Cugat, O., Delamare, J. and Gimeno, L., Combined pyroelectric, piezoelectric and shape memory effects for thermal energy harvesting, Journal of Physics: Conference Series Vol. 476, 2013, Paper Id:012012, 5. doi:10.1088/1742-6596/476/1/012021.

[41] Fasangi, M. A. A., Cottone, F., Sayyaadi, H., Zakerzadeh, M. R., Orfei, F. and Gammaitoni, L., Energy harvesting from structural vibrations of magnetic shape memory alloys, Applied Physical Letters 110, 2017, Paper Id: 103905, 4.

[42] Adeodato, A., Duarte, B. T., Monteiro, L. L. S., Pacheco, P. M. C. L. and Savi, M. A., Synergistic use of piezoelectric and shape memory alloy elements for vibration-based energy harvesting, International Journal of Mechanical Sciences 194, 2021, 10, Paper Id:106206.

[43] Safari, O., Zakerzadeh, M. R. and Baghani, M., Study of a magnetic SMA-based energy harvester using a corrugated structure, Journal of Intelligent Material Systems and Structures 32, 2021, 12. doi:10.1177/1045389X20983903.

[44] Abramovich, H., Tsikhotsky, E. and Klein, G., An experimental determination of the maximal allowable stresses for high power piezoelectric generators, Journal of Ceramic Science and Technology 4(3), 2013, 131–136. doi:10.4416/JCST2013-00006.

[45] Abramovich, H., Tsikhotsky, E. and Klein, G., An experimental investigation on PZT behavior under mechanical and cycling loading, Journal of the Mechanical Behavior of Materials 22(3–4), 2013, 129–136.

[46] Kim, S., Shen, J. and Ahad, M., Piezoelectric-based energy harvesting technology for road sustainability, International Journal of Applied Science and Technology 5(1), 2015, 20–25.

[47] Duarte, F. and Ferreira, A., Energy harvesting on road pavements: state of the art, Proceedings of the Institution of Civil Engineers, Energy 169(EN2), 2016, 79–90, Paper 1500005.

[48] Kour, R. and Charif, A., Piezoelectric roads: energy harvesting method using piezoelectric technology, Innovative Energy and Research 5(1), 2016, 6, Paper Id: 10000132. doi:10.4172/2576-1463.1000132.

[49] Papagiannakis, A. T., Montoya, A., Dessouky, S. and Helffrich, J., Development and evaluation of piezoelectric prototypes for roadway energy harvesting, Journal of Energy Engineering, ASCE 143(5), 2017, 7, Paper Id: 04017034. doi:10.1061/(ASCE)EY.1943-7897.0000467.

[50] Yang, H., Wang, L., Hou, Y., Guo, M., Ye, Z., Tong, X. and Wang, D., Development in stacked-array-type piezoelectric energy harvester in asphalt pavement, Journal of Materials in Civil Engineering, ASCE 29(11), 2017, 9, Paper Id: 04017224.

[51] Qabur, A., Alshammari, K. and Systematic, A., Review of energy harvesting from roadways by using piezoelectric materials technology, Innovative Energy & Research 7(1), 2018, 6, Paper Id: 10000191. doi:10.4172/2576-1463.1000191.

[52] Walubita, L. F., Djebou, D. C. S., Faruk, A. N. M., Lee, S. I., Dessouky, S. and Hu, X., Prospective of societal and environmental benefits of piezoelectric technology in road energy harvesting, Sustainability 10, 2018, 383, 13. doi:10.3390/su10020383.

Index